Hall Effect Devices

Second Edition

Series in Sensors

Senior Series Editor: **B E Jones**
Series Co-Editor: **W D Spillman, Jr**

Other titles in the series

Thin Film Resistive Sensors
P Ciureanu and S Middelhoek (Eds)

Biosensors: Microelectrochemical Devices
M Lambrechts and W Sansen

Automotive Sensors
M H Westbrook and J D Turner

Thermal Sensors
G C M Meijer and SAW van Herwaarden

Advances in Actuators
A P Dorey and J H Moore (Eds)

Sensor Materials
P T Moseley and J Crocker

Intelligent Sensor Systems
J E Brignell and N M White

Ultrasonic Sensors: for Chemical and Process Plant
R C Asher

Electronic Noses and Olfaction 2000
J W Gardner and K C Persaud (Eds)

Thin Film Magnetoresistive Sensors
S Tumanski

Sensors and their Applications XI
K T V Grattan and S H Khan (Eds)

Forthcoming titles in the series

Structural Sensing, Health Monitoring and Performance Evaluation
D R Huston

Environmental Sensors and Monitoring
M Campbell

Series in Sensors

Hall Effect Devices

Second Edition

R S Popovic
Swiss Federal Institute of Technology Lausanne (EPFL)

IoP
Institute of Physics Publishing
Bristol and Philadelphia

British Library Cataloguing-in-Publication Data

A catalogue record for this book is available from the British Library.

ISBN 0 7503 0855 9

Library of Congress Cataloging-in-Publication Data are available

Commissioning Editor: Tom Spicer
Production Editor: Simon Laurenson
Production Control: Sarah Plenty and Leah Fielding
Cover Design: Victoria Le Billon
Marketing: Nicola Newey and Verity Cooke

Published by Institute of Physics Publishing, wholly owned by The Institute of Physics, London

Institute of Physics Publishing, Dirac House, Temple Back, Bristol BS1 6BE, UK

US Office: Institute of Physics Publishing, The Public Ledger Building, Suite 929, 150 South Independence Mall West, Philadelphia, PA 19106, USA

Typeset by Academic + Technical, Bristol
Printed in the UK by MPG Books Ltd, Bodmin, Cornwall

To my wife
and our daughters
Tanja and Dragana

Contents

Preface

The purpose of this book is to relate the best results in the development and in the application of Hall effect devices with the sound physical and technological basics. For this reason, the book is 'wide-band': it includes the relevant chapters of solid-state physics, a thorough treatment of Hall effect devices, an introduction to interface electronics and magnetic circuits, selected examples of realized Hall magnetic sensors, and generic examples of their applications. I hope that the book will help a reader understand why a particular device works so well; and stimulate him to search for even better basic concepts, device designs, sensor systems, and further exciting applications.

The book is intended as a reference book for graduate students, physicists and engineers

- who need a profound treatment of galvanomagnetic transport effects in solids,
- who are involved in research and development of magnetic sensors,
- who are using the Hall effect for characterization of semiconductor materials,
- who are involved in precise magnetic measurements, and
- who are developing new products based on magnetic sensors.

This is the thoroughly revised, extended and updated edition of a book with the same title published in 1991. Briefly, in this edition I made several amendments in order to facilitate the use of the book; improved the treatment of the Hall effect (notably, by explaining the planar Hall effect and its relation with the magnetoresistance effect); extended the treatment of the electronic system aspects of Hall elements and of interface electronics; and completed the book with several new subjects that proved important or appeared during the past 12 years, such as sub-micrometer Hall devices, 2- and 3-axis Hall devices, Hall devices integrated with magnetic flux concentrators, and some applications thereof.

Acknowledgments

For the second edition

While occupying myself with the research and development of magnetic sensors over the past 10 years, I had the great privilege and pleasure to work together with many exceptional people both in academia and industry. Most new Hall magnetic sensors and their applications presented in this book are the results of these cooperations.

I am particularly grateful to my senior collaborators at EPFL, P A Besse, G Boero, P Kejik and B Furrer, for many fruitful discussions, for critically reading parts of the manuscript and correcting errors, for helping with graduate students, and for other support.

I am equally grateful to my former PhD students, whose research results are also built into this book: H Blanchard, Ch Schott, F Burger, D Manic, Z Randjelovic, M Demmierre, V Schlageter, P M Drljaca, and E Schurig.

Our academic research would have no real impact but for the devoted work of my collaborators at Sentron AG, who developed industrial versions of several new Hall magnetic sensors and their applications, presented in this book. I thank for this R Racz, F Betschart, Ch Schott, S Huber and A Stuck.

I began writing this book in the USA during the spring of 2002 while being on sabbatical leave from EPFL. I had wonderful time and I thank my hosts, Prof. J Judy (University of California, Los Angeles) and Prof. S von Molnar (Florida State University, Tallahassee), both for their hospitality and for many stimulating discussions that I had with them and with their collaborators.

I also thank all other people with whom I had collaboration and/or stimulating discussions on various aspects of magnetic sensors and their applications. Particularly, I thank: B Berkes (MCS Magnet Consulting Services, Switzerland); C Reymond and P Sommer (Metrolab, Switzerland); I Walker and L Law (GMW, USA); J Petr (Siemens Metering, Switzerland); J R Biard and R B Foster (Honeywell, USA); P Ripka (Czech Technical University, Czech Republic); I Shibasaki, H Endo, M Ozaki and K Shibahara (Asahi Kasei, Japan); P Nikitin (General Physics Institute of Russian Academy of Science, Russia); V Kempe (AMS, Austria); V Makoveev

(Joint Institute for Nuclear Research, Russia); I Bolshakova (Lviv Polytechnical State University, Ukraine); H Schneider-Muntau (National High Magnetic Field Laboratory, USA); D Pantic and Z Prijic (University of Nis, Serbia); U Ausserlechner (Infineon Technologies, Austria); and V Mordkovich (Institute of Microelectronics Technology of Russian Academy of Science, Russia).

I thank Mrs I Modoux (EPFL, Switzerland) and E Jovanovic (University of Nis, Serbia) for the technical help with the preparation of the manuscript.

Radivoje S Popovic
Lausanne, July 2003

For the first edition

I would like to thank the many friends and colleagues who supported and encouraged my research into magnetic sensors, and thus helped this book.

I am obliged to H Leinhard and H Vonarburg, my superiors at the Corporate Research and Development of Landis & Gyr, Zug, Switzerland, who built up this research laboratory with its special ambience. It has been very motivating to know that they believed in what I was doing. H Lienhard also kept encouraging me to search for what didn't exist.

I am grateful to H Baltes (now at the Swiss Federal Institute of Technology, Zurich), who engaged me in 1982 to develop a magnetic sensor which had to be much better than any sensor known at that time. While attempting to fulfil this task, I learned enough to write this book.

I acknowledge the collaboration of my colleagues who worked with me during the past eight years and who were fascinated with the research, the development, and the applications of magnetic sensors: J Petr who asked us to provide him with an extraordinary magnetic sensor, and who built a wonderful instrument around this sensor; J L Berchier and A Krause, who managed to develop a matchless magnetic sensor; J Hrejsa, D Schnetzler, M Kolman, M Habluetzel, E Habluetzel and B Ehrer, who participated in the development work and manufacture of the unique devices; G Schneider and H U Stocker, who proved the sensors to be truly unparalleled; U Falk and B Haelg, who looked into fundamental physical phenomena and managed to improve the magnetic sensor further; H U Schwarzenbach and H Ungricht, who helped to model and simulate magnetic sensors numerically; and Th Seitz, who helped the magnetic sensors appear much better than they really are by properly concentrating magnetic field on the sensitive region.

I also thank the many people of other organizations with whom I had the privilege to collaborate or discuss galvanomagnetic effects, magnetic sensors, and related phenomena and devices: H Melchior (Zurich, Switzerland), S Middelhoek (Delft, The Netherlands), F Rudolf (Neuchatel,

Switzerland), R Widmer (Zurich), S Cristloveanu (Grenoble, France), L K J Vandamme (Eindhoven, The Netherlands), E Vittoz (Neuchatel), V Zieren (Delft), S Kordic (Delft), A G Andreou (Baltimore, USA), A Nathan (Waterloo, Canada), Lj Ristic (Edmonton, Canada), Ch S Roumenin (Sofia, Bulgaria) and W Heidenreich (Regensburg, Germany).

I am also grateful to the following for granting permission to reproduce figures and tables included in this book

The authors of all figures not originated by myself.

Pergamon Press PLC for figures 2.2, 2.7, 3.31–5, 4.7, 5.13, 5.14, 6.6 and 7.2.

John Wiley & Sons, Inc. for figures 2.5, 2.6, 2.8, 2.11 and 7.39.

Springer Verlag for figures 2.9, 2.10 and 5.5.

Academic Press Ltd for figure 3.24.

VCH Verlagsgesellschaft GmbH for figure 3.36.

American Institute of Physics for figures 3.38, 3.39, 4.7–9, 5.15 and 7.15.

Plenum Press for figures 4.24 and 5.4.

Verlag der Zeitschrift für Naturforschung for figures 4.6 and 5.1.

IEEE for figures 5.7, 7.1, 7.8, 7.9, 7.14, 7.16, 7.18–21, 7.27, 7.31.

Elsevier Sequoia S.A. for figures 5.9, 7.3, 7.22, 7.23, 7.25, 7.26, 7.37 and 7.38.

Philips Research Laboratories for table 6.1 and figure 6.5.

IEEE Publishing for figures 7.34–6.

And I finally thank Mrs E Reuland and C Wiener for typing the manuscript of this book.

R S Popovic
June 1990

Chapter 1

Introduction

The subject of this book is the physics and technology of solid-state electron devices whose principle of operation is based on the Hall effect [1, 2]. These are devices such as Hall plates, semiconductor magnetoresistors, magnetodiodes and magnetotransistors. We also include semiconductor magnetoresistors in this group, because, as we shall see below, the Hall effect and the magnetoresistance effect [3] in non-magnetic materials are just two appearances of the same physical phenomena. For simplicity, we name all these devices with the representative term 'Hall effect devices'. The Hall effect devices are used for exploring the basic electronic properties of solids [4, 5], characterizing semiconductor materials [6, 7], and as magnetic sensors [8–11].

The quantized Hall effect [12] and corresponding devices are beyond the scope of this book. A review of the quantized Hall effect can be found, for example, in [13].

In this book, we treat only the Hall effect in non-magnetic materials, particularly in semiconductors. This effect exists also in ferromagnetic materials, then called 'ordinary Hall effect'. In addition to the ordinary Hall effect, the ferromagnetic materials show also an apparently stronger 'extraordinary' Hall effect. A good treatment of the galvanomagnetic effects in ferromagnetic materials can be found, for example, in [14].

In this chapter we provide a general introduction for newcomers to the field of Hall effect devices. Based on a brief description of the discoveries of the Hall effect and the magnetoresistance effect, we first give a notion on the underlying physics. Then we discuss the most important devices that exploit these effects. Finally, we review the most important applications of Hall effect devices. Throughout, we identify a few problems or refer to particular subjects treated in the book, which may motivate a reader to continue reading the rest of the book.

1.1 The basic physical effects

The operation of Hall effect devices is based on the phenomena called galvanomagnetic effects. The galvanomagnetic effects are physical effects

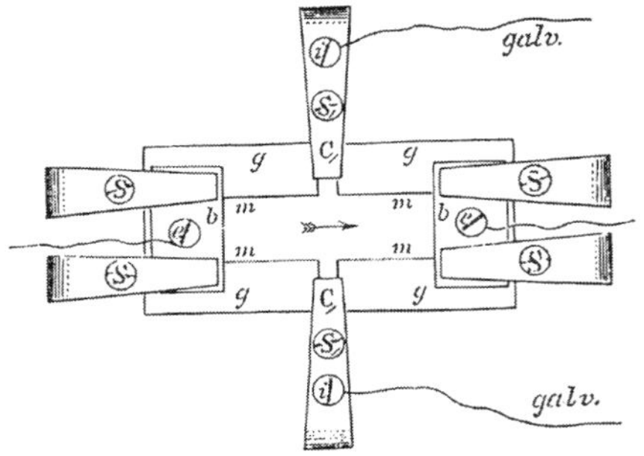

Figure 1.1. The very first Hall plate, the experimental device with which Mr Hall discovered the Hall effect. g g g g represents the plate of glass upon which the metal strip m m m m is mounted. The contacts for the main current (the arrow) are made by the two blocks of brass b b and the screws S S S S e e. A galvanometer is connected across the leaf using the clamps C C and the screws S S i i (reprinted from [2]).

arising in matter carrying electric current in the presence of a magnetic field. In the present context, the most important galvanomagnetic effects are the Hall effect and the magnetoresistance effect. In this section we shall briefly review these two effects.

1.1.1 The Hall effect

The effect is named after the American physicist *Edwin H Hall*, who discovered it in 1879. Then a graduate student, Hall was attempting to prove that a magnet directly affects current and not a wire bearing the current, as was believed at that time. His experimental device was a metal (gold, silver, iron) leaf mounted on a glass plate (figure 1.1). An electric current was passed through the leaf, and a sensitive galvanometer was connected across the leaf at two nearly equipotential points. The leaf was placed between the poles of an electromagnet (figure 1.2). Experimenting with this arrangement, Hall discovered 'a new action of the magnet on electric currents' [1, 2]. This action is now called the Hall effect.

What Hall actually observed was an electromotive force in the leaf, acting at right angles to the primary electromotive force and the magnetic field. It appeared as if the electric current was pressed, but not moved, towards one side of the leaf. Hall concluded that the new electromotive force was proportional to the product of the intensity of the magnetic field and the velocity of the electricity. In modern notation, his conclusion reads

$$\boldsymbol{E}_{\mathrm{H}} \sim [\boldsymbol{v} \times \boldsymbol{B}]. \tag{1.1}$$

Here, $\boldsymbol{E}_{\mathrm{H}}$ is the Hall electric field, $\boldsymbol{v}$ is 'the velocity of electricity' and $\boldsymbol{B}$ is the magnetic induction.

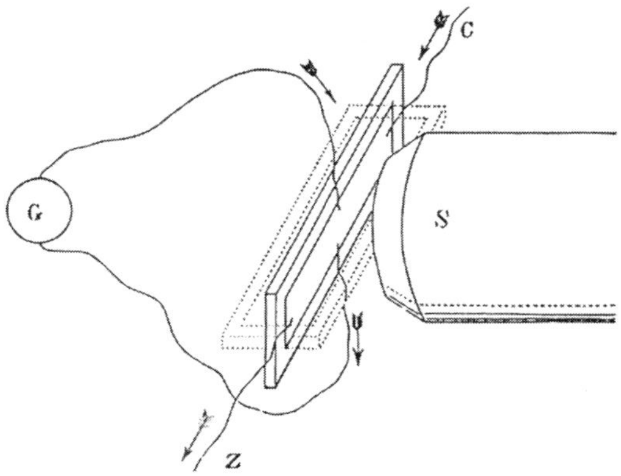

Figure 1.2. Mr Hall's key experiment. If the plate such as that shown in figure 1.1 is placed between the poles of an electromagnet in the position shown by the heavy lines, the galvanometer (G) measures a potential difference across the plate. If the plate is turned into the position indicated by the dotted lines, the potential difference disappears. S represents one pole of the electromagnet (reprinted from [2]).

Nowadays the Hall effect is an important and thoroughly elaborated topic of solid-state physics. In the modern theory of the Hall effect, Hall's conclusion (1.1) generally still holds. Of course, the velocity of electricity is now interpreted as the drift velocity of charge carriers:

$$\boldsymbol{v}_{\mathrm{d}} = \mu \boldsymbol{E}_{\mathrm{e}} \tag{1.2}$$

where μ denotes the carrier mobility and $\boldsymbol{E}_{\mathrm{e}}$ is the applied (external) electric field. Then, in accordance with (1.1), the Hall electric field is approximately given by

$$\boldsymbol{E}_{\mathrm{H}} \simeq \mu[\boldsymbol{E}_{\mathrm{e}} \times \boldsymbol{B}]. \tag{1.3}$$

The generation of the Hall electric field is not the only way the Hall effect appears. Under certain conditions, instead of the generation of the Hall electric field a current deflection in the sample may take place. In common with both modes of the Hall effect, a magnetic field introduces an angle between the total electric field and current density vectors. This angle is called the *Hall angle* and is given by

$$\Theta_{\mathrm{H}} \simeq \arctan(\mu B). \tag{1.4}$$

At weak magnetic inductions, this reduces to

$$\Theta_{\mathrm{H}} \simeq (\mu B). \tag{1.5}$$

It is interesting to note that Hall describes in his paper [2] an unsuccessful experiment of others with the goal to demonstrate the *current deflection effect*. On the page 186 we read: '(He) took a wire bifurcated throughout a part of its length and passed through it a current. He then endeavored by means of a magnet to divert the current somewhat from one branch of the wire and draw into the other branch more than its normal share.' We know now that the current deflection effect really exists, and that the described experiment was not successful only because of the crude means used at the time to detect it.

1.1.2 The magnetoresistance effect

In addition to the generation of a Hall voltage and/or a current deflection, a magnetic field causes a change in the electrical resistance of the sample. This is called the magnetoresistance effect. The magnetoresistance effect was discovered by *William Thomson* (*Lord Kelvin*) in 1856 [3]. Although Thomson describes in his paper the form of the magnetoresistance effect characteristic for ferromagnetic thin sheets, which is not the subject of this book, I can't help showing you one of his beautiful drawings, figure 1.3.

Thomson's experimental device is reminiscent of that of Hall (figure 1.1) (or, in view of the timing, vice-versa). Both devices were basically four-terminal thin metallic plates, fitted with two big input (current) contacts and two smaller output (voltage) contacts. In addition, Thomson's plate had the provision for offset cancelling—see figure 5.29 in §5.6.3. The essential difference is that Thomson's plate was ferromagnetic, whereas this was not relevant for Hall's plate.

Briefly, Thomson's experiment run as follows: while the electromagnet was switched off, he adjusted the position of the sliding contact K so that the output voltage V vanished. Then, he switched the electromagnet on, and the voltage V reappeared. He concluded that 'the electric conductivity

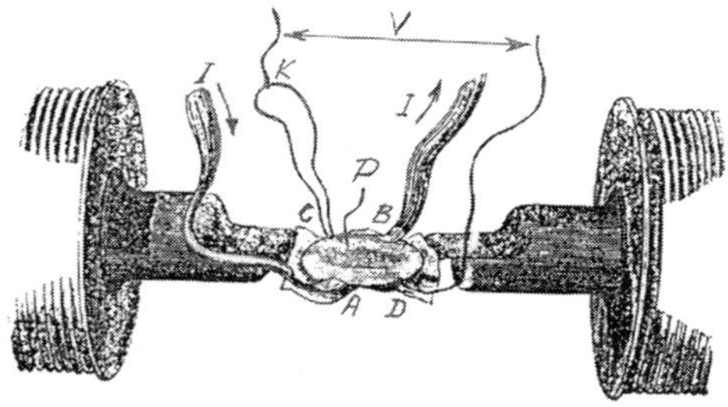

Figure 1.3. Thomson's experimental set-up with which he discovered the magnetoresistance effect in ferromagnetic thin plates. The device under test is a thin ferromagnetic plate P, fitted by four electrical contacts A, B, C, D. The plate is placed between the poles of an electromagnet so that the magnetic field is parallel to the plate and the line joining the electrodes A and B makes an angle of 45° to the magnetic field. A current I is forced between the contacts A and B, and the voltage difference V is measured between the contacts C and D. The sliding contact K is used to zero the output voltage V before switching on the magnetic field (reprinted from [3]).

is in reality greater across than along the lines of magnetization in magnetized iron.' We know now that the effect Thomson observed is due to the combination of two effects: the so-called magneto-concentration effect (see figure 5.37 in §5.7) and a form of the galvanomagnetic effects that appear also in non-magnetic layers (see §3.3.11).

To the best of my knowledge, it was Hall who described for the first time (but not experimentally verified) the magnetoresistance effect in non-magnetic materials. At the first page of his paper [2] he depicts very precisely the device shown in figure 3.5, §3.1.2, which is now known as a *Corbino disk*. He describes his vision of the effects in the device as: 'In such an apparatus the flow of electricity would be along the radii of the disk, but if a strong magnetic force (i.e. field) were made to act perpendicularly to the face of the disk a new electromotive force would be set up, which would be always perpendicular to the direction of the magnetic force (i.e. field) and to the actual direction of flow of the electricity. The resulting path of the electricity from the center to the circumference of the disk would be, not a straight line as under normal conditions, but a spiral. This line being longer than straight line, we would expect an increase of electrical resistance in the disk of gold leaf.' This is a precise summary of the modern theory of the geometrical magnetoresistance effect that we treat in §§3.1.2 and 3.1.3!

Therefore, the conventional Hall effect, the current deflection effect, and the magnetoresistance effect in non-ferromagnetic materials, have the same physical origin. This is why we ascribe in this book a broader sense to the notion of the Hall effect and treat the current deflection effect and the magnetoresistance effect as forms of the Hall effect.

The physics of the Hall effect is one of the main issues of this book. In the detailed treatment of the Hall effect, we shall have to take into account the type of charge carriers present in material, their concentration and their kinetic properties. One of the questions that will arise during the analysis is: what is the influence of the thermal agitation and scattering of carriers on the Hall effect? And still another: what happens if a Hall device becomes so small, that charge carriers pass through it without any scattering?

1.2 Hall effect devices and their applications

Under the term 'Hall effect devices' we understand all solid-state electron devices whose principle of operation is based on the Hall effect with the above broader sense of this notion.

In this field the following *terminology* is used: Hall effect devices similar to that with which Hall discovered his effect (figure 1.1) are today called 'Hall plates'. This term reflects the conventional form of a Hall effect device. Another term that reflects the form is a 'Hall cross' (compare figures 1.1 and 4.15(d)). In physics literature a Hall cross is sometimes called a 'Hall junction'. A merged combination of at least two Hall crosses (figure 4.15(b)) is called a 'Hall bridge'. If the form of the device is irrelevant, we use the terms 'Hall device', 'Hall element' or 'Hall cell'. In Japanese literature, the term Hall element is used for a discrete Hall device applied as a magnetic sensor. If we want to stress the application of a Hall device, we usually say a Hall magnetic sensor or just a 'Hall sensor'. In German literature we often see the expression 'Hall generator', which gives a hint on an aspect of its operation. In the magnetic field measurement community, a Hall magnetic field sensor, packaged in a suitable case, is normally referred to as a 'Hall probe'. An integrated circuit, incorporating a combination of a Hall device with some electronic circuitry, is usually called an integrated Hall effect magnetic sensor, 'Hall ASIC' (application-specific integrated circuit) or just a 'Hall IC'.

1.2.1 Hall plates

A simple-geometry plate-like Hall device suitable for our further analysis is shown in figure 1.4. This is a thin plate of conducting material fitted with four electrical contacts at its periphery. A bias current I is supplied to the device via two of the contacts, called the current contacts. The other two contacts are placed at two equipotential points at the plate boundary. These contacts are called the voltage contacts or the sense contacts. If a perpendicular magnetic field is applied to the device, a voltage appears between the sense contacts. This voltage is called the Hall voltage.

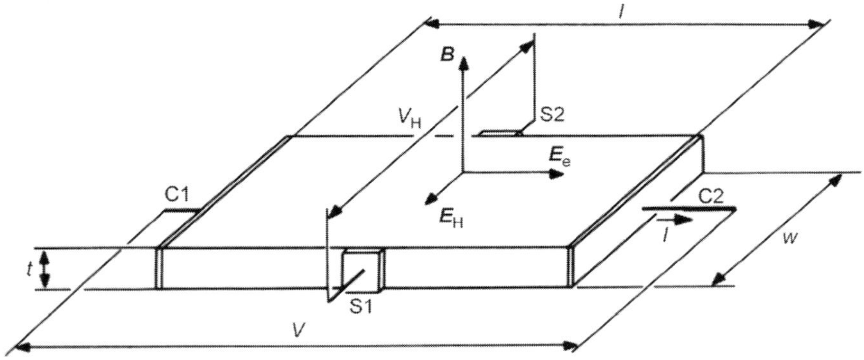

Figure 1.4. Hall device in the form of a rectangular plate. Modern Hall plates are usually of microscopic dimensions. For example, the thickness might be $t = 10\,\mu m$. the length $l = 200\,\mu m$ and the width $w = 100\,\mu m$. A bias voltage V is applied to the plate via the two current contacts C1 and C2. The bias voltage creates an electric field $\boldsymbol{E}_e$ and forces a current I. If the plate is exposed to a perpendicular magnetic induction $\boldsymbol{B}$, the Hall electric field $\boldsymbol{E}_H$ occurs in the plate. The Hall electric field gives rise to the appearance of the Hall voltage V_H between the two sense contacts S1 and S2.

The Hall voltage can be calculated by integrating the Hall electric field over the distance between the sense contacts:

$$V_{\mathrm{H}} = \int_{\mathrm{S1}}^{\mathrm{S2}} \boldsymbol{E}_{\mathrm{H}}\, \mathrm{d}\boldsymbol{w}. \tag{1.6}$$

For the configuration of figure 1.4, the external field is given by

$$\boldsymbol{E}_{\mathrm{e}} = (V/l)\boldsymbol{i} \tag{1.7}$$

where V denotes the voltage across the current contacts, l is the device length, and $\boldsymbol{i}$ is the unity vector collinear with the device longitudinal axis. The external field is collinear with the device longitudinal axis. Therefore, from (1.3), we see that the Hall field is collinear with the line connecting the two sense contacts. From (1.3), (1.6) and (1.7), we obtain the Hall voltage in the form

$$V_{\mathrm{H}} \simeq \mu \frac{w}{l} VB. \tag{1.8}$$

Here w denotes the plate width.

The Hall voltage can also be expressed in terms of the bias current I:

$$V_{\mathrm{H}} \simeq \frac{1}{qnt} IB. \tag{1.9}$$

Here q denotes the magnitude of the electron charge, n is the carrier concentration in the plate and t is the plate thickness.

The above simple formulae approximate reasonably well the Hall voltage of a very long rectangular Hall plate, with very small sense contacts. However, Hall devices suitable for practical applications usually have different shapes and large sense contacts. The question now arises, how to cope with the different geometries of Hall plates? Even more generally, are non-plate-like Hall devices feasible and tractable? These are some of the issues treated in this book.

1.2.2 Other Hall effect devices

Generally, the Hall effect takes place in any solid-state electron device exposed to a magnetic field. But the corresponding magnetic modulations of the characteristics of 'normal' electronic devices, such as diodes and transistors, are negligible. It turns out, however, that the structure and operating conditions of most semiconductor devices can be optimized with respect to the magnetic sensitivity of their characteristics. The optimization of the structures yields somewhat-modified conventional semiconductor devices, whose current–voltage characteristics become magnetically sensitive. In the literature, such magnetically sensitive devices are referred to as *magneto-transistors*, *magneto-diodes*, *MAG-FETs* (FET: field-effect transistor) and so on. Since the magnetic sensitivity of all these devices is a consequence of the Hall effect, we also include them in the family of Hall effect devices.

Note the difference in the terms we use: a Hall element is a device reminiscent of Hall's gold leaf with four contacts; a Hall effect device is any solid-state device whose principle of operation is based on the Hall effect.

The first Hall effect device, which was not similar to a Hall plate, was devised by Suhl and Shockley [15]. In 1949, Suhl and Shockley studied carrier transport phenomena in a point-contact transistor exposed to an external magnetic field. By varying the magnetic field, they were able to modulate the conductance of a semiconductor filament used as the base region of the transistor. Another early report concerns a change in the current gain of a transistor arising from a change in a magnetic field [16]. Later a large number of various magnetically sensitive semiconductor devices were proposed [17]: any known semiconductor device has probably got a 'magneto' brother. We shall discuss their structure and operation in an appropriate chapter.

In the current academic research on magnetic sensors based on the Hall effect, the following two approaches can be distinguished. First, one tries to build better sensors based on conventional Hall devices. The innovations come through a better understanding of the details of the operating principle, secondary effects, new applications of the mainstream microelectronics technologies, and through combination of Hall devices with other elements of a magnetic sensor system, such as interface electronics and magnetic circuit. In the second approach, one hopes to build better sensors by making use of the Hall effect in active devices, such as magneto-transistors and MAG-FETS. Many interesting magnetic sensors of this kind have been proposed during the past 40 years. However, none of them has (yet) become a competitor to the Hall devices [18, 19]. Why not? Is there any fundamental reason?

1.2.3 Applications of Hall effect devices

In more than a hundred years of their history, Hall effect devices have been used to demonstrate the basic laws of physics, to study details of carrier transport phenomena in solids, to detect the presence of a magnet, and as measuring devices for magnetic fields.

Consider briefly the principles of application of a Hall plate. The Hall voltage of a Hall plate may be regarded as a signal carrying information. If we know the material properties, device geometry and biasing conditions, the Hall voltage (1.9) can give us information about the magnetic induction $\boldsymbol{B}$. In this case the Hall device is applied as a magnetic sensor. Alternatively, we may control the biasing conditions and magnetic induction of a Hall device with a known geometry. From the measured Hall voltage, we may deduce some important properties of the material the device is made of (see equations (1.8) and (1.9)). In this case the Hall device is applied as a means of characterizing material. Similarly, other Hall effect devices can also be applied either as magnetic sensors or as material analysis tools.

The sensor applications of Hall effect devices became important only with the development of semiconductor technology. For one thing, the Hall effect is only strong enough for this purpose in some semiconductors. Therefore, the first Hall effect magnetic sensors became commercially available in the mid-1950s, a few years after the discovery of high-mobility compound semiconductors. Since then, the development of Hall effect devices has taken advantage of using high-quality materials and sophisticated, highly productive fabrication methods available in the microelectronics industry. On the other hand, advances in microelectronics have also created the need for cheap and good sensors of various non-electrical signals, including magnetic field.

Today, Hall effect magnetic sensors form the basis of a mature and important industrial activity. They are mostly used as key elements in contact-less sensors for linear position, angular position, velocity, rotation, electrical current, and so on. We estimated that more that two billion Hall magnetic sensors were sold worldwide in the year 2000 [18]. Currently (2003), the global *market of Hall magnetic sensors* is estimated to be about $600 million [20]. Moreover, these products have a big leverage effect: Hall sensors are the key enabling technology for much higher volume business—and this grows at more than 10% per year. Most currently produced Hall magnetic sensors are discrete elements, but an ever-increasing portion comes in the form of integrated circuits. Integrated Hall magnetic sensors are 'smart': they incorporate electronic circuits for biasing, offset reduction, compensation of temperature effects, signal amplification, and more. The integration helps improve sensor system performance at moderate costs, which allows a continuous penetration of Hall magnetic sensors into new application areas.

In the field of scientific research, Hall magnetic sensors have been extremely valuable tools for accurate measurement of magnetic fields in particle accelerators, for characterizing superconductors, for detecting and characterizing small magnetic particles, and so on.

One of the aims of this book is to establish clear relationships between the basics of the Hall effect and the most important applications of Hall effect devices. To this end, in the appropriate chapters, we shall be particularly interested in the system aspects of the incorporation of a Hall device into an electronic circuit and/or in combining a Hall magnetic sensor with other components of a sensor system, such as a magnets and magnetic circuit.

References

[1] Hall E H 1879 On a new action of the magnet on electric current *Am. J. Math.* **2** 287–92

[2] Hall E H 1880 On a new action of magnetism on a permanent electric current *Am. J. Sci.* series 3, **20** 161–86

[3] Thomson W 1856 On the effects of magnetisation on the electric conductivity of metals *Philos. Trans. R. Soc.* **A146** 736–51

[4] Putley E H 1960 *The Hall Effect and Semi-conductor Physics* (New York: Dover)

[5] Beer A C 1983 Galvanomagnetic effects in semiconductors *Solid State Phys.* suppl. 4 (New York: Academic Press)

[6] Kuchis E V 1974 *Metody Issledovanija Effekta Holla* (Moscow: Sovetskoe Radio)

[7] Wieder H H 1979 *Laboratory Notes on Electrical and Galvanomagnetic Measurements* (Amsterdam: Elsevier)

[8] Kuhrt F and Lippmann H J 1968 *Hallgeneratoren* (Berlin: Springer)

[9] Weiss H 1969 *Physik und Anwendung galvanomagnetischer Bauelemente* (Braunschweig: Vieweg); 1969 *Structure and Application of Galvanomagnetic Devices* (Oxford: Pergamon)

[10] Wieder H H 1971 *Hall Generators and Magnetoresistors* (London: Pion)

[11] Homeriki O K 1986 *Poluprovodnikovye Preobrazovateli Magnitnogo Polja* (Moscow: Energoatomizdat)

[12] von Klitzing K, Dorda G and Pepper M 1980 New method for high-accuracy determination of fine-structure constant based on quantized Hall resistance *Phys. Rev. Lett.* **45** 494–7

[13] Yennie D R 1987 Integral quantum Hall effect for nonspecialists *Rev. Mod. Phys.* **59** 781–824

[14] O'Handley R C 2000 *Modern Magnetic Materials* (New York: Wiley)

[15] Suhl H and Shockley W 1949 Concentrating holes and electrons by magnetic fields *Phys. Rev.* **75** 1617–18

[16] Bradner Brown C 1950 High-frequency operation of transistors *Electronics* **23** 81–3

[17] Baltes H P and Popovic R S 1986 Integrated semiconductor magnetic field sensors *Proc. IEEE* **74** 1107–32

[18] Popovic R S, Schott C, Shibasaki I, Biard J R and Foster R B 2001 *Hall-effect Magnetic Sensors*, in: P Ripka (ed), *Magnetic Sensors and Magnetometers*, Chapter 5 (Boston: Artech House)

[19] Popovic R S, Flanagan J A and Besse P A 1996 The future of magnetic sensors *Sensors and Actuators* **A56** 39–55

[20] Popovic D R, Fahrni F and Stuck A 2003 Minimizing investments in production of sensor micro systems *The 8th International Conference on the Commercialization of Micro and Nano Systems* 8–11 September 2003, Amsterdam, Holland

Chapter 2

Semiconductor physics: a summary

Before we start discussing details of the Hall effect, the magnetoresistance effect, and of the corresponding devices, we shall pass here quickly through a few general sections of semiconductor physics. Our aim is to refresh our understanding of some basic properties of semiconductors. For example, we shall see how and why quasi-free electron and hole concentrations in semiconductors depend on temperature and doping, and how charge transport phenomena, such as drift and diffusion, operate in the absence of a magnetic field. The physical phenomena that underline these properties are also basic to the galvanomagnetic effects.

There are many good books that treat the details of subjects mentioned here. For example, [1–4] on general semiconductor physics, and [5–7] on the physics of semiconductor devices.

2.1 Energy bands

All characteristic properties of semiconductors are the consequence of a basic physical phenomenon: the existence of certain energy bands in the energy spectrum of electrons. In this section we shall consider the basic relationships between the electrical properties of solid-state materials and the structure of energy bands.

2.1.1 Conductor, semiconductor, insulator

According to their electrical conductivity, materials are conventionally classified into three groups: conductors, semiconductors and insulators. Conductors, which include all metals, have high conductivities of $\sigma = (10^7 - 10^6)\,\mathrm{S\,m^{-1}}$, semiconductors show intermediate conductivities of $\sigma = (10^{-8} - 10^6)\,\mathrm{S\,m^{-1}}$, and insulators have low conductivities of $\sigma = (10^{-8} - 10^{-16})\,\mathrm{S\,m^{-1}}$. The distinction between semiconductors and insulators is only quantitative, whereas the distinction between semiconductors and metals is more profound.

The electrical conductivity of a metal is a rather stable property. For example, its dependence on temperature is weak. At about room temperature, the resistance of a metal sample increases with the increase in temperature:

$$R(T) \simeq R_0[1 + \alpha(T - T_0)]. \tag{2.1}$$

Here $R(T)$ is the resistance at an absolute temperature T, $T_0 = 273\,\mathrm{K}$, R_0 is the resistance at T_0, and α is the temperature coefficient of resistance. The value of α is of the order of $0.003\,\mathrm{K}^{-1}$.

On the other hand, the electrical conductivity of semiconductors and insulators is strongly dependent on the environmental conditions, such as temperature and illumination, and on the purity of the material. The resistance of a pure semiconductor sample generally decreases rapidly with an increase in temperature:

$$R(T) \simeq R_0 \exp\left(\frac{E_a}{kT}\right). \tag{2.2}$$

Here $R(T)$, T and R_0 have the same meaning as in (2.1), E_a is a quantity called the activation energy, and k is the Boltzmann constant. The activation energy is a characteristic parameter of each semiconductor. It is approximately constant within a limited temperature range. In different temperature ranges, its value and physical meaning might be different. Impure semiconductors may also show the resistance and temperature dependence of the type (2.1), but only in a limited temperature range.

The reason for such quite different behaviour of metals and semiconductors is that metals contain a constant number of mobile charge carriers at all temperatures and semiconductors do not. To become free to move, charge carriers in a semiconductor must first be activated. The activation requires some energy; this energy can be provided, for example, by thermal agitation. We shall see later that the activation energy is related to the energy band structure of the material.

The semiconductor materials of greatest practical importance are the elements silicon (Si) and germanium (Ge), and many III–V and II–VI compounds, such as gallium arsenide (GaAs), indium antimonide (InSb), indium arsenide (InAs) and cadmium sulphide (CdS).

The fundamental charge carriers in these materials are electrons. For this reason they are called electronic semiconductors. There are also semiconductor materials where conduction is due to the transport of ions. Such materials are called ionic semiconductors. In this book we shall be concerned only with electronic semiconductors.

2.1.2 Permitted and forbidden energy bands

Electrons in a crystalline solid interact with all other particles in the solid. Owing to the laws of quantum mechanics, they may then possess only the

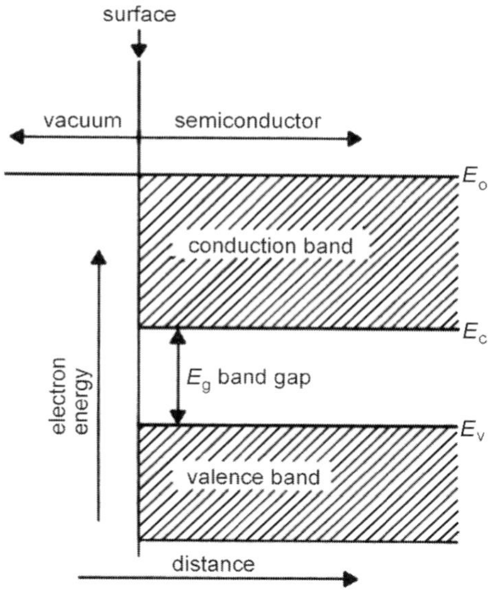

Figure 2.1. Energy band diagram of a semiconductor.

energies that belong to some permitted energy regions. In a diagram distance versus energy, the permitted energy regions appear as bands. Between the permitted energy bands are forbidden energy bands, or energy gaps. In an energy gap there are no allowed quantum states of electrons.

Theoretically, there is an unlimited number of energy bands. We are interested here only in those bands that determine the electron transport properties of solids. These bands contain the electrons with highest energies that still do not leave the solid material. High-energy electrons are not strongly bound to their native atoms; under certain conditions, they may wander through a solid.

The relevant part of the energy band diagram of a semiconductor is shown in figure 2.1. Let us define the zero electron energy as the energy of an electron at rest outside the sample, in vacuum. At and below the zero electron energy is a permitted energy band, called the *conduction band.* A semiconductor can conduct electricity only if there are some electrons in its conduction band (hence the name). The energy level at the bottom of the conduction band is designated by E_c. The next lower permitted energy band is called the *valence band.* It is so named because it is occupied by the valence electrons of the atoms the sample consists of. The energy at the top of the valence band is denoted by E_v. Between the two permitted bands is an *energy gap.* Its width

$$E_g = E_c - E_v \tag{2.3}$$

is called the band gap. This is one of the most important parameters in semiconductor physics.

The band gaps of pure silicon and gallium arsenide at room temperature are 1.12 and 1.43 eV, respectively. The band gaps of some other semiconductors are listed in Appendix C. The bandgaps of most semiconductors decrease with an increase in temperature. For pure silicon and gallium arsenide, the temperature coefficients dE_g/dT at room temperature are -2.55×10^{-4} and $-4.52 \times 10^{-4}\,\text{eV K}^{-1}$, respectively. Near room temperature the band gap of silicon decreases with pressure, and that of gallium arsenide increases with pressure.

Briefly, a *conductor* is a material which has plenty of electrons in the conduction band; a *semiconductor* needs an activation to get electrons in its conduction band; an *insulator* can be understood as a semiconductor with a big energy gap, say larger than 5 eV, so that electrons cannot be easily activated.

2.1.3 Electrons in solids

Recall that an electron in vacuum can be represented by a de Broglie plane wave. In terms of the wave vector $\boldsymbol{k}$ of the de Broglie wave, the electron momentum is given by

$$\boldsymbol{p} = \hbar \boldsymbol{k} \tag{2.4}$$

its velocity by

$$\boldsymbol{v} = \hbar \boldsymbol{k}/m \tag{2.5}$$

and its energy by

$$E = E_p + \frac{\hbar^2 (\boldsymbol{k})^2}{2m}. \tag{2.6}$$

Here $\hbar = h/2p$ (h being the Planck constant), m is the electron rest mass and E_p is the potential energy of the electron.

Similarly, a loosely bound electron in a crystalline solid can be represented by a somewhat-modified plane wave. The corresponding wavefunction is called the *Bloch function* and has the form

$$\psi_k(\boldsymbol{r}) = \exp(i\boldsymbol{k} \cdot \boldsymbol{r})\varphi_k(\boldsymbol{r}). \tag{2.7}$$

Here $\boldsymbol{k}$ is the wavevector, $\boldsymbol{r}$ is the position vector and $\varphi_k(\boldsymbol{r})$ is a periodic function with the period of the lattice potential field. The function (2.7) may be regarded as a plane wave $\exp(i\boldsymbol{k} \cdot \boldsymbol{r})$ with a variable amplitude $\varphi_k(\boldsymbol{r})$ modulated with the period of the crystal lattice. The electron energy E is a function of the wavevector $\boldsymbol{k}$: $E(\boldsymbol{k})$. All other parameters of the motion of electrons are related to $\boldsymbol{k}$ and the functions $E(\boldsymbol{k})$.

The band structure of the electron energy spectrum in a crystalline solid is a consequence of the interference of the Bloch waves. In particular, whenever the Wulf–Bragg condition between the de Broglie wavelength of an electron and the period of the crystal lattice is fulfilled, the electron wave will be totally reflected from the lattice. Owing to the interference of the incident and reflected waves, a standing wave arises. An electron with the corresponding wavevector $\boldsymbol{k}_0$ cannot propagate in the crystal. At such values of $\boldsymbol{k}$ the discontinuities in the electron energy spectrum appear. This corresponds to the energy levels at the limits of the energy bands, such as E_c and E_v in figure 2.1.

2.1.4 Electrons in the conduction band

In the $\boldsymbol{k}$ space, a discontinuity in the electron energy spectrum corresponds to an extreme point of the function $E(\boldsymbol{k})$. At the extreme point, $\mathrm{d}E/\mathrm{d}\boldsymbol{k}|_{k_0} = 0$. Around this point the electron energy can be approximated by the quadratic function

$$E(\boldsymbol{k}) \simeq E(\boldsymbol{k}_0) + \frac{1}{2}\sum_{i=1}^{3} \frac{\partial^2 E}{\partial k_i^2}(k_i - k_{0i})^2 \tag{2.8}$$

where k_i denotes the appropriate coordinate in the $\boldsymbol{k}$ space and $\boldsymbol{k}_0$ is the position vector of the extreme point. The term $E(\boldsymbol{k}_0)$ may be either a minimum or a maximum energy in a permitted band. At the minimum of the conduction band, $E(\boldsymbol{k}_0) = E_c$, while at the maximum of the valence band, $E(\boldsymbol{k}_0) = E_v$.

Consider the electrons in the conduction band of a semiconductor. As we shall see in the next section, most of these electrons occupy the energy states slightly above the bottom of the conduction band. Hence (2.8) describes accurately enough their $E(\boldsymbol{k})$ relationships. Besides, the number of electrons in the conduction band is normally much less than the number of available quantum states. Therefore, the electrons 'feel' practically no restriction due to the exclusion principle to take any of the energy states. Within a crystal, they can move and can be accelerated. Thus in many respects, such electrons behave like free electrons in vacuum. They are said to be quasi-free. For this reason, we may take the relations (2.6) and (2.8) to be analogous. Then the first term in (2.8) can be interpreted as the potential energy of an electron in the conduction band, and the second term as its kinetic energy.

By comparing the kinetic energy terms in (2.6) and (2.8), the following two physical quantities of electrons in solids have been introduced: the quasi-momentum, or crystal momentum,

$$\boldsymbol{P} = \hbar \boldsymbol{k} \tag{2.9}$$

and the effective masses

$$\frac{1}{m_i^*} = \frac{1}{\hbar^2} \frac{\partial^2 E(\boldsymbol{k})}{\partial k_i^2}\bigg|_{k_{0i}} \qquad i = 1, 2, 3. \tag{2.10}$$

Equation (2.8) can now be rewritten as

$$E(\boldsymbol{k}) = E_c + \frac{1}{2}\sum_{i=1}^{3}\frac{(\hbar k_i - \hbar k_{0i})^2}{m_i^*}. \tag{2.11}$$

Thus an electron in the conduction band of a semiconductor behaves like a free electron with a potential energy E_c, a momentum $(\hbar\boldsymbol{k} - \hbar\boldsymbol{k}_0)$ and an *effective mass* m_i^*.

The physical meaning of the existence of the three generally different values of the effective mass is that the acceleration of an electron depends on the direction of the force acting on the electron: the periodicity of the crystal lattice may be different in various directions. The values m_i^* are called the components of the effective mass. A convenient way to visualize the influence of the three components of the effective mass is to look at the constant energy surfaces $E(\boldsymbol{k}) = \text{const}$ (see figure 2.2). If all three effective

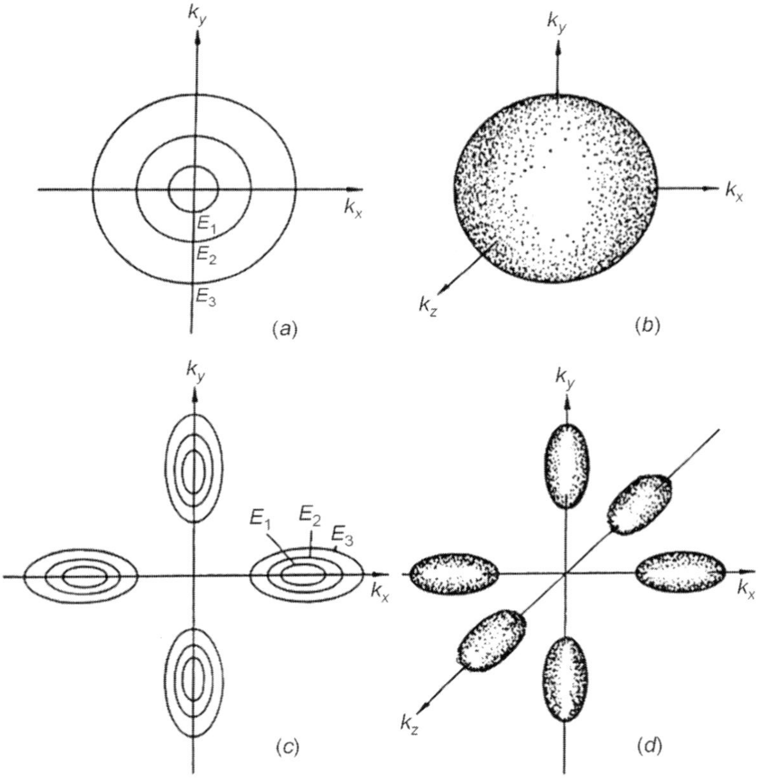

Figure 2.2. Constant energy surfaces in the k-space: (a) and (b), spherical constant energy surfaces; (c) and (d), ellipsoidal constant energy surfaces. In (a) and (c) three constant energy contours are shown, with $E_1 < E_2 < E_3$. (Adapted from [3].)

masses are equal, $m_1^* = m_2^* = m_3^*$, the constant energy surfaces reduce to spheres around the point $\boldsymbol{k}_0$. For example, this is the case with GaAs, where also $\boldsymbol{k}_0 = 0$, as shown in figure 2.2(a) and (b). If $m_1^* = m_2^* \neq m_3^*$, then the constant energy surfaces take the form of ellipsoids of revolution, as shown in figure 2.2(c) and (d). For example, this is the case with silicon. The rotation axes of the constant energy ellipsoids in silicon are parallel with the (100) crystal directions and there are six equivalent energy minima $E(\boldsymbol{k}) = E_c$. The effective mass along the rotation axis is called the *longitudinal effective mass* m_l^*; that perpendicular to the rotation axis is called the *transverse effective mass* m_t^*. In silicon, $m_l^* > m_t^*$. The constant energy surfaces are extended in the direction of the larger effective mass.

When we made the analogy between a free electron and an electron in the conduction band, we tacitly assumed that the effective mass of the latter was a positive quantity. This assumption is correct. The conduction band electrons occupy the energy levels slightly above the conduction band minima E_c. At a minimum of a function, the second derivatives in (2.10) are positive.

The effective mass of conduction band electrons is usually smaller than the normal mass of an electron m. In a crystal, electrons are lighter to accelerate than in vacuum. For example, the effective mass of conduction band electrons in GaAs is $0.068m$. The effective masses of carriers in some semiconductors are listed in Appendix C.

2.1.5 Holes in the valence band

Let us now turn to the electrons in the valence band. The extreme point $\boldsymbol{k}_0$ in (2.8) denotes a maximum of the $E(\boldsymbol{k})$ function, $E(\boldsymbol{k}_0) = E_v$. Typically, this maximum appears at $\boldsymbol{k}_0 = 0$. At a maximum of a function, the second derivatives are negative. Thus the effective mass (2.10) of electrons at the top of the valence band is negative. This means that an electric field in a crystal should accelerate these electrons along the field and thus generate a current in a wrong direction. Yet this does not happen, since the electrons in the valence band cannot be accelerated at all.

The acceleration of an electron means its successive transfer to the next energy levels, that is the next quantum states. However, for a valence band electron, the next quantum states are almost always occupied. (The number of electrons in the valence band is almost equal to the number of the available quantum states.) According to the Pauli exclusion principle, multiple occupancy of a quantum state is impossible. Therefore, a field cannot affect the acceleration of a valence electron. However, this does not mean that valence electrons do not take part in the transport phenomena: they do, but indirectly, via holes.

Holes are vacancies in the almost fully occupied matrix of valence electrons. At such a vacancy, the electrical neutrality of the crystal is disturbed: the positive charge prevails. The charge surplus is exactly equal to the absolute value of the charge of the missing electron. Holes 'feel' practically no restriction in exchanging their quantum states with the surrounding valence band electrons. So within a crystal, holes can wander and can be accelerated. Consequently, in many respects holes behave like positive quasi-free electrons.

The $E(\boldsymbol{k})$ relation of holes is the same as that of valence band electrons: holes are the absent valence electrons. However, contrary to the former case, the effective mass of holes is a positive quantity. This is clear from the following reasoning.

Consider a valence electron at an energy level E_e slightly beneath the top of the valence band E_v. The relative energy of this electron, with reference to E_v, is $\Delta E_e = E_e - E_v$. Since E_v is an energy maximum, ΔE_e is negative, $\Delta E_e < 0$. Here we are only interested in the electrodynamic behaviour of electrons and holes. It is therefore reasonable to express the relative energy of the electron in terms of an equivalent electric potential difference $\Delta\varphi$. This is given by $\Delta\varphi = \Delta E_e/(-q)$, where $-q$ is the negative charge of the electron. Imagine now that we substitute this electron by a hole. The hole possesses a relative energy $\Delta E_h = E_h - E_v$ and this must also be equivalent to the electric potential difference $\Delta\varphi$. Thus the relative energy of the hole is $\Delta E_h = q\Delta\varphi = -\Delta E_e$, where q is the positive charge of the hole. Since $\Delta E_e < 0$, $\Delta E_h = -\Delta E_e > 0$. This means that the relative energy of the hole is positive, $\Delta E_h = E_h - E_v > 0$. Consequently, the top of the valence band E_v is an energy minimum for holes. At an energy minimum, the second derivatives (2.10) are positive. Thus the effective mass of holes is also positive: an electric field accelerates a hole in the correct direction. The same result can be also obtained formally by stating that an electron (a negatively charged particle) with a negative effective mass is equivalent to a positively charged particle (that is, a hole) with a positive effective mass.

In conclusion, relation (2.11), which is analogous to the corresponding relation for a free electron, is valid for electrons in the conduction band as well as for holes in the valence band of semiconductors. The potential energy for electrons is the bottom of the conduction band E_c, while the potential energy for holes is the top of the valence band E_v. Simplified $E(\boldsymbol{k})$ relations for electrons and holes are illustrated in figure 2.3. In accordance with the above discussion, the positive energies of electrons and holes are measured in opposite directions. Conventionally, the electron energy is measured upwards, as before in figure 2.1. The hole energy is then measured downwards. This is compatible with the downward positive direction for displaying electric potential.

In some semiconductors there are two different paraboloids $E(\boldsymbol{k})$ at the top of the valence band: the valence band is degenerate. The two $E(\boldsymbol{k})$ functions have different second derivatives. Thus in such semiconductors there are holes with two different effective masses: they are known as light and heavy holes.

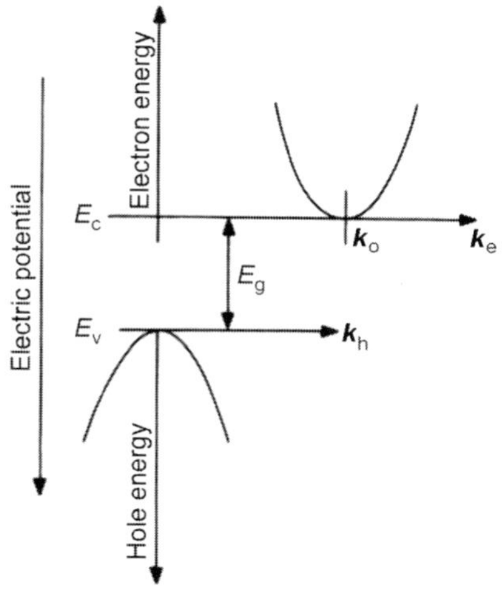

Figure 2.3. Two-dimensional $E(\boldsymbol{k})$ diagram for electrons and holes. $\boldsymbol{k}_e$ and $\boldsymbol{k}_h$ are the wavevectors of electrons and holes, respectively. The electron energy is measured upwards, and the hole energy and the electrical potential downwards.

2.2 Distribution and concentration of carriers

A solid can conduct electricity only if it contains some mobile charge carriers. As we have seen in the preceding section, in semiconductors there are two types of mobile or quasi-free charge carriers. These are electrons in the conduction band and holes in the valence band. The appurtenance to the band is often omitted, and the charge carriers are simply referred to as electrons and holes.

Our goal in this section is to find out how the concentrations of electrons and holes depend on the characteristics of a semiconductor and on temperature. To do this we need to know the density of states in each of the bands and how the charge carriers are distributed over the available states. Throughout this section, we shall assume the conditions of thermodynamic equilibrium.

2.2.1 Density of states

The number of quantum states in an energy band is finite. This is a direct consequence of the wave nature of electrons. Only those waves with precisely defined wavevectors can fit into a crystal. Hence the spectrum of the permitted wavevectors $\boldsymbol{k}$ is discrete. Owing to the $E(\boldsymbol{k})$ relation, the spectrum of the permitted energy values, or energy levels, is also discrete.

Consider a Bloch wave (2.7) along a main crystal axis i. Let the crystal along this axis have N_i cells of length a_i. The possible wavelengths of the Bloch wave are then restricted to the set of discrete values $\lambda_i = N_i a_i / n$, $n = 1, 2, \ldots$. Here n is an integer, which denotes the number of whole wavelengths that fit the crystal. Thus the permitted values of the wavenumber k_{pi} are

$$k_{pi} = \frac{2\pi}{\lambda_i} = 2\pi \frac{n}{N_i a_i} \qquad n = \pm 1, \pm 2, \ldots. \tag{2.12}$$

By including both signs of k_{pi}, we take into account both directions of wave propagation. The points k_{pi} are uniformly distributed along the k_i axis. Since a similar result holds for each k axis, we conclude that the points associated with the permitted $\boldsymbol{k}$ values are uniformly distributed in the $\boldsymbol{k}$ space. The density of these points in the $\boldsymbol{k}$ space per unit crystal volume can be obtained from (2.12), by substituting $i = 1, 2, 3$. This is

$$g_k = 2/(2\pi)^3. \tag{2.13}$$

Here the number 2 in the numerator stands for the two electron spin orientations.

Let us now denote by ΔG the number of states in the energy interval $\Delta E = E_2 - E_1$. ΔG is obviously equal to the number of permitted points enclosed between the two constant energy surfaces E_1 and E_2 in the $\boldsymbol{k}$ space (see figure 2.2). Since the density of the permitted points in the $\boldsymbol{k}$-space is constant (see 2.13)), $\Delta G = g_k \Delta V_k$. Here ΔV_k denotes the volume of the shell between the two constant energy surfaces E_1 and E_2.

The density of states per unit crystal volume and per unit energy interval is then

$$g(E) = \lim_{\Delta E \to 0} \frac{\Delta G}{\Delta E} = g_k \lim_{\Delta E \to 0} \frac{\Delta V_k}{\Delta E}. \tag{2.14}$$

$g(E)$ is referred to in short as the density of states. The density of states near the bottom of the conduction band can be found with the aid of (2.11) and (2.14) to be

$$g_c(E) = \frac{\sqrt{2}}{\pi^2 \hbar^3} M_c (m_{de}^*)^{3/2} (E - E_c)^{1/2}. \tag{2.15}$$

Here M_c is the number of equivalent energy minima in the conduction band and m_{de}^* is the density-of-states effective mass for electrons

$$m_{de}^* = (m_1^* m_2^* m_3^*)^{1/3}. \tag{2.16}$$

For example, for silicon $M_c = 6$ and $m_{de}^* = (m_1^* m_2^{*2})^{1/3}$; for GaAs, $M_c = 1$ and $m_{de}^* = m^*$.

A relation similar to (2.15) also holds for the density of states $g_v(E)$ near the top of the valence band.

In semiconductors, the total number of states in the valence band and in the conduction band is equal to the number of valence electrons. In metals, the total number of states in the valence band is larger than the number of valence electrons. This is the reason why the electrons in metals need no activation to become mobile.

2.2.2 Distribution function

Owing to the universal tendency of all particles in nature to take the lowest energy position available, the bottom of the valence band should be overcrowded with electrons. Yet this does not happen. The reasons are the following: the density of states at each energy is limited and, according to the Pauli exclusion principle, multiple occupancy of a quantum state is impossible. Therefore, the electrons have to occupy the energy levels over the whole valence band.

The probability of occupancy is practically equal to unity for all lower levels. Only those levels near the top of the valence band might not for sure be occupied by electrons. Because of the thermal agitation, some valence electrons might get enough energy to jump to the conduction band. The greatest probability to find an electron there is, of course, at the bottom of the conduction band. The higher the temperature, the more likely it is to find an electron in the conduction band. The probability distribution function, which we have just described, is called the *Fermi–Dirac distribution function*:

$$F(E) = \frac{1}{1 + \exp[(E - E_F)/kT]}. \tag{2.17}$$

Here $F(E)$ is the probability for an energy level E to be occupied by an electron, E_F is a parameter known as the *Fermi* (energy) *level*, k is the Boltzmann constant, and T is the absolute temperature.

Figure 2.4 shows plots of the Fermi–Dirac distribution function for various temperatures. The curves are central symmetrical around the point $(E_F, \frac{1}{2})$. Therefore, we may take the Fermi level E_F as a reference point for positioning the distribution function. It is intuitively clear from the preceding discussion that in a semiconductor the Fermi level should be somewhere near the middle of the band gap.

According to the Fermi–Dirac distribution function, the probability of occupancy drops from almost unity to almost zero within an energy interval of a few kT around the Fermi level E_F. The high-energy tail of the function (2.17), namely the region where $E - E_F \gg kT$, can be approximated by the function

$$F(E) \simeq \exp[-(E - E_F)/kT]. \tag{2.18}$$

This is known as the *Maxwell–Boltzmann distribution function*. In classical physics, it describes the kinetic energy distribution of gas particles.

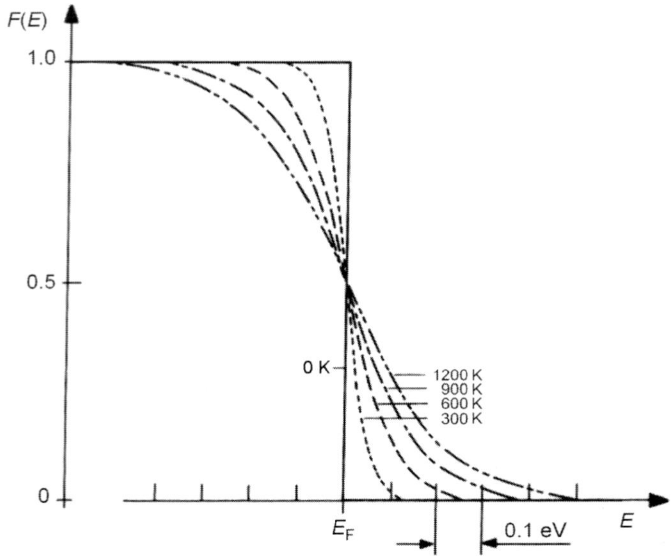

Figure 2.4. The Fermi–Dirac distribution function $F(E)$ for various values of temperature. Energy (horizontal) is measured in electron volts. E_F denotes the Fermi level.

If a gas of quasi-free electrons can be accurately described by the Maxwell–Boltzmann function, it is said to be non-degenerate (because then, from the point of view of classical physics, its distribution function is not degenerate). The conventional condition for the application of the non-degenerate distribution function is

$$E_c - E_F > 4kT. \tag{2.19}$$

If this condition is fulfilled, the average error due to the approximation (2.18) is less than 1%. Conversely, if condition (2.19) is not fulfilled, the electron gas is said to be degenerate.

At energies far below the Fermi level, where $E_F - E \gg kT$, the Fermi–Dirac distribution function can be approximated as

$$F(E) \simeq 1 - \exp[-(E_F - E)/kT]. \tag{2.20}$$

If the Fermi level is well above the top of the valence band, then this function describes accurately the distribution of the valence-band electrons. In this case, the probability for an energy level to be non-occupied is

$$F_h(E) = 1 - F(E) \simeq \exp[-(E_F - E)/kT]. \tag{2.21}$$

Non-occupied levels in the valence band are nothing more than holes. Thus we may interpret the function $F_h(E)$ (2.21) as the probability that a level is occupied by a hole. In analogy with the electron gas, one also speaks about hole gas. A hole gas is non-degenerate if

$$E_F - E_v > 4kT. \tag{2.22}$$

Otherwise, it is degenerate.

A semiconductor material is *non-degenerate* if none of the carrier gases is *degenerate*. Otherwise, it is called a degenerate semiconductor. It follows from (2.19) and (2.22) that in non-degenerate semiconductors the Fermi level is located in the band gap and is separated by more than $4kT$ from either band edge. At room temperature, $4kT \simeq 0.1$ eV. In this book, we shall mostly be concerned with non-degenerate semiconductors.

2.2.3 Carrier concentration

The concentration of quasi-free electrons is equal to the number of occupied energy levels in the conduction band per unit crystal volume:

$$n = \int_{E_c}^{E_{top}} g(E)F(E)\,dE \tag{2.23}$$

where E_c and E_{top} are the energies at the bottom and the top of the conduction band, g is the density of states (2.14) and F is the distribution function (2.17). Since the distribution function rapidly decreases at higher energies, we may approximate g by the density-of-states function valid near the bottom of the conduction band g_c (2.15), and take the upper integration limit as infinity. For a non-degenerate electron gas, we may also approximate F by the Maxwell–Boltzmann function (2.18). Thus (2.23) can be evaluated to be

$$n \simeq N_c \exp[-(E_c - E_F)/kT]. \tag{2.24}$$

Here N_c is the *effective density of states in the conduction band*

$$N_c = 2\left(\frac{m_{de}^* kT}{2\pi\hbar^2}\right)^{3/2} M_c. \tag{2.25}$$

It is so termed because (2.24) can be obtained from (2.23) if the density-of-states function g_c is replaced by N_c states at the bottom of the conduction band. M_c and m_{de}^* were defined by (2.15) and (2.16).

Analogously, the concentration of holes in a non-degenerate hole gas is given by

$$p \simeq N_v \exp[-(E_F - E_v)/kT] \tag{2.26}$$

where N_v is the *effective density of states in the valence band*

$$N_v = 2\left(\frac{m_{dh}^* kT}{2\pi\hbar^2}\right)^{2/3}. \tag{2.27}$$

Table 2.1. Effective density of states in germanium, silicon and GaAs at 300 K.

Semiconductor	N_c (cm^{-3})	N_c (cm^{-3})
Ge	1.04×10^{19}	6.0×10^{18}
Si	2.8×10^{19}	1.04×10^{19}
GaAs	4.7×10^{17}	7.0×10^{18}

Here m_{dh}^* denotes the density-of-states effective mass for holes. For silicon,

$$m_{dh}^* = (m_{li}^{*3/2} + m_{he}^{*3/2})^{2/3} \tag{2.28}$$

where m_{li}^* and m_{he}^* are the effective masses of light and heavy holes, respectively.

The numerical values of the effective density of states in some semiconductors are listed in table 2.1.

2.2.4 Intrinsic semiconductors

In an intrinsic semiconductor most of the charge carriers are created by the thermal ionization of its own atoms. To be intrinsic, a semiconductor must be chemically very pure. Since each electron, which appears in the conduction band, leaves a hole in the valence band, the concentrations of electrons and holes in an intrinsic semiconductor are equal:

$$n_i = p_i. \tag{2.29}$$

Here the subscript i stands for intrinsic.

The position of the Fermi level $E_F = E_i$ in intrinsic semiconductors can be found by substituting (2.24) and (2.26) into (2.29):

$$E_i = \frac{E_c + E_v}{2} + \frac{kT}{2} \ln\left(\frac{N_v}{N_c}\right) = \frac{E_c + E_v}{2} + \frac{3}{4} \ln\left(\frac{m_{dh}^*}{m_{de}^* M_c^{2/3}}\right). \tag{2.30}$$

E_i is often referred to in short as the intrinsic level. Since generally $N_v \simeq N_c$, the intrinsic level is situated near the middle of the band gap. The same conclusion was inferred before from qualitative arguments.

The result (2.30) justifies the use of the Maxwell–Boltzmann distribution functions for intrinsic semiconductors in most cases. If the band gap of a semiconductor is large enough, namely $E_g > 8kT$, then both relations (2.19) and (2.22) are fulfilled when this semiconductor is intrinsic.

To calculate the *intrinsic carrier concentration*, we multiply (2.24) and (2.26) and obtain

$$np = n_i^2 = N_c N_v \exp(-E_g/kT). \tag{2.31}$$

By substituting here the relations for N_c (2.25) and N_v (2.27), and the numerical values of the physical constants, we obtain

$$n_i = 4.9 \times 10^{15} \left(\frac{m^*_{de}}{m} \frac{m^*_{dh}}{m} \right)^{3/4} M_c^{1/2} T^{3/2} \exp(-E_g/2kT) \qquad (2.32)$$

in units of cm^{-3}. Here m denotes the free electron mass and the other notation is the same as before. The intrinsic carrier concentrations for silicon and GaAs at 300 K are 1.45×10^{10} and $1.79 \times 10^6\ cm^{-3}$, respectively.

2.2.5 Extrinsic semiconductors

Dopants

In an extrinsic semiconductor most of the quasi-free charge carriers are provided by some foreign atoms dissolved in the crystal. One then says that an extrinsic semiconductor is doped with foreign atoms. Foreign atoms capable of strongly affecting the electrical characteristics of a semiconductor are called *dopants*. The usual concentrations of dopants in semiconductors are rather small, normally less than 1%. For this reason dopants are also-called *impurities*.

Typical dopants are substitutional impurities: when dissolved in a semiconductor, dopant atoms substitute the host atoms in the crystal lattice. If a dopant atom has one valence electron more than the original lattice atom on that position had, the extra electron can easily become free. Such a dopant is called a *donor*, since it 'donates' a quasi-free electron to the crystal. For example, a phosphorus atom embedded in silicon is a donor. In a semiconductor doped with donors, there are more electrons than holes. Thus electrons are majority carriers and holes are minority carriers. Since in such an extrinsic semiconductor negatively charged carriers prevail, the material is called an *n-type semiconductor*.

Conversely, if a dopant atom has one valence electron less than the original lattice atom had, the missing electron is likely to be taken from the neighbourhood. Such a dopant is called an *acceptor*. For example, a boron atom in silicon is an acceptor. When a valence electron is 'accepted' by a dopant, a hole is created in the valance band. Hence in this case positively charged carriers prevail, and the material is called a *p-type semiconductor*.

The phenomenon of donating and accepting electrons by dopant atoms is obviously analogous to the phenomenon of ionizing isolated atoms. The ionization energies are

$$\Delta E_D = E_c - E_D \qquad \Delta E_A = E_A - E_v \qquad (2.33)$$

for donors and acceptors, respectively. Here E_D and E_A are the energy levels of donors and acceptors, respectively. These energy levels are situated in the forbidden gap, E_D near the bottom of the conduction band and E_A near the top of the valence band. The values of the ionization energies can be assessed by applying the hydrogen atom model on the impurity atoms. The ionization energy of an isolated hydrogen atom is 13.6 eV. Since electrons in a semiconductor have smaller effective masses and move in an environment with a much higher permittivity, the ionization energies of impurity atoms are much smaller than that. Calculations based on this model give dopant ionization energies of less than 0.05 eV for both silicon and GaAs. Dopants with such small ionization energies are called shallow centres. There are also impurities that produce deep centres. Figure 2.5 shows the measured ionization energies of various impurities in germanium, silicon and GaAs.

The probabilities for impurity energy states to be occupied by electrons are given by the following functions:

$$F_D(E_D) = \frac{1}{1 + \frac{1}{2}\exp[(E_D - E_F)/kT]} \tag{2.34}$$

for donors, and

$$F_A(E_A) = \frac{1}{1 + 2\exp[(E_A - E_F)/kT]} \tag{2.35}$$

for acceptors. These two functions are similar to the Fermi–Dirac distribution function (2.17). This was to be expected, since the impurities create some additional permitted states in an energy spectrum where the Fermi–Dirac distribution function holds. The only differences appear in the pre-exponential factors, which are due to the discrete nature of the dopant centres.

Carrier concentration

Let us now consider the carrier concentration in extrinsic semiconductors. Suppose a semiconductor is doped with both donors and acceptors, the concentrations of which are N_D and N_A, respectively. Generally, some of the impurity energy levels are occupied by electrons and some are not. If a donor level is not occupied, it means that the donor atom is ionized and thus positively charged. If an acceptor level is occupied by an electron, the acceptor atom is also ionized and thus negatively charged. Let us denote the concentrations of such ionized impurity atoms by N_D^+ and N_A^- respectively for donors and acceptors. The difference of these two ion concentrations creates a space charge in the semiconductor. This space charge must be compensated by the space charge of free carriers. The neutrality condition is

$$n + N_A^- = p + N_D^+ \tag{2.36}$$

where n and p are the concentrations of electrons and holes, respectively.

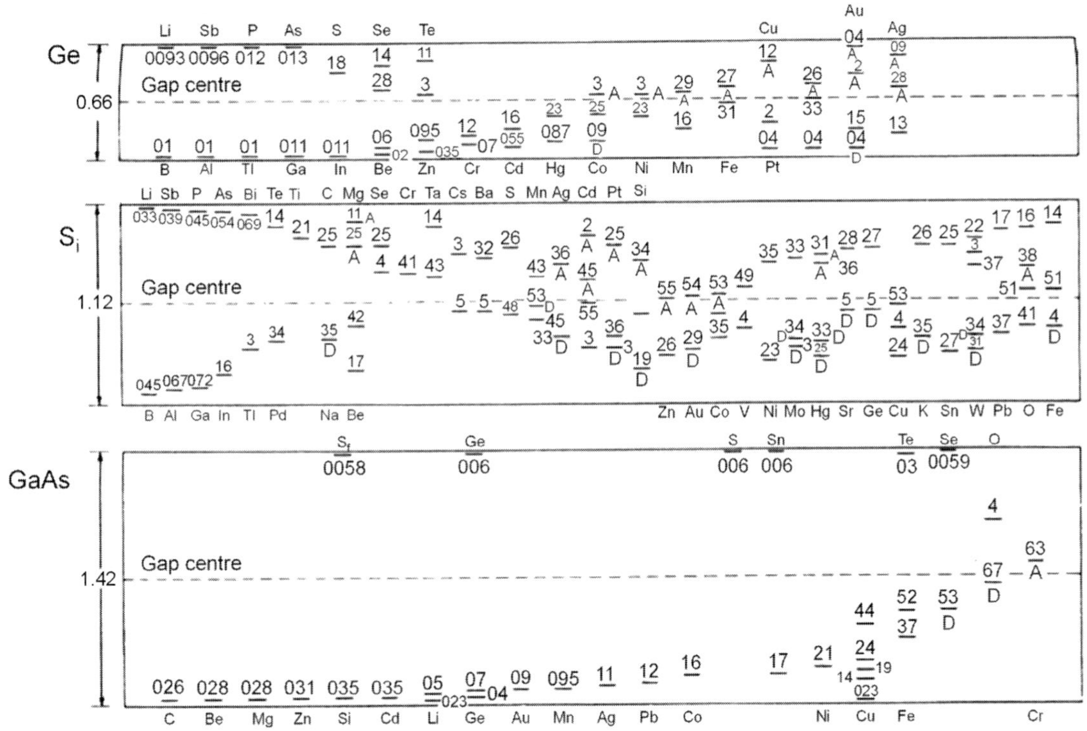

Figure 2.5. Measured ionization energies for various impurities in germanium, silicon and GaAs at 300 K. The values are in eV. Notation: for example, 0.26 means 0.026 eV, 44 means 0.44 eV, and so on. The levels below the gap centres are measured from the top of the valence band and are acceptor levels unless indicated by D for donor level. The levels above the gap centres are measured from the bottom of the conduction band and are donor levels unless indicated by A for acceptor level (adapted from [7]).

To calculate the electron and hole concentrations, we shall also make use of equation (2.31). (Note that we derived this equation from (2.24) and (2.26), both of which hold for any non-degenerate semiconductor.) Consequently, the np product does not depend on doping as long as the carriers remain non-degenerate. In an n-type semiconductor, $N_D > N_A$. By eliminating p from (2.31) and (2.36), we obtain the electron concentration as

$$n_n = \frac{1}{2}\{(N_D^+ - N_A^-) + [(N_D^+ - N_A^-)^2 + 4n_i^2]^{1/2}\}. \tag{2.37}$$

Here the subscript n stands for an n-type semiconductor.

Saturation range

To simplify further analysis, we shall make a few approximations at this point. First of all, at relatively low temperatures,

$$(N_D^+ - N_A^-)^2 \gg 4n_i^2 \tag{2.38}$$

and equation (2.37) reduces to

$$n_n \simeq N_D^+ - N_A^-. \tag{2.39}$$

Second, we shall find approximate relations for the ion densities N_D^+ and N_A^-. To do so we need to know the position of the Fermi level. We note first that, since in an n-type semiconductor $n > p$, the Fermi level moves from the intrinsic level E_i upwards, towards the conduction band (see (2.24) and (2.26)). The distance between the bottom of the conduction band and the Fermi level can be found from (2.24) and (2.39) to be

$$E_c - E_F = kT \ln\left(\frac{N_c}{N_D^+ - N_A^-}\right). \tag{2.40}$$

The separation of the Fermi level from the donor level is then

$$E_D - E_F = kT \ln\left(\frac{N_c}{N_D^+ - N_A^-}\right) - \Delta E_D \tag{2.41}$$

where $\Delta E_D = E_c - E_D$ is the donor ionization energy. We can now introduce, in analogy to (2.19), the condition for non-degenerate donor levels as

$$E_D - E_F > 5kT. \tag{2.42}$$

If this condition is fulfilled, then the distribution function for donor levels (2.34) differs by less than 1.5% from its Maxwell–Boltzmann-like approximation. At the same time, more than 98.5% of all donors will be ionized. Since the Fermi level is now even much more distant from the acceptor level, acceptors are also almost fully ionized. Therefore, if in an n-type semiconductor the condition (2.42) is fulfilled, then almost all dopant atoms are ionized, and

$$N_D^+ \simeq N_D \qquad N_A^- \simeq N_A. \tag{2.43}$$

Now (2.39) becomes

$$n_n \simeq N_D - N_A \tag{2.44}$$

or, if $N_D \gg N_A$,

$$n_n \simeq N_D. \tag{2.45}$$

The hole concentration can be obtained from (2.31) and (2.45):

$$p_n = n_i^2/n_n \simeq n_i^2/N_D. \tag{2.46}$$

Therefore, under the assumed conditions, the majority carrier concentration, which is the electron concentration in this case, is almost equal to the net impurity concentration. In particular, the majority carrier concentration practically does not depend on temperature. (All impurities are ionized, and the thermal activation of the valence electrons is negligible.) For this reason, the temperature range where the above conditions are met is called the saturation range of a semiconductor. Since the dopant atoms cannot then supply more carriers, this temperature range is also called the *exhaustion range*.

Intrinsic range

With an increase in temperature, the intrinsic carrier concentration rapidly increases, and the inequality (2.38) will eventually be reversed. Then $n_n \simeq p_n \simeq n_i$, and the semiconductor becomes intrinsic. Taking as a criterion for the onset of the intrinsic conditions

$$\frac{n_n - N_D}{N_D} = \frac{1}{100} \tag{2.47}$$

we obtain with the aid of (2.37)

$$\left(\frac{n_i(T_i)}{N_D}\right)^2 \simeq \frac{1}{100}. \tag{2.48}$$

Here we have assumed that $N_D \gg N_A$ and T_i denotes the temperature at the onset of intrinsic conditions.

Freeze-out range

On the other hand, with a decrease in temperature the thermal agitations eventually become insufficient to keep donors fully ionized. Some of the quasi-free electrons are then trapped by donors and become immobile. This phenomenon is called the freeze-out effect. To assess the temperature T_f at which the freeze-out effect begins, we can take (2.42) as an equality

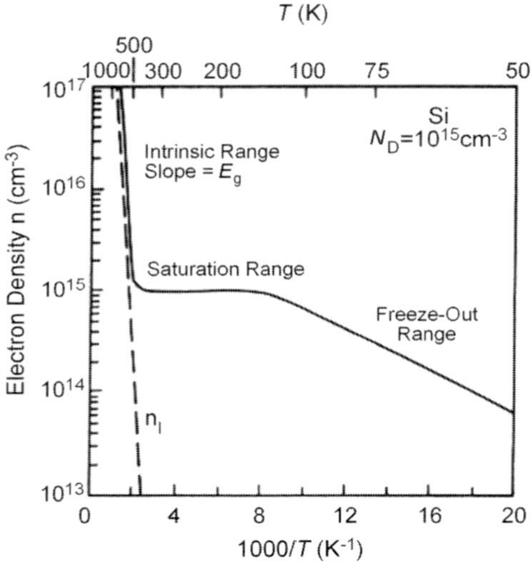

Figure 2.6. Electron density as a function of temperature for a silicon sample with donor impurity concentration of $10^{15}\,\mathrm{cm}^{-3}$ (reprinted from [7]).

and substitute it in (2.41). Thus

$$\frac{\Delta E_{\mathrm{D}}}{kT_{\mathrm{f}}} + 5 = \ln\left(\frac{N_{\mathrm{c}}(T_{\mathrm{f}})}{N_{\mathrm{D}} - N_{\mathrm{A}}}\right). \tag{2.49}$$

Note that the effective density of states in the conduction band N_{c} (2.25) depends on temperature as well.

A typical example of the temperature dependence of electron concentration in an n-type semiconductor is shown in figure 2.6. Note the two different slopes of the curve in the intrinsic and freeze-out ranges. These slopes correspond to the activation energies mentioned in §2.1.

In a p-type semiconductor, everything is similar but with the n and p notation reversed. Briefly,

$$p_{\mathrm{p}} \simeq N_{\mathrm{A}} - N_{\mathrm{D}} \tag{2.50}$$

and if $N_{\mathrm{A}} \gg N_{\mathrm{D}}$

$$p_{\mathrm{p}} \simeq N_{\mathrm{A}} \tag{2.51}$$

$$n_{\mathrm{p}} = n_{\mathrm{i}}^2/p_{\mathrm{p}} \simeq n_{\mathrm{i}}^2/N_{\mathrm{A}} \tag{2.52}$$

$$E_{\mathrm{F}} - E_{\mathrm{v}} = kT\ln(N_{\mathrm{v}}/N_{\mathrm{A}}). \tag{2.53}$$

Example

Let us find the range of donor concentrations in silicon under the condition that the saturation range in the electron concentration holds from −40 to 125 °C. The material is doped with phosphorus.

Consider first the onset of intrinsic conditions at the temperature $T_i = 398\,\mathrm{K}$. According to (2.47) and (2.48), the electron concentration at the temperature T_i will not exceed the donor concentration for more than 1% if $N_D > 10n_i(T_i)$. To find the intrinsic concentration $n_i(T_i)$, we first note that (see (2.32))

$$\frac{n_i(T_i)}{n_i(T)} = \left(\frac{T_i}{T}\right)^{3/2} \exp\left(\frac{E_g(T)}{2kT} - \frac{E_g(T_i)}{2kT_i}\right). \qquad (2.54)$$

At $T = 300\,\mathrm{K}$, the precise values of n_i and E_g are known: $n_i = 1.45 \times 10^{10}\,\mathrm{cm}^{-3}$ (§2.3.4) and $E_g = 1.12\,\mathrm{eV}$ (Appendix C). The band gap at the temperature T_i can be obtained as

$$E_g(T_i) = E_g(T) + \frac{\mathrm{d}E_g}{\mathrm{d}T}(T_i - T). \qquad (2.55)$$

Using the value of $\mathrm{d}E_g/\mathrm{d}T$ quoted in §2.2.1, we find that $E_g(T_i) = 1.10\,\mathrm{eV}$. By substituting these values in (2.54), we obtain n_i (125 °C) $\cong 6.1 \times 10^{12}\,\mathrm{cm}^{-3}$. Therefore, the donor concentration has to be $N_D > 6.1 \times 10^{13}\,\mathrm{cm}^{-3}$.

To ensure that not more than about 1.5% of electrons are frozen out at the temperature $T_f = 233\,\mathrm{K}$, the following condition must be fulfilled (see (2.49)):

$$\frac{N_c(T_f)}{N_D} > \exp\left(\frac{\Delta E_D}{kT_f} + 5\right). \qquad (2.56)$$

Using (2.25), we have

$$\frac{N_c(T_f)}{N_c(T)} = \left(\frac{T_f}{T}\right)^{3/2} \qquad (2.57)$$

and by substituting this in (2.56), we obtain

$$N_D < N_c(T)\left(\frac{T_f}{T}\right)^{3/2} \exp\left[-\left(\frac{\Delta E_D}{kT_f} + 5\right)\right]. \qquad (2.58)$$

From figure 2.5, we find the ionization energy of phosphorus in silicon to be $\Delta E_D = 0.045\,\mathrm{eV}$. Substituting this in (2.58), and with $T = 300\,\mathrm{K}$ and $N_c = 2.8 \times 10^{19}\,\mathrm{cm}^{-3}$ (from table 2.1), we obtain $N_D < 1.4 \times 10^{16}\,\mathrm{cm}^{-3}$.

Therefore, the electron concentration in n-type silicon will be constant to within about +1% in the temperature range from −40 °C to 125 °C if the donor concentration is in the range from $10^{14}\,\mathrm{cm}^{-3}$ to $10^{16}\,\mathrm{cm}^{-3}$.

2.2.6 Heavy doping effects

In the previous sections we have considered only pure and lightly doped semiconductors. In such semiconductors the energy bands and the density of states in the bands are determined solely by the host atoms of the crystal lattice. However, if more than roughly 10 per 10^6 host atoms are substituted by dopant atoms, the dopant atoms start to affect the band structure of the crystal. Then we say that a semiconductor is heavily doped. We shall now briefly discuss the heavy doping effects [8–10].

A qualitative picture that summarizes the changes that occur in the band structure of an n-type semiconductor due to the doping is given in figure 2.7. As the doping level increases, the band gap narrows, the conduction and valence band edges cease to be sharp and start developing tails, and the donor level broadens and approaches the conduction band.

The physical cause of the band-gap narrowing is the interaction between the carriers. If many electrons are present in a crystal, they start to affect the periodic potential field in which each of the electrons moves. The result is a reduction of the electron potential energy and thus a downward shift of the conduction band edge. At the same time the dense electrons screen the electrostatic field of holes. This causes a reduction of the hole potential energy and thus a shift of the valence band edge towards the conduction band. The two effects add up and produce a band-gap narrowing. It is called rigid gap narrowing, since the effects described so far do not affect the sharpness of the band edges.

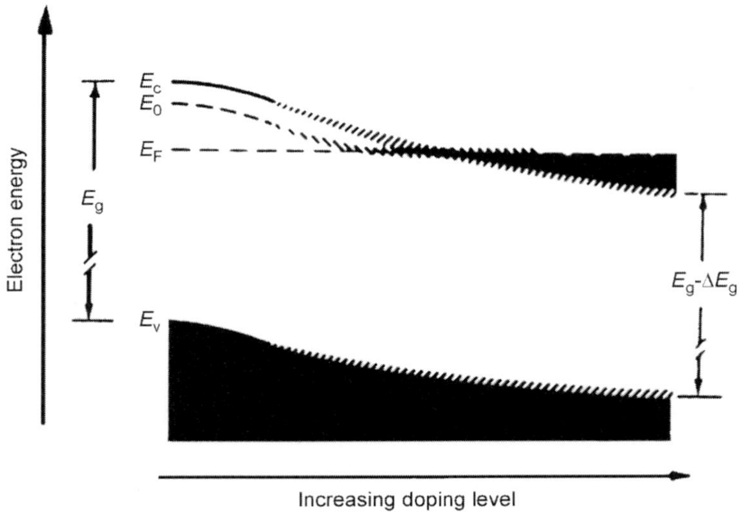

Figure 2.7. Influence of the doping level on the energy band diagram of an n-type semiconductor (adapted from [9]).

The tails in the density-of-states functions are due to the random impurity distribution. The spatial fluctuation in the doping density produces a fluctuation in the electrostatic potential energy. As a consequence, the average density-of-states functions show the tails that penetrate into the forbidden gap. The tails can be accounted for by an equivalent additional band-gap narrowing, while still assuming a parabolic density-of-state function. Such a model is adequate for treating the electrical characteristics of semiconductor devices. The overall band-gap narrowing, as determined by electrical measurements, is the sum of the rigid band-gap narrowing and the effective narrowing due to the band tails. For example, in heavily doped n-type silicon, the electrical band-gap narrowing amounts to about 30, 70 and 150 meV at the donor concentrations of 10^{18}, 10^{19} and 10^{20} cm^{-3}, respectively.

The random impurity distribution also causes the broadening of the impurity energy level into an impurity band. However, the formation of the impurity band has no essential influence on the effective band-gap narrowing. The impurity level shifts towards the appropriate energy band and merges with it before becoming broad enough to influence the gap narrowing. The shifting of the impurity level is due to the screening effect.

The screening of the impurity ions by the dense carrier cloud causes a reduction of the dopant ionization energy. At a critical carrier concentration, the dopant ionization energy reduces to zero. Then all the carriers from the dopant states become available for conduction at any temperature. In particular, the carrier freeze-out effect no longer appears. In this respect, a heavily doped semiconductor behaves like a metal. The conversion of a material by an increase in doping concentration from semiconductor-like to metal-like behaviour is called the *semiconductor metal transition*, or the *Mott transition*. In n-type silicon, the Mott transition arises at a donor density of about 3×10^{18} cm.

2.3 Scattering of carriers

As we have seen in §2.1.3, for a quasi-free electron a periodic crystal lattice is a perfectly transparent medium: the Bloch waves propagate through space with a periodic electric field without any attenuation. Regarded as a particle, a quasi-free electron thus travels through a perfect crystal along a straight line with a constant velocity. However, a crystal is never perfect. Its periodicity is disturbed by foreign atoms embedded in the lattice, dislocations and thermal vibrations.

When coming suddenly into the region without periodicity or with a different periodicity, an electron can no longer propagate in the same way. It has to adjust its motion to the new conditions, and does so by interacting with the lattice. During the interaction, the electron changes its momentum, and may lose or acquire some energy. Typically, this interaction takes place within a very small volume and lasts for a very short period of time. For these reasons, the process of electron–lattice interaction may be considered as a collision event. Owing to collisions, electrons change the directions of their propagation; they are scattered. The localized regions of the crystal lattice where electrons collide with the lattice and are scattered are called the collision or scattering centres.

2.3.1 Free transit time

The collision of an electron with the lattice is a random event. Hence the time elapsed between two successive collisions, called the free transit time, randomly assumes different values. For an ensemble of carriers possessing energy in the interval $(E, E + \Delta E)$, the probability of free transit time to lie within the time interval $(t, t + \mathrm{d}t)$ is given by the expression

$$w(t)\,\mathrm{d}t = \frac{1}{\tau}\exp(-t/\tau)\,\mathrm{d}t. \tag{2.59}$$

Here $w(t)$ is the free transit time distribution function, and τ is the *mean free transit time*. The mean free transit time generally depends on the energy of the carriers involved. There are many random processes in nature that can be described by equations similar to (2.59). All of them have in common the probability of an event that is proportional to a time interval or a path interval. The best-known examples of the kind are radioactive decay and light absorption.

The mean free transit time is related to the *mean free path* of the carriers l by

$$\tau = l/v \tag{2.60}$$

where v is the velocity of the carriers. The velocity of carriers is, in turn, related to their energy. This relation is fully analogous to the well-known relation that holds for a free particle. For example, the velocity of a quasi-free electron is given by (see (2.5), (2.6) and (2.11))

$$v = |\boldsymbol{v}| = \left|\frac{\hbar\boldsymbol{k} - \hbar\boldsymbol{k}_0}{m^*}\right| = \left(\frac{2(E - E_c)}{m^*}\right)^{1/2} = \left(\frac{2E_K}{m^*}\right)^{1/2} \tag{2.61}$$

where E_K is the kinetic energy of the electron. From (2.59) and (2.60) we obtain the probability for the free path of a carrier to be between x and $x + \mathrm{d}x$:

$$w(x)\,\mathrm{d}x = \frac{1}{l}\exp(-x/l)\,\mathrm{d}x \tag{2.62}$$

where l is the mean free path between collisions of the carriers with the selected energy.

Both the mean free transit time and the mean free path generally depend on carrier energy and crystal temperature. Let us consider, for example, the scattering of electrons by thermal vibrations and ionized impurities.

There are two types of elementary thermal vibrations of the crystal lattice: acoustic and optical *phonons*. The names come from the frequency ranges involved: they are in the acoustic and the optical range, respectively. The low-frequency acoustic thermal vibrations have the property that the neighbouring atoms in a crystal oscillate almost in phase with each other. Thus, they are similar to ordinary sound waves. However, being the quanta of thermal energy, phonons also have the properties of particles. In particular, one may define the density of such quasi-particles in a crystal. The spatial density of acoustic phonons N_p is proportional to the crystal absolute temperature.

Whenever an electron encounters a phonon in its way, it may be scattered. The probability of the scattering event is proportional to the spatial density of acoustic phonons. Hence the mean free path of an electron is proportional to N_p^{-1}, and thus to T^{-1}. From (2.60) and (2.61) we now find that the mean free transit time between the scattering of electrons by acoustic phonons is $\tau_p \approx T^{-1} E_K^{-1/2}$; it is short at higher temperatures and for high-energy carriers.

For ionized impurity scattering the mean free transit time is greater for higher-energy carriers. These carriers move faster, remain near the ion for a shorter time, and their paths are therefore less effectively deflected. For this reason, whenever the average velocity of carriers increases, for example because of an increase in temperature, scattering by ionized impurities becomes less significant.

If several scattering mechanisms characterized by the transit times τ_i act simultaneously, the resultant free transit time τ is given by

$$\frac{1}{\tau} = \sum_i \frac{1}{\tau_i}. \tag{2.63}$$

Of course, τ is shorter than any of the component transit times τ_i. The above equation is a consequence of the fact that τ^{-1} is proportional to the probability of carrier scattering. Indeed, according to (2.59), $\mathrm{d}t/\tau$ is the probability that the free transit time lies in the interval $(0, \mathrm{d}t)$. Thus this is also the probability for a carrier to scatter during this time interval. The probability for a carrier to scatter on any of the centres i, $\mathrm{d}t/\tau$, is equal to the sum of the probabilities of scattering on the different centre types, $\mathrm{d}t/\tau_i$; hence equation (2.63).

It turns out that the mean free transit time for various collision processes can be represented in a similar form, namely

$$\tau(E) \simeq \tau_0 E_K^p \tag{2.64}$$

Here T_0 is a coefficient, which depends on the effective mass of carriers and may also be a function of temperature, E_K is the kinetic energy of the considered ensemble of carriers, and p is a parameter, the value of which depends on the scattering mechanism involved.

Since the mean free transit time differs for carriers with various energies, it is useful to find some averages of the mean free transit times over all carriers. In particular, by the analysis of transport phenomena, we shall need the energy-weighted average free transit time. Weighting by energy should be done since the contribution of a carrier to a transport process is proportional to its energy, as we shall see in chapter 3. The energy-weighted average of the free transit time of electrons is given by

$$\langle \tau \rangle = \frac{\int_{E_c}^{E_{top}} \tau(E) \cdot E \cdot g(E) \cdot F(E) \, dE}{\int_{E_c}^{E_{top}} E \cdot g(E) \cdot F(E) \, dE}. \quad (2.65)$$

Here the angular brackets denote the energy-weighted average, E is the electron energy, g is the density of states (2.14), F is the distribution function (2.17), and the integration is done over the conduction band. Note that the expression in the denominator of (2.65) equals the total energy possessed by all electrons (compare the expression (2.23) for the electron concentration). We shall call $\langle \tau \rangle$ the *average free transit time* for short.

If a $T(E)$ dependence like (2.64) is assumed then, for a non-degenerate semiconductor, (2.65) reduces to

$$\langle \tau \rangle = \tau_0 (kT)^p \frac{\Gamma(\frac{5}{2} + p)}{\Gamma(\frac{5}{2})} \quad (2.66)$$

where $\Gamma(x)$ denotes the gamma function. The functional dependences of the mean and the average free transit times τ and $\langle \tau \rangle$ for various scattering mechanisms are summarized in table 2.2.

One can also define and calculate the average free path of carriers $\langle l \rangle$ in a similar manner as the average free transit time $\langle \tau \rangle$. In semiconductors, the average free path of carriers is of the order of 10 nm.

Table 2.2. Functional dependences of the mean free transit time τ and the average free transit time $\langle \tau \rangle$ for some carrier scattering mechanisms.

Scattering mechanism	τ	$\langle \tau \rangle$
Acoustic phonon	$\sim T^{-1} E_K^{-1/2}$	$\sim T^{-3/2}$
Ionized impurity	$\sim E_K^{3/2}$	$\sim T^{3/2}$
Neutral impurity	$\sim T^0 E_K^0$	$\sim T^0$

2.3.2 Relaxation time

Up to now we have considered scattering as individual events. However, the scattering process plays a major role in the collective behaviour of quasi-free carriers: it results in a frictional effect by the transport phenomena, and acts towards re-establishing equilibrium in the carrier distribution. We shall now discuss this second effect.

The frictional effect will be considered later in connection with electrical resistivity. In §2.3.2 we discussed the carrier distribution as a function of carrier energy, $F(E)$. But carrier energy is related to wavevector $\boldsymbol{k}$, momentum $\boldsymbol{p}$ and velocity $\boldsymbol{v}$ of a carrier. Hence it is reasonable also to consider carrier distribution as a function of, say, carrier velocity, $F(v)$.

In a sample with no macroscopic electric fields, carriers are in the state of thermodynamic equilibrium. The equilibrium distribution function $F_0(v)$, where the subscript 0 stands for equilibrium, is symmetrical around the point $\boldsymbol{v} = \boldsymbol{0}$. In other words, $F_0(v)$ is an even function, $F_0(v) = F_0(-v)$. If an electric field is established in the sample, a force acts on each quasi-free charge carrier and accelerates it. To be definite, let us take electrons. Their velocity components in the direction of the field decrease, and those against the field increase. As a result, the distribution function is disturbed and it loses its symmetry: $F_0(v) \neq F_0(-v)$. Moreover, since the electrons accumulate energy from the electric field, on average they tend to occupy higher energy levels than in the case of thermodynamic equilibrium: the average kinetic energy of electrons exposed to an electric field is higher.

If we suddenly turn the electric field off, the disturbed distribution function will start to recover. At each scattering event, a carrier will get a new velocity at random and, on average, lose some of its extra energy. After most of the carriers have been scattered at least once, the carrier gas will reassume its symmetrical distribution. Everything happens as if the carrier distribution function were under stress and relaxes afterwards. For this reason, the process of re-establishing equilibrium is called relaxation. The time constant of the (exponential) decay of the difference

$$\delta F(v) = F(v) - F_0(v) \tag{2.67}$$

during the relaxation is called the *relaxation time*. Thereby we have to distinguish between two relaxation times: the momentum relaxation time and the energy relaxation time.

The *momentum relaxation time* is the time constant of re-establishing symmetry in the carrier distribution function. Since by each scattering event a carrier completely 'forgets' the direction of its previous movement, we expect that the momentum relaxation time should be close to the mean free transit time discussed in §2.2.1. A detailed analysis shows that the two quantities are exactly equal. Therefore, we shall not specify any difference between them. Moreover, we will use the same symbol τ for both the mean free transit time and the momentum relaxation time.

The *energy relaxation time* is the time constant of re-establishing the original average energy of charge carriers. In other words, this is the time constant of re-establishing the complete thermodynamic equilibrium of charge carriers with the crystal lattice. Since most of the scattering events are almost elastic, i.e. energetically very inefficient, it takes many scattering events until the charge carriers reassume their thermodynamic equilibrium. Therefore, the energy relaxation time is usually much longer than the momentum relaxation time. We shall see in §2.4.3 that this difference in the magnitudes of the momentum relaxation time and the energy relaxation time has a dramatic consequence on the transport of carriers in very small semiconductor devices.

Since the mean free transit time generally depends on carrier energy, so too do the two relaxation times. Thus we also need to define various averages of the relaxation times. In particular, we shall often use the energy-weighted average momentum relaxation time $\langle\tau\rangle$. Of course, this quantity is identical to the average mean free transit time. Thereby we can directly use the results given by (2.65), (2.66) and in table 2.2. The energy-weighted average momentum relaxation time is often referred to as the average relaxation time for short, or even simply as the relaxation time.

The relaxation time of charge carriers in semiconductors is of the order of 10^{-13} s.

2.4 Charge transport

Quasi-free charge carriers in a crystal are subject to perpetual thermal motion. The carriers move at random: the magnitudes and directions of their velocities are steadily randomized by their collisions with crystal lattice imperfections. At thermal equilibrium, the thermal motion of carriers does not produce any unidirectional charge transport. When averaged over a long enough period of time, there is almost always an equal number of carriers passing any cross section of the crystal in one or another direction.

The situation changes if an electric field is applied to the crystal. Now a force acts on each charge carrier, so in addition to the random thermal velocity, the quasi-free carriers acquire a directed velocity component. As a consequence, the electron gas as a whole starts to move against the direction of the field and the hole gas starts to move in the direction of the field. This transport phenomenon is called *drift*. The electric current, associated with the drift of charged carriers, is called the *drift current*. Another term in use for the drift current is the conductivity current. The transport of charge carriers may also appear if a carrier concentration gradient or a temperature gradient exists in a sample. These phenomena are called the *diffusion* effect and the *thermoelectric* effect, respectively.

A magnetic field cannot cause the transport of charge carriers by itself. A magnetic force always acts perpendicularly to the direction of carrier velocity. Therefore the magnetic force cannot alter the absolute value of a carrier velocity, nor can it disturb the symmetry of the distribution function $F(v)$. But a magnetic field can affect, or modulate, an already-existing transport of carriers, irrespective of its origin. The influence of a magnetic field on drift and diffusion of carriers is a major issue in this book. We shall discuss the basics of this phenomenon in the next chapter.

In this section, we proceed with the discussion of the two basic charge transport effects: drift and diffusion.

2.4.1 Drift in a weak electric field

If an electric field $\boldsymbol{E}$ is established in a semiconductor sample, the force $\boldsymbol{F} = e\boldsymbol{E}$ acts on each charge carrier, where e denotes the charge. Owing to this force, each quasi-free charge carrier is accelerated. The acceleration is

$$\boldsymbol{a} = \frac{e}{m^*}\boldsymbol{E} \tag{2.68}$$

where m^* is the effective mass of the carrier. During its free transit time τ, a carrier then attains a velocity increment δv given by

$$\delta v = \boldsymbol{a}t = \frac{e}{m^*}\boldsymbol{E}t. \tag{2.69}$$

In the case of electrons $e = -q$, q being the elementary charge, and this velocity component is directed against the electric field. The thermal motion of quasi-free electrons is therefore altered: as explained before, the electron distribution function loses its symmetry, and the electron gas drifts.

In order to make further analysis of the drift phenomenon tractable, we have to make at this point some simplifying assumptions. These are as follows:

(i) The energy accumulated by an electron moving in the electric field during its free transit time is completely transferred to the crystal in the next collision. Put another way, in the shock of the collision an electron 'forgets' everything about its previous acceleration. After the collision, the acceleration of the electron starts all over again.

(ii) The disturbance in the distribution function (2.67) caused by the electric field is small, so that $F(v) \simeq F_0(v)$.

Both of these assumptions can be summarized by stating that quasi-free charge carriers stay essentially in thermodynamic equilibrium with the crystal lattice in spite of the action of the electric field.

Let us consider the background of the above assumptions. The first assumption is reasonable only if the condition

$$\delta E \le \Delta E_{\max} \tag{2.70}$$

is fulfilled. Here δE is the energy acquired from the field by an electron during its free transit time, and ΔE_{max} is the maximum energy that an electron can transfer to the lattice in a single collision event. We shall now estimate these two energies.

The energy acquired by an electron from the electric field during free flight is given by

$$\delta E(t) = (\delta \boldsymbol{P})^2/2\boldsymbol{m}^* = (\boldsymbol{F}t)^2/2m^* \tag{2.71}$$

where $\delta \boldsymbol{P}$ is the increase in the electron quasi-momentum, $\boldsymbol{F}$ is the force acting on the electron, and t is the duration of the free flight of the electron. By substituting $\boldsymbol{F} = e\boldsymbol{E}$ and $t \simeq \langle \tau \rangle$ here, we obtain the average increase in electron energy:

$$\langle \delta E \rangle \simeq \frac{q \langle \tau \rangle^2}{2m^*} (\boldsymbol{E})^2. \tag{2.72}$$

To assess how much energy an electron can lose in a collision with the lattice, we shall take a representative example: the electron–acoustic phonon interaction. The impurity collisions are less effective energy consumption mechanisms, and optical phonon scattering becomes significant only at very high electric fields.

An electron–phonon interaction means that the electron either creates or absorbs a phonon. During their random motion, quasi-free carriers continuously create and absorb phonons. If there is no electric field in the sample, the average rates of phonon creation and absorption are equal; in the presence of a field, however, the rate of phonon creation becomes greater. In this way the electrons dissipate energy obtained from the field.

At an electron–phonon interaction, both the total energy and the total momentum of the system must be conserved:

$$\Delta E = E' - E'' = \pm \hbar \omega_{\mathrm{p}} \tag{2.73}$$

$$\Delta \boldsymbol{P} = \boldsymbol{P}' - \boldsymbol{P}'' = \hbar \boldsymbol{q}_{\mathrm{p}}. \tag{2.74}$$

Here E' and E'' are the energies, and $\boldsymbol{P}'$ and $\boldsymbol{P}''$ the quasi-momenta of the electron before and after the interaction, respectively; $\hbar\omega_{\mathrm{p}}$ and $\hbar\boldsymbol{q}_{\mathrm{p}}$ are the energy and the quasi-momentum of the phonon, respectively; ω_{p} and $\boldsymbol{q}_{\mathrm{p}}$ are the frequency and the wavevector of the phonon, respectively; and the signs + and − correspond to the cases of creation and annihilation of a phonon, respectively. The energy of an acoustic phonon is related to its quasi-momentum as:

$$\hbar\omega_{\mathrm{p}} = \hbar |\boldsymbol{q}_{\mathrm{p}}| c. \tag{2.75}$$

Here c denotes the velocity of phonons, which equals the velocity of sound in the solid.

The largest change in the electron energy ΔE (2.73) corresponds to the largest possible value of the phonon quasi-momentum $\hbar\boldsymbol{q}_{\mathrm{p}}$. According to (2.74), a maximum value of $\hbar\boldsymbol{q}_{\mathrm{p}}$ is reached if $\boldsymbol{P}'' = -\boldsymbol{P}'$, that is for an electron collision similar to a frontal collision with a rigid barrier. Then $\hbar \boldsymbol{q}_{\mathrm{p,max}} = 2\boldsymbol{P}'$, and with the aid of (2.73) and (2.75) we get

$$\Delta E_{\max} = c2|\boldsymbol{P}'|. \tag{2.76}$$

Since for electrons $|\boldsymbol{P}'| = 2E'/v'$, v' being the electron velocity before scattering, (2.76) can be rewritten as

$$\Delta E_{\max} = 4\frac{c}{v'}E'. \tag{2.77}$$

Substituting $v' \simeq v_t$ and $E' \simeq \langle E_t \rangle$ here, we obtain an estimate of the maximum energy change of an electron in a collision with an acoustic phonon as

$$\langle \Delta E_{\max} \rangle \simeq 4\frac{c}{v_t}\langle E_t \rangle. \tag{2.78}$$

Here $\langle E_t \rangle$ is the average kinetic energy of electrons due to the thermal motion, and $v_t = \langle v^2 \rangle^{1/2}$ is the corresponding average thermal velocity of electrons. For a non-degenerate carrier gas,

$$\langle E_t \rangle = m^* \langle v^2 \rangle /2 = \tfrac{3}{2}kT. \tag{2.79}$$

At room temperature, $v_T \simeq 10^7\ \mathrm{cm\,s^{-1}}$. Since $c \simeq 10^5\ \mathrm{cm\,s^{-1}}$, it follows from (2.78) that $\Delta\langle E_{\max} \rangle \ll \langle E_t \rangle$.

Returning now to (2.70), we may interpret this condition in the following way:

$$\langle \delta E \rangle \ll \langle E_t \rangle. \tag{2.80}$$

Therefore, assumption (i) above is realistic if the average energy acquired by quasi-free charge carriers from the electric field during their free transits is much smaller than the average thermal energy of carriers. In this case the average carrier velocity increment $\langle \partial v \rangle$ due to the field will also be small relative to the average thermal velocity, and assumption (ii) sounds very reasonable, too.

From (2.70), (2.72), (2.78) and (2.79), we finally conclude that the above conditions can be fulfilled if the electric field in the sample is weak enough:

$$|\boldsymbol{E}| < 2\frac{m^*}{q\langle \tau \rangle}(cv_t)^{1/2}. \tag{2.81}$$

At room temperature, in most semiconductors electric fields up to some $100\ \mathrm{V\,cm^{-1}}$ can be considered as weak enough.

Drift velocity

If assumption (i) holds, then the mean drift velocity of a group of carriers having energies between E and $E + \mathrm{d}E$, $v_d(\boldsymbol{E})$, can be found in the following way. We calculate first the total distance L_n drifted per carrier during a long period of time t_n; then $v_d(E) = L_n/t_n$. The total distance drifted per carrier is given by $L_n = nl$. Here n is the number of free transits involved, $n = t_n/t$, and l is the mean path travelled by a carrier due to the field during a free transit time. Thus

$$v_d(\boldsymbol{E}) = \frac{\boldsymbol{L}_n}{t_n} = \frac{\boldsymbol{l}}{\tau} = \frac{1}{\tau}\int_0^\infty \frac{1}{2}\boldsymbol{a}t^2 w(t)\,dt. \tag{2.82}$$

where $w(t)$ is the probability given by (2.59). After substituting the expressions (2.68) and (2.59) in the above integral, we can evaluate it and obtain

$$\boldsymbol{v}_d(\boldsymbol{E}) = \frac{e}{m^*}\tau(E)\boldsymbol{E} \tag{2.83}$$

where τ is the mean free transit time.

The average drift velocity can easily be found, provided assumption (ii) above holds. According to the present simple model, the averaging should be done over all electrons, without any additional weighting. However, a more accurate analysis, presented in chapter 3, yields an expression equivalent to (2.83), which must be averaged over electron energies. To obtain a correct result, we shall also find an energy-weighted average. Replacing $\tau(E)$ in (2.65) by $v_d(E)$ (2.18), we readily obtain the average drift velocity:

$$\boldsymbol{v}_d = \frac{e}{m^*}\langle\tau\rangle\boldsymbol{E} \tag{2.84}$$

where $\langle\tau\rangle$ denotes the average relaxation time (2.65). For the sake of simplicity, we shall call the average drift velocity the drift velocity for short and denote it simply by v_d.

Mobility

The above analysis applies for each class of quasi-free charge carriers, namely for both electrons and holes. Therefore, in weak electric fields, the drift velocity of carriers (2.84) is proportional to the electric field,

$$v_{dn} = -\mu_n\boldsymbol{E} \qquad v_{dp} = -\mu_p\boldsymbol{E} \tag{2.85}$$

where the signs $-$ and $+$, and the subscripts n and p apply for electrons and holes, respectively. The proportionality coefficients

$$\mu_n = \frac{q}{m_n^*}\langle\tau_n\rangle \qquad \mu_p = \frac{q}{m_p^*}\langle\tau_p\rangle \tag{2.86}$$

are called mobilities. Mobility is by definition a positive quantity. If there is a danger of mixing up this mobility with the Hall mobility, which will be

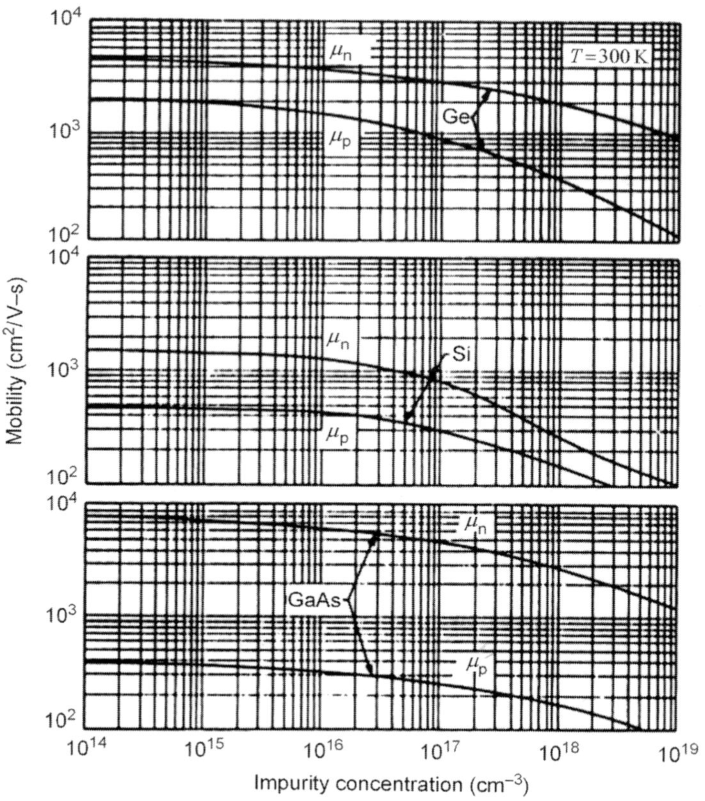

Figure 2.8. Drift mobility of charge carriers in germanium, silicon and GaAs at 300 K against impurity concentration (reprinted from [7]).

defined in chapter 3, the former is also referred to as *drift mobility* or *conductivity mobility*. The numerical values of carrier mobility are very different for various materials (see Appendix C).

For a given material, the mobility generally depends on temperature and doping concentration. This is because the mobility is proportional to the relaxation time $\langle \tau \rangle$, and the relaxation time depends on the scattering mechanism and temperature (see table 2.2).

The measured mobilities of carriers in germanium, silicon and GaAs are shown in figure 2.8. We note that at low impurity concentrations the curves are practically flat: here mobility does not depend on the impurity concentration. In this impurity concentration range the phonon scattering dominates. This is in agreement with a strong temperature dependence of mobility found experimentally in low-doped samples. However, the measured temperature dependences of $\mu_n \approx T^{-2.4}$ and $\mu_p \approx T^{-2.2}$ in low-doped silicon, shown in

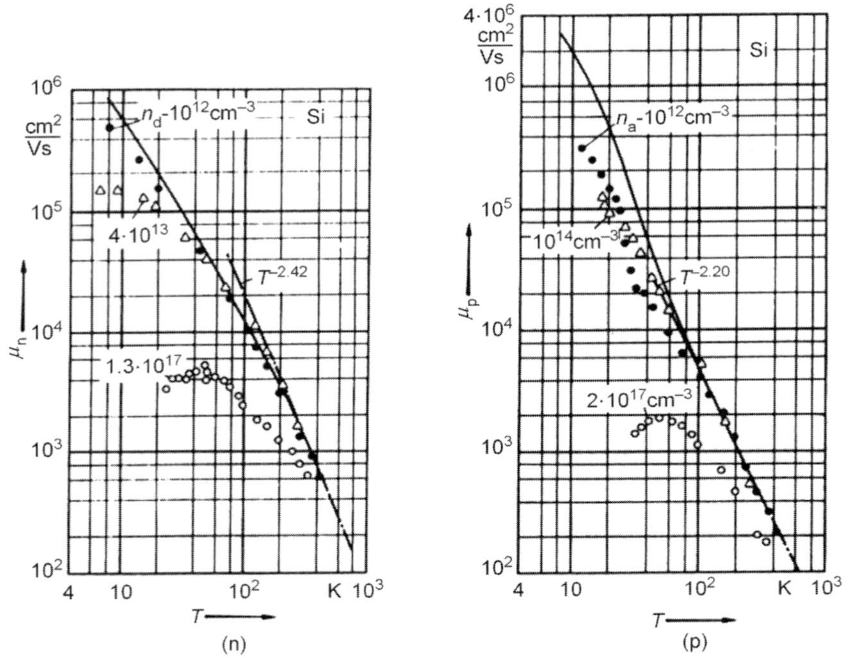

Figure 2.9. Drift mobility of carriers in silicon as a function of temperature: (*n*), electrons; (*p*), holes; points, experimental. The labels indicate the impurity concentration in the sample. Full lines indicate theoretical predictions for pure phonon scattering (adapted from [12]).

figure 2.9, do not agree in detail with the theoretical prediction $\mu \approx \langle \tau \rangle \approx T^{-3/2}$ for acoustic phonon scattering. Some additional scattering mechanisms are also involved [11]. At higher impurity concentrations the ionised impurity scattering is also effective, and the temperature dependence accordingly becomes weaker or even changes sign. A lot of data on mobility and other properties of various semiconductors can be found in [12].

Mobility of charge carriers is one of the most important parameters of materials used to build magnetic field sensors based on the galvanomagnetic effects. As we shall see in the later chapters, generally, the higher mobility, the easier is to achieve high sensor performance.

The semiconductor material with the highest electron mobility at room temperature is single-crystal indium-antimonide (InSb)—see Appendix D. In large and pure InSb crystals, the electron mobility is as high as $80\,000\ \mathrm{cm^2\,V\,s^{-1}}$. However, it is difficult to achieve this value of mobility in practical devices. Figure 2.10 shows the measured electron mobility in InSb single-crystal thin films, such as used for practical Hall magnetic field sensors [13]. The highest achieved mobility for a non-doped film is

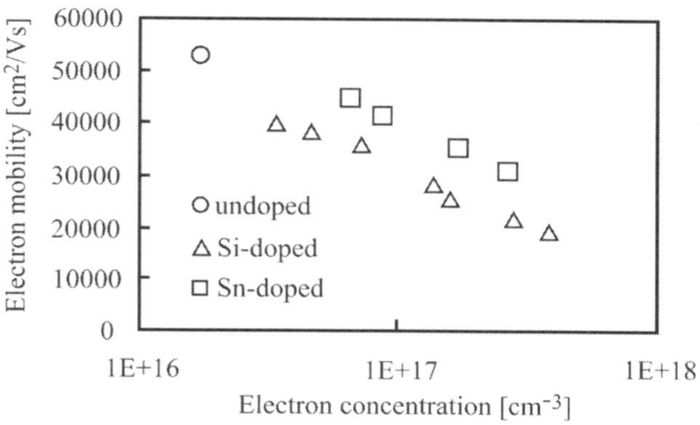

Figure 2.10. Drift mobility of electrons in InSb single crystal thin film as function of doping at 300 K. Experimental. The 1 μm thick InSb film was grown directly on a semi-insulating GaAs substrate using MBE (molecular beam epitaxy) (reprinted from [13]).

54 000 $cm^2\,V\,s^{-1}$. The mobility decreases with the doping density roughly in the same way as in the case of GaAs (figure 2.8).

The semiconductor material with the second-highest electron mobility at room temperature is indium arsenide (InAs). Thin films of a related ternary compound semiconductor InGaAs, epitaxially grown on InP substrate, are now often used in optoelectronics. Briefly, there are at least two good reasons to use such *hetero-epitaxial layers* for Hall devices instead of simple InSb thin films: first, it is possible to make a layer with much lower electron density than in the case of InSb; and second, the electron density is much less temperature-dependent than in the case of InSb. Figure 2.11 shows the measured electron *Hall mobility* and *sheet electron* density in two different $In_{0.53}Ga_{0.47}As/InP$ thin films [14]. Hall mobility is approximately equal with drift mobility—see §3.4.3. One of the structures (2DEG)

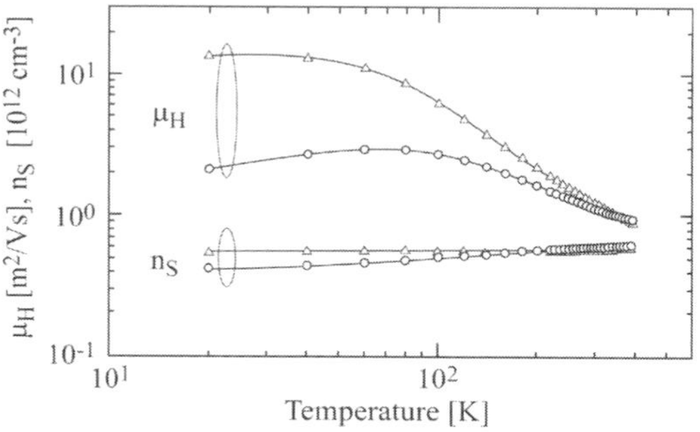

Figure 2.11. Temperature dependences of electron Hall mobility μ_H and sheet electron density n_S for hetero-epitaxial $In_{0.53}Ga_{0.47}As/InP$ thin films: doped channel (circles) and 2DEG-channel (triangles) (reprinted from [14]).

is the so-called *two-dimensional electron gas*, situated in a two-dimensional quantum well—see §5.3.3. At room temperature, the electron mobility is clearly limited by phonon scattering in both structures. But at low temperatures, the mobility in the quantum-well 2DEG strongly increases, reaching about 135 000 $cm^2\,V\,s^{-1}$ at 20 K.

Current density

The charge carriers that move collectively make electrical current. Let us denote the concentrations of electrons and holes by n and p, and their drift velocities by v_{dn} and v_{dp}, respectively. The current densities are then given by

$$\boldsymbol{J}_n = -qn\boldsymbol{v}_{dn} \qquad \boldsymbol{J}_p = qp\boldsymbol{v}_{dp} \tag{2.87}$$

for strongly extrinsic n-type and p-type semiconductors, respectively. Using (2.85), (2.87) can be rewritten as

$$\boldsymbol{J}_n = q\mu_n n\boldsymbol{E} \qquad \boldsymbol{J}_p = q\mu_p p\boldsymbol{E} \tag{2.88}$$

or

$$\boldsymbol{J}_n = \sigma_n\boldsymbol{E} \qquad \boldsymbol{J}_p = \sigma_p\boldsymbol{E}. \tag{2.89}$$

These equations express Ohm's law in differential form: the current density in a sample is proportional to the applied electric field. The proportionality coefficients

$$\sigma_n = q\mu_n n \qquad \sigma_p = q\mu_p p \tag{2.90}$$

are called the *electric conductivities*, and their inverse

$$\rho_n = \frac{1}{\sigma_n} = \frac{1}{q\mu_n n} \qquad \rho_p = \frac{1}{\sigma_p} = \frac{1}{q\mu_p p} \tag{2.91}$$

the *electric resistivities*.

If both electrons and holes exist in a sample, the total current is given by the sum of electron and hole currents,

$$\boldsymbol{J} = \boldsymbol{J}_n + \boldsymbol{J}_p. \tag{2.92}$$

From (2.89) and (2.92) Ohm's law again follows

$$\boldsymbol{J} = \sigma\boldsymbol{E} \tag{2.93}$$

where

$$\sigma = \sigma_n + \sigma_p \tag{2.94}$$

is the conductivity, and

$$\rho = \frac{1}{\sigma} \tag{2.95}$$

is the resistivity of a mixed conductor.

Resistivity characterizes the intensity of friction to which the flow of electricity is exposed in a material. According to what we said above, the friction is a consequence of the carrier scattering phenomena.

The resistivity of semiconductors depends on both temperature and impurity concentration. The temperature dependences are quite different in various temperature and doping ranges. In the intrinsic range (see §2.3.4), the dominant influence is that of carrier concentration, which is an exponential function of temperature (2.32); hence the exponential dependence of the resistance on temperature (2.2). In the extrinsic saturation range, the dominant influence has the mobility, since the carrier concentration practically equals the doping concentration (2.45). Semiconductors in the saturation range show a temperature dependence of resistance (2.1), which is typical for metals. Resistivities as functions of impurity concentration for some semiconductors are shown in figure 2.12.

Thermal agitation may activate additional conduction mechanisms in some materials. The influence of temperature on conduction in various solids is reviewed in [15].

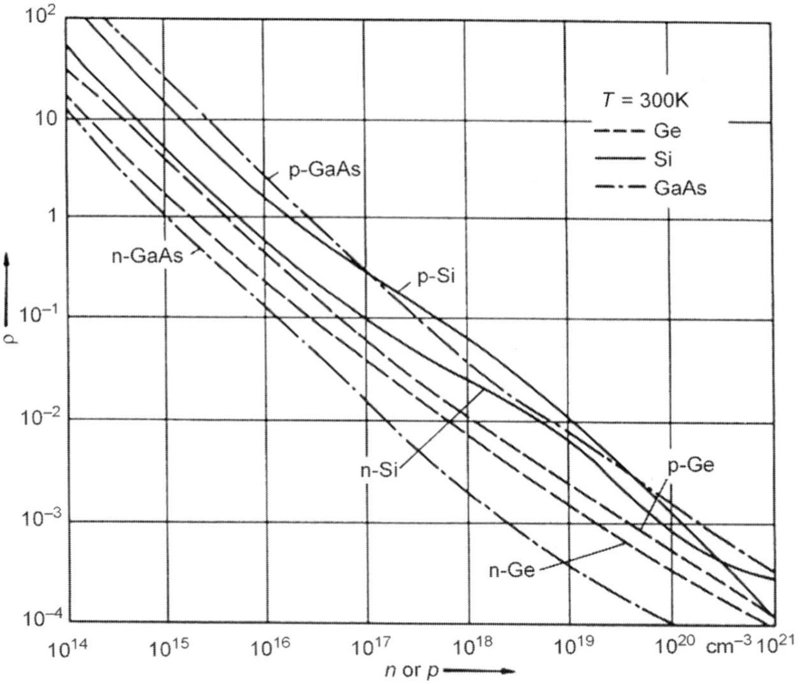

Figure 2.12. Resistivity against carrier concentration for germanium, silicon and GaAs at $T = 300$ K (reprinted from [12]).

Throughout the above analysis of electric conduction phenomena we tacitly assumed that the sample dimensions were large. If this assumption is not fulfilled, a number of new physical effects may arise. For instance, in very thin films, where the film thickness is comparable with the average free path of carriers, surface scattering plays an important role. Transport phenomena in very thin films are reviewed in [16].

2.4.2 High electric fields

If the electric field in a sample is very strong, so that the inequality (2.81) reverses, the thermodynamic equilibrium of the system lattice/carriers will be noticeably disturbed. During their free flights the carriers obtain on average more energy from the field than they can transfer to the lattice, so they start to accumulate energy. Their average kinetic energy becomes greater than that in thermodynamic equilibrium. A higher average kinetic energy corresponds to a higher temperature (2.79). Therefore, in a strong electric field charge carriers have their own temperature, which is higher than the lattice temperature. Such an ensemble of charge carriers is called *hot carriers*.

How strong the electric field must be to start heating carriers can be estimated from (2.81). Taking this as an equality, and substituting in it (2.86), we obtain

$$|\boldsymbol{E}_{\mathrm{h}}| \simeq \frac{2}{\mu}(cv_{\mathrm{t}})^{1/2} \tag{2.96}$$

where the subscript h stands for 'hot'. Thus the higher the carrier mobility, the lower the field is required to observe the high-field phenomena. Substituting $\mu \simeq 0.1\,\mathrm{m^2\,V^{-1}}$ Si (silicon), $c \simeq 10^5\,\mathrm{cm\,s^{-1}}$ and $v_{\mathrm{t}} \simeq 10^7\,\mathrm{cm\,s^{-1}}$, we obtain $|E_{\mathrm{h}}| \simeq 2 \times 10^3\,\mathrm{V\,cm^{-1}}$.

With an increase in the electric field, the carrier temperature at first also increases, but later eventually tends to saturate. This is because a high enough carrier temperature gives rise to some additional collision mechanisms, with a much higher energy transfer efficiency. These new scattering mechanisms are the generation of optical phonons and impact ionization.

Optical phonons are the elementary thermal vibrations of solids with the property that the neighbouring atoms oscillate with opposite phases. Because of a strong coupling of the neighbouring atoms, these are high-frequency and thus high-energy vibrations. For example, in silicon the energy of optical phonons is $E_{\mathrm{op}} = 0.063\,\mathrm{eV}$. (Note that at 300 K, $kT \simeq 0.026\,\mathrm{eV}$.) For this reason, optical phonons are very little excited at room temperature, and their density in a crystal is very low. Consequently, the electron–optical phonon interaction near room temperature and at low electric fields is a scarce event. This is why we have neglected so far the scattering of carriers on optical phonons.

Let us specifically consider now the gas of hot electrons. Of course, holes behave fully analogously.

Hot electrons

With an increase in the electric field, the electron temperature increases, too. Owing to the $T^{-3/2}$ law for the acoustic phonon scattering, the electron relaxation time and mobility start to decrease. In addition, more and more electrons acquire enough energy to create an optical phonon. This additional scattering mechanism causes a further reduction of the total mean free transit time (2.63) and the relaxation time, and thus a further reduction of mobility (2.86). At very high electric fields, the most frequent scattering event is that associated with the creation of an optical phonon. Now most of the electrons are involved in the following cycle: with little random motion, each electron is accelerated against the field and gains energy; this goes on until its kinetic energy becomes equal to the energy of optical phonons; then, with a high probability, the electron creates an optical phonon, and thereby loses almost all of its kinetic energy; the acceleration then starts all over again, and so on.

Saturation of drift velocity

According to this model, the drift velocity should saturate at about half of the velocity corresponding to the energy of optical phonons:

$$v_{\mathrm{d,sat}} \simeq \frac{1}{2}\left(\frac{2E_{\mathrm{op}}}{m^*}\right)^{1/2}. \tag{2.97}$$

For most semiconductors this expression yields a value for the carrier saturation velocity of about $v_{\mathrm{d,sat}} \simeq 10^7\ \mathrm{cm\,s^{-1}}$. By chance, this value equals the average thermal velocity of carriers at room temperature (see (2.79)).

An idea about the magnitude of the electric field at which we may expect the onset of velocity saturation can be obtained with the aid of (2.85), by substituting there $v_{\mathrm{d}} = v_{\mathrm{d,sat}}$ (2.97) and the low-field value of the mobility. Thus

$$|\boldsymbol{E}_{\mathrm{sat}}| \simeq v_{\mathrm{d,sat}}/\mu. \tag{2.98}$$

For example, for n-type silicon one obtains in this way a value of $|E_{\mathrm{sat}}| \simeq 10^4\ \mathrm{V\,cm^{-1}}$.

The results of the above simple model of the drift of hot carriers agree surprisingly well with the experimental results (see figure 2.13). The peculiar maximum of the curve for GaAs and other compound semiconductors is due to the so-called intervalley transfer mechanism, which is beyond the scope of this book.

Therefore, at electric fields stronger than that limited by (2.96), carrier drift velocity becomes gradually less dependent on the electric field and

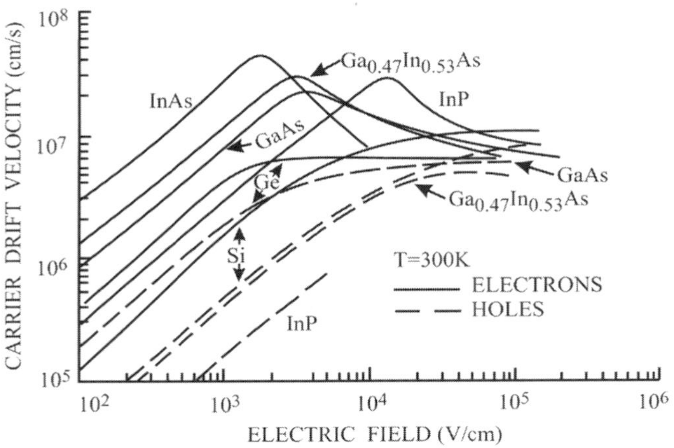

Figure 2.13. Measured carrier velocity against electric field for common high-purity semiconductors. For highly doped samples, the initial lines are lower than indicated here. In a high-field region, however, the velocity is essentially independent of doping (reprinted from [17]).

eventually saturates at a constant value. Ohm's law does not hold at high electric fields. Formally, one may still retain the notion of mobility as in (2.85); but mobility now depends on electric field. At very high electric fields, mobility tends to vanish:

$$\mu_h \simeq v_{d,sat}/|\boldsymbol{E}|. \tag{2.99}$$

At even higher electric fields than those at which the drift velocity saturates, the hottest carriers may attain enough energy to ionize the host atoms of the crystal. This phenomenon is called impact ionization. The threshold energy for impact ionization is roughly $E_I \simeq \frac{3}{2} E_g$. The threshold electric field for impact ionization in a semiconductor is, consequently, roughly proportional to its band gap. In silicon, the threshold field for impact ionization is about $2 \times 10^5\ \mathrm{V\,cm^{-1}}$.

Under certain conditions, impact ionization may result in an unlimited cumulative increase in carrier density. This phenomenon is called avalanche breakdown.

A wealth of literature on hot carrier phenomena is available. The interested reader may continue studying the subject with the aid of [18] and [19].

2.4.3 Diffusion current

Diffusion is a kind of collective transport of micro-particles, which is due exclusively to the random motion of individual particles. For example, the random motion of particles is due to thermal agitation and then no external field is necessary. The direction of the net particle transport due to diffusion is always such as to decrease any spatial non-uniformity in particle density. The rate of the particle transport is proportional to the gradient of the particle density,

$$\boldsymbol{j} = -D\nabla n. \tag{2.100}$$

Here $\boldsymbol{j}$ denotes the particle current density, D is the diffusion coefficient and ∇n is the gradient of the particle density. The negative sign reflects the fact that diffusion always goes 'downhill', that is towards lower concentrations.

Diffusion of quasi-free charge carriers in a semiconductor produces an electrical current. The expressions for the corresponding current densities are obtained by multiplying (2.100) by $-q$ or q for electrons and holes, respectively:

$$\boldsymbol{J}_{\mathrm{n}} = qD_{\mathrm{n}}\nabla n \qquad -\boldsymbol{J}_{\mathrm{p}} = qD_{\mathrm{p}}\nabla p. \tag{2.101}$$

The diffusion coefficients are related to the corresponding carrier mobilities. For non-degenerate carrier gases and diffusion due to thermal agitation,

$$D_{\mathrm{n}} = \frac{kT}{q}\mu_{\mathrm{n}} \qquad D_{\mathrm{p}} = \frac{kT}{q}\mu_{\mathrm{p}}. \tag{2.102}$$

These expressions are known as the *Einstein relationships*.

Let us consider now, without taking too much care with rigour, how the random thermal motion brings about the diffusion effect.

Imagine a semiconductor sample in which a non-uniform concentration of electrons is somehow established and kept. Let the electron concentration in a region monotonously decrease in the x direction, as shown in figure 2.14. We shall find the net rate of electrons passing through a unit area of the plane $x = x_0$, placed perpendicularly to the electron concentration gradient. We assume a constant temperature throughout the sample and no electric field.

Consider first the wandering of a single electron. This electron moves along a piecewise straight line, making a sharp turn at each collision. During a time interval t_{N}, the electron makes $N = t_{\mathrm{N}}/t$ straight-line steps, t being the mean transit time, and arrives at a distance L_{N} away from its initial position. According to the random walk model [20], the most probable value for this distance is given by

$$\langle L_{\mathrm{N}}^2\rangle = \frac{1}{3}Nl^2 = \frac{1}{3}\frac{t_{\mathrm{N}}}{\tau}l^2. \tag{2.103}$$

Here t_{N} is the average number of steps the electron makes along one coordinate axis, and l^2 is the squared mean free path of the electron.

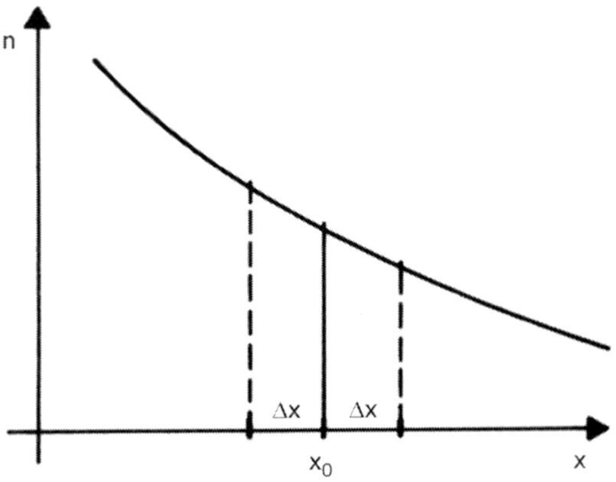

Figure 2.14. Non-uniform concentration of electrons in a sample along the x axis. The plate $x = x_0$ is placed at the point x_0. Electrons diffuse through this from left to right.

Notice now in our semiconductor sample the two slices of thickness $\Delta x = (\langle L_N^2 \rangle)^{1/2}$ on both sides of the plane $x = x_0$ (see figure 2.14). All the electrons present in a slice at the moment $t = 0$ will probably leave the slice during the time interval t_N. Half of the electrons from each slice leave it by wandering to the left, and the other half by wandering to the right. The net number of electrons crossing a unit surface of the plane $x = x_0$ during the time t_N is then

$$\phi_x = \tfrac{1}{2} n_L \, \Delta x - \tfrac{1}{2} n_R \, \Delta x. \tag{2.104}$$

Here n_L and n_R are the mean electron densities in the left and right slice, respectively. Substituting here

$$n_L = n(x_0) - \frac{dn}{dx}\frac{\Delta x}{2} \qquad n_R = n(x_0) + \frac{dn}{dx}\frac{\Delta x}{2}$$

we obtain

$$\phi_x = -\frac{1}{2}(\Delta x)^2 \frac{dn}{dx}. \tag{2.105}$$

Since $(\Delta x)^2 = \langle L_N^2 \rangle$, from (2.103) and (2.105) it follows that

$$j_x = \frac{\phi_x}{t_N} = -\frac{1}{6}\frac{l^2}{\tau}\frac{dn}{dx} = -D\frac{dn}{dx}. \tag{2.106}$$

This expression is the one-dimensional version of (2.100) for a concentration gradient along the x axis.

Equation (2.106) gives a hint of a relation between the diffusion coefficient, which is a macroscopic quantity, and the microscopic parameters of the system. To elaborate this relation further, consider first a sub-ensemble of electrons with energies in a narrow range around a value E. The diffusion coefficient of these electrons is

$$D = \frac{1}{6}\frac{l^2}{\tau} = \frac{1}{6}\frac{v^2\overline{t^2}}{\tau} = \frac{1}{3}v^2\tau \tag{2.107}$$

where v is the electron velocity (2.61) and $\overline{t^2}$ is the mean square value of the electron transit times. With the aid of (2.59), we obtain $\overline{t^2} = 2\tau^2$. After averaging (2.107) over all electrons in a non-degenerate electron gas, we obtain for the diffusion coefficient the expression

$$D = \frac{1}{3}v_t^2\langle\tau\rangle = \frac{kT}{m^*}\langle\tau\rangle. \tag{2.108}$$

Here we also made use of (2.79). By comparing the expressions for the diffusion coefficient (2.108) with that for mobility (2.86), we obtain the Einstein relationship (2.102).

If in addition to an electron concentration gradient an electric field is also present in the sample, then both drift and diffusion current will flow. The total electron current density at any point of the sample is the sum of the two current components:

$$\boldsymbol{J}_{\mathrm{n}} = q\mu_{\mathrm{n}}n\boldsymbol{E} + qD_{\mathrm{n}}\nabla n. \tag{2.109}$$

Analogously, for the total hole current

$$\boldsymbol{J}_{\mathrm{p}} = q\mu_{\mathrm{p}}p\boldsymbol{E} - qD_{\mathrm{p}}\nabla p. \tag{2.110}$$

In mixed conductors, the total current density is the sum of the above two.

2.4.4 Non-stationary transport

In all our discussions of the carrier transport up to now, we assumed that carriers stay under given conditions (electric field, density of scattering centres, temperature) during enough time to reach steady-state distributions both in terms of momentum and energy. 'Enough time' means enough to make many collisions with scattering centres. But (i) if a device is very small or (ii) if an electric field appears in a region as a very short pulse, so that electrons experience just a few or even no collisions while passing a region with a high electric field, then electrons cannot settle in the steady state. Then we have non-stationary transport. The most important phenomena of non-stationary transport are velocity overshoot and ballistic transport [21].

Velocity overshoot means a swift temporary increase in the drift velocity of charge carriers well above the value of stationary saturation drift velocity. Velocity overshoot occurs during a very short period of time to electrons, which suddenly enter a high electric field region of a semiconductor device. For example, for electrons in GaAs, this happens after the first 0.1 ps or so in a field of 10 kV/cm at 300 K. At these conditions, the mean distance that electrons travel before reaching a steady state is about 100 nm in GaAs and 10 nm in Si. In GaAs, the overshoot velocity may peak at a value five times higher than the stationary saturation velocity.

Briefly, velocity overshoot is a consequence of the difference in the momentum and energy relaxation times, see §2.3.2. It takes many collisions with phonons to reach the steady-state in energy distribution of electrons. And while the electron mean energy is still low, their mean transit time between collisions is that of non-hot electrons, see §2.4.2. Consequently, during the first few collisions in a region with very high electric field, the average mobility of electrons is the same as that in a low electric field.

The term *ballistic transport* implies that an electron 'flies' trough a region of a solid-state device without any collision, similarly as an electron flies between the two electrodes in a vacuum tube. This may occur in a solid-state device if the length of the region under consideration is (i) shorter than the mean free path between scatterings, but (ii) still much larger than the wavelength of electrons. Then we can treat electrons as little charged billiard balls of a mass equal to the effective mass of electrons in the material under consideration. We call such electrons *ballistic electrons*.

Let us estimate the length of a device in which we may have ballistic transport. We calculate the average distance l that an electron travels without scattering in a similar way as (2.82) and (2.83), and obtain

$$l = \frac{q}{m^*}\tau^2 E. \tag{2.111}$$

Using (2.86) to eliminate the dependence on τ, we obtain

$$l \simeq \mu^2 \frac{m^*}{q} E. \tag{2.112}$$

This shows that ballistic transport is most probable in high-mobility materials. But note that we now use the notion of mobility merely as a criterion for judging if in a material we may find ballistic transport or not. Otherwise, the movement of ballistic electrons exposed to electric field is determined by their effective mass, not by mobility.

As an example, let us take the mobility of 2DEG at low temperatures from figure 2.11, $\mu \approx 10^5\,\mathrm{cm^2\,V\,s^{-1}}$, the effective mass of electrons in InAs $m^* = 0.023m$ (Appendix C), and the electric field of $10^4\,\mathrm{V/cm} = 1\,\mathrm{V/\mu m}$. Then with (2.112) we calculate $l \approx 13\,\mu\mathrm{m}$. Therefore, under these conditions, a device will be predominantly ballistic if it is smaller than, say, $10\,\mu\mathrm{m}$.

2.5 Other relevant phenomena

In this section we shall briefly mention a few phenomena that are not directly connected with the Hall effect. However, these phenomena may define general conditions for the operation of some Hall effect devices, or appear as parasitic effects that essentially limit the performance of Hall effect devices.

2.5.1 Recombination and generation

Quasi-free charge carriers that are present in a semiconductor in thermodynamic equilibrium are a product of the thermal excitation of either host or impurity atoms. When a valence electron is excited into the conduction band, we say that it has been generated; when an electron and a hole annihilate each other, we say that an electron-hole pair has recombined. In semi-conductors, carriers are continuously generated and recombined. At thermodynamic equilibrium, the average rates of carrier generation and recombination are equal, and the average carrier concentrations, found in §2.2.3, are constant.

If the equilibrium in carrier concentration is somehow disturbed, for instance by illumination, the equality of generation and recombination rates no longer holds. The net generation rate assumes such a value as to drive the system towards equilibrium. If the disturbed system is left alone again, the excess carrier concentration will start to decay:

$$\delta n(t) = \delta n(0) \exp(-t/\tau_r). \tag{2.113}$$

Here $\delta n(t) = n(t) - n_0$ is the excess carrier density, $n(t)$ and n_0 are the instantaneous and the equilibrium carrier concentrations, respectively, and τ_r is a parameter called the *lifetime*. From this equation, we obtain the net generation rate

$$U = \frac{d(\delta n)}{dt} = -\frac{\delta n(t)}{\tau_r}. \tag{2.114}$$

The lifetime of excess carriers in semiconductors may have very different values. For example, in pure and perfect silicon crystals it may be greater than 10^{-3} s; and in heavily doped material it may be as short as 10^{-9} s. Some impurity atoms promote recombination and are called recombination centres. Especially efficient as recombination centres are impurity atoms that produce energy levels at the middle of the band gap. Such impurities are silver and gold in silicon (see figure 2.5).

The surface of a semiconductor device may also contain a lot of recombination centres. The net recombination rate per unit surface area can be expressed in this way:

$$U_s = s\delta n_s. \tag{2.115}$$

Here δn_s is the excess carrier density at the surface, and s is a parameter called the *surface recombination velocity*. The name comes about because of the dimensions of s, which are in metres per second.

2.5.2 PN junction

A pn junction is a merger of the semiconductor p- and n-type regions. In a transition layer between the two regions the carrier densities are much lower than in the neutral regions away from the junction. For this reason, this transition layer is also called the depletion region. The simplest pn region is a junction where both sides are homogeneously doped and the transition between the acceptor and donor doping is abrupt. Such a junction is called an abrupt pn junction.

The width of the *depletion region* of an abrupt pn junction is given by

$$W \simeq \left[\frac{2\varepsilon_s\varepsilon_0}{q}\left(\frac{N_A + N_D}{N_A N_D}\right)(V_{bi} - V)\right]^{1/2}. \qquad (2.116)$$

Here ε_s is the relative dielectric constant of the semiconductor, ε_0 is the permittivity of free space, N_A and N_D are the acceptor and donor concentrations at the p- and n-side of the junction, respectively, $V_{bi}(\leq E_g/q)$ is the built-in potential, and V is the externally applied bias voltage. $V > 0$ if the p-side is positively biased.

A larger part of the depletion region appears at the lower doped side of the junction. If the doping is very asymmetric, for instance $N_A \gg N_D$, almost the whole depletion region will be at the n-type side. In this case, (2.116) reduces to

$$W \simeq \left(\frac{2\varepsilon_s\varepsilon_0}{qN_D}(V_{bi}/V)\right)^{1/2}. \qquad (2.117)$$

Such a junction is called a one-sided abrupt pn junction.

A pn junction exhibits highly non-linear, rectifying *voltage-current characteristics*. The idealized current density-voltage relationship of a pn junction diode is given by the Shockley equation

$$J = J_0[\exp(qV/kT) - 1] \qquad (2.118)$$

where J_0 is the saturation current density and V the biasing voltage. The saturation current density is given by

$$J_0 = qn_i^2\left(\frac{D_n}{N_A L_n} + \frac{D_p}{N_D L_p}\right). \qquad (2.119)$$

Here D_n and D_p are the diffusion coefficients of electrons and holes, respectively. If the neutral regions on either side of the junction are very thick, then L_n and L_p denote the diffusion lengths: $L_n = (D_n\tau_{rn})^{1/2}$ and $L_p = (D_p\tau_{rp})^{1/2}$ for the p- and n-side of the junction, respectively. τ_{rn} and τ_{rp} denote the lifetimes of electrons and holes, respectively. If the width of a neutral region is much shorter than the diffusion length, then either L_n or L_p denote this width. Such a device is called a *short-base diode*.

Equation (2.118) is a good approximation of real diode characteristics if the generation-recombination effect in the depletion layer is negligible and the diode operates under low-level injection conditions. The meaning of the term *injection level* is explained below.

At zero bias voltage, between the n-region and the p-region of a diode there is a potential barrier, which prevents the majority carriers from one region from entering the other region. Under forward bias, this barrier decreases and the carriers diffuse into the neighbouring regions. Take, for example, electrons at the n-side of the junction. Owing to the barrier lowering, they can enter into the p-region. Then we say that the electrons are injected into the p-region. At the p-side of the junction, the injected electrons are minority carriers. Their excess concentration at the p-side boundary of the depletion layer ($x = 0$) is given by

$$\delta n_{\mathrm{p}}(x=0) = n_{\mathrm{p0}}[\exp(qV/kT) - 1] \tag{2.120}$$

and decreases with the distance x from the junction. Here n_{p0} is the equilibrium electron concentration at the p-type side. From (2.120) it follows that more minority carriers are injected into the lower doped side of the junction: if $N_{\mathrm{D}} > N_{\mathrm{A}}$, then $p_{\mathrm{n0}} < n_{\mathrm{p0}}$.

Let us suppose the p-side is the lower doped side of the junction. *Low-level injection* means

$$\delta n_{\mathrm{p}} \ll p_{\mathrm{p0}} \simeq N_{\mathrm{A}} \tag{2.121}$$

where p_{p0} is the equilibrium hole (majority carriers) concentration in the injected region. Under low-level injection, the majority carrier concentration stays practically constant.

At higher bias voltages the inequality (2.121) eventually reverses. Owing to the neutrality condition (2.36), the majority carrier density must also increase, and

$$\delta n_{\mathrm{p}} = \delta p_{\mathrm{p}} \simeq p_{\mathrm{p}} \gg p_{\mathrm{p0}}. \tag{2.122}$$

Then we have *high-level injection*. At high-level injection, the concentration of the injected electrons at the p-side boundary of the depletion layer is given by

$$\delta n_{\mathrm{p}}(x=0) \simeq n_{\mathrm{i}} \exp(qV/2kT) \tag{2.123}$$

and the current density by

$$J \simeq J_{\mathrm{h}} \exp(qV/2kT). \tag{2.124}$$

Note that at a high injection level, the carrier density in the injected region is proportional to the diode current density.

2.5.3 Noise

Electrical noise is a spontaneous random fluctuation of the voltage or current of an electron device. A common description of a *noise* quantity is in terms of its mean square, or *root-mean-square (RMS)*, value. Take, for instance, the output voltage of a sensor $v(t)$. Generally, it can be represented in the form

$$v(t) = v_s(t) + v_N(t). \tag{2.125}$$

Here v_s is a signal voltage, which may contain information on a measured quantity, and v_N is a noise voltage. In most cases, the time average of the noise voltage is zero; but the time average of noise power is not. A descriptor of this power is the mean square noise voltage, defined as

$$\overline{v_N^2} = \lim_{T\to\infty} \frac{1}{T} \int_0^T [v_N(t)]^2 \,dt. \tag{2.126}$$

The RMS noise voltage is then $v_N = (\overline{v_N^2})^{1/2}$.

Noise voltage and its RMS value depend on the frequency range involved. The harmonic content of a noise voltage is usually expressed by its spectral density function $S_{NV}(f)$ .The voltage spectral density is proportional to the power per unit frequency range that a voltage signal could deliver to a resistor at a given frequency. Given the noise voltage spectral density, the RMS noise voltage can be calculated as

$$v_N = \left(\int_{f_1}^{f_2} S_{NV}(f)\,df \right)^{1/2} \tag{2.127}$$

where f_1 and f_2 are the boundaries of the frequency range in which the noise voltage is considered.

Generally, two kinds of noise can be distinguished: generating noise and modulating noise. A generating noise is associated with an internal fluctuating electromotive force, and is consequently independent of the device biasing conditions. The well-known thermal noise is a generating noise. A modulating noise is due to the fluctuations of a device parameter, such as conductance. Hence modulating noise appears only when a device carries a current. Shot noise, recombination-generation noise, and $1/f$ noise belong to this category.

Thermal noise, also known as Johnson–Nyquist noise, is due to the thermal agitation of charge carriers. The voltage spectral density of thermal noise is given by

$$S_{NV} = 4kTR \tag{2.128}$$

where R is the resistance of the noise source. The spectral density is independent of frequency up to very high frequencies. Such a spectrum is called a white spectrum.

Shot noise is due to the discrete nature of electricity. It is noticeable when current in a circuit is limited by the passing of charge carriers over a potential barrier. A typical example is the collector current of a transistor. The current noise spectral density of shot noise is given by

$$S_{NI} = 2qI \tag{2.129}$$

where I is the device current. This spectrum is also white up to very high frequencies.

Generation-recombination (GR) noise is a result of the fluctuation in number of quasi-free carriers in a device. As we saw earlier in §2.5.1, the equilibrium carrier concentrations are maintained through a balance in the generation and recombination of charge carriers. But generation and recombination are independent random processes, and therefore the instantaneous carrier concentrations fluctuate around their equilibrium values. The noise current spectral density of GR noise for one type of GR process is given by

$$S_{NI} = AI^2 \frac{4\tau_r}{1 + (2\pi f \tau_r)^2} \tag{2.130}$$

where A is a parameter whose value depends on details of the GR process and device parameters, and τ_r is the lifetime of carriers for this GR process. This noise might be significant at low frequencies, where its spectrum is also white.

1/f noise is a conductivity-modulating noise whose physical origin is still under dispute [22, 23]. Experimentally it was found that the current spectral density of this noise can be described by the following expression [24]:

$$S_{NI} \simeq I^2 \frac{\alpha}{N} \frac{1}{f^\beta}. \tag{2.131}$$

Here I is the device current, N is the total number of charge carriers in the device, α is a dimensionless parameter called the Hooge parameter, and $b \simeq 1 \pm 0.1$ (typically).

It was long believed that the Hooge parameter α is approximately a constant of about 2×10^{-3}. Later, for semiconductors α values of 10^{-7} [25] and even as low as 10^{-9} [26] were reported. This indicates that the α parameter might rather characterize a specific device processing than the basic material itself.

In a semiconductor device, all of the above-mentioned noise sources may act simultaneously. The total noise spectral density is then the sum of the particular contributions. Figure 2.15 illustrates the noise spectrum of a unipolar device, such as a resistor, carrying a current. Each of the three relevant noise mechanisms dominates in a frequency range.

For further reading, several books on noise are available [27].

2.5.4 Piezoresistance

Piezoresistance is the change of the electrical resistance of a sample upon the application of a mechanical force. Thus the relative change of resistivity is

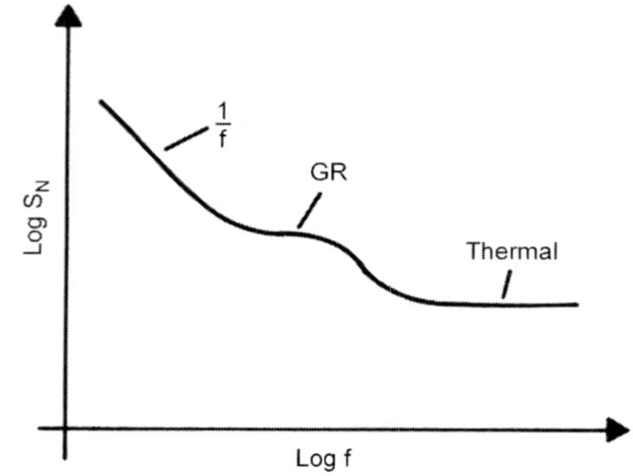

Figure 2.15. A typical current noise spectrum of a unipolar semiconductor device, such as a resistor or a Hall device.

proportional to stress X as

$$\Delta\rho/\rho_0 = \Pi X \tag{2.132}$$

where Π is the coefficient of piezoresistance.

In some crystalline materials, piezoresistance is an anisotropic effect: it depends on the directions of the stress and the current density relative to the crystal axes. (Note that the current density lines define the paths along which the change in resistivity is considered.) In particular, one can distinguish the longitudinal piezoresistance effect, when the vectors of stress $\boldsymbol{X}$ and current density $\boldsymbol{J}$ are collinear, and the transverse piezoresistance effect, when $\boldsymbol{X}$ and $\boldsymbol{J}$ are perpendicular to each other. Table 2.3 shows some experimental values of the piezoresistance coefficients of n-type silicon. Detailed data on piezoresistance in silicon can be found in [28]. The piezoresistance coefficients of GaAs are more than an order of magnitude smaller than those of silicon [29].

The physical origin of the piezoresistance effect is the change of the interatomic distances in a crystal under stress. A shift in the interatomic

Table 2.3. Piezoresistance coefficient of low-doped n-type silicon at room temperature.

	Direction of		
Effect	$\boldsymbol{X}$	$\boldsymbol{J}$	$\Pi \times 10^{12}(\mathrm{Pa}^{-1})$
Longitudinal	⟨100⟩	⟨100⟩	−10.2
	⟨111⟩	⟨111⟩	−0.7
Transverse	⟨100⟩	⟨010⟩	+5.3

distances produces a change in the periodic field of the lattice, and hence also alterations in the band gap and the effective masses of carriers. A change in the effective mass directly causes a change in the mobility (2.86). This is the main mechanism underlying the piezoresistance effect in semiconductors with spherical constant energy surfaces, like GaAs (see figure 2.2(a,b)).

In semiconductors with ellipsoidal constant energy surfaces at $\boldsymbol{k} \neq 0$, such as silicon (figure 2.2(c,d)), there is an additional effect that even prevails. This is the repopulation of energy valleys. As a consequence of the alternation in the lattice periodic field, the minimum energy (bottom of an energy valley) of some ellipsoids increases, and that of the other ellipsoids decreases. Now the available conduction band electrons increasingly populate the lower-energy valleys. If the ellipsoids corresponding to the more populated valleys are elongated in the direction of the current, the average electron effective mass increases, and so also does the resistivity, for the constant energy surfaces are extended in the direction of the larger effective mass (see §2.1.2).

References

[1] Kireev P S 1974 *Semiconductor Physics* (Moscow: MIR)

[2] Seeger K 1989 *Semiconductor Physics* (Berlin: Springer)

[3] Nag B R 1972 *Theory of Electrical Transport in Semiconductors* (Oxford: Pergamon)

[4] Smith R A 1987 *Semiconductors* (Cambridge: Cambridge University Press)

[5] Grove A S 1967 *Physics and Technology of Semiconductor Devices* (New York: Wiley)

[6] Muller R and Kamins T I 1977 *Device Electronics for Integrated Circuits* (New York: Wiley)

[7] Sze S M 2002 *Semiconductor Devices* (New York: Wiley)

[8] Van Overstraeten R J and Mertens R P 1987 Heavy doping effects in silicon *Solid-State Electron.* **30** 1077–87

[9] del Alamo J A and Swanson R M 1987 Modeling of minority-carrier transport in heavily doped silicon emitters *Solid-State Electron.* **30** 1127–36

[10] Lang D and Tufte O N 1980 The Hall effect in heavily doped semiconductors *The Hall Effect and its Applications* ed. C L Chien and C R Westate (New York: Plenum)

[11] Jacobini C, Canoli C, Oltaviani G and Alberigi Quaranta A 1977 A review of some charge transport properties of silicon *Solid-Stllte Electron.* **20** 77–89

[12] Landolt-Bornstein 1982 Numerical and functional relationships in science and technology *Semiconductors* vol. III/17a (Berlin: Springer)

[13] Shibasaki I, Okamoto A, Ashihara A and Suzuki K 2001 Properties and applications of InSb single crystal thin film Hall elements, *Technical Digest of the 18th Sensor Symposium*, pp 233–238

[14] Kyburz R, Schmid J, Popovic R S and Melchior H 1994 High-performance $In_{0.53}Ga_{0.47}As/InP$ Hall sensors with doped and 2DEG channels and screening front and back gate layers *Sensors and Materials* **6** 279–291

[15] Roberts G G, Apsley N and Munn R W 1980 Temperature dependent electronic conduction in semiconductors *Phys. Rep.* **60** 59–150

[16] Askerov B M, Kuliev B I and Figarova S R 1984 Electron transport phenomena in conductive films *Phys. Status Solidi* (b) **121** 11–37

[17] Brennan K F and Brown A S 2002 *Theory of Modern Electronic Semiconductor Devices* (New York: Wiley)

[18] Ridley B K 1986 Hot electrons in semiconductors *Sci. Prog.* **70** 425–59

[19] Reggiani L (ed.) 1985 *Hot-electron Transport in Semiconductors* (Berlin: Springer)

[20] For example: MacDonald D K C 1962 *Noise and Fluctuations* (New York: Wiley) pp 12–14

[21] Ref. 17, pp 105–109

[22] Hooge F N, Kleinpenning T G M and Vandamme L K J 1981 Experimental studies on $1/f$ noise *Rep. Prog. Phys.* **44** 479–532

[23] Weissman M B 1988 $1/f$ noise and other slow, nonexponential kinetics in condensed matter *Rev. Mod. Phys.* **60** 537–71

[24] Hooge F N 1969 $1/f$ noise is no surface effect *Phys. Lett.* **29A** 139–40

[25] Vandamme L K J 1983 Is the $1/f$ noise parameter constant? *Noise in Physical Systems and 1/f Noise* ed. M Savelli, G Lecoy and J-P Nougier (Amsterdam: Elsevier) pp 183–92

[26] van der Ziel A, Handel P H, Zhu X and Duh K H 1985 A theory of the Hooge parameters of solid-state devices *IEEE Trans. Electron Devices* **ED-32** 667–71

[27] Bell D A 1960 *Electrical Noise* (London: Van Nostrand)

[28] Kanda Y 1982 A graphical representation of the piezoresistance coefficients in silicon *IEEE Trans. Electron Devices* **ED-29** 64–70

[29] Ref. 12, p 241

Chapter 3

Galvanomagnetic effects

This chapter is devoted to the physical effects basic to the operation of Hall and magnetoresistance devices. These effects are called the Hall effect and the magnetoresistance effect.

The Hall effect is the generation of a transverse electromotive force in a sample carrying an electric current and exposed to a perpendicular magnetic field. Depending on the sample geometry, this electromotive force may cause the appearance of a transverse voltage across the sample, or a current deflection in the sample. The generation of this transverse voltage, called the Hall voltage, is the generally known way for the appearance of the Hall effect [1].

The magnetoresistance effect is an increase in sample resistance under the influence of a magnetic field.

Both the Hall effect and the magnetoresistance effect belong to the more general class of phenomena called galvanomagnetic effects. Thence the title of this chapter. Galvanomagnetic effects are the manifestations of charge transport phenomena in condensed matter in the presence of a magnetic field. Taking the direction of the electric field as a reference, the Hall effect could be defined as a transverse isothermal galvanomagnetic effect. The magnetoresistance effect is a longitudinal isothermal galvanomagnetic effect. Galvanomagnetic effects also include non-isothermal galvanomagnetic effects, known as galvanothermomagnetic effects. A general treatment of galvanomagnetic effects can be found in several books, for example [2–8].

Our goal in this chapter is to understand the physical background of the Hall and the magnetoresistance effects. As we shall see, the two effects have a common origin and always coexist in a sample. We shall begin by discussing a simplified model of the Hall effect. This will give us a rough idea of the Hall effect and help to identify more detailed issues. Then we shall dig more into the physics of carrier transport in solid state and so proceed with a more accurate analysis.

In accordance with the definition of the Hall and the magnetoresistance effects, we assume isothermal conditions. To simplify the analysis, we also assume equilibrium carrier concentrations in samples, and thus neglect diffusion currents. However, at the end of the chapter we shall extend our results to account for diffusion effects. Throughout this chapter, we shall analyse the galvanomagnetic effects in samples of particularly convenient and idealized geometrical forms, called very short and very long plates. This will allow us to see selectively the effects of interest in their pure forms. We shall also explain the reasons for choosing a plate as a basic device form. The issue of the influence of realistic sample geometry on the device characteristics we leave for the next chapters.

3.1 Approximate analysis

All galvanomagnetic effects come about as a manifestation of the action of the Lorentz force on quasi-free charge carriers in condensed matter. The *Lorentz force* is the force acting on a charged particle in an electromagnetic field. Moreover, the existence of the Lorentz force is the fundamental indication of a very presence of an electric and/or magnetic fields. It is given by

$$\boldsymbol{F} = e\boldsymbol{E} + e[\boldsymbol{v} \times \boldsymbol{B}]. \tag{3.1}$$

Here e denotes the particle charge (for electrons $e = -q$ and for holes $e = q$, where q is the magnitude of the electron charge), $\boldsymbol{E}$ is the electric field, $\boldsymbol{v}$ the carrier velocity and $\boldsymbol{B}$ the magnetic induction. The first term on the right-hand side of (3.1) is often referred to as electrostatic force, and the second term as the Lorentz force. In this book, we shall refer to the two terms as the electric part and the magnetic part of the Lorentz force, respectively.

In order to be able to immediately introduce the magnetic part of the Lorentz force (3.1) into the familiar carrier drift equations, we shall use in the first few paragraphs of this section the following simplifying approximation.

Smooth-drift approximation

This approximation consists of the assumption that charge carriers move uniformly as a result of an electric field or other driving forces, the velocity of this movement being the same for all carriers and equal to the appropriate drift velocity. So we can replace the carrier velocity of each individual carrier, $\boldsymbol{v}$ in (3.1), by the average drift velocity of all carriers. We neglect completely the thermal motion of carriers and approximate the energy dissipation effect of scattering by a smooth friction. In a way, we model our charge carriers by charged macroscopic particles immersed in a viscous liquid (the role of which is here played by the crystal lattice). As we shall see in §3.1.5, the smooth drift approximation is equivalent to the assumption of a very weak magnetic field. Thanks to this approximation, we shall avoid occupying ourselves with fine details, and quickly get an overall insight into the galvanomagnetic effects.

In §§3.1.5 and 3.1.6, we shall drop the smooth-drift approximation, and represent carrier scattering with another simple model. We shall see that the two models give a consistent picture of the galvanomagnetic effects.

Although very useful, the simple models have their limits. In §3.1.7 we shall identify a few issues beyond these models, and so prepare ourselves for a more profound analysis of the galvanomagnetic effects, given in §§3.2–3.5.

3.1.1 The (original) Hall effect

We shall first study the Hall effect in the form in which Hall discovered it [1]. Hall's experimental device was a long gold leaf. We shall also consider a long flat sample; but in order to quickly come close to modern applications of the Hall effect, we shall study it in semiconductor samples.

Let us consider the transport of carriers in long and thin semiconductor strips, such as those shown in figure 3.1. 'Long' here means that the length of a strip l is much larger than its width w. In such a sample we can certainly neglect the influence of the current-supplying contacts on the phenomena

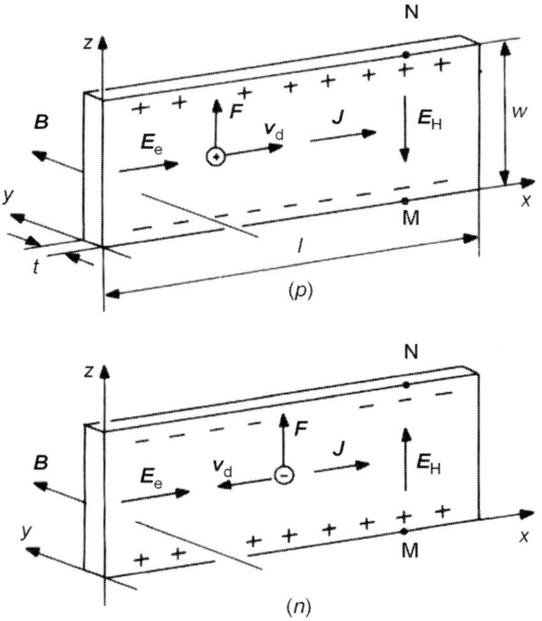

Figure 3.1. The Hall effect in long plates of p-type (p) and n-type (n) material. $\boldsymbol{E}_e$ is the external electric field, $\boldsymbol{B}$ the magnetic induction, $\boldsymbol{v}_d$ the drift velocity of carriers, $\boldsymbol{F}$ the magnetic force, $\boldsymbol{J}$ the current density and $\boldsymbol{E}_H$ the Hall electric field. The magnetic forces press both positive and negative charge carriers towards the upper boundary of a strip. The Hall voltage appears between the charged edges of the strips.

under examination. Let the strips be strongly extrinsic and of different conductivity types, one p-type (p) and the other n-type (n). For the time being, we neglect minority carriers. We choose the coordinate system so that the x axis is parallel with the strip axis, and the xz plane leans on one of the larger faces of the strip.

Suppose that in the strips, along the x direction, an external electric field $\boldsymbol{E}_e = (E_x, 0, 0)$ is established, and the magnetic field is zero. Then in the Lorentz force (3.1) only the first term, which is the electrical force, exists. As we have seen in §2.6.1, the electrical force causes the drift of charge carriers. In the present case, charge carriers drift along the strips in two opposite directions. The drift velocities are given by (2.85):

$$\boldsymbol{v}_{dp} = \mu_p \boldsymbol{E}_e \qquad \boldsymbol{v}_{dn} = \mu_n \boldsymbol{E}_e \tag{3.2}$$

where μ_p and μ_n are the mobilities of holes and electrons, respectively. The associated current densities are given by (2.88):

$$\boldsymbol{J}_p = q\mu_p p \boldsymbol{E}_e \qquad \boldsymbol{J}_n = q\mu_n n \boldsymbol{E}_e \tag{3.3}$$

where p and n denote the hole and the electron density in the p-type and the n-type strip, respectively.

Hall electric field

Let us now expose the current-carrying strips to a magnetic field. Let the magnetic induction $\boldsymbol{B}$ be collinear with the y axis (figure 3.1). Now on each charge carrier in the samples both parts of the Lorentz force (3.1) act. Since we assumed a uniform directional velocity of all carriers in a sample, the magnetic forces acting on the carriers are given by

$$\boldsymbol{F}_p = e[\boldsymbol{v}_{dp} \times \boldsymbol{B}] \qquad \boldsymbol{F}_n = e[\boldsymbol{v}_{dn} \times \boldsymbol{B}] \tag{3.4}$$

in the samples p and n, respectively. These forces have the same direction in both strips: since $e = q$ and $e = -q$ for holes and electrons, respectively, from (3.2) and (3.4) it follows that

$$\boldsymbol{F}_p = q\mu_p[\boldsymbol{E}_e \times \boldsymbol{B}] \qquad \boldsymbol{F}_n = q\mu_n[\boldsymbol{E}_e \times \boldsymbol{B}]. \tag{3.5}$$

For our particular geometry the magnetic forces are collinear with the z axis:

$$\boldsymbol{F}_p = (0, 0, q\mu_p E_x B_y) \qquad \boldsymbol{F}_n = (0, 0, q\mu_n E_x B_y). \tag{3.6}$$

These forces push carriers towards the upper edges of the strips. Consequently the carrier concentrations at these upper edges of the strips start to increase, while the carrier concentrations at the lower edges start to decrease. Then the charge neutrality condition (2.36) is disturbed, and an electric field appears between the strip edges. This field, $\boldsymbol{E}_H$ in figure 3.1, acts on the moving carriers too, and pushes them in such a direction as to decrease the excess charges at the edges. In both samples, this is downwards.

Eventually the transverse electrical force becomes strong enough to balance the magnetic force (3.4):

$$e[\boldsymbol{v}_\mathrm{d} \times \boldsymbol{B}] + e\boldsymbol{E}_\mathrm{H} = 0. \tag{3.7}$$

From this moment on, the charge carriers again move parallel to the strip axes, as if only the external electric field were acting on them.

The transverse electric field $\boldsymbol{E}_\mathrm{H}$, which counterbalances the action of the magnetic force, is called the *Hall electric field*. From (3.7), the Hall electric field is

$$\boldsymbol{E}_\mathrm{H} = -[\boldsymbol{v}_\mathrm{d} \times \boldsymbol{B}]. \tag{3.8}$$

Using (3.2), we also have

$$\boldsymbol{E}_\mathrm{Hp} = -\mu_\mathrm{p}[\boldsymbol{E} \times \boldsymbol{B}] \qquad \boldsymbol{E}_\mathrm{Hn} = \mu_\mathrm{n}[\boldsymbol{E} \times \boldsymbol{B}] \tag{3.9}$$

and in our particular case (figure 3.1)

$$\boldsymbol{E}_\mathrm{Hp} = (0, 0, -\mu_\mathrm{p} E_x B_y) \qquad \boldsymbol{E}_\mathrm{Hn} = (0, 0, \mu_\mathrm{n} E_x B_y). \tag{3.10}$$

Therefore, in a sample carrying electrical current and exposed to a magnetic field, the magnetic part of the Lorentz force has a tendency to 'press' electricity against one edge of the sample. ('Pressing electricity' is a notion used by Hall.) Electricity, however, behaves somewhat like an incompressible fluid. It reacts by developing an electric field, the Hall field, which exactly counterbalances the magnetic pressure. The magnetic force has the direction of $[\boldsymbol{E}_\mathrm{e} \times \boldsymbol{B}]$ irrespective of the carrier's charge sign. Hence the Hall electric field has to have opposite directions in p-type and n-type semiconductors.

Hall voltage

A more tangible effect associated with the Hall field is the appearance of a transverse voltage between the edges of a strip. This voltage is known as the *Hall voltage*. Let us choose two points M and N at the opposite edges of a strip, under the condition that both points lie in the same equipotential plane when $\boldsymbol{B} = 0$ (figure 3.1). Then the Hall voltage is given by

$$V_\mathrm{H} = \int_M^N \boldsymbol{E}_\mathrm{H}\,\mathrm{d}z \tag{3.11}$$

and

$$V_\mathrm{Hp} = \mu_\mathrm{p} E_x B_y w \qquad V_\mathrm{Hn} = \mu_\mathrm{n} E_x B_y w \tag{3.12}$$

for our p-type and n-type strips, respectively. Here the negative sign has been omitted and w denotes the width of the strips.

The generation of the transverse Hall electric field and the associated Hall voltage under the experimental conditions similar to that shown in figure 3.1 are usually referred to as the Hall effect. However, there is another, more fundamental feature of the Hall effect: the Hall angle.

Hall angle

In the presence of a magnetic field, the total electric field in the sample $\boldsymbol{E} = \boldsymbol{E}_e + \boldsymbol{E}_H$ is not collinear with the external electric field $\boldsymbol{E}_e$. In the present case the current in the sample is confined to the direction of the external electric field, and the current density is collinear with the external electric field (see figure 3.1). Hence the total electric field is not collinear with the current density either. Therefore, the Hall effect in a long sample shows up through the tilting of the total electric field relative to the external electric field and the current density in the sample. The angle of inclination Θ_H is called the *Hall angle*. With the aid of figure 3.2, the magnitude of the Hall angle is given by

$$\tan \Theta_H = |\boldsymbol{E}_H|/|\boldsymbol{E}_e|. \tag{3.13}$$

For a reason which will become clear later, it is convenient to measure the Hall angle with respect to the direction of the total electric field, as shown in figure 3.2. Then we may define the Hall angle as the angle of inclination of the current density $\boldsymbol{J}$ with respect to the total electric field $\boldsymbol{E}$. According to this definition and (3.13), we obtain for the two particular cases from figures 3.1 and 3.2

$$\tan \Theta_{Hp} = \mu_p B_y \qquad \tan \Theta_{Hn} = -\mu_n B_y \tag{3.14}$$

for the p-type and n-type semiconductor strip, respectively. The value of the Hall angle depends only on the applied magnetic induction and the mobility of the charge carriers. The sign of the Hall angle coincides with the sign of the charge carriers.

Hall coefficient

From (3.3) and (3.9) we can find a relationship between the current density and the Hall field:

$$\boldsymbol{E}_{Hp} = -\frac{1}{qp}[\boldsymbol{J} \times \boldsymbol{B}] \qquad \boldsymbol{E}_{Hn} = \frac{1}{qn}[\boldsymbol{J} \times \boldsymbol{B}]. \tag{3.15}$$

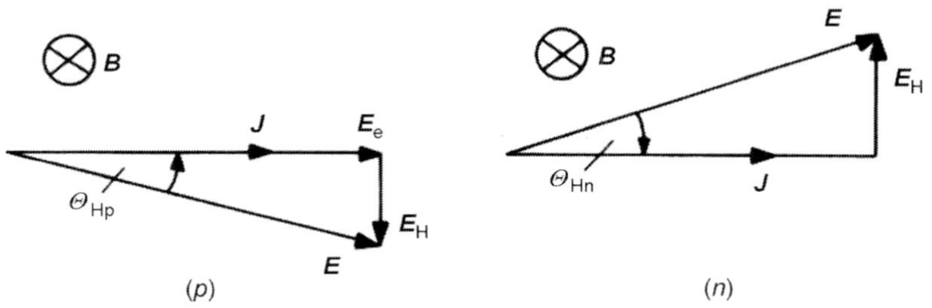

Figure 3.2. The vector diagrams of electric fields and current densities in long samples: (p), in a p-type material; (n), in an n-type material. $\boldsymbol{J}$ is the current density, $\boldsymbol{E}_e$ the external electric field, $\boldsymbol{E}_H$ the Hall electric field, $\boldsymbol{E}$ the total electric field and Θ_H is the Hall angle.

These equations can be rewritten as

$$\boldsymbol{E}_{\mathrm{H}} = -R_{\mathrm{H}}[\boldsymbol{J} \times \boldsymbol{B}] \tag{3.16}$$

where R_{H} is a parameter called the *Hall coefficient*. Equation (3.16) is a modern representation of the conclusions made by Hall after his experimental findings.

The Hall coefficient is a material parameter that characterizes the intensity and sign of the Hall effect in a particular material. The unit of the Hall coefficient is $\mathrm{V\,m\,A^{-1}\,T^{-1}}$ (volt metre per ampere tesla), which is sometimes expressed in a more compact form as $\Omega\,\mathrm{m\,T^{-1}}$ (ohm metre per tesla) or, equivalently, $\mathrm{m^3\,C^{-1}}$ (metre^3 per coulomb).

From a comparison of (3.15) and (3.16), we can find the Hall coefficients of strongly extrinsic semiconductors

$$R_{\mathrm{Hp}} = \frac{1}{qp} \qquad R_{\mathrm{Hn}} = -\frac{1}{qn} \tag{3.17}$$

for p-type and n-type material, respectively. The sign of the Hall coefficient coincides with the sign of the majority carriers, and the magnitude of the Hall coefficient is inversely proportional to the majority carrier concentration.

In the practical applications of the Hall effect it is convenient to operate with pure macroscopic and integral quantities characterizing a Hall device. To this end, for a long Hall device we obtain from (3.11) and (3.16) the *Hall voltage*:

$$V_{\mathrm{H}} = \frac{R_{\mathrm{H}}}{t} I B_{\perp}. \tag{3.18}$$

Here t is the thickness of the strip, I is the device current given by $I = Jwt$, and $B_{\perp}$ is the component of the magnetic induction perpendicular to the device plane. We omit the negative sign.

Equation (3.18) shows why a plate is a preferential shape of a Hall device with voltage output: for a given biasing current and magnetic induction, the thinner the sample, the higher the resulting Hall voltage.

Example

Let us find the numerical values of the characteristic quantities for the Hall device shown in figure 3.3. The device is made of homogeneously doped n-type silicon of resistivity $\rho = 1\,\Omega\,\mathrm{cm}$ at 300 K. The dimensions are: length $l = 300\,\mu\mathrm{m}$, width $w = 100\,\mu\mathrm{m}$, thickness $t = 20\,\mu\mathrm{m}$. The biasing voltage is $V = 10\,\mathrm{V}$, and the perpendicular magnetic induction $B_{\perp} = 1\,\mathrm{T}$.

Hall coefficient (3.17): From figure 2.10 we find the doping concentration corresponding to $\rho = 1\,\Omega\,\mathrm{cm}$ to be $N_{\mathrm{D}} \simeq 4.5 \times 10^{15}\,\mathrm{cm}^{-3}$. According to the result of the example in §2.3.5, the material is at room temperature

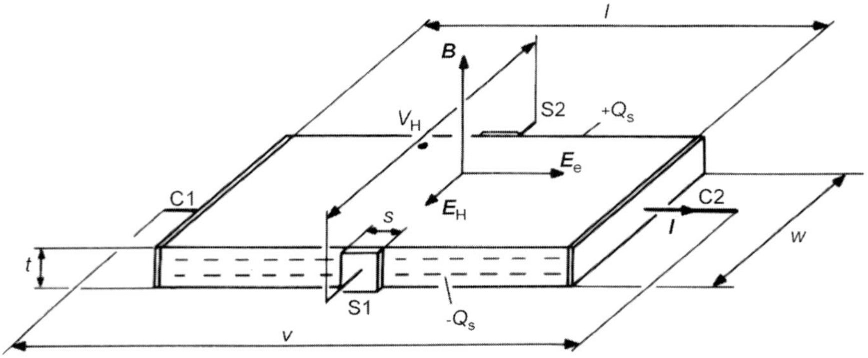

Figure 3.3. A rectangular Hall device. It can be understood as a portion of a strip from figure 3.1, fitted by four contacts.

at saturation, that is $n \simeq N_D$. Thus

$$R_H \simeq \frac{1}{qn} \simeq \frac{1}{qN_D} = -1.39 \times 10^{-3}\,\mathrm{C}^{-1}\,\mathrm{m}^3. \tag{3.19}$$

Device resistance:

$$R = \rho\frac{l}{wt} \simeq 1.5 \times 10^3\,\Omega. \tag{3.20}$$

Current:

$$I = \frac{V}{R} \simeq 6.67 \times 10^{-3}\,\mathrm{A}. \tag{3.21}$$

Hall voltage (3.18):

$$V_H = \frac{R_H}{t} IB_\perp \simeq 0.46\,\mathrm{V}. \tag{3.22}$$

Hall angle (3.14): From figure 2.8 we find the electron mobility which corresponds to $N_D = 4.5 \times 10^{15}\,\mathrm{cm}^{-3}$ to be $\mu_n \simeq 1400\,\mathrm{cm}^2\,\mathrm{V}^{-1}\,\mathrm{s}^{-1}$. Thus

$$\tan\Theta_{Hn} = -\mu_n B_\perp \simeq -0.140 \qquad \Theta_{Hn} = -0.139\,\mathrm{rad} \simeq -7.97°. \tag{3.23}$$

Even at high magnetic inductions such as $B = 1\,\mathrm{T}$, the Hall angle in silicon is relatively small, and $\Theta_H \simeq \tan\Theta_H$.

Excess charge: The Hall voltage stems from the excess charges kept separated by the magnetic force at the corresponding device boundaries. The surface density of these charges is given by

$$Q_s = C_s V_H \tag{3.24}$$

where C_s is the capacitance per unit area between the charged faces (perpendicular to the Hall field $\boldsymbol{E}_H$; see figure 3.3). To assess C_s, we may assume that the permittivity of the Hall device material $\varepsilon_0\varepsilon_s$ is much greater than the permittivity of the surrounding space, and that there are no other conductive bodies in the vicinity of the device. Then $C_s \simeq \varepsilon_0\varepsilon_s/w$. For silicon, $\varepsilon_s \simeq 12$ and $C_s \simeq 1 \times 10^{-6}\,\mathrm{F_m^{-2}}$. Using this value, from (3.24) and (3.22) we obtain $Q_s \simeq 0.46 \times 10^{-6}\,\mathrm{C\,m^{-2}}$. This corresponds to an electron surface density of

$$N_s = \frac{Q_s}{q} \simeq 2.9 \times 10^{12}\,\mathrm{m^{-2}}. \tag{3.25}$$

This is a very small surface charge density. However, the situation changes drastically if the surface is strongly capacitively coupled with other bodies.

3.1.2 The current deflection effect

Consider now the Hall effect in a very short Hall plate. A sample for a Hall effect experiment is called short if its dimension in the current direction is much smaller than that in the direction of the magnetic force acting on carriers. An example of a short sample is the short strip shown in figure 3.4. The basic shape of this device, its position relative to the coordinate system, and the notation are the same as in the case of the long strips in figure 3.1. Now, however, the strip is short, $l \ll w$, and it is laterally sandwiched between two large current contacts. In such short samples the Hall effect takes a form which is sometimes called the current deflection effect.

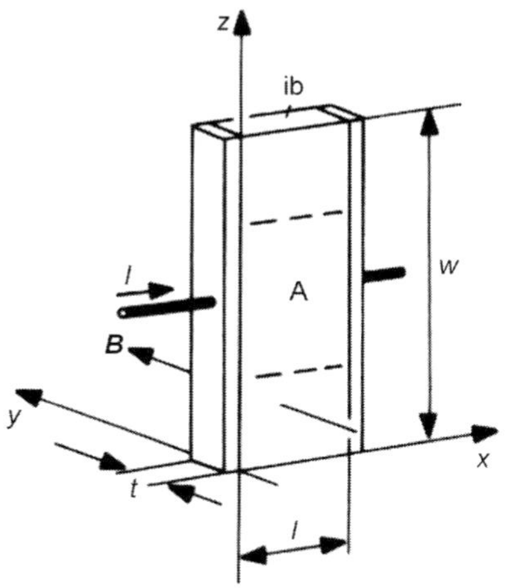

Figure 3.4. A short Hall plate. The distance between the current contacts is much smaller than the distance between the insulating boundaries (ib).

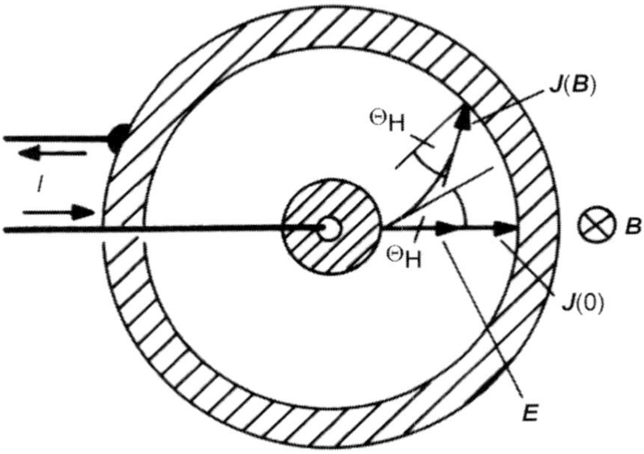

Figure 3.5. A Corbino disc. This is a round semiconductor plate with one electrode at the centre and the other around the circumference. In a Corbino disc, the electric field $\boldsymbol{E}$ is radial irrespective of the magnetic field. The current density $\boldsymbol{J}$ is tilted by the Hall angle Θ_H relative to the electric field.

Corbino disc

To understand the peculiarities of the Hall effect in short samples, let us consider first a special case: the ring-shaped short sample shown in figure 3.5. Such a sample is called a *Corbino disc* [9], which is a disc or a ring of a low-conductivity material, such as a semiconductor, fitted with two highly conductive, concentric electrodes. If a voltage is applied between the electrodes, the equipotential lines are also concentric, and the electric field is radial. Then in the absence of a magnetic field, a radially oriented current flows in the disc. In a perpendicular magnetic field, the current density tilts relative to the radial direction (see figure 3.5). In a Corbino disc there are no isolating boundaries where charge could accumulate, as in the previous case of long strips. Therefore, the Hall electric field cannot appear, there is nothing to counterbalance the action of the magnetic force, and carriers are forced to deflect from their shortest paths.

The short strip in figure 3.4 can be viewed as a section of a Corbino disc with an infinitely large radius. It is intuitively clear that the inner portion of this sample (A), far away from the insulating boundaries (ib), should behave exactly as a portion of a real Corbino disc. Consequently, in the largest part of a short sample there is no Hall field; the current deflection takes place there.

The absence of Hall voltage in short samples can also be explained by using an electric circuit model. According to this model, the Hall voltage does not appear because the magnetic electromotive force is short-circuited by the large contacts. Instead, a transverse current flows. The resultant of the 'normal' and the transverse current density is skewed respective to the electric field.

Current deflection

Let us now find a quantitative relationship between the electric field and the current density in short samples. We shall again use the approximation which we introduced above: we neglect the thermal motion of carriers, and assume that carriers move uniformly under the action of the Lorentz force. Equation (3.1) can then be rewritten as

$$\boldsymbol{F}_{\rm p} = q\boldsymbol{E} + q[\boldsymbol{v}_{\rm p} \times \boldsymbol{B}] \qquad \boldsymbol{F}_{\rm n} = -q\boldsymbol{E} - q[\boldsymbol{v}_{\rm n} \times \boldsymbol{B}] \tag{3.26}$$

for holes and electrons, respectively. Here $\boldsymbol{v}_{\rm p}$ and $\boldsymbol{v}_{\rm n}$ denote the directional velocities of holes and electrons due to the action of the pertinent Lorentz forces $\boldsymbol{F}_{\rm p}$ and $\boldsymbol{F}_{\rm n}$, and friction forces, respectively.

To make the further steps in our analysis more obvious, we shall formally substitute the Lorentz forces $\boldsymbol{F}_{\rm p}$ and $\boldsymbol{F}_{\rm n}$ in (3.26) by the equivalent electrical forces $q\boldsymbol{E}_{\rm p}$ and $-q\boldsymbol{E}_{\rm n}$:

$$q\boldsymbol{E}_{\rm p} = q\boldsymbol{E} + q[\boldsymbol{v}_{\rm p} \times \boldsymbol{B}] \qquad -q\boldsymbol{E}_{\rm n} = -q\boldsymbol{E} - q[\boldsymbol{v}_{\rm n} \times \boldsymbol{B}]. \tag{3.27}$$

Carriers would drift due to the equivalent electric fields $\boldsymbol{E}_{\rm p}$ and $\boldsymbol{E}_{\rm n}$ exactly in the same way as they do under the action of the Lorentz forces. Multiplying now the two equations (3.27) by $\mu_{\rm p}p$ and $-\mu_{\rm n}n$, we obtain

$$\begin{aligned} \boldsymbol{J}_{\rm p}(B) &= \boldsymbol{J}_{\rm p}(0) + \mu_{\rm p}[\boldsymbol{J}_{\rm p}(\boldsymbol{B}) \times \boldsymbol{B}] \\ \boldsymbol{J}_{\rm n}(B) &= \boldsymbol{J}_{\rm n}(0) - \mu_{\rm n}[\boldsymbol{J}_{\rm n}(\boldsymbol{B}) \times \boldsymbol{B}]. \end{aligned} \tag{3.28}$$

Here $\boldsymbol{J}_{\rm p}(B)$ and $\boldsymbol{J}_{\rm n}(B)$ are the hole and electron current densities respectively in the presence of a magnetic induction $\boldsymbol{B}$,

$$\begin{aligned} \boldsymbol{J}_{\rm p}(\boldsymbol{B}) &= q\mu_{\rm p}p\boldsymbol{E}_{\rm p} = \mu_{\rm p}p\boldsymbol{F}_{\rm p} = qp\boldsymbol{v}_{\rm p} \\ \boldsymbol{J}_{\rm n}(\boldsymbol{B}) &= q\mu_{\rm n}n\boldsymbol{E}_{\rm n} = \mu_{\rm n}n\boldsymbol{F}_{\rm n} = -qn\boldsymbol{v}_{\rm n} \end{aligned} \tag{3.29}$$

and $\boldsymbol{J}_{\rm p}(0)$ and $\boldsymbol{J}_{\rm n}(0)$ are the drift current densities due to the electric field $\boldsymbol{E}$ when $\boldsymbol{B} = 0$:

$$\boldsymbol{J}_{\rm p}(0) = q\mu_{\rm p}p\boldsymbol{E} \qquad \boldsymbol{J}_{\rm n}(0) = q\mu_{\rm n}n\boldsymbol{E}. \tag{3.30}$$

Equations (3.26), (3.28) and (3.30) are graphically represented in figure 3.6. The orientations of the vectors $\boldsymbol{E}$ and $\boldsymbol{B}$ correspond to those in figure 3.4: the external electric field $\boldsymbol{E}$ is collinear with the x axis, and the magnetic induction $\boldsymbol{B}$ points into the drawing plane. The external field is assumed to be kept constant. The essential features of the diagrams in figure 3.6 are that the vectors of the external electric field $\boldsymbol{E}$ and the resultant current densities $\boldsymbol{J}_{\rm p}(B)$ and $\boldsymbol{J}_{\rm n}(B)$ are not collinear: in a short sample, a magnetic field deflects the current from its usual way along the electrical field.

Hall angle

Therefore, in short samples the Hall effect shows up through the tilting of the current density $\boldsymbol{J}(B)$ with respect to the external electric field $\boldsymbol{E}$. The angle of

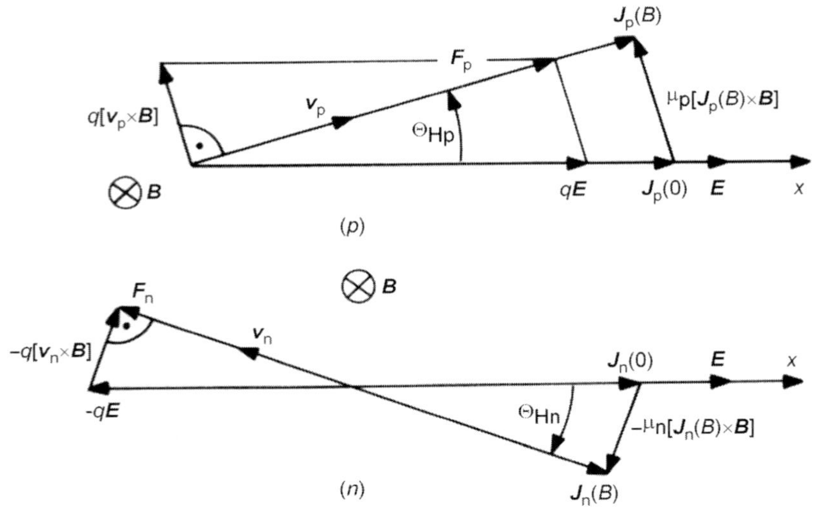

Figure 3.6. The graphical representations of the vector equations (3.26), (3.28) and (3.30) that hold in short Hall plates: (p) of a p-type semiconductor; (n) of an n-type semiconductor.

inclination is the Hall angle Θ_H. In the extreme case of a Corbino disc, there is no Hall electric field, and the external electric field is also the total electric field. Hence the Hall angle is the angle of inclination of the current density with respect to the total electric field in a sample. This definition of the Hall angle is identical with the one that we made before (see the text following (3.13)). Moreover, using the diagrams in figure 3.6, we find

$$\tan\Theta_{Hp} = \mu_p B_\perp \qquad \tan\Theta_{Hn} = -\mu_n B_\perp \tag{3.31}$$

for holes and electrons, respectively. The relationships defining the values and signs of the Hall angles in long and short samples, (3.14) and (3.31) respectively, are identical.

Note that in the above discussion of the effects in short plates we never mentioned the thickness of the plate, t in figure 3.4. This means that, unlike the case of the conventional Hall effect (§3.1.1), a device utilizing the current deflection effect does not have to be plate-like. However, the thinner the device, the higher its resistance, and the less current is needed for its practical operation. For this reason, a plate-like shape is preferable also for a current-deflection device.

3.1.3 The magnetoresistance effect

By inspecting the diagrams in figure 3.6, we notice that the current density vectors $\boldsymbol{J}(B)$ are smaller than $\boldsymbol{J}(0)$. This means that in short Hall plates a

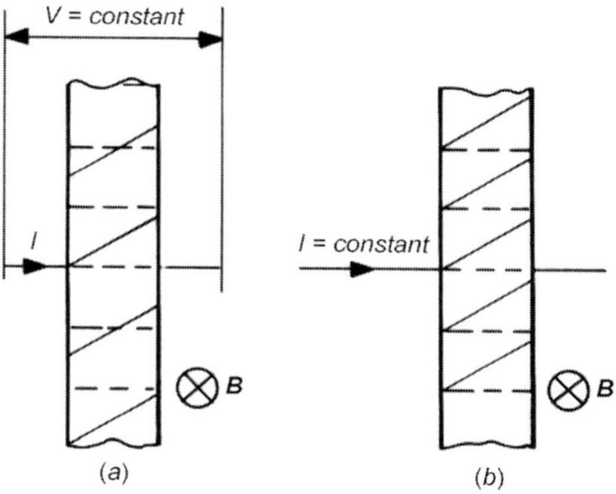

Figure 3.7. The current deflection effect in a short Hall plate, biased by a constant voltage (a) and a constant current (b). The broken lines are the current lines at $\boldsymbol{B} = 0$, whereas the full inclined lines are the current lines at $\boldsymbol{B} \neq 0$.

magnetic field also has an additional effect: it causes a reduction in current density.

The current deflection effect is illustrated in figure 3.7. As shown in figure 3.7(a), current density in the presence of a magnetic field is smaller than that with no magnetic field. The attenuation in the current density is a consequence of the current deflection effect: the deflected current lines between the sample contacts are longer (see figure 3.7(a)). The longer current path means also greater effective resistance of the sample. (Recall that we assume a constant electric field $\boldsymbol{E}$ in the sample, and consequently, a constant biasing voltage V, as shown in figure 3.7(a). Should the device be biased by constant current, the current density in it would increase in the presence of a magnetic field, as illustrated in figure 3.7(b). But this would give rise to an increase in the voltage drop over the device.)

To find how much the current density decreases due to a magnetic induction, we have to solve (3.28) with respect to $\boldsymbol{J}(B)$. We can solve this equation with the aid of (E.8) of Appendix E. For a perpendicular magnetic field, such that $\boldsymbol{E} \cdot \boldsymbol{B} = 0$, the solutions of equations (3.28) are

$$\begin{aligned} \boldsymbol{J}_{\mathrm{p}}(\boldsymbol{B}) &= \sigma_{\mathrm{pB}}\boldsymbol{E} + \sigma_{\mathrm{pB}}\mu_{\mathrm{p}}[\boldsymbol{E} \times \boldsymbol{B}] \\ \boldsymbol{J}_{\mathrm{n}}(\boldsymbol{B}) &= \sigma_{\mathrm{nB}}\boldsymbol{E} - \sigma_{\mathrm{nB}}\mu_{\mathrm{n}}[\boldsymbol{E} \times \boldsymbol{B}] \end{aligned} \tag{3.32}$$

where

$$\sigma_{\mathrm{pB}} = \sigma_{\mathrm{p0}}/[1 + (\mu_{\mathrm{p}}B)^2] \qquad \sigma_{\mathrm{nB}} = \sigma_{\mathrm{n0}}/[1 + (\mu_{\mathrm{n}}B)^2] \tag{3.33}$$

are the effective conductivities of p-type and n-type materials in the presence of a magnetic induction $\boldsymbol{B}$, and σ_{p0} and σ_{n0} are the conductivities at $\boldsymbol{B} = \mathbf{0}$ (2.90), respectively. Note that, at a given electric field, the coefficients σ_B (3.33) determine the longitudinal current density in an infinitely short sample. Since a materialization of an infinitely short sample is a Corbino disc (figure 3.5), we shall call the coefficients σ_B the *Corbino conductivity*. We shall further discuss the Corbino conductivity in §§3.3.5 and 3.4.3.

The first terms on the right-hand side of (3.32) represent the lateral current densities. The integral of this current density over the contact surface is the device current. From (3.32) and (3.33) we deduce that the device current decreases as the magnetic field increases. Therefore, if a sample is exposed to a magnetic field, everything happens as if the conductivity of the sample material were decreased. The corresponding relative increase in the effective material resistivity, from (3.37) and (2.91), is given by

$$\frac{\rho_{pB} - \rho_{p0}}{\rho_{p0}} = (\mu_p B)^2 \qquad \frac{\rho_{nB} - \rho_{n0}}{\rho_{n0}} = (\mu_n B)^2. \tag{3.34}$$

The increase in material resistivity under the influence of a magnetic field is called the *magnetoresistance effect*. In particular, the just described increase in resistivity due to the current deflection effect in short samples is called the *geometrical magnetoresistance effect*. The attribute 'geometrical' reflects the fact that the effect is related to the change in geometry of the current lines, as illustrated in figure 3.7. We shall discuss this effect further in §3.1.4 and still more thoroughly in §3.3.3 and §3.4. In §3.3.4 we shall learn about another, more subtle magnetoresistance effect, which is not related to any geometrical change in current distribution.

For a similar reason as in the case of current-deflection devices (see the end of §3.1.2: the thinner the device, the higher its resistance), a magnetoresistor usually has a plate-like shape.

3.1.4 Galvanomagnetic current components

Up to now, in order to simplify our analysis, we have always assumed that our sample is plate-like, and that the magnetic field was perpendicular to the plate. Let us drop now this assumption and consider a general case. Now we do not impose restrictions either on the sample geometry or on the mutual position of the vectors of electric field $\boldsymbol{E}$ and magnetic induction $\boldsymbol{B}$. We only assume that these two vectors are known so we seek the current density vector. The interpretation of the current density components in such a case will give us a further insight into the beauty of the galvanomagnetic effects, help us to better understand the magnetoresistance effect, and prepare us for the analysis of the so-called planar Hall effect.

The basic equations relating the resulting current density $\boldsymbol{J}(\boldsymbol{E}, \boldsymbol{B})$ with the electric field $\boldsymbol{E}$ and the magnetic induction $\boldsymbol{B}$ are given by (3.28) to

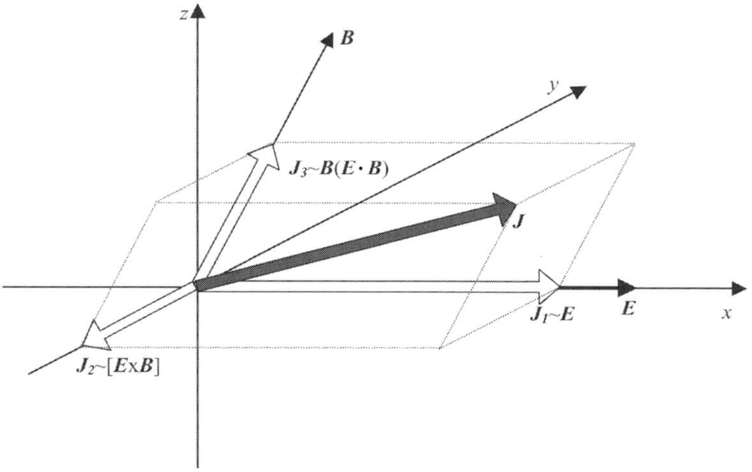

Figure 3.8. The current density components arising in matter in the presence of an electric field $\boldsymbol{E}$ and a magnetic induction $\boldsymbol{B}$ (3.36a and b). The coordinate system is chosen so that $\boldsymbol{E}$ is collinear with the x axis and $\boldsymbol{B}$ lies in the xz plane. The $\boldsymbol{E}$ and $\boldsymbol{B}$ vectors are not mutually orthogonal. The total current density is the sum $\boldsymbol{J} = \boldsymbol{J}_1 + \boldsymbol{J}_2 + \boldsymbol{J}_3$. $\boldsymbol{J}_1$ is collinear with $\boldsymbol{E}$, $\boldsymbol{J}_2$ is collinear with $[\boldsymbol{E} \times \boldsymbol{B}]$ (which is here collinear with the y axis), and $\boldsymbol{J}_3$ is collinear with B. Note that we tacitly assume positive charge carriers. In the case of negative charge carriers, the component $\boldsymbol{J}_2$, being proportional to the mobility, would have an inversed direction (along the $+y$ axis)

(3.30). We can rewrite these equations in the following compact form:

$$\boldsymbol{J}(\boldsymbol{B}) = \sigma \boldsymbol{E} + \mu[\boldsymbol{J}(\boldsymbol{B}) \times \boldsymbol{B}]. \tag{3.35}$$

The notations are as follows: $\boldsymbol{J}(\boldsymbol{E}, \boldsymbol{B}) \equiv \boldsymbol{J}(\boldsymbol{B})$, σ is the material conductivity at $\boldsymbol{B} = 0$ and μ is the mobility of carriers. Comparing (3.35) with (3.28) we see that (3.35) is applicable for both type of charge carriers if we state that μ carries the sign of the corresponding carriers.

We can solve (3.35) with the aid of Appendix E. According to (E.8), the solution is

$$\boldsymbol{J} = \frac{\sigma \boldsymbol{E} + \mu\sigma[\boldsymbol{E} \times \boldsymbol{B}] + \mu^2 \sigma \boldsymbol{B}(\boldsymbol{E} \cdot \boldsymbol{B})}{1 + \mu^2 B^2}. \tag{3.36}$$

This equation shows explicitly the total current density in matter, exposed to electric and magnetic fields, as a function of these fields and the material characteristics. With reference to (3.36) and figure 3.8, we notice the following essential facts:

- The total galvanomagnetic current density generally consists of three components:

$$\boldsymbol{J} = \boldsymbol{J}_1 + \boldsymbol{J}_2 + \boldsymbol{J}_3 \tag{3.36a}$$

$$J_1 = \left(\frac{\sigma}{1+\mu^2 B^2}\right) E;$$

$$J_2 = \left(\frac{\sigma}{1+\mu^2 B^2}\right) \mu[E \times B]; \qquad (3.36b)$$

$$J_3 = \left(\frac{\sigma}{1+\mu^2 B^2}\right) \mu^2 B[E \cdot B].$$

J_1 is proportional to the electric field E, J_2 is proportional to the vector $[E \times B]$ (which is perpendicular to both E and B), and J_3 is proportional to the vector $B(E \cdot B)$ (which is collinear with the magnetic induction B).

- If the vectors E and B are mutually orthogonal, then the current component collinear with the magnetic field vanishes. This case we had in (3.32).
- The current density component J_1, which is proportional to the electric field, decreases with increase in magnetic induction. In §3.1.3 we attributed this to the geometrical magnetoresistance effect. The other two current density components, being also proportional to B and B^2, respectively, are less affected by an increase in magnetic induction.

In order to thoroughly understand the above facts, we shall now try to find the solution of (3.35) in another way. We apply an *iterative procedure*, where each iteration step has a clear physical interpretation. The iterative process will be performed in cycles. We denote the symbols pertinent to the first iteration cycle by a prime ($'$) and those of the second cycle by a second ($''$). Figure 3.9 illustrates the first cycle of this iterative procedure.

Let us assume that carriers in a solid are exposed to an electric field E and a weak magnetic induction B. Then we imagine the current genesis process as follows:

1st approximation (figure 3.9(a)): Since B is small, the current density is mostly due to conventional drift. According to (3.28) and (3.30), we write

$$J \simeq J_1' = \sigma E \qquad (3.37a)$$

2nd approximation (figure 3.9(b)): The existence of the current density (3.37a) means that the charge carriers move with the drift velocity $v_d = \mu E$. Then they also 'feel' the magnetic part of the Lorentz force (3.4). Therefore, they will also drift in response to this force, which produces a current density J_2'. The action of the magnetic force is equivalent to the action of an equivalent electric field (see (3.27)), given by $E_2 = [v_d \times B] = [\mu E \times B]$. Therefore, the current density is $J_2' = \sigma E_2$ or

$$J_2' = \mu\sigma[E \times B] = \mu[J_1' \times B] \qquad (3.37b)$$

which is collinear with the negative y axis.

3rd approximation (figure 3.9(c)): Now since the carriers move along the y axis, they also 'feel' a magnetic Lorentz force collinear with the vector

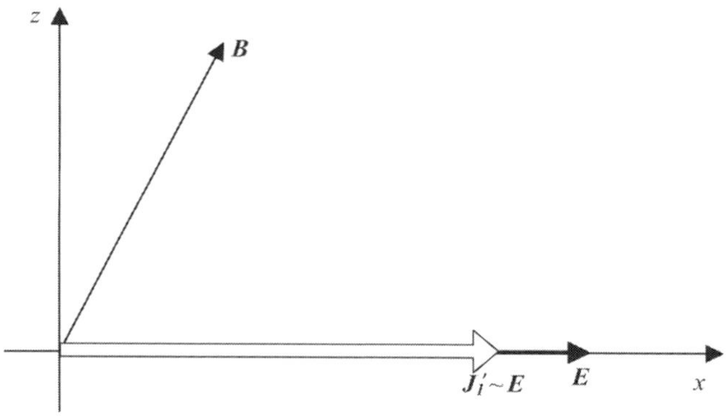

Figure 3.9. (a) Illustrating the first step in the current genesis process in matter exposed to an electric field $\boldsymbol{E}$ and a magnetic induction $\boldsymbol{B}$. The coordinate system is chosen so that the $\boldsymbol{E}$ is collinear with the x axis and $\boldsymbol{B}$ lies in the xz plane. The $\boldsymbol{E}$ and $\boldsymbol{B}$ vectors are not mutually orthogonal. To the first approximation, only a conventional drift current density $\boldsymbol{J}_1'$ arises, which is proportional to $\boldsymbol{E}$.

$[\boldsymbol{J}_2' \times \boldsymbol{B}]$. They move accordingly, producing a current density $\boldsymbol{J}_{3\mathrm{T}}'$, which is parallel with the xz plane. With an analogue reasoning as that in the 2nd approximation and using (3.37b), we find

$$\boldsymbol{J}_{3\mathrm{T}}' = \mu[\boldsymbol{J}_2' \times \boldsymbol{B}] = \mu[\mu\sigma[\boldsymbol{E} \times \boldsymbol{B}] \times \boldsymbol{B}]. \tag{3.37c}$$

Being in the same plane as the vectors $\boldsymbol{E}$ and $\boldsymbol{B}$ (the xz plane), the vector $\boldsymbol{J}_{3\mathrm{T}}'$ can be represented by its two components $\boldsymbol{J}_{3\mathrm{E}}'$ and $\boldsymbol{J}_{3\mathrm{B}}'$, one collinear with $\boldsymbol{E}$ and the other with $\boldsymbol{B}$ (see figure 3.9(d)). We can find the two components by

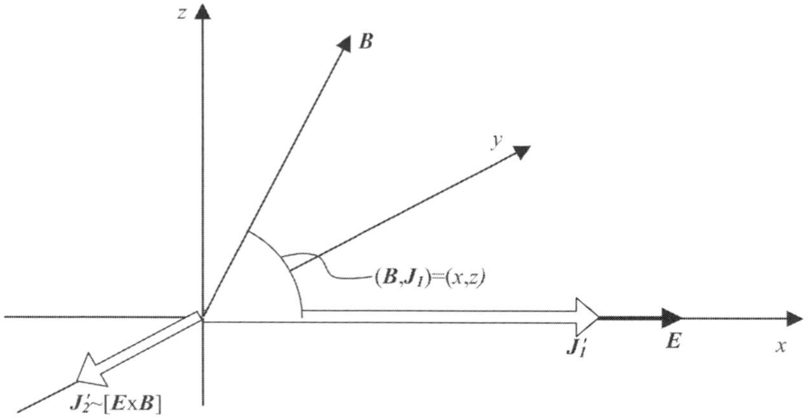

Figure 3.9. (b) The Lorentz force deflects the current component $\boldsymbol{J}_1'$, which gives rise to the generation of the current component $\boldsymbol{J}_2'$. $\boldsymbol{J}_2'$ is perpendicular to both $\boldsymbol{J}_1'$ and $\boldsymbol{B}$ (and thus collinear with the z axis), and proportional to $[\boldsymbol{E} \times \boldsymbol{B}]$.

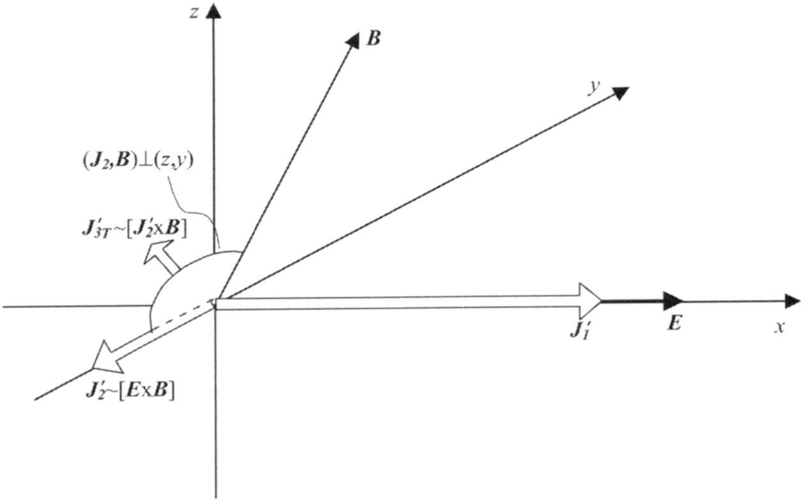

Figure 3.9. (c) The Lorentz force deflects the current component $\boldsymbol{J}_2'$, which gives rise to the generation of the current component $\boldsymbol{J}_{3\mathrm{T}}'$. $\boldsymbol{J}_{3\mathrm{T}}'$ is perpendicular to both $\boldsymbol{J}_{12}'$ and $\boldsymbol{B}$ (and therefore lies in the xz plane), and proportional to $[\boldsymbol{J}_2' \times \boldsymbol{B}]$.

transforming (3.37c) according to (E.2) of Appendix E, which gives

$$
\begin{aligned}
\boldsymbol{J}_{3\mathrm{T}}' &= \mu^2\sigma(\boldsymbol{B}(\boldsymbol{E}\cdot\boldsymbol{B}) - \boldsymbol{E}(\boldsymbol{B}\cdot\boldsymbol{B})) \\
\boldsymbol{J}_{3\mathrm{B}}' &= \mu^2\sigma\boldsymbol{B}(\boldsymbol{E}\cdot\boldsymbol{B}) \\
\boldsymbol{J}_{3\mathrm{E}}' &= -\mu^2\sigma\boldsymbol{E}(\boldsymbol{B}\cdot\boldsymbol{B}).
\end{aligned}
\tag{3.37d}
$$

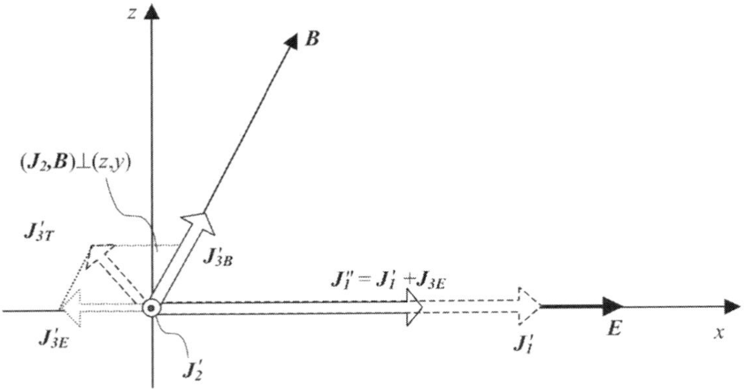

Figure 3.9. (d) The vector $\boldsymbol{J}_{3\mathrm{T}}'$ can be replaced by its components $\boldsymbol{J}_{3\mathrm{E}}'$ and $\boldsymbol{J}_{3\mathrm{B}}'$. The current component $\boldsymbol{J}_{3\mathrm{B}}'$ is collinear with the magnetic field. The sum of $\boldsymbol{J}_{3\mathrm{E}}'$ and $\boldsymbol{J}_1'$ gives $\boldsymbol{J}_1'$, which is the second approximation of the drift current density $\boldsymbol{J}_1$.

Now the part of $\boldsymbol{J}'_{3\mathrm{T}}$ collinear with $\boldsymbol{E}$ (3.37d), which we denote by $\boldsymbol{J}'_{3\mathrm{E}}$, corrects the first approximation of $\boldsymbol{J}_1$ (3.37). This closes our first iteration cycle and starts the second. The first step in the second cycle is the second approximation of $\boldsymbol{J}_1$, given by

$$\boldsymbol{J}''_1 = \boldsymbol{J}'_1 + \boldsymbol{J}'_{3\mathrm{E}} = \sigma \boldsymbol{E}(1 - \mu^2 B^2). \tag{3.37e}$$

The second approximation cycle can be continued in the same manner as above, while using (3.37e) as an approximation of $\boldsymbol{J}_1$ instead of (3.37a), and so on. The iterative process converges if $\mu \boldsymbol{B} < 1$.

From (3.38a–e), we can write now the total galvanomagnetic current density as

$$\boldsymbol{J} \simeq \sigma \boldsymbol{E}(1 - \mu^2 B^2) + \mu\sigma[\boldsymbol{E} \times \boldsymbol{B}] + \mu^2 \sigma \boldsymbol{B}(\boldsymbol{E} \cdot \boldsymbol{B}). \tag{3.38}$$

So in this way we come to the same conclusion as above (3.36): The total galvanomagnetic current density consists of three components: $\boldsymbol{J} = \boldsymbol{J}_1 + \boldsymbol{J}_2 + \boldsymbol{J}_3$. (For the sake of simplicity, we denote now the above current component $\boldsymbol{J}_{3\mathrm{B}}$ by $\boldsymbol{J}_3$.) Briefly, $\boldsymbol{J}_1$ is essentially due to conventional drift driven by the electric field $\boldsymbol{E}$; $\boldsymbol{J}_2$, which is proportional to the vectors $[\boldsymbol{J}_1 \times \boldsymbol{B}]$ and $[\boldsymbol{E} \times \boldsymbol{B}]$, appears because of the Lorentz deflection of the current component $\boldsymbol{J}_1$; and $\boldsymbol{J}_3$, which is proportional to the vector $\boldsymbol{B}(\boldsymbol{E} \cdot \boldsymbol{B})$, comes about as a consequence of the Lorentz deflection of the current component $\boldsymbol{J}_2$. We can illustrate the total galvanomagnetic current density (3.38) in the same way as (3.36) (see figure 3.8).

Equation (3.38) shows also the first approximation of the magnetoresistance effect: for the term $(1 - \mu^2 B^2)$ in (3.38) is the first approximation of the term $1/(1 + \mu^2 B^2)$ figuring in (3.36). Now we see intuitively how the magnetoresistance effect comes about: The magnetoresistance effect can be explained as an opposition of the current density component $\boldsymbol{J}_{3\mathrm{E}}$ to the main drift current density $\boldsymbol{J}_1$ (see figure 3.9(d) and equation (3.37e)). Note that the current density component $\boldsymbol{J}_3$ is a consequence of the existence of the current density component $\boldsymbol{J}_2$. Therefore, within the limits of the present 'smooth drift' approximation, the magnetoresistance effect exists only if the current deflection effect exists. We shall see in §3.3.8 that the planar Hall effect is related to the current deflection phenomenon in a similar way.

3.1.5 Charge carrier transport in very high magnetic fields

As we stated at the beginning of §3.1, up to now we have assumed that our samples were exposed to a very weak magnetic field. Consider now the other extreme: let us assume that a sample is exposed to a very high magnetic field. Qualitatively, a high magnetic field in the present context means that a charge carrier suffers a dramatic influence of the magnetic field already during a single free flight, between two successive collisions. Later in this section we shall be able to make precise this vague definition.

Free charged particle

Since we talk about a free flight, it is useful to briefly recall two simple cases of the motion of a completely free charged particle in vacuum. A more detailed description of these examples can be found in any physics or electromagnetics textbook.

The classical equation of motion of a free charged particle exposed to combined electric and magnetic fields is

$$\boldsymbol{F} = e\boldsymbol{E} + e[\dot{\boldsymbol{r}} \times \boldsymbol{B}] = m\ddot{\boldsymbol{r}} \tag{3.39}$$

where $\boldsymbol{F}$ is the Lorentz force (3.1), m is the mass of the particle and $\boldsymbol{r}$ is its position vector.

If only an electric field acts on the particle, the particle attains a constant acceleration. We analysed such a case in §2.6.1.

The movement of a charged particle in vacuum in the presence of a magnetic field is illustrated in figure 3.10.

Consider first the case when only a magnetic field is present and a particle has a velocity $\boldsymbol{v}$ at an instant t. Let us take the direction of the magnetic field $\boldsymbol{B}$ as a reference direction, which is the z axis in figure 3.10(a). The velocity of the particle can be divided into two components: $\boldsymbol{v} = \boldsymbol{v}_{\parallel} + \boldsymbol{v}_{\perp}$, where $\boldsymbol{v}_{\parallel}$ is collinear with $\boldsymbol{B}$ and $\boldsymbol{v}_{\perp}$ is perpendicular to $\boldsymbol{B}$. The velocity component parallel with the magnetic field $\boldsymbol{v}_{\parallel}$ is not affected by the magnetic field. The particle moves with this velocity along the direction of the magnetic field irrespective of the strength of the magnetic field.

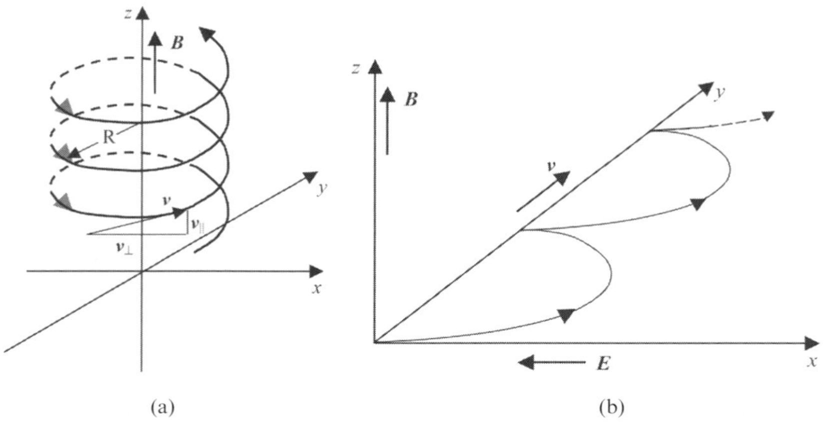

Figure 3.10. Trajectories of a (negatively) charged particle moving in vacuum: (a) The helicoidal trajectory of a particle with a velocity $\boldsymbol{v}$ in a magnetic field. By and large, the particle moves in the direction of the magnetic induction $\boldsymbol{B}$ with velocity $\boldsymbol{v}_{\parallel}$. (b) The cycloidal trajectory of a particle in crossed electric and magnetic fields. By and large, the particle moves in the direction perpendicular to both electric and the magnetic fields with velocity $v = E/B$.

On the other hand, according to (3.39), the component of the velocity perpendicular to the magnetic field $\boldsymbol{v}_\perp$ is steadily modified by the magnetic field. The result is a rotation of the particle around the z axis. The charged particle appears trapped by a magnetic field line and moves in a helix (see figure 3.10(a)). On average, the particle moves only along the direction of the magnetic field with velocity $\boldsymbol{v}_\parallel$. Note, however, that a magnetic field alone cannot produce a transport of carriers. In a system of charge carriers, there will be on average equal numbers of carriers moving in two opposite directions along a magnetic field line.

The radius of the helix is given by

$$R = mv_\perp/eB. \tag{3.40}$$

The angular velocity of the particle, also called the *cyclotron frequency*, is given by

$$\omega_c = eB/m. \tag{3.41}$$

Figure 3.10(b) shows the trajectory of a charged particle of zero initial velocity in crossed electric and magnetic fields. The more the particle is accelerated along the electric field, the more it is deviated from this direction by the magnetic field. The resulting trajectory is a cycloid, a curve consisting of an array of half circles. The radius of the half circles is

$$R = 2m|\boldsymbol{E}|/eB^2. \tag{3.42}$$

On average, the particle travels in a direction perpendicular to both electric and magnetic fields, with a mean velocity

$$\bar{\boldsymbol{v}} = |\boldsymbol{E}|/B. \tag{3.43}$$

Note that the transport along the direction perpendicular to electric and magnetic fields corresponds to a Hall angle of $\pi/2$.

Charge carrier in a solid

A charge carrier in a solid behaves, between collisions, almost as a charged particle in vacuum: as we saw in §2.2.2, the influence of the crystal lattice can be taken into account by introducing the appropriate effective mass of carriers. Therefore, the above relations (3.40)–(3.43) also apply to a charge carrier in a solid, provided the carrier can make several cycloid circles before scattering again. This condition can be fulfilled if the magnetic field is strong enough, that is if

$$T_c = 2\pi/\omega_c \ll \langle\tau\rangle \tag{3.44}$$

where T_c is the period of a carrier cycle in the magnetic field and $\langle\tau\rangle$ is the average free transit time or the relaxation time (see §2.5.2). From (3.41) and (3.44), we obtain the condition for a very high magnetic field in relation to the cyclotron effect in a solid:

$$\frac{e}{m^*}\langle\tau\rangle\frac{B}{2\pi} = \frac{\mu B}{2\pi} \gg 1 \tag{3.45}$$

where we also used the relation (2.86) for the mobility μ. For example, for electrons in silicon with $\mu = 1000\,\text{cm}^2\,\text{V s}^{-1}$ at 300 K, (3.45) gives $\boldsymbol{B} \gg 60\,\text{T}$. This is a hardly accessible high field. But in a high-mobility material at low temperature we may have $\mu = 100\,000\,\text{cm}^2\,\text{V s}^{-1}$. Then (3.45) gives $\boldsymbol{B} \gg 0.6\,\text{T}$, which is easy to realize.

3.1.6 The role of collisions

Consider now qualitatively the galvanomagnetic transport of charge carriers in a solid while taking into account collisions of carriers. We suppose that carriers are subject to crossed electric and weak magnetic fields. With reference to the above discussion (3.45), we define a *weak magnetic field* by

$$\mu B/2\pi \ll 1. \tag{3.46}$$

To get an idea about carrier motion in this case, we shall again make use of assumption (i) of §2.6.1: we suppose that a carrier loses all of its energy in a collision. Then the trajectories of charge carriers look like those illustrated in figure 3.11. Generally, carriers move along cycloidal trajectories, but cannot complete a half cycle before a collision. After each collision, a carrier has zero velocity and starts a new cycloid in the direction of the electric field (see figure 3.11(a)); but after following only a small portion of the cycloid, the carrier collides, loses its energy, and the cycle starts all over again. Compare with figure 3.10(b)! On average, the carrier now moves with a velocity $\boldsymbol{v}_\text{d}$, the direction of which is inclined by an angle $\Theta_\text{H} < \pi/2$ with respect to the electric field. We recognize $\boldsymbol{v}_\text{d}$ as the drift velocity and Θ_H as the Hall angle.

Note that the drift velocity can be represented by two components (see figure 3.10(a)): one ($\boldsymbol{v}_{\text{d}\parallel}$) parallel with the electric field, and the other ($\boldsymbol{v}_{\text{d}\perp}$) perpendicular to both electric and the magnetic fields. This reminds us of the two components of the current density in (3.32).

According to this model, the Hall angle should depend on the collision frequency $f_\text{c} = l/\langle\tau\rangle$, as illustrated in figure 3.11(b). The two curves represent schematically the paths of two carriers with different free transit times, $\tau_1 > \tau_2$. For simplicity, we assume that the free transit time of each carrier is constant. The cycloid portions making the curve 1 are longer than those making the curve 2 and, consequently, $\Theta_{\text{H}1} > \Theta_{\text{H}2}$. Moreover, it can be shown that the Hall angle is proportional to the free transit time, $\Theta_\text{H} \sim \tau$. Recall that, according to our previous analysis, $\Theta_\text{H} \simeq \mu\boldsymbol{B}$ (3.31). Our present qualitative considerations corroborate this result, for $\mu \sim \langle\tau\rangle$ (2.86).

In reality, the free transit time is not constant, as assumed in figure 3.11(a) and (b). It attains randomly different values, the probabilities of

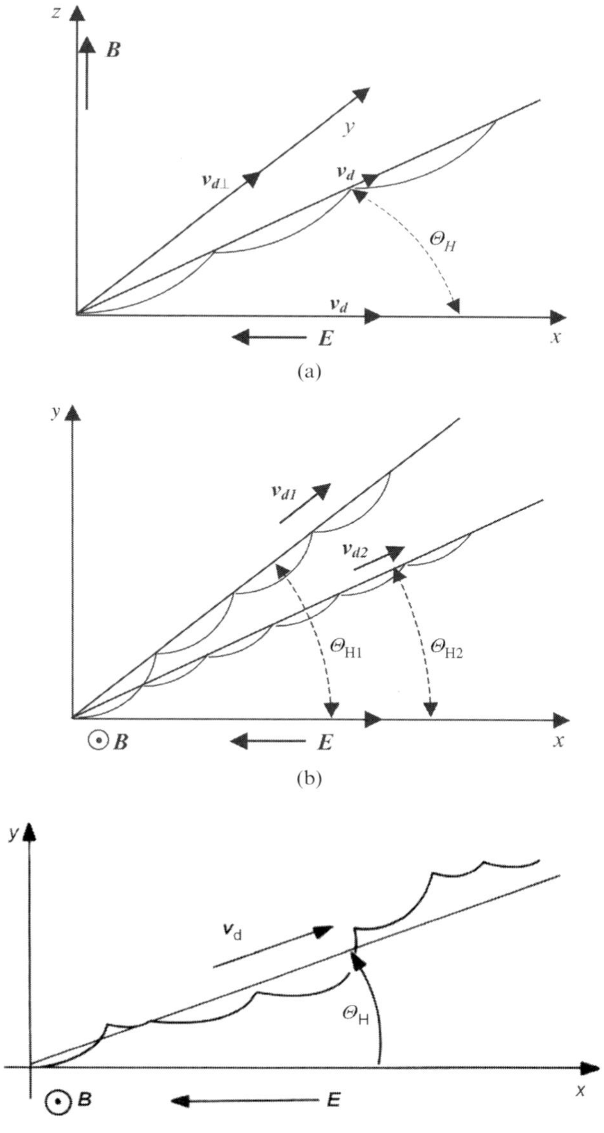

Figure 3.11. Schematic representations of the trajectory of a charged particle in a solid in crossed electric and weak magnetic fields. (a) The average trajectory is inclined with respect to the electric field $\boldsymbol{E}$ by the Hall angle θ_H. (b) Curves 1 and 2 represent the trajectories of two carriers with different free transit times, $\tau_1 > \tau_2$. The longer the free transit path of a carrier, the larger the Hall angle. (c) The trajectory of a charge carrier with randomly varying free transit times.

which are given by (2.59). Therefore, the trajectory of a carrier might look something like the curve shown in figure 3.11(c). During a longer free transit time, the carrier moves along a longer portion of a cycloid and deflects more from the direction of the electric field; conversely, during a shorter free transit time, the deflection is smaller. The Hall angle, being a macroscopic quantity, is the angle of inclination of the average drift velocity of all carriers with respect to the electric field.

How do collisions affect the helicoidal motion of carriers shown in figure 3.10(a)? Similarly to the previous case, collisions usually break the movement long before a circle is completed. However, a tendency of the movement of a carrier around and along a magnetic field line persists. If the $\boldsymbol{E}$ and $\boldsymbol{B}$ vectors are not mutually orthogonal, this will give rise to an additional current component collinear with $\boldsymbol{B}$, as illustrated in figure 3.8. So the present intuitive analysis gives a result coherent with that obtained in §3.1.4. The picture becomes more complicated if one also takes into account the thermal motion of carriers. The analysis of the galvanomagnetic effects then becomes more involved, and we postpone it until §3.2.

3.1.7 Summary and further questions

In most of §3.1, we studied the galvanomagnetic effects in a current-caring plate exposed to a perpendicular magnetic field. The Hall effect shows up in long and short plates in two apparently different forms: in long plates (as a transverse voltage) and in short plates (as a transverse current component). In short plates, in addition, arises a magnetoresistance effect. All three phenomena are consequences of the action of the Lorentz force on quasi-free charge carriers. The electric component of the Lorentz force (3.1) provides the lateral transport of carriers. The magnetic component of the Lorentz force, which is perpendicular to the velocity of carriers, is responsible for the transverse effects.

If the current in a sample is confined to the lateral direction, as in a long plate, a transverse voltage appears. This voltage, known as the *Hall voltage*, provides a counterbalance to the lateral magnetic pressure on carriers. The Hall voltage is given by (3.18)

$$V_{\mathrm{H}} = \frac{R_{\mathrm{H}}}{t} I B_{\perp}. \tag{3.47}$$

If the current is not confined to the lateral direction, which is the case in a short sample, the transverse magnetic force produces a transverse current component, and hence a *current deflection*.

Both forms of the Hall effect can be regarded as consequences of a transverse macroscopic electromotive force. Depending on the sample geometry, this electromotive force may cause the appearance of a Hall voltage, or a current deflection in the sample.

Owing to the Hall effect, the current density in a sample is not collinear with the electric field. The angle of inclination between the electric field and the current density is called the *Hall angle*. The Hall angle depends only on carrier mobility and magnetic induction, and its sign on the type of carriers and the direction of the magnetic field. For weak magnetic inductions, the Hall angles are given by (3.14) and (3.31):

$$\Theta_{\mathrm{Hp}} \simeq \mu_{\mathrm{p}} B \qquad \Theta_{\mathrm{Hn}} \simeq -\mu_{\mathrm{n}} B \tag{3.48}$$

for holes and electrons, respectively. Specifically, the Hall angle does not depend on sample geometry.

The current deflection in a short sample causes an increase in the sample resistivity. This effect is called the *geometrical magnetoresistance effect*. The relative increase in resistance due to the geometrical magnetoresistance effect is given by (3.34)

$$\frac{\Delta\rho}{\rho_0} = (\mu B)^2. \tag{3.49}$$

In a general case, when the magnetic field and electric field in a sample are not mutually perpendicular, the Hall deflection current gives rise to the creation a third current component, which is collinear with the magnetic field. We have seen that the geometrical magnetoresistance effect is related with this third current component. But how about magnetoresistance effect in samples in which, due to their geometry, one or the other lateral current density component cannot exist? Does a magnetoresistance effect still appear or not? And is there any other consequence of the current confinement in this case?

In the analysis of the Hall effect in §§3.1.1–3.1.4, we applied the smooth drift approximation: we assumed that the velocities of all carriers were the same and equal to the appropriate drift velocity. In reality, carriers are subject to random thermal motion, which is, in most cases, even much more intensive than directed or drift motion; moreover, carriers have a large spread in directional velocities, and scatter at each collision without losing much of their total energy. The question now naturally arises: what is the influence of the thermal agitation and scattering of carriers on the Hall and the magnetoresistance effects?

To answer these questions, we must learn how to cope with the action of the Lorentz force on randomly moving charge carriers. The next sections, §§3.2 and 3.3, are devoted to this issue.

Figure 3.12 illustrates the relationships between electric field and current density in a long and a short sample. The current density is also represented by the current lines, and the electric field by the equipotential lines. (Electric field, being the gradient of electric potential, is perpendicular to equipotential lines.) In the central part of a long sample (l), the current lines always remain parallel with the long insulating boundaries, whereas the equipotential lines are inclined by the Hall angle with respect to their position at $\boldsymbol{B} = 0$. In the

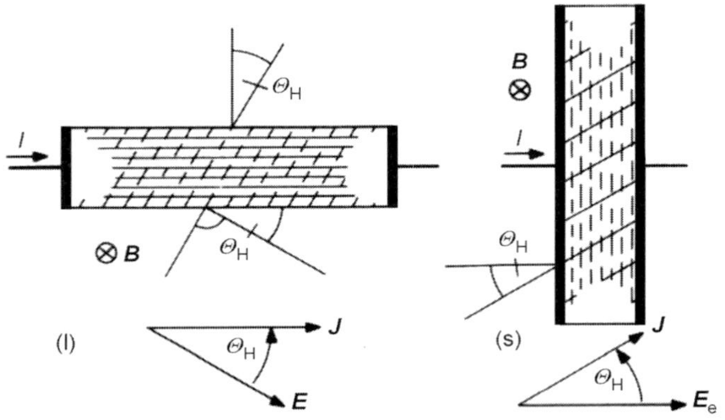

Figure 3.12. The current density (full lines) and the equipotential lines (broken lines) in a long sample (l) and a short sample (s). Also shown are the corresponding vector diagrams: current density (***J***)–electric field (***E***).

central part of a short sample (s), the equipotential lines always remain parallel to the biasing contacts, irrespective of the strength of the magnetic field, but the current lines rotate by the Hall angle respective to their position at $\boldsymbol{B} = 0$.

What happens in the clear areas of the two samples in figure 3.12? A thorough analysis [10] shows that, even in a long sample, close to a contact, the current deflection takes place; and even in a short sample, close to a sample insulating boundary, the Hall field develops. So both things happen irrespective of the overall sample geometry. Depending on the sample geometry, one or the other effect may prevail, but both of them are always present. In chapter 4, we shall discuss a method for taking both these effects into account in real Hall devices.

3.2 Kinetic equation approach

In the present section, we shall apply a method of analysis of the galvanomagnetic transport phenomena in solids, which does not require the crude approximations we had to introduce before. This approach is based on the solution of an equation, called the Boltzmann kinetic equation. Essentially, the kinetic equation relates the disturbance in the carrier distribution function caused by external fields to the equilibrium-restoring effect of scattering. As we have noted several times already, a disturbance in the symmetry of the distribution function is a necessary condition for a transport process to appear. The kinetic equation method is, therefore, quite general and powerful. In particular, it allows for the thermal motion of carriers, their scattering, and the action of the Lorentz force.

3.2.1 Kinetic equation

Consider the distribution function of electrons in a semiconductor sample exposed to external electric and magnetic fields.

In thermodynamic equilibrium, the distribution of electrons is only a function of electron energy: $F(E)$. But when the sample is exposed to external fields, electrons are no longer in the state of thermodynamic equilibrium. Their distribution function may then depend on the electron wavevector $\boldsymbol{k}$, position in the sample $\boldsymbol{r}$, and time t: $F(\boldsymbol{k}, \boldsymbol{r}, t)$. We shall now construct an equation which will enable us to find the non-equilibrium distribution function $F(\boldsymbol{k}, \boldsymbol{r}, t)$.

Let us select a volume element of the sample, ΔV_r, defined by the position vectors $\boldsymbol{r} + \Delta\boldsymbol{r}$, and also a subpopulation of electrons therein, with wavevectors $\boldsymbol{k} + \Delta\boldsymbol{k}$. These wavevectors define a volume element ΔV_{k} in the $\boldsymbol{k}$ space. The concentration of electrons with wavevectors $\boldsymbol{k} + \Delta\boldsymbol{k}$ in the volume element ΔV_r is given by

$$n_{kr} = g_k F(\boldsymbol{k}, \boldsymbol{r}, t). \tag{3.50}$$

This is the density of occupied states in the $\boldsymbol{k}$ space in the volume element ΔV_k that belong to the volume element ΔV_r of the $\boldsymbol{r}$ space. g_k is the density of permitted states in the $\boldsymbol{k}$ space per unit crystal volume (2.13) and F is the required probability distribution function.

The electron concentration (3.50) may vary in time. Formally, the rate of change is given by its time derivative, which is

$$\frac{\mathrm{d}n_{kr}}{\mathrm{d}t} = g_k \frac{\partial F}{\partial t}. \tag{3.51}$$

Physically, the change in the electron concentration (3.51) can arise only because of the following three processes: (i) generation and recombination of carriers, which we shall neglect; then the number of electrons and corresponding occupied states in the $\boldsymbol{k}$ space is conserved; (ii) the transport of the electrons with the selected values of wavevector between the selected volume element ΔV_r of the $\boldsymbol{r}$ space and its surroundings; (iii) the change of the wavevectors of the electrons that are already in the volume element ΔV_r. This change of the wavevectors can be represented by a transport of the occupied points in the $\boldsymbol{k}$ space between the selected volume element ΔV_k and its surroundings. The transport of electrons and occupied points can be represented by the appropriate particle current densities:

$$\boldsymbol{J}_r = n_{kr}\dot{\boldsymbol{r}} \qquad \boldsymbol{J}_k = n_{kr}\dot{\boldsymbol{k}} \tag{3.52}$$

where $\dot{\boldsymbol{r}} = \boldsymbol{v}$ is the velocity of electrons in the $\boldsymbol{r}$ space and $\dot{\boldsymbol{k}}$ is the velocity of the occupied points in the $\boldsymbol{k}$ space. The latter can also be represented by

$$\dot{\boldsymbol{k}} = \frac{1}{\hbar}\dot{\boldsymbol{P}} = \frac{1}{\hbar}\boldsymbol{F}_{\mathrm{t}} \tag{3.53}$$

where $\boldsymbol{P}$ is the electron quasi-momentum (2.9) and $\boldsymbol{F}_t$ is the total force acting on an electron.

Applying now the continuity equation to the transport of particles in the $\boldsymbol{r}$ and $\boldsymbol{k}$ spaces, we have

$$\frac{dn_{kr}}{dt} = -\nabla_r \boldsymbol{J}_r - \nabla_k \boldsymbol{J}_k. \tag{3.54}$$

Here ∇_r and ∇_k are the nabla operators for the $\boldsymbol{r}$ space and the $\boldsymbol{k}$ space, respectively. Substituting (3.50), (3.52) and (3.53) into (3.54), we obtain

$$\frac{dn_{kr}}{dt} = \left(-\nabla_r F \cdot \boldsymbol{v} - \nabla_k F \cdot \frac{\boldsymbol{F}_t}{\hbar}\right) g_k. \tag{3.55}$$

Comparing this equation with (3.51), we find

$$\frac{\partial F}{\partial t} = -\nabla_r F \cdot \boldsymbol{v} - \nabla_k F \cdot \frac{\boldsymbol{F}_t}{\hbar}. \tag{3.56}$$

The total force acting on a carrier, $\boldsymbol{F}_t$, consists of two quite different components: the external force $\boldsymbol{F}$, which is due to the external macroscopic electric and magnetic fields; and the internal force $\boldsymbol{F}_c$, which is due to the internal localized fields associated with the crystal defects. Thus

$$\boldsymbol{F}_t = \boldsymbol{F} + \boldsymbol{F}_c. \tag{3.57}$$

The force $\boldsymbol{F}$ tends to cause a macroscopic transport of carriers; the force $\boldsymbol{F}_c$ is associated with the collision and scattering of carriers, and thus represents a friction reaction to the transport of carriers.

Using (3.57), we may rewrite (3.56) as

$$\frac{\partial F}{\partial t} = -\nabla_r F \cdot \boldsymbol{v} - \nabla_k F \cdot \frac{\boldsymbol{F}}{\hbar} - \nabla_k F \cdot \frac{\boldsymbol{F}_c}{\hbar}. \tag{3.58}$$

This is a form of the *Boltzmann kinetic equation*. It states that the electron distribution function may vary in time as a result of the following three physical effects: the redistribution or diffusion of electrons in the sample (the first term on the right-hand side of (3.58), which is proportional to the gradient of the distribution function); the action of the external fields (the second term); and the action of internal forces during collisions (the third term). Therefore, the kinetic equation can also be expressed in the form

$$\frac{\partial F}{\partial t} = \left(\frac{\partial F}{\partial t}\right)_d + \left(\frac{\partial F}{\partial t}\right)_{f'} + \left(\frac{\partial F}{\partial t}\right)_c \tag{3.59}$$

where the subscript d stands for diffusion, f$'$ for external forces, and c for collisions. The diffusion term is also a consequence of an external influence, for a gradient in the distribution function can be produced by the injection of carriers or by a temperature gradient. Regarding generally all external influences as the results of external fields, we may summarize the terms d and f′ into a single external field term f, and rewrite (3.59) as

$$\frac{\partial F}{\partial t} = \left(\frac{\partial F}{\partial t}\right)_{\mathrm{f}} + \left(\frac{\partial F}{\partial t}\right)_{\mathrm{c}}. \tag{3.60}$$

Obviously, equations (3.58) and (3.60) hold also for holes.

The Boltzmann kinetic equation cannot be solved in its general form given by (3.58). For one thing, the collision forces cannot be known and handled in detail for all electrons. In order to render the kinetic equation applicable for the treatment of transport phenomena, we must introduce some approximations into it. First of all, we shall simplify the collision term in (3.58) by using the so-called relaxation time approximation.

3.2.2 Relaxation time

We have already introduced the notion of relaxation time in §2.5.2. Now we shall define it more precisely, in relation to the kinetic equation.

Suppose a solid is exposed to external fields, and the distribution of its electrons is not at equilibrium. We shall be interested here only in the external influences that do not produce any change in the spatial distribution of electrons, so we assume that the disturbed electron distribution function depends only on the electron wavevector $\boldsymbol{k}$ and time t: $F(\boldsymbol{k}, \tau)$.

If at time $t = 0$ the external fields suddenly disappear, then the relaxation process begins: the system of electrons approaches its equilibrium and the electron distribution function tends to its equilibrium value $F_0(\boldsymbol{k})$. During the relaxation, the field term in (3.60) vanishes, the distribution function changes only because of the collisions, and (3.60) reduces to

$$\frac{\partial F}{\partial t} = \left(\frac{\partial F}{\partial t}\right)_{\mathrm{c}}. \tag{3.61}$$

This equation shows that the collision processes, taken as a whole, define the rate of change of the distribution function. Thus if we could determine this rate, we might avoid treating the details of the scattering events.

We may assume that the equilibrium restoration rate is proportional to the displacement from equilibrium, for many other equilibrium-restoring processes in nature behave that way. In this case, (3.61) becomes

$$\frac{\partial F}{\partial t} = -\frac{F(\boldsymbol{k}, t) - F_0(\boldsymbol{k})}{\tau(\boldsymbol{k})} \tag{3.62}$$

where F and F_0 are the disturbed and the equilibrium distribution functions, respectively, and τ is a parameter. The minus sign in this equation reflects the fact that the system of electrons, if left alone, always tends towards equilibrium. The solution of (3.62) is

$$F(\boldsymbol{k}, t) - F_0(\boldsymbol{k}) = [F(\boldsymbol{k}, 0) - F_0(\boldsymbol{k})] \exp[-t/\tau(\boldsymbol{k})]. \tag{3.63}$$

Thus the parameter τ is the time constant of the return of the electron system to the equilibrium state. For this reason, τ is called the *relaxation time*.

The substitution of the collision term in the kinetic equation by the expression on the right-hand side of (3.62) is called the *relaxation time approximation*.

The relaxation time generally depends on the magnitudes of the wavevectors of the group of electrons under consideration, and hence also on their energy. We shall see later that this dependence determines some important coefficients of transport processes. According to our qualitative discussion in §2.5.2, the relaxation time of the subpopulation of electrons with the wavevectors around a value $\boldsymbol{k}$ is equal to the mean free transit time of these electrons. Since this is so for each value of energy, the energy-weighted average relaxation time over all electrons $\langle\tau\rangle$ also equals the average free transit time (2.65). Then equation (2.66) and table 2.2 also hold for the relaxation time. If there is no danger of misunderstanding, we shall refer to the average relaxation time as the relaxation time in short, as we did in §2.5.2.

3.2.3 Solution for the steady state

We shall now find the electron distribution function in a solid exposed to slowly varying electric and magnetic fields. 'Slowly varying' here means slow with respect to the duration of the relaxation process. Since the relaxation time is of the order of 10^{-13} s, fields with frequencies below 10^{12} Hz can be considered as slowly varying. To simplify the mathematics, we shall also assume that: (i) electrons have scalar effective mass (i.e. spherical constant energy surfaces in the $\boldsymbol{k}$ space; see §2.2.2); (ii) the electric and magnetic fields are weak; and (iii) there are no other external influences.

Herring–Vogt transformation. In the case of anisotropic effective mass of carriers, one may proceed in the following way. An ellipsoidal constant energy surface around an energy minimum at $\boldsymbol{k} = \boldsymbol{k}_0$ can be transformed into a spherical surface around the point $\boldsymbol{K} = 0$ in a $\boldsymbol{K}$ space. To this end, the Herring–Vogt transformation [11] can be used:

$$K_i = \left(\frac{m_c^*}{m_i^*}\right)^{1/2} (k_i - k_{0i}) \qquad i = 1, 2, 3. \tag{3.64}$$

Here $\boldsymbol{K}_i$ is the ith coordinate in the $\boldsymbol{K}$ space and m_c^* is an average value of the three effective masses m_i^*, called the conductivity effective mass. A formula for m_c^* is given in §3.2.7. m_i^* is the effective mass in the k_i principal direction in the $\boldsymbol{k}$ space. From (3.63) and (2.11), we obtain the electron energy in the $\boldsymbol{K}$ space:

$$E(\boldsymbol{k}) = E_c + \frac{\hbar^2 (\boldsymbol{K})^2}{2m_c^*}. \tag{3.65}$$

Thus in the $\boldsymbol{K}$ space the constant energy surfaces are spheres around the point $\boldsymbol{K} = 0$. As a result of this fact, further analysis can be done in an analogous way to that given below.

In the steady state, the time derivative of the electron distribution function, $\partial F/\partial t$ in (3.58)–(3.60), vanishes. Then the kinetic equation (3.60) attains the form

$$-\left(\frac{\partial F}{\partial t}\right)_{\mathrm{f}} = \left(\frac{\partial F}{\partial t}\right)_{\mathrm{c}}. \tag{3.66}$$

This states the obvious fact that, in the steady state, the change in the distribution function affected by the external fields is exactly compensated by the influence of collisions. Substituting now the left-hand side of (3.66) by the external field term from (3.58) (note that in the present case the diffusion term reduces to zero), and the right-hand side by (3.62), we obtain

$$\frac{\boldsymbol{F}}{\hbar}\cdot\nabla_k F(\boldsymbol{k}) = -\frac{F(\boldsymbol{k}) - F_0(\boldsymbol{k})}{\tau(\boldsymbol{k})}. \tag{3.67}$$

This is the *Boltzmann kinetic equation* in the relaxation time approximation for the steady state.

Small disturbance of the distribution function

To solve the kinetic equation (3.67), we have to introduce an additional approximation: we shall assume that the disturbance of the distribution function due to the external fields is small. Then we may seek the distribution function in the form of the sum

$$F(\boldsymbol{k}) = F_0(\boldsymbol{k}) + F_1(\boldsymbol{k}) \tag{3.68}$$

where $F_1(\boldsymbol{k}) \ll F_0(\boldsymbol{k})$. F_0 denotes the distribution function at thermodynamic equilibrium, and F_1 is a small change in the distribution function caused by external fields. In §2.6.1 we already used the assumption of a small disturbance of the distribution function and found that this assumption is reasonable if the applied electric field is weak enough (2.81).

Substituting (3.68) into (3.67), we obtain

$$F_1(\boldsymbol{k}) = -\frac{\tau(\boldsymbol{k})}{\hbar}\boldsymbol{F}\cdot\nabla_k[F_0(\boldsymbol{k}) + F_1(\boldsymbol{k})]. \tag{3.69}$$

Before we start solving this differential equation, let us find an approximation of F_1, assuming that $\nabla_k F_0 \gg \nabla_k F_1$. This will give us an idea about the structure and magnitude of the correct solution for the function F_1. Retaining only the first term on the right-hand side of (3.69), we have

$$F_1^{(1)}(\boldsymbol{k}) = -\frac{\tau(\boldsymbol{k})}{\hbar}\boldsymbol{F}\cdot\nabla_k F_0(\boldsymbol{k}) \tag{3.70}$$

where the superscript (1) denotes the first approximation. The gradient of $F_0(\boldsymbol{k})$ is given by

$$\nabla_k F_0(\boldsymbol{k}) = \frac{\partial F_0}{\partial E}\frac{\partial E}{\partial \boldsymbol{k}} = \frac{\partial F_0}{\partial E}\hbar \boldsymbol{v} \tag{3.71}$$

where, by making use of (2.17),

$$\frac{\partial F_0}{\partial E} = -\frac{\exp[-(E - E_F)/kT]}{\{1 + \exp[(E - E_F)/kT]\}^2}\frac{1}{kT} \tag{3.72}$$

and $\boldsymbol{v}$ is the velocity of the selected group of electrons. For electrons with an isotropic effective mass, in analogy with (2.5), the velocity is given by

$$\boldsymbol{v} = \hbar \boldsymbol{k}/m^*. \tag{3.73}$$

By substituting in (3.70) the Lorentz force (3.1) and the gradient (3.71), we obtain

$$F_1^{(1)}(\boldsymbol{k}) = -\frac{\partial F_0}{\partial E}\tau(\boldsymbol{k})e\boldsymbol{E}\cdot\boldsymbol{v}. \tag{3.74}$$

This is the required rough approximation for the change in the distribution function. It does not contain the term with magnetic induction, since $[\boldsymbol{v} \times \boldsymbol{B}]\cdot\boldsymbol{v} = 0$. To a first approximation, the electron distribution function does not depend on the magnetic field. This conclusion justifies our first approximation for the galvanomagnetic current density in §3.1.4 and (3.37a).

Correction of the distribution function

It is reasonable to suppose that an accurate expression for the alteration of the distribution function due to the external fields might approximately equal (3.74). Thus we may seek the function F_1 in the form

$$F_1(\boldsymbol{k}) = -\frac{\partial F_0}{\partial E}\boldsymbol{X}\cdot\boldsymbol{v} \tag{3.75}$$

where $\boldsymbol{X}$ is an as-yet-unknown vector function; however, we know in advance that for $\boldsymbol{B} \to 0$, (3.75) must tend to (3.74), and hence

$$\boldsymbol{X} \simeq \tau(\boldsymbol{k})e\boldsymbol{E}. \tag{3.76}$$

It is interesting to note the physical meaning of the vector on the right-hand side of (3.76): it can be interpreted as the mean electron momentum associated with the drift motion of electrons (compare with (2.83)).

To find $F_1(\boldsymbol{k})$, we shall need its gradient in the k space. Using (3.74) and (3.76), we have

$$\nabla_k F_1(\boldsymbol{k}) = -\frac{\partial F_0}{\partial E}\boldsymbol{X}\frac{\hbar}{m^*} - \boldsymbol{v}\frac{\partial}{\partial \boldsymbol{k}}\left(\frac{\partial F_0}{\partial E}\boldsymbol{X}\right). \tag{3.77}$$

Substituting (3.1), (3.67), (3.75) and (3.77) into (3.69), we obtain

$$-\frac{\partial F_0}{\partial E}(\boldsymbol{X}\cdot\boldsymbol{v}) = -\frac{\partial F_0}{\partial E}\tau e(\boldsymbol{E}\cdot\boldsymbol{v}) + \frac{\partial F_0}{\partial E}\frac{\tau e}{m^*}(\boldsymbol{E}\cdot\boldsymbol{X}) + \frac{\tau e}{\hbar}(\boldsymbol{E}\cdot\boldsymbol{v})\frac{\partial}{\partial \boldsymbol{k}}\left(\frac{\partial F_0}{\partial E}\boldsymbol{X}\right) + \frac{\tau e}{m^*}\frac{\partial F_0}{\partial E}[\boldsymbol{v}\times\boldsymbol{B}]\cdot\boldsymbol{X} \tag{3.78}$$

where the two terms with $[\boldsymbol{v}\times\boldsymbol{B}]\cdot\boldsymbol{v}$ were omitted. Recalling that we have assumed a weak electric field, and taking into account (3.76), we can neglect the second and the third term with $\boldsymbol{E}$ in (3.78). Thus we obtain

$$\boldsymbol{X} \simeq e\tau(\boldsymbol{k})\boldsymbol{E} + \frac{e\tau(\boldsymbol{k})}{m^*}[\boldsymbol{X}\times\boldsymbol{B}]. \tag{3.79}$$

This vector equation has the same form as (3.35) and (E.4) of Appendix E. Thus in analogy with (3.36) and (E.8), the solution of (3.79) is given by

$$\boldsymbol{X} \simeq \frac{e\tau\boldsymbol{E} + (e^2\tau^2/m^*)[\boldsymbol{E}\times\boldsymbol{B}] + (e^3\tau^3/m^{*2})\boldsymbol{B}(\boldsymbol{E}\cdot\boldsymbol{B})}{1 + (e^2\tau^2/m^{*2})B^2} \tag{3.80}$$

which determines the required correction of the distribution function.

In summary, the electron distribution function in a solid exposed to a weak electric field and a magnetic field is slightly disturbed relative to its value at thermodynamic equilibrium. The non-equilibrium distribution function is given by

$$F(\boldsymbol{k}) = F_0(\boldsymbol{k}) + F_1(\boldsymbol{k}) \tag{3.81}$$

where F_0 is the equilibrium distribution function and F_1 is a small alteration or correction in the distribution function due to the fields. The latter is given by

$$F_1(\boldsymbol{k}) = -\frac{\partial F_0}{\partial E}\boldsymbol{X}\cdot\boldsymbol{v} \tag{3.82}$$

where $\boldsymbol{X}$ is the vector function (3.80) found above and $\boldsymbol{v}$ is the carrier velocity.

3.2.4 Non-equilibrium distribution function

To understand the physical meaning of the solution for the non-equilibrium distribution function found above, let us rewrite it in a simplified form. To this end, we assume that the electric and magnetic fields are perpendicular to each other, and thus $\boldsymbol{E}\cdot\boldsymbol{B} = 0$, and that the magnetic induction is low; hence the term with B^2 in the denominator of (3.80) can be neglected. Then, substituting (3.72), (3.82), (3.80) and $e = -q$ into (3.81), we obtain

$$F(\boldsymbol{k}) \simeq F_0(\boldsymbol{k}) + \left(-\frac{\partial F_0}{\partial E}\right)\mu(\boldsymbol{k})\hbar\boldsymbol{k}\cdot\{-\boldsymbol{E} + \mu(\boldsymbol{k})[\boldsymbol{E}\times\boldsymbol{B}]\} \tag{3.83}$$

where we introduce the notation

$$\mu(\boldsymbol{k}) = q\tau(\boldsymbol{k})/m^*. \tag{3.84}$$

This is the mobility of the electrons with the selected value of wavevector $\boldsymbol{k}$ (compare with (2.83)–(2.86)).

Equation (3.83) shows that the non-equilibrium function differs from the equilibrium distribution function mostly for the wavevectors which are collinear with the vector $-\boldsymbol{E} + \mu[\boldsymbol{E} \times \boldsymbol{B}]$. This vector is inclined with respect to the electric field for an angle determined by

$$\tan \theta_{\mathrm{H}} = -\mu(\boldsymbol{k})B. \tag{3.85}$$

Comparing (3.85) with (3.31), we infer that this angle θ_{H} is nothing more than the Hall angle for the subpopulation of electrons with the selected value of the wavevector $\boldsymbol{k}$. Note that the angle θ_{H} depends on the wavevector $\boldsymbol{k}$. Therefore, the macroscopic Hall angle Θ_{H} is an average value of the angles $\theta_{\mathrm{H}}(\boldsymbol{k})$ over all electron energies.

The distribution function is the probability that the permitted states in the $\boldsymbol{k}$ space are occupied. Since the density of the permitted states in the $\boldsymbol{k}$ space is constant, the density of the occupied states is proportional to the distribution function. Therefore, if in a drawing we denote the occupied states by points, the density of the points will represent the value of the distribution function. This is done in figure 3.13 for three characteristic cases. In the case

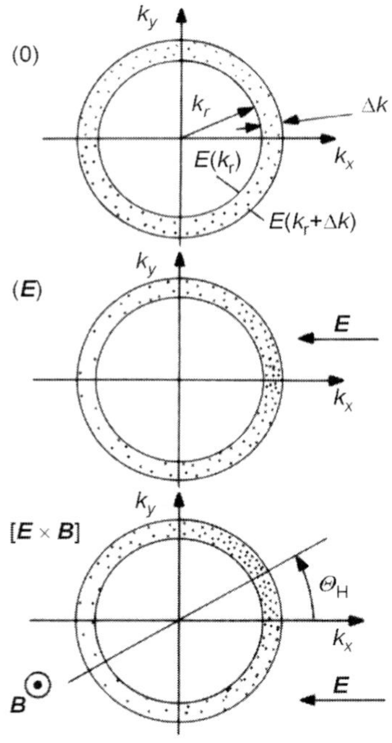

Figure 3.13. The distribution of charge carriers in the $\boldsymbol{k}$ space. Each point represents an occupied quantum state. (0), thermodynamic equilibrium; ($\boldsymbol{E}$), an electric field is present; $[\boldsymbol{E} \times \boldsymbol{B}]$, crossed electric and magnetic fields are present.

with no macroscopic fields (0), the distribution function (3.83) reduces to that of thermodynamic equilibrium, $F(\boldsymbol{k}) = F_0(\boldsymbol{k})$, and depends only on the absolute value of the wavevector $\boldsymbol{k}$. Hence the density of the occupied states between the two constant energy surfaces is constant. If an electric field ($\boldsymbol{E}$) is established in the sample, the distribution function (3.83) contains the term proportional to the scalar product $-\boldsymbol{k} \cdot \boldsymbol{E}$, and the density of the occupied states depends also on the direction of the $\boldsymbol{k}$ vectors. Similarly, if both electric and magnetic fields are present, the largest change in the density of the occupied states appears for the $\boldsymbol{k}$ vectors collinear with the field term in (3.83).

Each point in figure 3.13 represents an electron per unit volume of the sample. Since the velocity of electrons is related to their wavevectors (3.73), figure 3.13 also illustrates the directions of electron velocities. Therefore, external fields cause the appearance of a preferred direction for electron velocities, and hence the transport of electrons.

Obviously, everything above also holds analogously for holes.

3.2.5 Galvanomagnetic transport of carriers. Current density

Let us now apply the steady-state solution of the kinetic equation to a quantitative analysis of transport phenomena. Specifically, let us calculate the electric current density in a solid exposed to slowly varying electric and magnetic fields. Thereby we shall consider only one type of carrier and keep all the assumptions that we introduced above.

Consider first a subpopulation of carriers with the wave vectors between $\boldsymbol{k}$ and $\boldsymbol{k} + \mathrm{d}\boldsymbol{k}$. The concentration of these carriers is, according to (3.50) and (2.13), given by

$$n_k = \frac{2}{(2\pi)^3} F(\boldsymbol{k}). \tag{3.86}$$

The elementary current density due to the selected subpopulation of carriers is given by

$$\mathrm{d}\boldsymbol{J}_k = e\boldsymbol{v}\frac{\mathrm{d}V_k}{4\pi^3} F(\boldsymbol{k}). \tag{3.87}$$

Here e denotes the charge and $\boldsymbol{v}$ the velocity of a carrier, $\mathrm{d}V_k$ is the volume element of the $\boldsymbol{k}$ space which corresponds to the selected wavevector $\boldsymbol{k}$, and F is the carrier distribution function. The total current density, due to all carriers of the considered type, is given by

$$\boldsymbol{J} = \frac{e}{4\pi^3} \int_{V_k} \boldsymbol{v} F(\boldsymbol{k})\, \mathrm{d}V_k. \tag{3.88}$$

The integration should be done over the volume V_k of the $\boldsymbol{k}$ space, which contains the relevant $\boldsymbol{k}$ vectors. (This part of the $\boldsymbol{k}$ space is called the Brillouin zone.) Substituting (3.81) into (3.88), we obtain

$$\boldsymbol{J} = \frac{e}{4\pi^3}\left(\int_{V_k} vF_0(\boldsymbol{k})\,\mathrm{d}V_k + \int_{V_k} vF_1(\boldsymbol{k})\,\mathrm{d}V_k\right). \tag{3.89}$$

The first integral vanishes, since the equilibrium distribution function $F_0(\boldsymbol{k})$ is an even function, and thus $\boldsymbol{v}F_0(\boldsymbol{k})$ is an odd function of $\boldsymbol{k}$. This was to be expected, for there can be no transport at thermodynamic equilibrium. Substituting (3.82) into the second term of (3.89), we obtain

$$\boldsymbol{J} = \frac{e}{4\pi^3}\int_{V_k} \boldsymbol{v}\left(-\frac{\partial F_0}{\partial E}\right)(\boldsymbol{X}\cdot\boldsymbol{v})\,\mathrm{d}V_k. \tag{3.90}$$

Substituting here (3.80), we have

$$\begin{aligned}\boldsymbol{J} = e^2\int_{V_k} &\frac{\boldsymbol{v}(\boldsymbol{v}\cdot\boldsymbol{E})\tau}{4\pi^3[1+(e^2\tau^2/m^{*2})B^2]}\left(-\frac{\partial F_0}{\partial E}\right)\mathrm{d}V_k \\ &+ \frac{e^3}{m^*}\int_{V_k}\frac{\boldsymbol{v}(\boldsymbol{v}\cdot[\boldsymbol{E}\times\boldsymbol{B}])\tau^2}{4\pi^3[1+(e^2\tau^2/m^{*2})B^2]}\left(-\frac{\partial F_0}{\partial E}\right)\mathrm{d}V_k \\ &+ \frac{e^4}{m^{*2}}\int_{V_k}\frac{\boldsymbol{v}(\boldsymbol{v}\cdot\{\boldsymbol{B}(\boldsymbol{E}\cdot\boldsymbol{B})\})\tau^3}{4\pi^3[1+(e^2\tau^2/m^{*2})B^2]}\left(-\frac{\partial F_0}{\partial E}\right)\mathrm{d}V_k.\end{aligned} \tag{3.91}$$

Integrals determining the carrier transport

By inspecting the above integrals, we notice that all three of them are of the form

$$\boldsymbol{I}_s = \int_{V_k}\frac{\boldsymbol{v}(\boldsymbol{v}\cdot\boldsymbol{V}_s)\tau^s}{4\pi^3[1+(e^2\tau^2/m^{*2})B^2]}\left(-\frac{\partial F_0}{\partial E}\right)\mathrm{d}V_k \tag{3.92}$$

where $\boldsymbol{V}_s$, $s = 1, 2, 3$, are the vectors determined solely by the applied fields

$$\boldsymbol{V}_1 = \boldsymbol{E}, \qquad \boldsymbol{V}_2 = [\boldsymbol{E}\times\boldsymbol{B}], \qquad \boldsymbol{V}_3 = \boldsymbol{B}(\boldsymbol{E}\cdot\boldsymbol{B}). \tag{3.93}$$

We shall now rearrange this integral so as to take the vector $\boldsymbol{V}_s$ outside the integration. By doing so, we can put the expression (3.92) in the convenient form of the product of two terms, one depending only on the external fields, the other only on the sample material properties. We shall also substitute the integration over the $\boldsymbol{k}$ space by the integration over electron energy.

For a semiconductor with a scalar effective mass (i.e. spherical constant energy surfaces), we may substitute (3.73) into (3.92), and obtain

$$\boldsymbol{I}_s = \int_{V_k}\boldsymbol{k}(\boldsymbol{k}\cdot\boldsymbol{V}_s)f_s\,\mathrm{d}V_k. \tag{3.94}$$

Here we have introduced the notation

$$f_s = f_s(\boldsymbol{k}) = \frac{\hbar^2}{m^{*2}}\frac{1}{4\pi^3}\frac{\tau^s(\boldsymbol{k})}{(1+\mu^2(\boldsymbol{k})B^2)}\left(-\frac{\partial F_0}{\partial E}\right) \tag{3.95}$$

where $\mu(\boldsymbol{k})$ (3.84) is the mobility of the carriers with the wavevector $\boldsymbol{k}$.

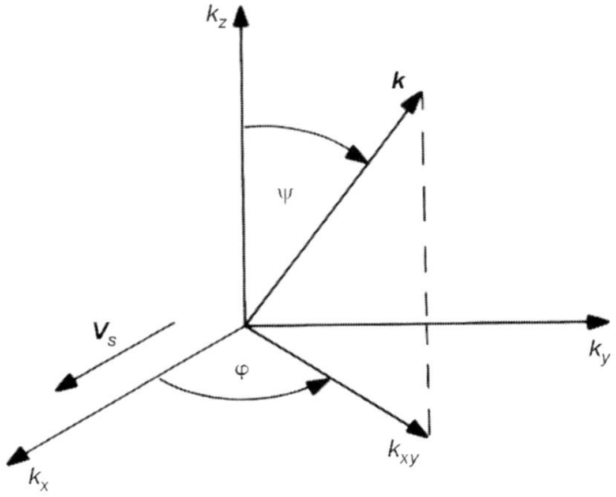

Figure 3.14. Coordinate system in the $\boldsymbol{k}$ space. $\boldsymbol{V}_s$ denotes a vector defined by equation (3.93).

To evaluate the integral (3.94), let us choose a coordinate system so that the k_x axis is collinear with the vector $\boldsymbol{V}_s$ (figure 3.14). Then the ith component of the vector $\boldsymbol{I}_s$ (3.94) is given by

$$I_{si} = |\boldsymbol{V}_s| \int_{V_k} k_x k_i f_s \, \mathrm{d}V_k. \tag{3.96}$$

In the spherical coordinate system (figure 3.14) a volume element in the $\boldsymbol{k}$ space is given by

$$\mathrm{d}V_k = k^2 \sin\psi \, \mathrm{d}\psi \, \mathrm{d}\varphi \, \mathrm{d}k \tag{3.97}$$

and the components of $\boldsymbol{k}$ by

$$k_x = k \sin\psi \cos\varphi \qquad k_y = k \sin\psi \sin\varphi \qquad k_z = k\cos\psi. \tag{3.98}$$

Substituting (3.97) and (3.98) into (3.96), we obtain

$$I_{si} = |\boldsymbol{V}_s| \int_{V_k} k^4 f_s \, \mathrm{d}k \int_{\psi,\varphi} \xi_{xi} \, \mathrm{d}\psi \, \mathrm{d}\varphi \tag{3.99}$$

where ξ_{xi} is a function of the two angles ψ and φ. The second integral in (3.99) is given by

$$\begin{aligned}
\int_{\psi,\varphi} \xi_{xx} \, \mathrm{d}\psi \, \mathrm{d}\varphi &= \int_0^{\pi}\int_0^{2\pi} \sin^2\psi \cos^2\varphi \sin\psi \, \mathrm{d}\psi \, \mathrm{d}\varphi = \frac{4\pi}{3} \\
\int_{\psi,\varphi} \xi_{xy} \, \mathrm{d}\psi \, \mathrm{d}\varphi &= \int_0^{\pi}\int_0^{2\pi} \sin^2\psi \cos\varphi \sin\varphi \sin\psi \, \mathrm{d}\psi \, \mathrm{d}\varphi = 0 \\
\int_{\psi,\varphi} \xi_{xz} \, \mathrm{d}\psi \, \mathrm{d}\varphi &= \int_0^{\pi}\int_0^{2\pi} \sin\psi \cos\varphi \cos\psi \, \mathrm{d}\psi \, \mathrm{d}\varphi = 0.
\end{aligned} \tag{3.100}$$

Substituting now (3.100) into (3.99), we have

$$I_{sx} = |\boldsymbol{V}_s| \frac{4\pi}{3} \int_k k^4 f_s \, \mathrm{d}k \qquad I_{sy} = I_{sz} = 0 \, . \tag{3.101}$$

Therefore, (3.94) attains the form

$$\boldsymbol{I}_s = \left(\frac{4\pi}{3} \int_k k^4 f_s \, \mathrm{d}k \right) \boldsymbol{V}_s . \tag{3.102}$$

For semiconductors with spherical constant energy surfaces, the $E(\boldsymbol{k})$ relation (2.11) reduces to

$$E(\boldsymbol{k}) = E_c + \frac{1}{2} \frac{(\hbar \boldsymbol{k})^2}{m^*} . \tag{3.103}$$

Changing the variable in (3.102) with the aid of (3.103), we obtain (3.102) in terms of electron energies:

$$\boldsymbol{I}_s = \left[\frac{4\pi}{3} \frac{m^*}{\hbar^2} \int_{E_c}^{E_{top}} \left(\frac{2m^*}{\hbar^2} (E - E_c) \right)^{3/2} f_s \, \mathrm{d}E \right] \boldsymbol{V}_s \tag{3.104}$$

where the integration is done over the conduction band. Comparing this expression with (2.15), we may rewrite (3.104) as

$$\boldsymbol{I}_s = \left(\frac{8\pi}{3} \frac{m^*}{\hbar^2} \int_{E_c}^{E_{top}} g_c(E)(E - E_c) f_s \, \mathrm{d}E \right) \boldsymbol{V}_s \tag{3.105}$$

where g_c is the density of states near the bottom of the conduction band. Substituting here (3.95) also, we obtain the required rearranged form of the expression (3.92):

$$\boldsymbol{I}_s = \left[\frac{2}{3m^*} \int_0^{E_{top}} \frac{E \tau^s(E)}{(1 + \mu^2(E) B^2)} g_c(E) \left(-\frac{\partial F_0}{\partial E} \right) \mathrm{d}E \right] \boldsymbol{V}_s \tag{3.106}$$

where we have taken the bottom of the conduction band as the energy reference: $E_c = 0$.

Galvanomagnetic current density

Returning now to the current density, we see that (3.91) can, using (3.92), (3.93) and (3.106), be expressed in the following compact form:

$$\boldsymbol{J} = e^2 K_1 \boldsymbol{E} + \frac{e^3}{m^*} K_2 [\boldsymbol{E} \times \boldsymbol{B}] + \frac{e^4}{m^{*2}} K_3 \boldsymbol{B} (\boldsymbol{E} \cdot \boldsymbol{B}). \tag{3.107}$$

The coefficients K_1, K_2 and K_3 are called the *kinetic coefficients* and are given by

$$K_s = \frac{2}{3m^*} \int_0^{E_{top}} \frac{E \tau^s(E)}{(1 + \mu^2(E) B^2)} g_c(E) \left(-\frac{\partial F_0}{\partial E} \right) \mathrm{d}E \tag{3.108}$$

with $s = 1, 2, 3$. Obviously, relations (3.107) and (3.108) hold also for holes.

Equation (3.107) is the main result of this section. This equation is similar to (3.36), which is illustrated in figure 3.8. Therefore, the present accurate analysis corroborates the result of our approximate analysis of the galvanomagnetic transport in §3.1.4. In both cases we conclude that **the macroscopic current density associated with the transport of change carriers in a solid, subject to electric and magnetic fields, has generally three components: one is collinear with the electric field, one is perpendicular to both the electric and the magnetic fields, and the third one is collinear with the magnetic field.**

Let us briefly recall (see §3.1.4) our insight about the physical origins of the three current components shown in figure 3.8. The current component $\boldsymbol{J}_1$, proportional to the electric field $\boldsymbol{E}$, is due to the conventional drift of charge carriers under the influence of the electrical field. The current component $\boldsymbol{J}_2$, proportional to the vector $[\boldsymbol{E} \times \boldsymbol{B}]$, is the Hall deflection current, due to the Lorentz force action on the drifting charge carriers in the direction perpendicular to both electric and magnetic fields. The current component $\boldsymbol{J}_3$, proportional to the vector $\boldsymbol{B}[\boldsymbol{E} \cdot \boldsymbol{B}]$, is essentially due to the Lorentz deflection of the current component $\boldsymbol{J}_2$. By careful examination of the analysis in §3.2, we see no reason why we should not adopt the same interpretation for the three current components figuring in (3.107).

Note also that the magnetoresistance effect term of (3.36), $1/(1 + \mu^2 B^2)$, exists also in (3.107), although in somewhat modified form: it appears under the averaging integrals of the kinetic coefficients; see (3.108).

We shall further comment on (3.107) in §3.3.1 and use it as a basic equation of the galvanomagnetic effects.

3.2.6 Kinetic coefficients

A general analysis of transport phenomena in solids leads to a relation of a similar form to the above expression (3.107), namely

$$\phi = \sum_{r,s} C_{rs} K_{rs} \boldsymbol{V}_{rs}. \tag{3.109}$$

Here ϕ may denote the electric current density or the thermal energy flux density associated with the transport of carriers; C_{rs} are the coefficients that depend only on the charge and the effective mass of the carriers involved; K_{rs} are the kinetic coefficients; and V_{rs} are the generalized field vectors causing the transport. In our case, the field vectors are given by (3.93). In the general case, the field vectors may include terms proportional to the gradients of temperature and the Fermi level.

Generally for semiconductors with non-spherical constant energy surfaces in the $\boldsymbol{k}$ space, the kinetic coefficients in (3.109) are tensors. However, if several non-spherical constant energy surfaces are symmetrically arranged in the $\boldsymbol{k}$ space, the transport coefficients reduce to scalars. This is the case, for example, with silicon and germanium. Thus (almost) everything happens as if the effective mass of carriers were isotropic. For simplicity, we treat in detail only the case of isotropic effective mass. We shall take into account the anisotropy in the effective mass only when the exact numerical values of kinetic coefficients are considered, or if it plays the essential role in an effect. We have already done so in connection with the piezoresistance effect in §2.5.4.

For semiconductors with spherical constant energy surfaces, the kinetic coefficients are scalars given by

$$K_{rs} = \frac{2}{3m^*}\int_0^{E_{\text{top}}} \frac{E^r \tau^s(E)}{(1+\mu^2(E)B^2)} g_c(E)\left(-\frac{\partial F_0}{\partial E}\right) \mathrm{d}E. \tag{3.110}$$

This is a generalization of expression (3.108). Therefore, the three kinetic coefficients relevant to the pure galvanomagnetic effects can be denoted by K_{1s}, with $s = 1, 2, 3$. Since we do not need the other kinetic coefficients, we can omit the first subscript in K_{1s} and denote it simply by K_s.

The kinetic coefficients can be expressed in a closed form only if some additional approximations are made. We shall now find the approximate expressions for the kinetic coefficients for non-degenerate and strongly degenerate semiconductors.

Non-degenerate carriers

For non-degenerate carriers, the electron distribution function at equilibrium is given by the Maxwell–Boltzmann function (2.18). Then the derivative (3.72) reduces to

$$\frac{\partial F_0}{\partial E} \simeq -\frac{F_0(E)}{kT}. \tag{3.111}$$

Substituting this expression into (3.108), we have

$$K_s \simeq \frac{2}{3m^* kT}\int_0^{\infty} \frac{E\tau^s(E)}{(1+\mu^2(E)B^2)} g_c(E) F_0(E)\,\mathrm{d}E. \tag{3.112}$$

Since the integrand is proportional to $F_0(E)$, which is a fast-decreasing function of energy, we substituted in this expression the upper integration limit by infinity.

For a reason that will become clear later, let us multiply and divide (3.112) by the carrier concentration n. Then we obtain

$$K_s = \frac{\dfrac{n}{m^*}\displaystyle\int_0^{\infty} \frac{E\tau^s(E)}{(1+\mu^2(E)B^2)} g_c(E) F_0(E)\,\mathrm{d}E}{\frac{3}{2}kTn}. \tag{3.113}$$

Note that the term $\frac{3}{2}kT$ in the denominator of (3.113) equals the average thermal energy of non-degenerate carriers. This average energy is given by

$$\frac{\int_0^\infty E g_c(E) F_0(E)\,\mathrm{d}E}{n} = \tfrac{3}{2}kT. \tag{3.114}$$

Using (3.114), (3.113) can be rewritten as

$$K_s = \frac{\dfrac{n}{m^*}\displaystyle\int_0^\infty \frac{E\tau^s(E)}{(1+\mu^2(E)B^2)} g_c(E)F_0(E)\,\mathrm{d}E}{\displaystyle\int_0^\infty E g_c(E)F_0(E)\,\mathrm{d}E}. \tag{3.115}$$

The ratio of integrals in this expression equals the energy-weighted average of the term $\tau^s/(1+\mu^2 B^2)$. Consequently, the kinetic coefficients can be expressed in the following compact form:

$$K_s = \frac{n}{m^*}\left\langle \frac{\tau^s}{1+\mu^2 B^2}\right\rangle \tag{3.116}$$

where the angular brackets denote averaging over electron energies.

For $\boldsymbol{B} = 0$ and $s = 1$, the term in the angular brackets of (3.116) reduces to the relaxation time, and its energy-weighted average to (2.65). If the energy dependence of the relaxation time can be expressed by (2.64), the average relaxation time is given by (2.66). The energy-weighted average of the powered relaxation time is given by

$$\langle \tau^s \rangle = \tau_0^s (kT)^{ps} \frac{\Gamma(\frac{5}{2}+ps)}{\Gamma(\frac{5}{2})}. \tag{3.117}$$

Note that (2.66) is a special case of (3.117) for $s = 1$.

The expression (3.117) is obtained in the following way. By substituting $\boldsymbol{B} = 0$, g_c (2.15), F_0 (2.18) and τ (2.64) into the ratio of integrals in (3.115), this becomes

$$\langle \tau^s \rangle = \frac{\int_0^\infty \tau_0^s E^{(3/2)+ps} \exp(-E/kT)\,\mathrm{d}E}{\int_0^\infty E^{3/2}\exp(-E/kT)\,\mathrm{d}E}. \tag{3.118}$$

(Compare with (2.65).) Introducing here the new variable $x = E/kT$, this equation attains the form

$$\langle \tau^s \rangle = \frac{\tau_0^s (kT)^{ps} \int_0^\infty x^{(3/2)+ps}\,\mathrm{e}^{-x}\,\mathrm{d}x}{\int_0^\infty x^{3/2}\,\mathrm{e}^{-x}\,\mathrm{d}x}. \tag{3.119}$$

Recall that the gamma function is defined by [12]

$$\Gamma(\alpha) = \int_0^\infty x^{\alpha-1} \mathrm{e}^{-x} \, \mathrm{d}x. \tag{3.120}$$

Therefore the integrals in (3.119) can be expressed by gamma functions. Thence (3.117).

Substituting now (3.117) into (3.116), we obtain the kinetic coefficients for $\boldsymbol{B} = 0$ in the form

$$K_s(\boldsymbol{B} = 0) = \frac{n}{m^*} \langle \tau^s \rangle = \frac{n}{m^*} \tau_0^s (kT)^{ps} \frac{\Gamma(\frac{5}{2} + ps)}{\Gamma(\frac{5}{2})}. \tag{3.121}$$

Strongly degenerate carriers

For strongly degenerate carriers, the derivative of the distribution function (3.72) features a sharp maximum at $E = E_{\mathrm{F}}$, E_{F} being the Fermi level. Therefore, we may approximate the derivative of the distribution function by the Dirac delta function:

$$\frac{\partial F_0}{\partial E} \simeq -\delta(E - E_{\mathrm{F}}). \tag{3.122}$$

This means that the transport properties of degenerate carriers are predominantly determined by the carriers with energies close to the Fermi level. Substituting (3.122) into (3.108), we obtain

$$K_s \simeq \frac{n}{m^*} \frac{\tau^s(E_{\mathrm{F}})}{1 + \mu^2(E_{\mathrm{F}})B^2}. \tag{3.123}$$

This expression could also be obtained formally from (3.116) by considering only those electrons with energies $E \sim E_{\mathrm{F}}$. Therefore, (3.116) holds for both non-degenerate and strongly degenerate carrier gases.

3.2.7 Magnetic field dependence of the kinetic coefficients

We shall now find the approximations of (3.116) for two extreme values of the magnetic field: very weak and very strong. We shall need these expressions in §3.3.

A *low magnetic induction* is defined by

$$\mu^2(E)B^2 \ll 1 \tag{3.124}$$

for all carrier energies. Note that this condition is milder than (3.46). The inequality (3.124) yields quite different limits for various semiconductors. Assume that (3.124) is fulfilled if $\mu^2 B^2 < 0.1$. Then, for n-type silicon with $\mu = 0.14\,\mathrm{m^2\,V^{-1}\,s^{-1}}$, an induction as high as $B = 2.2\,\mathrm{T}$ may be considered as a weak magnetic induction. In contrast, for electrons in InSb, with

$\mu = 8\,\mathrm{m^2\,V^{-1}\,s^{-1}}$, already the induction $B = 0.04\,\mathrm{T}$ cannot be considered as weak.

For a weak magnetic induction, we can expand the term $(1 + \mu^2 B^2)^{-1}$ in (3.116) in a series in $\mu^2 B^2$:

$$\frac{1}{1 + \mu^2 B^2} = \sum_{m=0}^{\infty} (-1)^m (\mu^2 B^2)^m. \tag{3.125}$$

Using (3.84), we can express the mobility in terms of the relaxation time. Then by substituting (3.125) into (3.116), we obtain

$$K_s = \frac{n}{m^*} \sum_{m=0}^{\infty} (-1)^m \left(\frac{q}{m^*}\right)^{2m} \langle \tau^{s+2m} \rangle B^{2m} \tag{3.126}$$

where $\langle \tau^x \rangle$ denotes the *energy-weighted average* value of the carrier *relaxation time* to the power x:

$$\langle \tau^x \rangle = \int_0^\infty E[\tau(E)]^x g(E) F(E)\,\mathrm{d}E \bigg/ \int_0^\infty E g(E) F(E)\,\mathrm{d}E \tag{3.127}$$

(compare with (2.65)).

Keeping only the first two terms of the series (3.125), we obtain the following approximations for the three kinetic coefficients:

$$K_1 \simeq \frac{n}{m^*} \left[\langle \tau \rangle - \left(\frac{q}{m^*}\right)^2 \langle \tau^3 \rangle B^2 \right] \tag{3.128}$$

$$K_2 \simeq \frac{n}{m^*} \left[\langle \tau^2 \rangle - \left(\frac{q}{m^*}\right)^2 \langle \tau^4 \rangle B^2 \right] \tag{3.129}$$

$$K_3 \simeq \frac{n}{m^*} \left[\langle \tau^3 \rangle - \left(\frac{q}{m^*}\right)^2 \langle \tau^5 \rangle B^2 \right]. \tag{3.130}$$

These relations hold for a weak magnetic field (3.124).

A *high magnetic induction* is defined by

$$\mu^2(E) B^2 \gg 1 \tag{3.131}$$

for all carrier energies. This is easier to fulfil than (3.45). At a high magnetic induction, the term $(1 + \mu^2 B^2)^{-1}$ can be expanded in the following series:

$$\frac{1}{1 + \mu^2 B^2} = \frac{1}{\mu^2 B^2} \sum_{m=0}^{\infty} (-1)^m \left(\frac{1}{\mu^2 B^2}\right)^m. \tag{3.132}$$

Substituting this into (3.116), we obtain

$$K_s = \frac{n}{m^*} \left(\frac{m^*}{q}\right)^s \left\langle \frac{1}{\mu^{2-s} B^2} \sum_{m=0}^{\infty} (-1)^m \left(\frac{1}{\mu^2 B^2}\right)^m \right\rangle. \tag{3.133}$$

Keeping only the first two terms of this series, we obtain the following approximations for the three kinetic coefficients:

$$K_1 \simeq \frac{nm^*}{q^2}\frac{1}{B^2}\left(\langle\tau^{-1}\rangle - \left(\frac{q}{m^*}\right)^{-2}\langle\tau^{-3}\rangle B^{-2}\right) \tag{3.134}$$

$$K_2 \simeq \frac{nm^*}{q^2}\frac{1}{B^2}\left(1 - \left(\frac{q}{m^*}\right)^{-2}\langle\tau^{-2}\rangle B^{-2}\right) \tag{3.135}$$

$$K_3 \simeq \frac{nm^*}{q^2}\frac{1}{B^2}\left(\langle\tau\rangle - \left(\frac{q}{m^*}\right)^{-2}\langle\tau^{-1}\rangle B^{-2}\right). \tag{3.136}$$

These relations hold for a strong magnetic field (3.131).

3.3 An accurate analysis

We shall now apply the results of the general treatment of isothermal galvanomagnetic transport phenomena, obtained in §3.2, to study the Hall and the magnetoresistance effects. This will enable us to make some important refinements in the simple models of these effects, developed in §3.1. The present analysis is quite general and physically accurate: the validity of the results is limited only by the simplifying assumptions that we made while composing and solving the kinetic equation. Let us recall these assumptions: the applicability of the relaxation time approximation (3.67), steady-state conditions (3.66), a small disturbance of the distribution function (3.68), and spherical constant energy surfaces in the $\boldsymbol{k}$ space (3.103).

3.3.1 Basic equations

The full set of basic equations for semiconductor device operation comprises the Maxwell equations, the current density equations, and the continuity equations [13]. In a limited range of operating conditions of a device, some of these equations may become simpler or even irrelevant. To this end, we shall also assume some reasonable limits of the parameters and operating conditions of Hall and magnetoresistance devices. In this way, we can greatly simplify their analysis. These assumptions and their consequences are listed below.

(i) Low frequencies of the biasing current and the magnetic field. We may then neglect dielectric displacement currents and inductive currents, and thus drop the first two Maxwell equations.

(ii) The neutrality condition (2.36) is fulfilled throughout the device active region. There is then no space charge in this region, and the third Maxwell equation reduces to the Laplace equation

$$-\nabla \cdot \boldsymbol{E} = \Delta\varphi = 0. \tag{3.137}$$

Here φ is the electric potential. Note, however, that at the boundary of a device, a space charge may exist (see §3.1.1).

(iii) There is no generation and recombination of carriers. The continuity equations then read

$$\nabla \cdot \boldsymbol{J} = 0 \tag{3.138}$$

for each kind of carrier present in the sample.

(iv) No considerable gradients in carrier densities. We may then neglect diffusion currents.

If these conditions are fulfilled, then the only equations left from the above-mentioned full set of equations are the current density equations (without diffusion terms). The current density arising from one kind of carrier in a sample exposed to electric and magnetic fields is accurately described by equation (3.107). For easy reference, we repeat this equation:

$$\boldsymbol{J} = e^2 K_1 \boldsymbol{E} + \frac{e^3}{m^*} K_2 [\boldsymbol{E} \times \boldsymbol{B}] + \frac{e^4}{m^{*2}} K_3 \boldsymbol{B}(\boldsymbol{E} \cdot \boldsymbol{B}). \tag{3.139}$$

It relates the electric current density $\boldsymbol{J}$ at any point of the sample with the local electric field $\boldsymbol{E}$ and the local magnetic induction $\boldsymbol{B}$. The notations are: e is the charge, m^* is the effective mass of the carriers, and K_1, K_2 and K_3 are the kinetic coefficients (3.116). For the physical interpretation of this equation, recall our discussion in §3.1.4 and that following (3.107).

If several kinds of charge carriers are present in a sample, then the total current density consists of the sum of the current densities due to each group of carriers:

$$\boldsymbol{J} = \sum_c \boldsymbol{J}_\mathrm{c} \tag{3.140}$$

where $\boldsymbol{J}_\mathrm{c}$ denotes the current density arising from carriers of the kind c. Each of the current components is determined by its own equation (3.139) with appropriate values of the parameters e, m^* and the kinetic coefficients. However, each group of carriers 'swims in the same pool': they are exposed to the common electric and magnetic fields, $\boldsymbol{E}$ and $\boldsymbol{B}$.

With the aid of Appendix E, (E.9)–(E.13), the function $\boldsymbol{J}(\boldsymbol{E})$ (3.139) can be transformed into its inverse function $\boldsymbol{E}(\boldsymbol{J})$:

$$\boldsymbol{E} = \frac{1}{1 + \left(\frac{e}{m^*}\frac{K_2}{K_1}\right)^2 B^2} \times \left\{ \frac{1}{e^2 K_1}\boldsymbol{J} - \frac{1}{em^*}\frac{K_2}{K_1^2}[\boldsymbol{J} \times \boldsymbol{B}] + \frac{\left(\frac{e}{m^*}\frac{K_2}{K_1}\right)^2 - \frac{e^2}{m^{*2}}\frac{K_3}{K_1}}{e^2 K_1 \left(1 + \frac{e^2}{m^{*2}}\frac{K_3}{K_1} B^2\right)} (\boldsymbol{J} \cdot \boldsymbol{B})\boldsymbol{B} \right\}. \tag{3.141}$$

Note the symmetry of (3.139) and (3.141). We shall use latter equation rather than the original one (3.139) when the local values of the current density are known.

Equations (3.137)–(3.141) are the basic equations of Hall devices and magnetoresistors. In principle, all variations of the isothermal galvanomagnetic effects can be elaborated by solving these equations, subject to appropriate boundary conditions. Note that the field vectors $\boldsymbol{E}$ and $\boldsymbol{B}$ figuring in (3.139) and (3.141) are the local values of these vectors. When deriving or importing the basic equations, we assumed nothing specific about the origins of the electric and magnetic fields. Consequently, these fields may be of any origin, including galvanic, inductive and electrostatic. Moreover, they may also be self-fields: a component of $\boldsymbol{B}$ may be associated with the current in the sample, and a component of $\boldsymbol{E}$ may arise as a consequence of the Hall effect itself. In fact, the latter case is essential for the appearance of the Hall effect in a long sample. As we have seen in §3.1 while performing the approximate analysis of the Hall effect, the sample geometry imposes boundary conditions, which may dramatically influence the local values of the electric field and of the current density. In the next sections we shall carefully revisit this effect and consider the application of the basic equations in a few particular device structures.

3.3.2 Drift current density at $B = 0$

Let us first refine our approximate results of §2.6.1 about the current density in the absence of a magnetic field. If an electric field is established in a sample having quasi-free charge carriers, a drift electric current appears in the sample. In the absence of the magnetic field, the current density $\boldsymbol{J}$ and the electric field $\boldsymbol{E}$ are related by Ohm's law (2.89):

$$\boldsymbol{J} = \sigma \boldsymbol{E} \tag{3.142}$$

where σ denotes the *electric conductivity* (2.90).

Ohm's law can also be derived as a special case from our present analysis of transport phenomena. In fact, for one kind of charge carrier and $\boldsymbol{B} = 0$, (3.139) reduces to

$$\boldsymbol{J} = e^2 K_1 \boldsymbol{E}. \tag{3.143}$$

Thus in terms of the kinetic coefficients the electric conductivity for one type of carrier is given by

$$\sigma = e^2 K_1(\boldsymbol{B} = 0). \tag{3.144}$$

Substituting here $e^2 = q^2$ and K_1 by (3.116) with $s = 1$, we obtain

$$\sigma = \frac{q^2}{m^*}\langle\tau\rangle n \tag{3.145}$$

from which comes the average value of the relaxation time $\langle\tau\rangle$, which we postulated in (2.84). Making use of the expression (2.86) for mobility, we obtain

$$\sigma = q\mu n. \tag{3.146}$$

This is formally identical with (2.90) for each type of carrier.

For semiconductors with many-valley band structure and non-spherical constant energy surfaces, Ohm's law (3.142) and the expression for conductivity (3.146) also hold. However, the *mobility* attains the form

$$\mu = q\frac{\langle\tau\rangle}{m_c^*} \tag{3.147}$$

where m_c^* denotes the *conductivity effective mass*. For example, for electrons in silicon the conductivity effective mass is given by

$$\frac{1}{m_c^*} = \frac{1}{3}\left(\frac{1}{m_l^*} + \frac{1}{m_t^*}\right) \tag{3.148}$$

where m_l^* and m_t^* are the *longitudinal* and *transverse effective masses*, respectively (see §2.2.2).

3.3.3 Current density in a plate under perpendicular magnetic field

As we have seen in §§3.1.1–3.1.3, a preferential form of a Hall device or a magnetoresistor is a plate. In typical applications of these devices, the magnetic field vector is approximately perpendicular to the plate plane. Let us consider qualitatively the current components in a plate in such a case. To this end, let us combine figures 3.3 and 3.8. The result is shown in figure 3.15.

The plate geometry imposes such boundary conditions that a macroscopic current component perpendicular to the plate plane, which is here the z component of $\boldsymbol{J}_3$ of figure 3.8, does not exist. Since the vector of the magnetic induction $\boldsymbol{B}$ is almost perpendicular to the plate, then the in-plate component of $\boldsymbol{J}_3$ is very small. So we may neglect this current component. This is equivalent to the approximation that the two vectors $\boldsymbol{E}$ and $\boldsymbol{B}$ are mutually exactly perpendicular. Then the current vector parallelepiped in figure 3.8 attains its simplest form: it reduces to a parallelogram in the xz plane, and the total current density is given by $\boldsymbol{J} \approx \boldsymbol{J}_{12} = \boldsymbol{J}_1 + \boldsymbol{J}_2$. Thus (3.139) reduces to

$$\boldsymbol{J} = \sigma_B\boldsymbol{E} + \sigma_B\mu_H[\boldsymbol{E}\times\boldsymbol{B}] \quad \text{if } \boldsymbol{B}\perp\boldsymbol{E} \tag{3.149}$$

where we have also introduced the new notation

$$\sigma_B = e^2K_1 \tag{3.150}$$

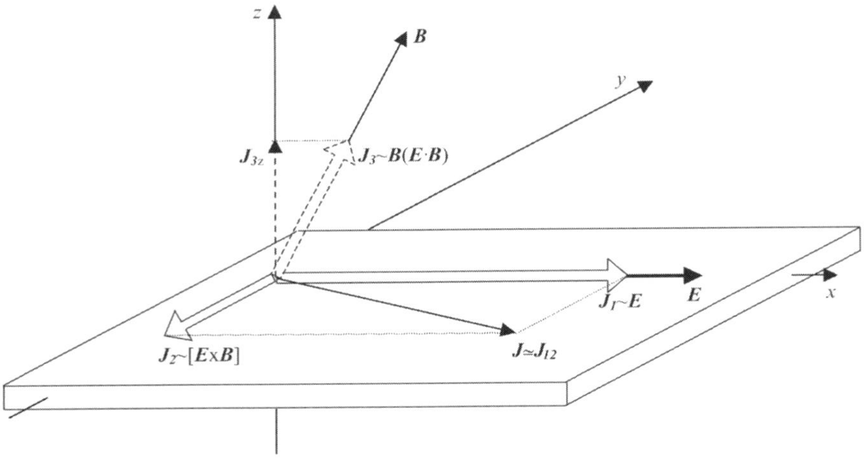

Figure 3.15. A plate exposed to an in plane electric field $\boldsymbol{E}$ and an almost perpendicular magnetic field. Compare with figures 3.3 and 3.8. The plate plane is the xy plane and the magnetic induction vector $\boldsymbol{B}$ is in the xz plane. A current component density perpendicular to the plate plane $\boldsymbol{J}_{3z}$ cannot exist.

$$\sigma_{\mathrm{B}}\mu_{\mathrm{H}} = \frac{e^3}{m^*}K_2. \tag{3.151}$$

The coefficient σ_{B} is an effective conductivity, and μ_{H} is called the Hall mobility. The physical meaning of these two coefficients will be explained in §3.3.5. If not stated otherwise, we shall *from now on assume* ***B*** *perpendicular to the plate and thus also to* **E**, and use (3.149) instead of (3.139) as the basic equation. A notable exception will be made in §3.3.9.

Figure 3.16 is a graphical representation of equation (3.149). The fundamental feature of this equation is the existence of an angle between the total current density $\boldsymbol{J}$ and the electric field $\boldsymbol{E}$. This angle, denoted by Θ_{H}, is obviously the Hall angle, which we introduced before (see (3.14)

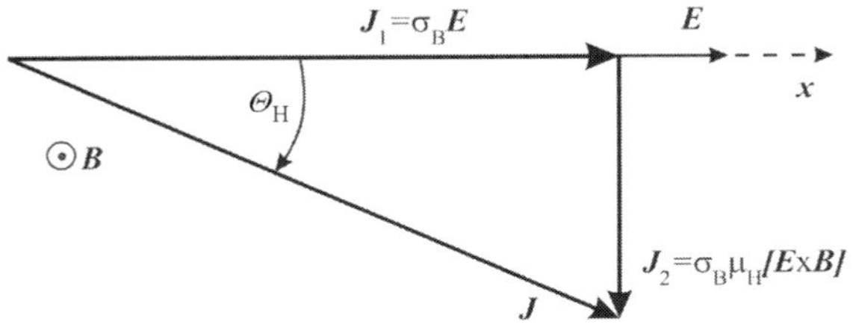

Figure 3.16. If the magnetic induction $\boldsymbol{B}$ is perpendicular to the plate, then the vector diagram from figure 3.15 reduces to this one. The current density vectors are in the plate plane. The total current density $\boldsymbol{J}$ is inclined with respect to the electric field $\boldsymbol{E}$ for the Hall angle Θ_{H}.

and (3.31)). With reference to (3.149) and figure 3.16, the Hall angle is given by

$$\tan \Theta_H = \mu_H B. \tag{3.152}$$

This is reminiscent of (3.13) and (3.31). However, (3.152) contains the proper sign of the Hall angle, which is the sign of the Hall mobility; and owing to the relation of the Hall mobility to the kinetic coefficients, the above equation also includes the subtle influence of the carrier scattering process on the Hall angle.

3.3.4 Boundary conditions for $\boldsymbol{B} \perp \boldsymbol{E}$

In the present context, the boundary conditions denote the relations among the macroscopic values of a magnetic induction $\boldsymbol{B}$, electric field $\boldsymbol{E}$, current density $\boldsymbol{J}$ and the electric potential φ that hold at the boundaries of a plate-like galvanomagnetic device. A boundary of such a device is an interface between its low-conductivity active region and its surroundings. According to this definition, high-conductivity metal electrodes do not belong to the device active region.

To simplify further discussions, we now have to choose a particular shape of the device for analysis. Let it be a plate with contacts fitted at vertical cylindrical side edges, as shown in figure 3.17. Such a device is called a Hall plate. In a Hall plate, both electric field and current density have only components parallel to the plane of the plate. If the plate is exposed to a perpendicular magnetic field, the above assumption $\boldsymbol{B} \perp \boldsymbol{E}$ is certainly fulfilled. We shall also assume that the contacts are ohmic, and the rest of the device surface makes a perfectly isolating boundary. Consider now the boundary conditions at an ohmic contact and at an insulating boundary of a Hall plate.

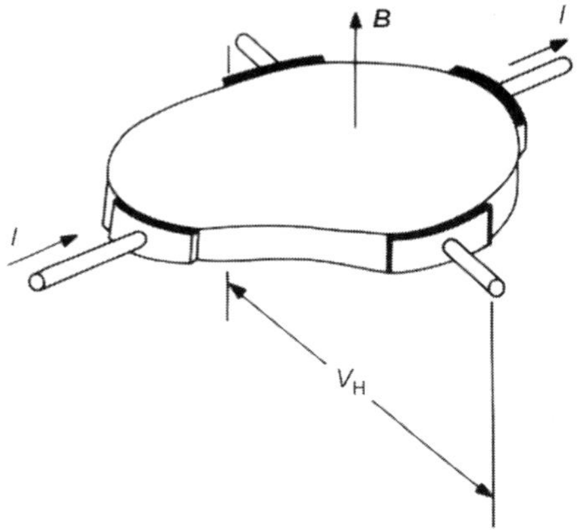

Figure 3.17. The Hall plate of arbitrary shape. This is a thin semiconductor slice with a cylindrical boundary, fitted with four electrical contacts.

Ohmic contact

An ohmic contact is a non-rectifying metal–semiconductor contact. A good ohmic contact also has the property that the voltage drop across the contact interface is negligible relative to the voltage drop in the semiconductor sample. Assume that the metal used for the contact is highly conductive. Then the contact is equipotential and, consequently, the contact boundary itself is an equipotential plane:

$$\varphi|_{\text{cb}} = \text{constant} \tag{3.153}$$

where cb refers to a (ohmic) contact boundary.

Electric field at an Ohmic contact

Recall that, generally, the tangential component of the electric field at an interface is conserved. Hence the tangential component of the electric field at the semiconductor side of an ohmic contact is zero. Consequently, the electric field at an ohmic contact of a Hall device is perpendicular to the contact interface:

$$\boldsymbol{E} \cdot \boldsymbol{u}_{\text{t}}|_{\text{cb}} = 0 \tag{3.154}$$

where $\boldsymbol{u}_{\text{t}}$ is the unit vector tangential to the boundary, see figure 3.18.

Current density at an ohmic contact

But the current density at an ohmic contact is not necessarily perpendicular to the interface. In particular, if a magnetic induction parallel to the contact

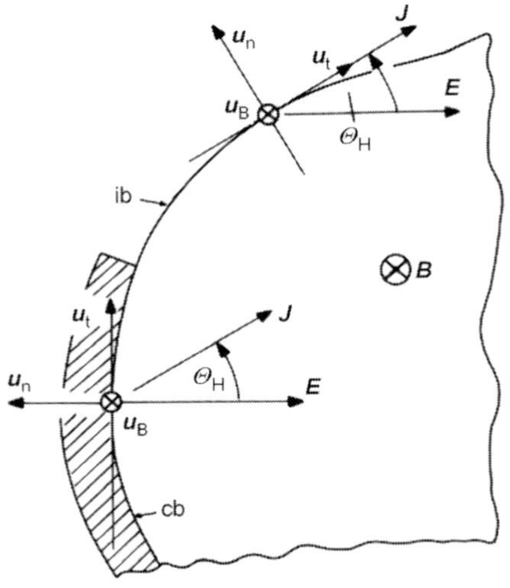

Figure 3.18. The boundary conditions in a Hall plate. cb denotes the contact boundary, and ib the insulating boundary.

interface is present, the current density is inclined by the Hall angle (3.152) with respect to the electric field. Thus the current density at an ohmic contact of a Hall device is also inclined by the Hall angle with respect to the normal on the contact.

To show this rigorously, let us first rewrite (3.149) so:

$$\boldsymbol{E} = \frac{1}{\sigma_B}\boldsymbol{J} - \mu_H[\boldsymbol{E} \times \boldsymbol{B}]. \tag{3.155}$$

By analogy with (E.4)–(E.8) of Appendix E, the solution of (3.155) with respect to $\boldsymbol{E}$ is given by

$$\boldsymbol{E} = \frac{1}{\sigma_B(1+\mu_H^2B^2)}\boldsymbol{J} - \frac{\mu_H}{\sigma_B(1+\mu_H^2B^2)}[\boldsymbol{J} \times \boldsymbol{B}] \tag{3.156}$$

where we have omitted the term with $(\boldsymbol{J} \cdot \boldsymbol{B})$, since $\boldsymbol{B} \perp \boldsymbol{J}$. Equation (3.156) is special case of (3.141).

Let us now introduce three mutually orthogonal unit vectors: $\boldsymbol{u}_t$, $\boldsymbol{u}_B$ and $\boldsymbol{u}_n$. $\boldsymbol{u}_t$ is tangential to the boundary and $\boldsymbol{u}_t \perp \boldsymbol{B}$; $\boldsymbol{u}_B \| \boldsymbol{B}$ and $\boldsymbol{u}_n$ is normal to the boundary. It holds that

$$\boldsymbol{u}_t = [\boldsymbol{u}_B \times \boldsymbol{u}_n]. \tag{3.157}$$

Multiplying (3.156) by (3.157), we obtain

$$\boldsymbol{J} \cdot \boldsymbol{u}_t|_{cb} = -\mu_H B(\boldsymbol{J} \cdot \boldsymbol{u}_n). \tag{3.158}$$

Here we have made use of (3.154) and the vector identity

$$(\boldsymbol{a} \times \boldsymbol{b}) \cdot (\boldsymbol{b} \times \boldsymbol{d}) = (\boldsymbol{a} \cdot \boldsymbol{c})(\boldsymbol{b} \cdot \boldsymbol{d}) - (\boldsymbol{a} \cdot \boldsymbol{d})(\boldsymbol{b} \cdot \boldsymbol{c}). \tag{3.159}$$

Equation (3.158) shows that the angle of inclination of $\boldsymbol{J}$ with respect to $-\boldsymbol{u}_n$ is $\Theta_H = \arctan \mu_H B$.

Insulating boundary

An insulating boundary of a Hall device is the interface between the active region of the device and its non-conductive surroundings. Since there is no current through such a boundary, only a tangential component of the current density may exist. Thus

$$\boldsymbol{J} \cdot \boldsymbol{u}_n|_{ib} = 0 \tag{3.160}$$

where ib refers to an insulating boundary. At the two insulating boundaries that define the plate and are perpendicular to the magnetic field, only the tangential component of an electric field exists, irrespective of the magnetic field.

If a magnetic induction perpendicular to the current density is established, the electric field will be tilted by the Hall angle with respect to the current density. Thus the electric field at an insulating boundary, which is parallel to the magnetic induction, is inclined by the Hall angle also with respect to the boundary (see figure 3.18).

This can also be shown by multiplying (3.149) by

$$\boldsymbol{u}_{\mathrm{n}} = [\boldsymbol{u}_{\mathrm{t}} \times \boldsymbol{u}_{\mathrm{B}}]. \tag{3.161}$$

Taking into account (3.159) and (3.160), we have

$$-\boldsymbol{E} \cdot \boldsymbol{u}_{\mathrm{n}}\big|_{\mathrm{ib}} = \mu_{\mathrm{H}} B(\boldsymbol{E} \cdot \boldsymbol{u}_{\mathrm{t}}). \tag{3.162}$$

This shows that the inclination of $\boldsymbol{E}$ with respect to $\boldsymbol{u}_{\mathrm{t}}$ is $\Theta_{\mathrm{H}} = \arctan \mu_{\mathrm{H}} B$. Thus in the presence of a magnetic field, a normal component of an electric field may also exist at an insulating boundary.

The above boundary conditions are sufficient for solving the basic equations of Hall plates (3.137), (3.138), (3.141) and (3.149). Moreover, if one of the vectors $\boldsymbol{E}$ or $\boldsymbol{J}$ is determined by some additional conditions, the solutions of the transport equations become especially simple. This is the case if a Hall plate is very long or very short. We have already treated these two device types approximately in §3.1. Now we are able to perform the analysis more rigorously.

3.3.5 Infinitely short plate with $\boldsymbol{B} \perp \boldsymbol{E}$. Corbino conductivity and Hall mobility

The structure and operation of short Hall plates are thoroughly described in §3.1.2. Briefly, in a short Hall plate the distance between the biasing contacts l is much smaller than the contact width w (see figure 3.4). With reference to this figure, we define the infinitely short sample by the condition

$$l/w \to 0. \tag{3.163}$$

Now we shall also assume that the device is fitted with two parallel contacts, and the contacts are ohmic. Since the electric field is perpendicular to an ohmic contact (3.154), here the electric field in the whole device must be perpendicular to the biasing contacts. The electric field is determined by the applied voltage V and the distance between the biasing contacts l:

$$\boldsymbol{E} = \frac{V}{l} \boldsymbol{u}_{\mathrm{n}} \tag{3.164}$$

where $\boldsymbol{u}_{\mathrm{n}}$ denotes the unit vector normal to the contact (see figure 3.18). We shall assume here that the sample is strongly extrinsic and contains only one kind of carrier. The non-extrinsic case is discussed in §3.4.7.

In a strongly extrinsic infinitely short sample, the current density is readily given by (3.149), which we repeat:

$$\boldsymbol{J} = \sigma_{\mathrm{B}} \boldsymbol{E} + \sigma_{\mathrm{B}} \mu_{\mathrm{H}} [\boldsymbol{E} \times \boldsymbol{B}]. \tag{3.165}$$

Note that this equation is of the same form as the approximate equations (3.32) that we derived in §3.1.3. Therefore, the solutions are formally the same. So the basic conclusion about the existence of the current deflection stays the same. However, the coefficients in the two equations are not identical: those in (3.165) contain more detailed information on the underlying physics. Therefore, in the rest of this section, we shall concentrate on the explanation of the physical nature of the coefficients σ_B and μ_H.

Corbino conductivity

We introduced in §3.1.3 (3.33) an approximation of the coefficient σ_B figuring in (3.165) as Corbino conductivity. It characterizes the effective conductivity of a sample in which a full *current deflection effect* takes place. Otherwise, σ_B *is the effective material conductivity in the direction of the electric field* $\boldsymbol{E}$ when $\boldsymbol{E} \perp \boldsymbol{B}$. (Note the difference relative to the definition of ρ_b in §3.3.6!) By inspecting (3.165) and figure 3.16, we see that σ_B relates the longitudinal current density component with the electric field in much the same way as the usual conductivity does (3.143) in the absence of a magnetic field (here longitudinal means collinear with $\boldsymbol{E}$). At $\boldsymbol{B} \sim 0$, σ_B reduces to the usual, zero magnetic field conductivity (3.144). Since the kinetic coefficient K_1 depends on the magnetic induction (3.116), so also does the conductivity σ_B: $\sigma_B(\boldsymbol{B})$. Taking an infinitely short plate as a black box, σ_B is its apparent conductivity. Therefore, the conductivity σ_B contains information on the *geometrical magnetoresistance effect* (see §3.1.3). We shall analyse the conductivity σ_B further in §3.4.

Hall mobility

To understand the physical nature of the Hall mobility, consider the second term in (3.165):

$$\boldsymbol{J}_t = \sigma_B \mu_H [\boldsymbol{E} \times \boldsymbol{B}]. \tag{3.166}$$

This is a transverse current density: it is transverse to both $\boldsymbol{E}$ and $\boldsymbol{B}$. The transverse current is a consequence of the action of magnetic forces on moving charge carriers—see also $\boldsymbol{J}_2$ in figure 3.15. The Hall mobility determines the average drift velocity of carriers due to the Hall's 'magnetic pressure'. To see this, we will formally replace the magnetic forces by equivalent electric forces. To this end, we have to introduce a fictitious transverse electric field $\boldsymbol{E}_t$. Under the action of this electric field, the carriers would drift in the transverse direction so as to produce a current density exactly equal to $\boldsymbol{J}_t$ (3.166).

Let us determine the equivalent transverse electric field $\boldsymbol{E}_t$. Assuming a low magnetic induction, we may substitute σ_B in (3.166) by its low-field value (3.146). Then equation (3.166) can be rewritten as

$$\boldsymbol{J}_t = qn\mu_H \mu [\boldsymbol{E} \times \boldsymbol{B}]. \tag{3.167}$$

The term $\mu[\boldsymbol{E} \times \boldsymbol{B}]$ has the dimensions of electric field. Moreover, this term equals the negative of the Hall electric field in a corresponding very long sample (see (3.9)). It seems reasonable to interpret this term as the required equivalent electric field, $\boldsymbol{E}_t = \mu[\boldsymbol{E} \times \boldsymbol{B}]$, for in a very long sample $\boldsymbol{E}_t$ compensates the action of the Hall electric field $\boldsymbol{E}_H$ and there is no transverse current. In the present case, there is no Hall field, and the field $\boldsymbol{E}_t$ produces the transverse current (3.167). Thus the latter equation can be rewritten as

$$\boldsymbol{J}_t = qn\mu_H \boldsymbol{E}_t. \tag{3.168}$$

Formally, this equation expresses Ohm's law for the transverse direction (compare with (2.88)). However, in (3.168) the role of the drift mobility plays the Hall mobility μ_H.

Therefore, under the action of the magnetic forces, the carriers drift in the transverse direction with a drift velocity given by $\mu_H \boldsymbol{E}_t$. Consequently, the Hall mobility is an effective mobility of charge carriers under the action of transverse magnetic force.

Let us now express the Hall mobility in terms of already-known coefficients. From (3.150) and (3.151), we find

$$\mu_H = \frac{e}{m^*} \frac{K_2}{K_1}. \tag{3.169}$$

Since the kinetic coefficients K_1 and K_2 depend on the magnetic field, so also does the Hall mobility. We shall discuss this dependence later in §3.4. For the moment, it is interesting to derive an expression for the low magnetic field value of the Hall mobility. To this end, we substitute into (3.169) the approximate values of the kinetic coefficients K_1 and K_2 (3.116) for $\boldsymbol{B} \sim 0$. Hence we obtain

$$\mu_H(B \simeq 0) = \frac{e}{m^*} \frac{\langle \tau^2 \rangle}{\langle \tau \rangle}. \tag{3.170}$$

Using (2.86), (3.170) can be rewritten as

$$\mu_H(B \simeq 0) = \text{sign}[e] r_{H0} \mu \tag{3.171}$$

where sign$[e]$ denotes the sign of the charge of the carriers involved, μ is the drift mobility of those carriers, and r_{H0} is a quantity called the *Hall scattering factor*, given by

$$r_{H0} = \frac{\langle \tau^2 \rangle}{\langle \tau \rangle^2}. \tag{3.172}$$

For a reason which will become clear later, we shall denote r_{H0} also by r_H. The Hall scattering factor is a number (a dimensionless quantity) of about unity. Thus the Hall mobility (3.171) approximately equals the corresponding drift mobility. However, the drift mobility is a positive quantity, whereas the Hall mobility, by definition, carries the sign of the corresponding carriers. For holes, the Hall mobility is positive, but for electrons it is negative.

3.3.6 Infinitely long plate with $\boldsymbol{B} \perp \boldsymbol{E}$. Physical magnetoresistance and Hall effect

Let us also check and refine the results on the Hall effect in very long plates that we found in §3.1.1. Recall that, in a long sample, the distance between the biasing contacts l is much larger than their width w. We shall again assume a strip-like sample (see figure 3.1). With reference to this figure, we define an infinitely long sample as a strip satisfying the condition

$$l/w \rightarrow \infty. \tag{3.173}$$

In addition, we assume that the side contacts for sensing the Hall voltage are infinitely small, that is point-like.

In an infinitely long sample, the current is perfectly confined by the long isolating boundaries and the condition (3.160) holds. The magnitude of the current density is determined by the biasing current I and the strip cross section lw. Thus the current density is given by

$$\boldsymbol{J} = \frac{I}{lw}\boldsymbol{u}_{\mathrm{t}} \tag{3.174}$$

where $\boldsymbol{u}_{\mathrm{t}}$ is the unit vector tangential to the insulating boundary (see figure 3.18) and parallel to the x axis in figure 3.1.

Assume again a strongly extrinsic material. Since the current density is known, the electric field in the very long plate is readily given by (3.141). If the magnetic induction is perpendicular to the larger strip face (and thus to the $\boldsymbol{J}$ vector), the third term on the right-hand side vanishes. Let us rewrite the rest of (3.141) in a more compact form:

$$\boldsymbol{E} = \rho_{\mathrm{b}}\boldsymbol{J} - R_{\mathrm{H}}[\boldsymbol{J} \times \boldsymbol{B}]. \tag{3.175}$$

The coefficients r_{b} and R_{H} are given by

$$\rho_{\mathrm{b}} = \frac{1}{1 + \left(\dfrac{e}{m^*}\dfrac{K_2}{K_1}\right)^2 B^2}\,\frac{1}{e^2 K_1} \tag{3.176}$$

$$R_{\mathrm{H}} = \frac{1}{1 + \left(\dfrac{e}{m^*}\dfrac{K_2}{K_1}\right)^2 B^2}\,\frac{1}{em^*}\frac{K_2}{K_1^2}. \tag{3.177}$$

We shall explain below the nature of these two coefficients. Note first that the total electric field $\boldsymbol{E}$ (3.175) consists of two mutually perpendicular components: the longitudinal component $r_{\mathrm{b}}\boldsymbol{J}$, which is collinear with the current density; and the transverse component $-R_{\mathrm{H}}[\boldsymbol{J} \times \boldsymbol{B}]$, which is perpendicular to the current density. We had a similar situation in §3.1.1 (see figure 3.2).

The physical magnetoresistance effect

The longitudinal electric field component $\rho_b \boldsymbol{J}$ in (3.175) is obviously due to the voltage drop along the sample. Thus the coefficient ρ_b plays the role of the usual resistivity. Moreover, at very low magnetic inductions, the coefficient ρ_b reduces to the usual resistivity—this is easy to see using (3.128), (3.129) and (3.176). But ρ_b generally depends on magnetic induction. The change in resistivity of an infinitely long Hall plate under the influence of a perpendicular magnetic field is called the *physical* (or, also, *intrinsic*) *magnetoresistance effect*. The attribute 'physical' is used to stress the fact that this effect is related to some deep physical phenomenon, and to distinguish it from the more obvious geometrical magnetoresistance effect, described in §3.1.3.

The coefficient ρ_b (3.176) we shall call *intrinsic magnetoresistivity* or, in short, *magnetoresistivity*. This coefficient denotes the effective material resistivity of the infinitely long Hall plate along its length, when exposed to a perpendicular magnetic field. Otherwise, this is the material resistivity of a sample *collinear with the current density vector* $\boldsymbol{J}$ when $\boldsymbol{J} \perp \boldsymbol{B}$. Note the difference relative to the definition of σ_B in §3.3.5!

In §3.1.4 we concluded that the magnetoresistance effect exists only if the current deflection effect exists. But in an infinitely long sample, the two transverse components of the Lorentz force, namely the magnetic force and the electric force due to the Hall electric field, are at balance. Hence there is no macroscopic current deflection. So how, in spite of this fact, does a magnetoresistance effect also show up in very long samples?

In order to resolve this apparent paradox, we shall perform an iterative analysis in a similar way as we did in §3.1.4 (see figure 3.9). Now we analyse a very long plate subject to a longitudinal external electric field $\boldsymbol{E}$ and a weak perpendicular magnetic induction $\boldsymbol{B}$. For a reason that will become clear later, we now take into account the fact that charge carriers drift with different drift velocities. To roughly model this, we are assuming that there are just two groups of charge carriers with two different mobilities, μ_+ and μ_-. Let $\mu_+ > \mu_-$. Figure 3.19 illustrates the first cycle of the present iterative procedure. We imagine the current genesis process in the plate as follows (compare with that of §3.1.4, equations (3.37a)–(3.37e)):

1st approximation (figure 3.19(a)): Since $\boldsymbol{B}$ is small, the current density is mostly due to conventional drift, much as in (3.37a). In order to get an idea about the effect of a spread in drift velocities, we model this by representing the drift current by the sum of two components:

$$\boldsymbol{J}_1' = \sigma \boldsymbol{E} = \boldsymbol{J}_{1+} + \boldsymbol{J}_{1-} \qquad \boldsymbol{J}_{1+} = \boldsymbol{J}_{1-} = \tfrac{1}{2}\sigma \boldsymbol{E}. \tag{3.178a}$$

$\boldsymbol{J}_{1+}$ is the current density due to fast drifting carriers (with the mobility μ_+) and $\boldsymbol{J}_{1-}$ is the one due to slow-drifting carriers (with the mobility μ_-).

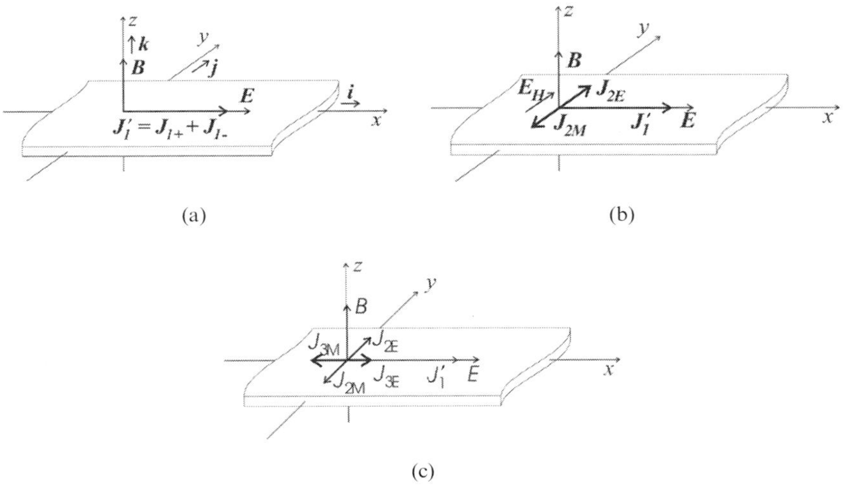

Figure 3.19. (a) Illustrating the first step in the current genesis process in a very long plate exposed to a longitudinal electric field $\boldsymbol{E}$ and a perpendicular magnetic induction $\boldsymbol{B}$. The plate is in the xy plane, $\boldsymbol{E}$ is collinear with the x axis and $\boldsymbol{B}$ is collinear with the z axis. $\boldsymbol{i}$, $\boldsymbol{j}$, $\boldsymbol{k}$ are the unit vectors along the coordinate axes. To the first approximation, only a conventional drift current density $\boldsymbol{J}'_1$ arises. $\boldsymbol{J}_{1+}$ and $\boldsymbol{J}_{1-}$ denote the contributions of fast and slow carriers, respectively. (b) The Lorentz force deflects the current components $\boldsymbol{J}_{1+}$ and $\boldsymbol{J}_{1-}$, which gives rise to the generation of the current components $\boldsymbol{J}_{2\mathrm{M}}$ and $\boldsymbol{J}_{2\mathrm{E}}$. Due to the Hall electric field $\boldsymbol{E}_\mathrm{H}$, the sum $\boldsymbol{J}_2 = \boldsymbol{J}_{2\mathrm{M}} + \boldsymbol{J}_{2\mathrm{E}} = 0$. (c) The Lorentz force deflects the current components $\boldsymbol{J}_{2\mathrm{M}}$ and $\boldsymbol{J}_{2\mathrm{E}}$, which gives rise to the generation of the current components $\boldsymbol{J}_{3\mathrm{M}}$ and $\boldsymbol{J}_{3\mathrm{E}}$. Although $\boldsymbol{J}_{2\mathrm{M}}$ and $\boldsymbol{J}_{2\mathrm{E}}$ are of equal amplitudes, $\boldsymbol{J}_{3\mathrm{M}}$ and $\boldsymbol{J}_{3\mathrm{E}}$ are not. The result is an attenuation of the drift current density $\boldsymbol{J}_1$, which is a magnetoresistance effect.

2nd approximation (figure 3.19(b)): The drifting charge carriers, which make up $\boldsymbol{J}'_1$, also 'feel' the magnetic part of the Lorentz force (3.4). This force tends to produce a current density $\boldsymbol{J}'_2$ (3.37b), shown in figure 3.9(b). But now the action of the magnetic force is balanced by the electric force due to the Hall electric field $\boldsymbol{E}_\mathrm{H}$, so that the transverse current $\boldsymbol{J}'_2$ does not exist:

$$\boldsymbol{J}_2 = \mu_+[\boldsymbol{J}_{1+} \times \boldsymbol{B}] + \mu_-[\boldsymbol{J}_{1-} \times \boldsymbol{B}] + \sigma \boldsymbol{E}_\mathrm{H} = 0. \tag{3.178b-1}$$

We can transform this equation into the following:

$$\begin{aligned} \boldsymbol{J}_2 &= \boldsymbol{J}_{2\mathrm{M}} + \boldsymbol{J}_{2\mathrm{E}} = 0 \\ \boldsymbol{J}_{2\mathrm{M}} &= \mu_+[\boldsymbol{J}_{1+} \times \boldsymbol{B}] + \tfrac{1}{2}\sigma \boldsymbol{E}_\mathrm{H} \\ \boldsymbol{J}_{2\mathrm{E}} &= \mu_-[\boldsymbol{J}_{1-} \times \boldsymbol{B}] + \tfrac{1}{2}\sigma \boldsymbol{E}_\mathrm{H}. \end{aligned} \tag{3.178b-2}$$

So we represent $\boldsymbol{J}'_2$ by two current components, $\boldsymbol{J}_{2\mathrm{M}}$ and $\boldsymbol{J}_{2\mathrm{E}}$. The subscript M means that the magnetic force dominates, and the subscript E means that the electric force dominates. With the aid of (3.9), (3.178b-2) becomes

$$\begin{aligned} \boldsymbol{J}_{2M} &= -\boldsymbol{j}\tfrac{1}{2}\sigma EB\Delta\mu \\ \boldsymbol{J}_{2E} &= \boldsymbol{j}\tfrac{1}{2}\sigma EB\Delta\mu \\ \Delta\mu &= \tfrac{1}{2}(\mu_+ - \mu_-). \end{aligned} \tag{3.178b-3}$$

The two forces, magnetic and electric (due to the Hall field), are really only equal for some 'average' carriers. Faster carriers are more affected by the magnetic force, and slower carriers by the electric force. Because of their distribution over energies, most of the carriers are still deflected from their straight free transits: a microscopic current deflection takes place.

3rd approximation (figure 3.19(c)): Now since the carriers move along the y axis ($\boldsymbol{J}_{2M}$ and $\boldsymbol{J}_{2E}$), they also 'feel' a magnetic Lorentz force collinear with the vector $[\boldsymbol{J} \times \boldsymbol{B}]$, which is collinear with the x axis. They move accordingly, producing a current density $\boldsymbol{J}_3$. In analogy with (3.37c), we find

$$\begin{aligned} \boldsymbol{J}_3 &= \boldsymbol{J}_{3M} + \boldsymbol{J}_{3E} \\ \boldsymbol{J}_{3M} &= \mu_+[\boldsymbol{J}_{2M} \times \boldsymbol{B}] \\ \boldsymbol{J}_{3E} &= \mu_-[\boldsymbol{J}_{2E} \times \boldsymbol{B}]. \end{aligned} \tag{3.178c}$$

Substituting here (3.178b-3), we obtain

$$\boldsymbol{J}_3 = -\mathrm{i}\sigma\boldsymbol{E}(\Delta\mu\boldsymbol{B})^2. \tag{3.178d}$$

Now since $\boldsymbol{J}_3$ is collinear with the x axis, it corrects the first approximation of $\boldsymbol{J}_1$ (3.178a). This closes our first iteration cycle and starts the second. The first step in the second cycle is the second approximation of $\boldsymbol{J}_1$, given by:

$$\boldsymbol{J}_1'' = \boldsymbol{J}_1' + \boldsymbol{J}_3 = \sigma\boldsymbol{E}(1 - (\Delta\mu\boldsymbol{B})^2). \tag{3.178e}$$

This equation can be re-written in the form of Ohm's law:

$$\boldsymbol{E} \simeq \rho_b\boldsymbol{J} \tag{3.179}$$

where resistivity is given by

$$\rho_b \simeq \frac{1}{\sigma}[1 + (\Delta\mu\boldsymbol{B})^2]. \tag{3.180}$$

Equations (3.178e) and (3.180) show the first approximation of the intrinsic magnetoresistance effect. Now we see intuitively how the *physical magnetoresistance* effect comes about: the physical magnetoresistance effect can be explained as an opposition of the current density component $\boldsymbol{J}_3$ to the main drift current density $\boldsymbol{J}_1$—see figure 3.19(c) and (3.178e). The current density component $\boldsymbol{J}_3$ is a consequence of the existence of a spread in carrier mobility. Slower carriers are deflected in one direction and faster carriers in the other, see figure 3.19(b). Although the deflection currents $\boldsymbol{J}_{2M}$ and $\boldsymbol{J}_{2E}$ on average cancel, together with the magnetic field they generate a non-zero reaction current $\boldsymbol{J}_3$ (3.178d). A consequence is an attenuation of the drift current component dependent on magnetic induction (3.178e) and an increase in resistivity (3.180). This is the physical magnetoresistance effect.

A more accurate (but less instructive) description of the physical magnetoresistance effect is contained in the expression for ρ_b (3.176). In order to see why ρ_b depends on magnetic field, let us find an approximation of (3.176) for very low magnetic fields. To this end, we can substitute in (3.176) the low-field values of the kinetic coefficients (3.128) and (3.129). By neglecting all the terms of higher orders than $\boldsymbol{B}^2$ and using (3.144), we obtain

$$\rho_b \simeq \frac{1}{\sigma}\left[1 + \left(\frac{\langle\tau^3\rangle}{\langle\tau\rangle^3} - \frac{\langle\tau^2\rangle^2}{\langle\tau\rangle^4}\right)(\mu\boldsymbol{B})^2\right]. \tag{3.181}$$

This equation shows that, if the relaxation time depends on electron energy, then the resistivity ρ_b depends on the magnetic field. Otherwise, this is not the case. Note the similarity between (3.180) and (3.181). We see that the term

$$\left(\frac{\langle\tau^3\rangle}{\langle\tau\rangle^3} - \frac{\langle\tau^2\rangle^2}{\langle\tau\rangle^4}\right)^{1/2}\mu \sim \Delta\mu \tag{3.182}$$

figuring in (3.181) expresses the spread of carrier mobilities over energy.

If the relaxation time does not depend on electron energy, then, to the first approximation, the resistivity ρ_b does not depend on the magnetic field either. For example, in a metal, conductivity depends mostly on the electrons at the Fermi level (see (3.122)). Within a small energy domain around the Fermi level, the mobility cannot vary much. Therefore, we conclude that the *intrinsic magnetoresistance effect in metals must be very weak*. Indeed, by substituting (3.123) into (3.176), we obtain a ρ_b of the same form as in (2.91), which means that then (within the present approximations) resistivity does not depend on magnetic field. Experimental facts corroborate this conclusion.

Note that we treated here the physical magnetoresistance effect under the assumption that the magnetic field is perpendicular to the plate plane. Otherwise, the magnetoresistance effect depends also on the magnetic field direction (see §3.3.11).

The Hall effect

The transverse component of the electric field in (3.175), the term $-R_H[\boldsymbol{J} \times \boldsymbol{B}]$, is nothing more than the *Hall electric field*:

$$\boldsymbol{E}_H = -R_H[\boldsymbol{J} \times \boldsymbol{B}]. \tag{3.183}$$

This becomes obvious by a comparison of the diagrams in figures 3.19 and 3.2, and equations (3.183) and (3.65). Moreover, our earlier expression for the Hall coefficient R_H (3.17) is an approximation of the present accurate expression (3.177). In fact, using (3.176) and (3.169), the *Hall coefficient* (3.177) can be expressed as

$$R_{\mathrm{H}} = \mu_{\mathrm{H}}\rho_{\mathrm{b}}. \tag{3.184}$$

At $\boldsymbol{B} \sim 0$, we may substitute here the zero-field expression for the Hall mobility (3.171) and resistivity to obtain

$$R_{\mathrm{H}}(B \simeq 0) \simeq \mathrm{sign}[e]\frac{r_{\mathrm{H}}}{qn} \tag{3.185}$$

where r_{H} is the Hall scattering factor (3.172). Since $r_{\mathrm{H}} \sim 1$, (3.185) is approximately equal to (3.17). We shall discuss further details of the Hall coefficient in §3.4.

Another useful relation for the Hall electric field follows from (3.183) by substituting the Hall coefficient there by (3.184). Since the product $\rho_{\mathrm{b}}\boldsymbol{J}$ determines the longitudinal electric field, we may write

$$\boldsymbol{E}_{\mathrm{H}} = -\mu_{\mathrm{H}}[\boldsymbol{E}_{\mathrm{e}} \times \boldsymbol{B}]. \tag{3.186}$$

The electric field E_{e} is parallel to the long strip axis and is simply the externally applied field. Thus equation (3.186) corresponds to the approximate equation (3.9).

As shown in figure 3.19, the total electric field is not collinear here with the current density. Taking as before the total electric field direction as a reference, the *Hall angle* is given by

$$\tan\Theta_{\mathrm{H}} = R_{\mathrm{H}}B/\rho_{\mathrm{b}} = \mu_{\mathrm{H}}B \tag{3.187}$$

where we have made use of (3.184). In the earlier analysis, this expression was approximated by (3.14). By comparing (3.187) with (3.152), which holds for an infinitely short sample, we see that the two expressions are identical. This confirms that the Hall angle is a local galvanomagnetic quantity. It does not depend on sample geometry.

The integral quantities pertinent to a very long Hall device, such as the Hall voltage, the bias voltage and the device resistance, can be calculated in similar way as we did in §3.1.1: see Chapter 4.

3.3.7 The galvanomagnetic effects in a mixed conductor

Up to now we have always assumed that the galvanomagnetic devices we analysed contained only one kind of charge carrier. Now we shall extend the analyses to the case when a material contains several kinds of carrier. Such a material might be a non-extrinsic semiconductor, containing both electrons and holes, or a material with carriers of a single charge type, but with different effective masses. This appears, for example, in p-type silicon: in silicon, there are light and heavy holes.

The current density in a mixed conductor is given by (3.140), which we repeat:

$$\boldsymbol{J} = \sum_c \boldsymbol{J}_{\mathrm{c}} \tag{3.188}$$

where $\boldsymbol{J}_{\mathrm{c}}$ is the current density component due to carriers of the kind c. Substituting $\boldsymbol{J}_{\mathrm{c}}$ here by (3.149), we obtain

$$\boldsymbol{J} = \left(\sum_c \sigma_{\mathrm{Bc}}\right)\boldsymbol{E} + \left(\sum_c \sigma_{\mathrm{Bc}}\mu_{\mathrm{Hc}}\right)[\boldsymbol{E} \times \boldsymbol{B}] \tag{3.189}$$

where σ_{Bc} and μ_{Hc} are the coefficients, given by (3.150) and (3.151), of the c kind of carriers. This is a general equation for the galvanomagnetic effects in substances with several kinds of carriers. All particular forms of the Hall effect and the magnetoresistance effect can be elaborated by solving (3.189) subject to the appropriate boundary conditions. We shall apply this equation to analyse the Hall effect in samples of the two idealized geometries discussed above. For the sake of simplicity, we assume a very weak magnetic field.

Infinitely short samples. Average Corbino conductivity and mobility

Equation (3.189) can be written as

$$\boldsymbol{J} = \bar{\sigma}_{\mathrm{B}}\boldsymbol{E} + \bar{\sigma}_{\mathrm{B}}\bar{\mu}_{\mathrm{H}}[\boldsymbol{E} \times \boldsymbol{B}] \tag{3.190}$$

where we have introduced the notation

$$\bar{\sigma}_{\mathrm{B}} = \sum_c \sigma_{\mathrm{Bc}} \tag{3.191}$$

$$\bar{\mu}_{\mathrm{H}} = \left(\sum_c \sigma_{\mathrm{Bc}}\mu_{\mathrm{Hc}}\right) \Big/ \left(\sum_c \sigma_{\mathrm{Bc}}\right). \tag{3.192}$$

The coefficient $\bar{\sigma}_{\mathrm{B}}$ is the total Corbino conductivity and $\bar{\mu}_{\mathrm{H}}$ is the average Hall mobility of the mixed conductor. We see that equation (3.190) is of the same form as (3.165). Therefore, in a mixed conductor, macroscopically everything happens as in a material with a single kind of carrier; only the coefficients are given by (3.191) and (3.192). In particular, the deflection angle of the total current is given by

$$\tan \bar{\Theta}_{\mathrm{H}} = \bar{\mu}_{\mathrm{H}} B. \tag{3.193}$$

However, the deflection of each current component is fully independent of each other and might be drastically different than that of the total current.

Consider specifically a semiconductor containing electrons and holes, one kind of each. Let us denote their concentrations by n and p, respectively. At $\boldsymbol{B} \sim 0$, the conductivity $\bar{\sigma}_{\mathrm{B}}$ reduces to (2.94):

$$\bar{\sigma}_{\mathrm{B}} \simeq qn\mu_{\mathrm{n}} + qp\mu_{\mathrm{p}} \tag{3.194}$$

and the average Hall mobility is given by

$$\bar{\mu}_{\mathrm{H}} \simeq \frac{\mu_{\mathrm{n}}\mu_{\mathrm{Hn}}n + \mu_{\mathrm{p}}\mu_{\mathrm{Hp}}p}{\mu_{\mathrm{n}}n + \mu_{\mathrm{p}}p}. \tag{3.195}$$

Substituting here (3.171), we have

$$\bar{\mu}_{\mathrm{H}} \simeq \frac{1 - sb^2x}{1 + bx}\mu_{\mathrm{Hp}} \tag{3.196}$$

with

$$s = \frac{r_{\mathrm{Hn}}}{r_{\mathrm{Hp}}} \qquad b = \frac{\mu_{\mathrm{n}}}{\mu_{\mathrm{p}}} \qquad x = \frac{n}{p} \tag{3.197}$$

where r_{Hp} and r_{Hn} are the Hall scattering factors for holes and electrons, respectively. The average Hall mobility is smaller than that of faster carriers. Consequently, the deflection of the total current in a mixed conductor is always inferior in comparison with that of the fastest carriers. Moreover, if in an extrinsic semiconductor the condition

$$\frac{n}{p} = \frac{r_{\mathrm{Hn}}}{r_{\mathrm{Hp}}}\left(\frac{\mu_{\mathrm{n}}}{\mu_{\mathrm{p}}}\right)^2 \tag{3.198}$$

is fulfilled, the deflection of the total current reduces to zero.

Infinitely long samples. Average magnetoresistivity and Hall coefficient

Suppose that the total current density is known. To find the electric field, we have to solve equation (3.189) with respect to $\boldsymbol{E}$. We can do this in the same way as before (see (3.155) and (3.156)) and obtain

$$\boldsymbol{E} = \frac{\dfrac{1}{\sum_c \sigma_{\mathrm{Bc}}}\boldsymbol{J} - \dfrac{\sum_c \sigma_{\mathrm{Bc}}\mu_{\mathrm{Hc}}}{(\sum_c \sigma_{\mathrm{Bc}})^2}[\boldsymbol{J} \times \boldsymbol{B}]}{1 + \left(\dfrac{\sum_c \sigma_{\mathrm{Bc}}\mu_{\mathrm{Hc}}}{\sum_c \sigma_{\mathrm{Bc}}}\right)^2 B^2}. \tag{3.199}$$

In analogy with (3.175), we can rewrite this equation as

$$\boldsymbol{E} = \bar{\rho}_{\mathrm{B}}\boldsymbol{J} - \bar{R}_{\mathrm{H}}[\boldsymbol{J} \times \boldsymbol{B}] \tag{3.200}$$

where

$$\bar{\rho}_{\mathrm{b}} = \left\{\sum_c \sigma_{\mathrm{Bc}}\left[1 + \left(\frac{\sum_c \sigma_{\mathrm{Bc}}\mu_{\mathrm{Hc}}}{\sum_c \sigma_{\mathrm{Bc}}}\right)^2 B^2\right]\right\}^{-1} \tag{3.201}$$

is the *average magnetoresistivity*, and

$$\bar{R}_{\mathrm{H}} = \frac{\sum_c \sigma_{\mathrm{Bc}}\mu_{\mathrm{Hc}}}{\left(\sum_c \sigma_{\mathrm{Bc}}\right)^2\left[1 + \left(\dfrac{\sum_c \sigma_{\mathrm{Bc}}\mu_{\mathrm{Hc}}}{\sum_c \sigma_{\mathrm{Bc}}}\right)^2 B^2\right]} \tag{3.202}$$

is the *average Hall coefficient of the mixed-conductivity material.*

Consider again a non-extrinsic semiconductor with one kind of each carrier type. At $\boldsymbol{B} \sim 0$, $\bar{\rho}_0$ (3.201) reduces to the usual resistivity (2.95), namely

$$\bar{\rho}_{\mathrm{b}}(B \simeq 0) = \frac{1}{qn\mu_{\mathrm{n}} + qp\mu_{\mathrm{p}}} \tag{3.203}$$

and the average Hall coefficient (3.202) becomes

$$\bar{R}_{\mathrm{H}}(B \simeq 0) = \frac{\sigma_{\mathrm{Bn}}\mu_{\mathrm{Hn}} + \sigma_{\mathrm{Bp}}\mu_{\mathrm{Hp}}}{(\sigma_{\mathrm{Bn}} + \sigma_{\mathrm{Bp}})^2}. \tag{3.204}$$

With reference to (3.195) and (3.203), this can also be put in the following compact form:

$$\bar{R}_{\mathrm{H}}(B \simeq 0) = \bar{\mu}_{\mathrm{H}}\bar{\rho}_{\mathrm{b}} \tag{3.205}$$

which is analogous to (3.184). Using (3.196), the Hall coefficient can also be put in this form:

$$\bar{R}_{\mathrm{H}}(B \simeq 0) = \frac{1 - sb^2x}{(1 + bx)^2} \frac{r_{\mathrm{Hp}}}{qp} \tag{3.206}$$

where the parameters s, b and x are given by (3.197).

Obviously, the Hall coefficient of a mixed conductor is smaller than if all the carriers were of the faster type. This is because the magnetic force deflects both holes and electrons towards the same boundary of a sample (see figure 3.1), so carriers of the opposite types compensate each other's charge, and the Hall field is diminished. If the condition (3.198) is fulfilled, the Hall field even vanishes altogether. In fact, according to (3.206),

$$\bar{R}_{\mathrm{H}}(B \simeq 0) = 0 \quad \text{if } sb^2x = \frac{r_{\mathrm{Hp}}}{r_{\mathrm{Hn}}}\left(\frac{\mu_{\mathrm{p}}}{\mu_{\mathrm{n}}}\right)^2 \frac{p}{n} = 1 \tag{3.207}$$

and then also $\boldsymbol{E}_{\mathrm{H}} = 0$.

3.3.8 The Hall effect in a many-valley semiconductor

In the study of the transport phenomena we have used up to now a simple model of a semiconductor: we assumed a single energy minimum in an energy band, and an isotropic effective mass of charge carriers. However, for some important semiconductors, such as silicon and germanium, this model gives only a rough approximation of their real behaviour. In the conduction band of these semiconductors there are several equivalent energy minima, or valleys. The electrons are equally distributed among these valleys. The effective mass of an electron associated with a valley depends on the acceleration direction (see §2.2.2): the effective mass of an electron in a many-valley semiconductor is anisotropic. Moreover, the scattering probability of an electron depends on the direction of its motion: electron scattering in a many-valley semiconductor is also anisotropic. In spite of this, a transport property of a many-valley semiconductor may be isotropic. This is the case with cubic crystals, such as silicon and germanium. A quantity averaged over cubic symmetrically arranged valleys turns out to be direction independent.

Transport phenomena in many-valley semiconductors were first studied by Herring [14] and Herring and Vogt [11]. Their results were later confirmed and refined by several people [15–17]. Some results of the Herring–Vogt transport theory for electrons in silicon [11] are summarized below.

Mobility, Hall factor, Hall coefficient

The drift mobility of electrons in silicon is given by

$$\mu = q\frac{1}{3}\left(2\frac{\langle\tau_t\rangle}{m_t^*} + \frac{\langle\tau_l\rangle}{m_l^*}\right) \tag{3.208}$$

where m_l and m_t are the longitudinal and transverse effective masses of an electron, respectively, defined in §2.2.2, and $\langle\tau_l\rangle$ and $\langle\tau_t\rangle$ are the energy-weighted averages of the relaxation times of electrons moving in the longitudinal and transverse directions, respectively. The longitudinal direction is parallel to the rotation axis of a constant energy ellipsoid, and the transverse direction is perpendicular to the rotation axis. In silicon, $\langle\tau_l\rangle \sim \langle\tau_t\rangle$. By substituting these two relaxation times by their average value $\langle\tau\rangle$, (3.208) reduces to

$$\mu = q\langle\tau\rangle\frac{1}{3}\left(\frac{2}{m_t^*} + \frac{1}{m_l^*}\right). \tag{3.209}$$

We have already cited this equation before (see (3.147) and (3.148)).

The Hall mobility of electrons in silicon can be expressed, in analogy with (3.171), as

$$\mu_H = -r_H\mu. \tag{3.210}$$

Here μ is the drift mobility (3.208) and the term r_H is called simply the *Hall factor* (without 'scattering'). Thus the Hall factor now replaces the Hall scattering factor. The two factors are approximately equal numbers of about unity. The Hall factor is given by

$$r_H = 3\left(\frac{\langle\tau_t\tau_l\rangle}{m_t^* m_l^*} + \frac{\langle\tau_t^2\rangle}{m_t^{*2}}\right) \Bigg/ \left(\frac{\langle\tau_l\rangle}{m_l^*} + 2\frac{\langle\tau_t\rangle}{m_t^*}\right)^2 \tag{3.211}$$

with the same notation as above. This indicates why the attribute 'scattering' is omitted in the name of r_H: it differs from unity not only because of scattering, but also because of anisotropy.

If $\langle\tau_l\rangle \sim \langle\tau_t\rangle \sim \langle\tau\rangle$, this expression attains the form

$$r_H = r_{H0}a \tag{3.212}$$

where r_{H0} is the Hall scattering factor for the isotropic model given by (3.172), and a is the anisotropy factor of the Hall effect. We shall call it the anisotropy factor for short. The anisotropy factor is given by

$$a = \frac{3K(2+K)}{(1+2K)^2} \tag{3.213}$$

with

$$K = m_l^*/m_t^*. \tag{3.214}$$

For electrons in silicon, $K \sim 4.7$ and $a \sim 0.87$ [18]. At a high magnetic induction, the anisotropy factor approaches unity.

The Hall coefficient of strongly extrinsic n-type silicon at low fields is given by

$$R_H = -r_{H0}a/qn. \tag{3.215}$$

Thus the Hall mobility of electrons in silicon and the Hall coefficient of n-type silicon should be slightly smaller than those predicted by the simple isotropic model. The reduction is due to the anisotropy effects and amounts to about 13%.

Equations (3.210), (3.212) and (3.215) indicate why up to now we have omitted the subscript 0 in r_{H0}. In this way, equations involving r_H such as (3.215) are generally valid, irrespective of whether r_H is due only to scattering or also to anisotropy.

Modelling anisotropy

To obtain a physical insight of the influence of anisotropy on the Hall effect, we shall now construct a simple hybrid model of electrons in silicon. As in §3.1, we shall model electrons by charged macroscopic particles immersed in a viscous liquid: we assume that electrons move only because of the action of the Lorentz forces, and we substitute the effect of the thermal motion and scattering of electrons by a smooth friction. But we also take into account that the electrons are distributed among the six energy valleys, and that the mobility of each electron depends on the direction of its acceleration. Specifically, in analogy with isotropic mobility (2.86), we introduce a longitudinal and transverse mobility, defined by

$$\mu_l = q\frac{\langle\tau_l\rangle}{m_l^*} \qquad \mu_t = q\frac{\langle\tau_t\rangle}{m_t^*} \tag{3.216}$$

respectively.

Consider now the Hall effect in an infinitely long, strongly extrinsic n-type silicon strip. We adopt the geometry, which we used in §3.1.1, shown

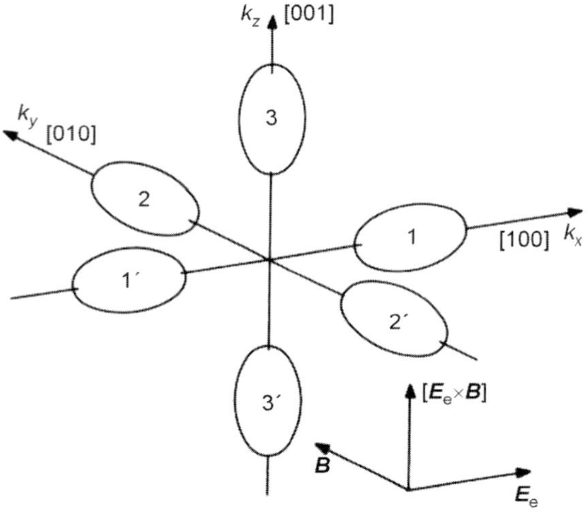

Figure 3.20. Constant energy surfaces of electrons in silicon. 1 and 1′, 2 and 2′, and 3 and 3′ are the pairs of equivalent constant energy ellipsoids. The crystal is exposed to crossed electric (***E***) and magnetic (***B***) fields of the shown directions.

in figure 3.1. In addition, we assume that the large face of the strip (the xz plane) coincides with the (010) crystal plane. The corresponding arrangement of the ellipsoidal constant energy surfaces of electrons in the $\boldsymbol{k}$ space is sketched in figure 3.20. We shall refer to the twin ellipsoids 1 and 1′ and the corresponding valleys as ellipsoids 1 or valleys 1, and so on: ellipsoids 2 and 2′ as 2, and 3 and 3′ as 3. The three coordinate axes k_x, k_y and k_z are parallel to the three main crystal directions ⟨100⟩. The rotation axis of the ellipsoids 1 is parallel to [100], the rotation axis of 2 is parallel to [010], and the rotation axis of 3 is parallel to [001]. The directions of the driving fields are also shown: $\boldsymbol{E} \parallel [100]$ and $\boldsymbol{B} \parallel [010]$.

As in the case of a mixed conductor (3.188), the total current density in the sample is given by the sum of the current densities due to various groups of electrons:

$$\boldsymbol{J} = \sum_{1}^{M_v} \boldsymbol{J}_v. \tag{3.217}$$

Here $\boldsymbol{J}_v$ is the current density due to the electrons pertinent to the v valley, and M_v is the number of valleys. For silicon, $M_v = 6$. In analogy with (2.87), the drift current density due to the electrons of the vth valley is given by

$$\boldsymbol{J}_v = -qn_v\boldsymbol{v}_v \tag{3.218}$$

where n_v is the concentration of electrons pertinent to the v valley, and $\boldsymbol{v}_v$ is their drift velocity. In silicon, all six valleys are equivalent, and consequently $n_v = \frac{1}{6}n$, n being the total electron density.

Consider now the drift current density in the direction of the external field $\boldsymbol{E}_e$. In analogy with (2.85), and with reference to (3.216) and figure 3.20, the drift velocities in the x direction are given by

$$\boldsymbol{v}_{1x} = -\mu_l \boldsymbol{E}_e \qquad \boldsymbol{v}_{2x} = \boldsymbol{v}_{3x} = -\mu_t \boldsymbol{E}_e \tag{3.219}$$

for the electrons of valleys 1, 2 and 3, respectively. From (3.217)–(3.219) we readily obtain the total drift current density along the x axis:

$$\boldsymbol{J}_x = q\frac{n}{3}(\mu_l + 2\mu_t)\boldsymbol{E}_e. \tag{3.220}$$

If we introduce an average value of the anistropic mobilities, defined by

$$\mu = \tfrac{1}{3}(\mu_l + 2\mu_t) \tag{3.221}$$

equation (3.220) becomes fully analogous with that for carriers in an isotropic band (2.88). When we substitute (3.216) into (3.221), the average mobility attains the form

$$\mu = \frac{2}{3}\left(\frac{\langle\tau_l\rangle}{m_l^*} + 2\frac{\langle\tau_t\rangle}{m_t^*}\right) \tag{3.222}$$

which is identical with the result (3.209) of the accurate analysis. It is easy to show that the average mobility μ (3.221) is invariant with the transport direction.

The current in the transverse direction is due to the Lorentz force component parallel with the z axis. When we substitute the appropriate terms into (3.1), the Lorentz force attains the form

$$\boldsymbol{F}_z = e\boldsymbol{E}_H + e[\boldsymbol{v}_x \times \boldsymbol{B}]. \tag{3.223}$$

Here $\boldsymbol{E}_H$ is the Hall electric field, which is as yet unknown, and $\boldsymbol{v}_x$ is the drift velocity of an electron in the x direction. It is convenient to substitute the Lorentz force $\boldsymbol{F}_z$ by an equivalent electrical force $e\boldsymbol{E}'_z$, $\boldsymbol{E}'_z$ being an equivalent electric field parallel with the z axis. Using (3.223), the equivalent electric field is given by

$$\boldsymbol{E}'_z = \boldsymbol{E}_H + [\boldsymbol{v}_x \times \boldsymbol{B}]. \tag{3.224}$$

Substituting here (3.219), we obtain

$$\begin{aligned} \boldsymbol{E}'_{z1} &= \boldsymbol{E}_H - \mu_l[\boldsymbol{E}_e \times \boldsymbol{B}] \\ \boldsymbol{E}'_{z2} &= \boldsymbol{E}'_{z3} = \boldsymbol{E}_H - \mu_t[\boldsymbol{E}_e \times \boldsymbol{B}]. \end{aligned} \tag{3.225}$$

Using again the analogy with (2.85), and with reference to figure 3.19, the drift velocities of electrons in the direction of the z axis are given by

$$\boldsymbol{v}_{1z} = -\mu_t \boldsymbol{E}'_{z1} \qquad \boldsymbol{v}_{2z} = -\mu_t \boldsymbol{E}'_{z2} \qquad \boldsymbol{v}_{3z} = -\mu_l \boldsymbol{E}'_{z3} \tag{3.226}$$

for the electrons of valleys 1, 2 and 3, respectively. From (3.218), (3.226) and (3.225), we obtain the current components along the z axis:

$$\begin{aligned} \boldsymbol{J}_{z1} &= q\frac{n}{3}\{\boldsymbol{E}_{\mathrm{H}} - \mu_{\mathrm{l}}[\boldsymbol{E}_{\mathrm{e}} \times \boldsymbol{B}]\} \\ \boldsymbol{J}_{z2} &= q\frac{n}{3}\{\boldsymbol{E}_{\mathrm{H}} - \mu_{\mathrm{t}}[\boldsymbol{E}_{\mathrm{e}} \times \boldsymbol{B}]\} \\ \boldsymbol{J}_{z3} &= q\frac{n}{3}\{\boldsymbol{E}_{\mathrm{H}} - \mu_{\mathrm{t}}[\boldsymbol{E}_{\mathrm{e}} \times \boldsymbol{B}]\} \end{aligned} \tag{3.227}$$

due to the electrons of valleys 1, 2 and 3, respectively. But the total current density parallel with the z axis must vanish:

$$\boldsymbol{J}_z = \boldsymbol{J}_{z1} + \boldsymbol{J}_{z2} + \boldsymbol{J}_{z3} = 0 \tag{3.228}$$

for there can be no transverse macroscopic current in an infinitely long strip. By substituting (3.227) into (3.228), we obtain the *Hall electric field*,

$$\boldsymbol{E}_{\mathrm{H}} = \frac{\mu_{\mathrm{t}}(2\mu_{\mathrm{l}} + \mu_{\mathrm{t}})}{\mu_{\mathrm{l}} + 2\mu_{\mathrm{t}}}[\boldsymbol{E}_{\mathrm{e}} \times \boldsymbol{B}]. \tag{3.229}$$

Comparing this equation with (3.186), we conclude that the term in front of the vector product in (3.229) plays the role of the Hall mobility. Thus the *Hall mobility* is given by

$$\mu_{\mathrm{H}} = \frac{\mu_{\mathrm{t}}(2\mu_{\mathrm{l}} + \mu_{\mathrm{t}})}{\mu_{\mathrm{l}} + 2\mu_{\mathrm{t}}}. \tag{3.230}$$

It is customary to relate the Hall mobility to the drift mobility. In analogy with (3.171), we may write

$$\mu_{\mathrm{H}} = -r_{\mathrm{H}}\mu \tag{3.231}$$

where r_{H} is the Hall factor. Using (3.230) and (3.222), the *Hall factor* may be expressed as

$$r_{\mathrm{H}} = \frac{3\mu_{\mathrm{t}}(2\mu_{\mathrm{l}} + \mu_{\mathrm{t}})}{(\mu_{\mathrm{l}} + 2\mu_{\mathrm{t}})^2}. \tag{3.232}$$

By making use of (3.216), this becomes

$$r_{\mathrm{H}} = 3\left(2\frac{\langle\tau_{\mathrm{t}}\rangle\langle\tau_{\mathrm{l}}\rangle}{m_{\mathrm{l}}^* m_{\mathrm{t}}^*} + \frac{\langle\tau_{\mathrm{t}}\rangle^2}{m_{\mathrm{l}}^*}\right) \Big/ \left(\frac{\langle\tau_{\mathrm{l}}\rangle}{m_{\mathrm{l}}^*} + 2\frac{\langle\tau_{\mathrm{t}}\rangle}{m_{\mathrm{t}}^*}\right)^2. \tag{3.233}$$

This is very similar to the accurate expression (3.211). The only difference between the two expressions is in the order of the multiplication and the averaging of the terms $\tau_{\mathrm{t}}\tau_{\mathrm{l}}$ and $\tau_{\mathrm{t}}\tau_{\mathrm{t}}$. In the present simple model, we operate with the average mobilities. Hence the multiplication comes after the averaging.

The Hall electric field can also be expressed as a function of the current density. By making use of (3.220), (3.229) becomes

$$\boldsymbol{E}_{\mathrm{H}} = \frac{3\mu_{\mathrm{t}}(2\mu_{\mathrm{l}} + \mu_{\mathrm{t}})}{(\mu_{\mathrm{l}} + 2\mu_{\mathrm{t}})^2}\frac{1}{qn}[\boldsymbol{J} \times \boldsymbol{B}]. \tag{3.234}$$

Comparing this equation with (3.183), we conclude that the term multiplying the vector product in (3.234) is the Hall coefficient. Thus the *Hall coefficient* of strongly extrinsic *n-type silicon* is given by

$$R_H = -r_H/qn \tag{3.235}$$

where r_H denotes the Hall factor given by (3.232). This result is a good approximation of the accurate expression (3.215).

So far, we have assumed a low magnetic induction. With the increase in magnetic induction, the influence of the anisotropy decreases and eventually vanishes. This is because, at higher magnetic inductions, an electron travels along a longer portion of a cycloid (see §3.1.6). While travelling along a cycloid, the direction of the electron acceleration changes. Thus in a single free transit, the effective mass of an electron varies. The kinetic properties of each electron are then determined by an average effective mass, irrespective of the valley to which it belongs.

Physical insight

Let us summarize. The fundamental consequence of the existence of many valleys in an energy band of a semiconductor is the presence of several groups of charge carriers all of the same type but with different kinetic properties in various directions. When exposed to an electric field, each group of carriers drifts with a different drift velocity. In crossed electric and magnetic fields, the magnetic forces acting on carriers moving with different velocities are also different. On the other hand, the electrical forces are equal for all carriers. Therefore, in a long sample the Hall electric field cannot balance the magnetic forces for each group of carriers. For carriers drifting faster as a result of the external electric field, the Hall field is too weak to balance the magnetic forces, and these carriers also drift in the direction of resultant magnetic forces. For carriers drifting slower, the Hall field overcompensates the action of the magnetic forces, and these carriers drift also as a result of the Hall electric field. In a way, in crossed electric and magnetic fields, carriers are separated according to their effective masses. Therefore, the condition of a zero transverse current does not hold in detail. To be sure, the resultant transverse current does not exist, but the transverse current density components in the opposite directions do. The existence of the transverse current components means the existence of a kind of current deflection effect. The result is a contribution to the physical magnetoresistance effect, discussed in §3.3.6.

The question now arises, how is it that continuous transverse currents flow in an infinitely long sample in spite of the isolating boundaries? Where do the carriers come from and where do they go to? One can explain this paradox by taking into account the intervalley scattering of carriers. This scattering process always works towards the re-establishment of equilibrium. Let us assume that the carriers associated with a valley drift towards a sample boundary. In due course, this valley will be more populated than the other valleys in the vicinity of the boundary. Then, at a collision with a high-energy phonon, a carrier of the more populated valley is more likely to be transferred to a depleted valley than vice versa. Thus in a many-valley band of a semiconductor exposed to crossed electric and magnetic fields, a carrier generally moves along a zigzag route. In addition to the drift in the longitudinal direction arising from the external electric field, the carrier also drifts in the transverse direction. When it arrives at an insulating boundary, such a carrier will probably jump into another valley and start to transversely drift towards the opposite boundary.

3.3.9 The planar Hall effect

In 1954 Goldberg and Davis reported a 'new galvanomagnetic effect' [19], which they named the planar Hall effect. They observed the effect by measuring the induced voltage normal to the direction of current flow as in the conventional Hall effect, but with the magnetic field in the current–voltage plane.

Let us consider the planar Hall effect in a very long Hall plate—see figure 3.21). The plate boundaries confine the current distribution in the plate so that only the longitudinal current density component $\boldsymbol{J}_1$ exists, $\boldsymbol{J} = \boldsymbol{J}_1$. Then the electric field in the plate is readily given by (3.141). Accordingly, the electric field consists of three components. The first component, the term proportional with the current density $\boldsymbol{J}$ in (3.141), is obviously the longitudinal external electric field $\boldsymbol{E}$. The second component, the term proportional to the vector $[\boldsymbol{J} \times \boldsymbol{B}]$, is the conventional Hall electric field $\boldsymbol{E}_H$. In the present case, the Hall electric field is oriented across the thickness of the plate (i.e. it is collinear with the y axis in figure 3.21) and, therefore,

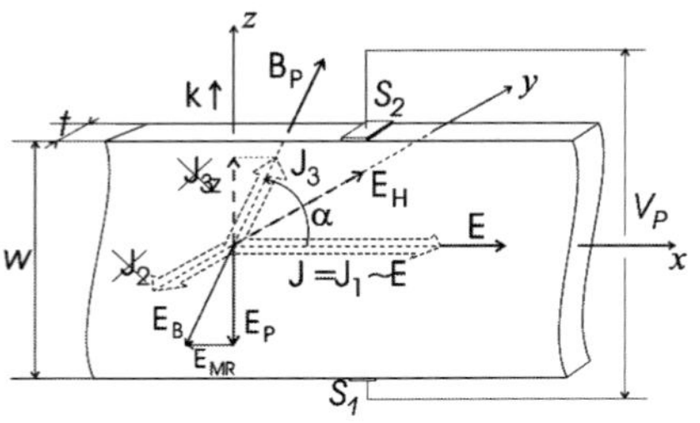

Figure 3.21. Illustrating the planar Hall effect in a very long plate. The plate is in the xz plane. The plate is exposed to a longitudinal external electric field $\boldsymbol{E}$ and an in-plane magnetic induction $\boldsymbol{B}_P$. Due to the plate boundaries, the current components $\boldsymbol{J}_2$ and $\boldsymbol{J}_{3Z}$ do not exist—compare with figure 3.15. The current component $\boldsymbol{J}_{3Z}$ is suppressed by the planar Hall electric field $\boldsymbol{E}_P$. The output voltage is the planar Hall voltage $V_P = \boldsymbol{w} \cdot \boldsymbol{E}_P$.

produces no voltage at the output terminals S1, S2 of the Hall plate. The third component of the total electric field (3.141), which is the third term, collinear with the magnetic induction $\boldsymbol{B}$, may be expressed in this form:

$$\boldsymbol{E}_{\mathrm{B}} = P_{\mathrm{H}}(\boldsymbol{J} \cdot \boldsymbol{B})\boldsymbol{B} \tag{3.236}$$

where we introduced notation

$$P_{\mathrm{H}} = \frac{1}{1 + \left(\dfrac{e}{m^*}\dfrac{K_2}{K_1}\right)^2 B^2} \frac{\left(\dfrac{e}{m^*}\dfrac{K_2}{K_1}\right)^2 - \dfrac{e^2}{m^{*2}}\dfrac{K_3}{K_1}}{e^2 K_1 \left(1 + \dfrac{e^2}{m^{*2}}\dfrac{K_3}{K_1}B^2\right)}. \tag{3.237}$$

The coefficient P_{H} is called the *planar Hall coefficient.* Note the analogy between the definitions of the 'normal' Hall coefficient R_{H} (3.16), (3.185), and that of the planar Hall coefficient P_{H} (3.236).

The vector $\boldsymbol{E}_{\mathrm{B}}$ (3.236) can be decomposed into two components: a transverse one and a longitudinal one. We shall discuss the longitudinal component E_{MR} later in §3.3.11. The transverse component is called the *planar Hall electric field* and is given by

$$\boldsymbol{E}_{\mathrm{P}} = (\boldsymbol{E}_{\mathrm{B}} \cdot \boldsymbol{k})\boldsymbol{k} = P_{\mathrm{H}} J B_{\mathrm{P}}^2 \sin\alpha \cos\alpha \boldsymbol{k} \tag{3.238}$$

where $\boldsymbol{k}$ is the unit vector along the plate width (here collinear with the z axis), B_{P} denotes the component of the magnetic induction parallel with the plate, and α is the angle enclosed by the vectors $\boldsymbol{B}_{\mathrm{P}}$ and $\boldsymbol{J}$. The planar Hall electric field $\boldsymbol{E}_{\mathrm{P}}$ suppresses the transverse current density component $\boldsymbol{J}_{3\mathrm{Z}}$ (see figure 3.21). The planar Hall electric field also generates a transverse output voltage called the planar Hall voltage (see §4.1.1).

The planar Hall coefficient

At small magnetic fields, we can neglect the terms with B^2 in (3.237), so that P_{H} can be simplified to

$$P_{\mathrm{H}} \simeq \frac{1}{m^{*2}}\left(\frac{K_2^2}{K_1^3} - \frac{K_3}{K_1^2}\right). \tag{3.239}$$

Substituting here (3.128)–(3.130) where we neglect again the terms with B^2, this becomes

$$P_{\mathrm{H}} \simeq \frac{\langle\tau\rangle}{nm^*}\left(\frac{\langle\tau^2\rangle^2}{\langle\tau\rangle^4} - \frac{\langle\tau^3\rangle}{\langle\tau\rangle^3}\right). \tag{3.240}$$

This equation shows that the planar Hall effect exists only if the carrier relaxation time depends on carrier energy. Recall that we had a similar situation with the physical magnetoresistance effect, §3.3.6. There we saw that we can explain the physical magnetoresistance effect by a deflection of carriers with different drift velocities, even though there is no deflection of the average current. In a similar way, it can also be demonstrated that the planar Hall effect is a consequence of a dependence of carrier drift velocity on carrier energy. Since drift velocity is proportional to mobility, and mobility to relaxation time, this model corroborates (3.240).

In §3.3.11 we shall demonstrate a quantitative relationship between the planar Hall effect and the physical magnetoresistance effect.

3.3.10 Alternative forms of the basic equations

Infinitely long samples

As we have seen in §3.3.6, in an infinitely long plate, on average there is no current deflection effect, and the current density vector is well defined by the total current and the geometry of the sample. In such a case, it is adequate to apply (3.141) as a basic equation for the treatment of the galvanomagnetic effects. We can find in the literature a few other ways of writing this equation. For historical or practical reasons, different forms of the basic equation are usually used in different scientific communities. In this section, we shall show the relationships between these alternative forms of the equation.

In view of the notation used in (3.175) and (3.236), equation (3.141) can be rewritten in the following compact form:

$$\boldsymbol{E} = \rho_b \boldsymbol{J} - R_H[\boldsymbol{J} \times \boldsymbol{B}] + P_H(\boldsymbol{J} \cdot \boldsymbol{B})\boldsymbol{B}. \tag{3.241}$$

Recall the meaning of the coefficients: ρ_b (3.176) is the intrinsic magnetoresistivity, R_H (3.177) is the Hall coefficient, and P_H (3.237) is the planar Hall coefficient. This form of the basic equation is the closest to the original form (3.107), derived by the kinetic equation approach. Therefore, it is preferably used when a clear relationship of the coefficients in the equation with their physical meaning is important.

Another form of this equation can be obtained if we write the magnetic induction vector in this way [20]:

$$\boldsymbol{B} = \boldsymbol{b}|\boldsymbol{B}| \tag{3.242}$$

where $\boldsymbol{b}$ denotes a unity vector collinear with $\boldsymbol{B}$. Then (3.241) can be rewritten as

$$\boldsymbol{E} = \rho_b \boldsymbol{J} - R_H|\boldsymbol{B}|[\boldsymbol{J} \times \boldsymbol{b}] + P_H \boldsymbol{B}^2(\boldsymbol{J} \cdot \boldsymbol{b})\boldsymbol{b}. \tag{3.243}$$

At the right-hand side of this equation, all non-vector coefficients have the dimensions of resistivity, and the vectors $\boldsymbol{J}$, $[\boldsymbol{J} \times \boldsymbol{b}]$, and $(\boldsymbol{J} \cdot \boldsymbol{b}) \cdot \boldsymbol{b}$ have the dimensions of current density. Therefore, we can consider (3.243) as a formal *generalization of Ohm's* law in differential form (2.89) for the case of the presence of a magnetic field.

Equation (3.243) attains a particularly interesting form in the case of small magnetic fields. For small magnetic fields, we can use for magnetoresistivity ρ_b the approximation (3.181) and for the planar Hall coefficient P_H the approximation (3.240). Then we can write (3.243) in the following form [20]:

$$\boldsymbol{E} = \rho_\perp \boldsymbol{J} + \rho_H [\boldsymbol{J} \times \boldsymbol{b}] + (\rho_\| - \rho_\perp)(\boldsymbol{J} \cdot \boldsymbol{b})\boldsymbol{b} \tag{3.244}$$

where we introduced the following notation:

$$\rho_\perp \equiv \rho_b \simeq \frac{1}{\sigma}\left[1 + \left(\frac{\langle \tau^3 \rangle}{\langle \tau \rangle^3} - \frac{\langle \tau^2 \rangle^2}{\langle \tau \rangle^4}\right)(\mu B)^2\right] \tag{3.245}$$

$$\rho_H \equiv -R_H B \tag{3.246}$$

$$\rho_\| \equiv \frac{1}{\sigma} = \rho(B = 0). \tag{3.247}$$

In the notation (3.245) we stress the fact that ρ_b (3.176) is the material resistivity of a sample collinear with the current density vector $\boldsymbol{J}$ when $\boldsymbol{J} \perp \boldsymbol{B}$. This is why we call $\rho_b = \rho_\perp$ also a *perpendicular magnetoresistivity*. In contrast, $\rho\|$ (3.247) is called a *parallel magnetoresistivity*. This term is used because, for the case $\boldsymbol{J} \| \boldsymbol{B}$, the equation (3.244) reduces to $\boldsymbol{E} = \rho_\| \boldsymbol{J}$. This means that in this case and within the limits of the present approximation, the effective material resistivity does not change with a magnetic field—see also the next section, §3.3.11. The term ρ_H in (3.246) is called the *Hall resistivity*. This is formally justified by the Ohm's law form of equation (3.244).

Equation (3.244) is used as a standard equation in the field of *anisotropic magnetoresistors* (AMR) made of thin ferromagnetic films [21]. A thin ferromagnetic film is normally spontaneously magnetized to its saturation. Therefore, an external magnetic field produces merely a rotation of the internal magnetic field. So the use of the unity vectors $\boldsymbol{b}$, representing the direction of the magnetization in the film, is very convenient.

Equation (3.244) is also often used if the background information on the equation's coefficients is not known and one has to determine the coefficients experimentally. Then one can consider a Hall plate as a 'black box' (see §4.1.2). Since one measures currents and voltages, one expresses the coefficients as resistances. Consider a very long Hall plate in the xy plane, with a current flow along the x axis, the Hall field along the y axis and a perpendicular magnetic field. Then one can write the basic equation in the matrix form:

$$\begin{bmatrix} E_x \\ E_y \end{bmatrix} = \begin{bmatrix} \rho_{xx} & \rho_{xy} \\ \rho_{yx} & \rho_{yy} \end{bmatrix} \begin{bmatrix} J_x \\ J_y \end{bmatrix}. \tag{3.248}$$

By comparing this equation with (3.244), we identify the two relevant elements of the ρ matrix as follows:

$$\rho_{xx} = \rho_b = \rho_\perp \qquad \rho_{yx} = \rho_H. \tag{3.249}$$

The other coefficients are not relevant if the current density component J_y is negligible, which is usually the case.

This way of writing the basic equation is common in the field of the quantum Hall effect.

Infinitely short samples

As we have seen in §3.3.5, in an infinitely short plate, on average there is no Hall electric field, and the electric field vector is well defined by the applied voltage and the geometry of the sample. In such a case, it is adequate to apply (3.139) as a basic equation for the treatment of the galvanomagnetic effects. Following the notation used in (3.149), equation (3.139) can be rewritten in the following compact form:

$$\boldsymbol{J} = \sigma_{\mathrm{B}}\boldsymbol{E} + \sigma_{\mathrm{B}}\mu_{\mathrm{H}}[\boldsymbol{E} \times \boldsymbol{B}] + Q_{\mathrm{H}}(\boldsymbol{E} \cdot \boldsymbol{B})\boldsymbol{B}. \tag{3.250}$$

Recall the meaning of the previously used coefficients: σ_{B} (3.150) is the Corbino resistivity and μ_{H} (3.151) is the Hall mobility. The coefficient Q_{H} is a new coefficient and is defined by (3.139):

$$Q_{\mathrm{H}} = \frac{e^4}{m^{*2}} K_3. \tag{3.251}$$

In analogy with the planar Hall coefficient P_{H}, figuring in (3.241), we call Q_{H} the *planar Hall-current coefficient*.

3.3.11 Relation between the magnetoresistance effect and the planar Hall effect. Magnetoresistivity for an arbitrary direction of *B*

In §3.3.6 and 3.3.9 we saw that both the physical magnetoresistance effect and the planar Hall effect exist only if the carrier relaxation time depends on carrier energy. This is because in this case a partial current deflection effect may exist even though there is now global current deflection. The common physical origin of these two effects gives rise to a clear quantitative relationship between the respective coefficients characterizing the two effects.

Consider first the approximate expressions for intrinsic magnetoresistivity (3.181) and the planar Hall coefficient (3.240) at small magnetic fields. By comparing these two equations it is easy to show that formally

$$\Delta\rho_{\mathrm{b}} \simeq -P_{\mathrm{H}}B^2 \tag{3.252}$$

where $\Delta\rho_{\mathrm{b}} = \rho_{\mathrm{b}}(B) - \rho(0)$ is the increase in resistivity of a very long sample due to a perpendicular magnetic field. Not the attribute ‘formally’: we use it because the coefficient $\rho_{\mathrm{b}}(\mathrm{B})$ and thus also $\Delta\rho_{\mathrm{b}}$ are defined for a perpendicular magnetic field, whereas the coefficient P_{H} can be measured only with a non-perpendicular magnetic field.

We can get an idea on the real relationship between the two effects, measured at a certain non-perpendicular magnetic field, by inspecting

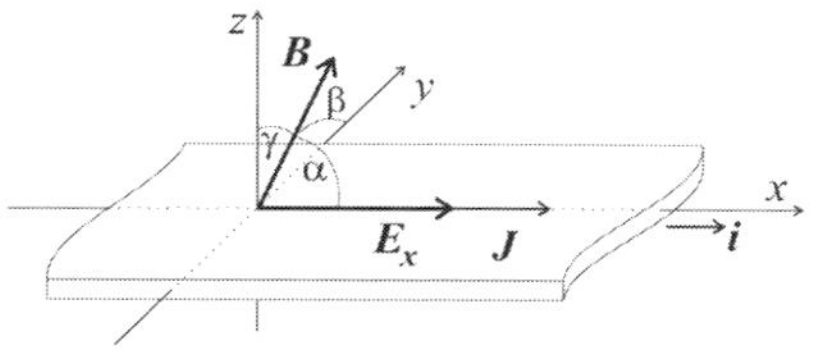

Figure 3.22. A very long Hall plate exposed to a magnetic induction $\boldsymbol{B}$ of arbitrary direction. The vector $\boldsymbol{B}$ encloses the angles α, β and γ with the axes x, y and z, respectively.

figure 3.21. Recall that in this figure the vector $\boldsymbol{E}_\mathrm{B}$ denotes the electric field component collinear with the magnetic induction $\boldsymbol{B}$ (3.236). We saw in §3.3.9 that the transverse component of this vector gives the *planar Hall electric field* (3.238). On the other hand, the longitudinal component of the vector $\boldsymbol{E}_\mathrm{B}$, being collinear with the external electric field, affects the *magnetoresistance effect*.

We may express mathematically the relationship between the two effects by calculating the total longitudinal component of the electric field in a long Hall plate. To this end, we multiply the equation (3.241) by the unity vector $\boldsymbol{i}$ collinear with the x axis and obtain:

$$E_x = \rho_\mathrm{b} J + P_\mathrm{H} B^2 J \cos^2 \alpha. \tag{3.253}$$

Here E_x denotes the component of the electric field along the plate and collinear with the current density $\boldsymbol{J}$, and α is the angle enclosed by the vectors $\boldsymbol{B}$ and $\boldsymbol{J}$ (see figure 3.22). Based on this equation, we can define the effective resistivity of a very long sample exposed to a magnetic induction of an arbitrary direction as follows:

$$\rho(B) = \rho_\mathrm{b} + P_\mathrm{H} B^2 \cos^2 \alpha. \tag{3.254}$$

For a small magnetic field, we can use instead (3.241) the approximation (3.244). Then (3.254) reduces to

$$\rho(B) \simeq \rho_\perp + (\rho_\parallel - \rho_\perp) \cos^2 \alpha. \tag{3.255}$$

According to (3.254) or (3.255), for a perpendicular magnetic field ($\alpha = \pi/2$), the field-dependent effective resistivity $\rho(B)$ reduces to the intrinsic magnetoresistivity ρ_b. However, for a magnetic field collinear with the current density ($\alpha = 0$), to the first approximation, the intrinsic magnetoresistance effect disappears altogether: then $\rho(B) \sim \rho_\parallel = \rho(0)$.

Why does the magnetoresistance effect vanish if the current density and magnetic field are collinear? We can get an intuitive hint on this while looking at figure 3.9(a). A magnetic field affects the velocity component of a charge carrier that is perpendicular to the magnetic field, and not that which is collinear with the magnetic field.

In conclusion, let us note again that the intrinsic magnetoresistance effect and the planar Hall effect have the same physical origin. The consequence is a functional relationship between the intrinsic magneto-resistivity and the planar Hall coefficient. For a similar reason, an analogue relationship exists also between the Corbino-resistivity and the planar Hall-current coefficient.

3.4 Galvanomagnetic coefficients

The basic galvanomagnetic transport coefficients are those coefficients figuring in the basic equations (3.139) and (3.141), that describe the galvanomagnetic transport of charge carriers in a solid. They are completely determined by the kinetic transport coefficients, discussed in §§3.2.6 and 3.2.7. For historical and convenience reasons, usually in practice a set of derived galvanomagnetic coefficients are used. These are the coefficients figuring in the equations (3.250) and (3.241), which we repeat:

$$\boldsymbol{J} = \sigma_{\mathrm{B}}\boldsymbol{E} + \sigma_{\mathrm{B}}\mu_{\mathrm{H}}[\boldsymbol{E} \times \boldsymbol{B}] + Q_{\mathrm{H}}(\boldsymbol{E} \cdot \boldsymbol{B})\boldsymbol{B} \tag{3.256}$$

$$\boldsymbol{E} = \rho_{\mathrm{b}}\boldsymbol{J} - R_{\mathrm{H}}[\boldsymbol{J} \times \boldsymbol{B}] + P_{\mathrm{H}}(\boldsymbol{J} \cdot \boldsymbol{B})\boldsymbol{B}. \tag{3.257}$$

The values of the galvanomagnetic transport coefficients depend on the material involved, carrier type and concentration, temperature, and magnetic field. In this section, we shall discuss these dependences and review the pertinent experimental results available in the literature.

While analysing some of the galvanomagnetic transport coefficients before, we often expressed them in terms of average values of the relaxation time $\langle\tau\rangle$ (see (3.181) or (3.240)). To make our mathematic expressions simpler, we shall now first derive some useful relations concerning these terms.

3.4.1 Scattering factors. Hall factor

The galvanomagnetic transport coefficients can be expressed in terms of the ratio

$$r_m = \langle\tau^m\rangle/\langle\tau\rangle^m. \tag{3.258}$$

Here m is an integer, $\langle\tau^m\rangle$ is given by (3.127), and τ denotes the relaxation time of the carriers involved. By making use of (3.84), (3.258) attains the form

$$r_m = \langle\mu^m\rangle/\langle\mu\rangle^m \tag{3.259}$$

where $\langle\mu^m\rangle = q\langle\tau^m\rangle/m^*$ is the averaged-powered mobility of the charge carriers.

The numerical value of the quantity r_m depends on the energy distribution of the relaxation times. This energy distribution is a function of the scattering process (see table 2.2). Thus the quantity r_m depends on the scattering process. In a formula for the transport coefficient, the terms r_m

Table 3.1. Numerical values of the scattering factors r_m (3.260) for $m = -1, 2, 3, 4$, and a few values of the parameter p—see (2.64) and table 2.2.

		Scattering factor			
Scattering mechanism	p	r_{-1}	r_2	r_3	r_4
Acoustic phonon	$-1/2$	1.13	1.18	1.77	4.16
Ionized impurities	$3/2$	3.40	1.93	5.90	19.14
Neutral impurities	0	1	1	1	1

express the influence of the scattering process on this transport coefficient. For this reason, we shall call the quantities r_m *scattering factors*.

Non-degenerated carriers

Let us consider the scattering factors of non-degenerated carriers. Suppose that the relaxation time can be expressed by (2.64). Then by using (3.117), the scattering factors (3.258), (3.259) attain the form

$$r_m = \frac{\Gamma(\frac{5}{2} + pm)[\Gamma(\frac{5}{2})]^{m-1}}{[\Gamma(\frac{5}{2} + p)]^m}. \tag{3.260}$$

The numerical values of some scattering factors for a few scattering processes are listed in table 3.1. The scattering factors are numbers of about unity; the higher the order of the factor (m), the larger the departure from unity.

For single scattering mechanisms, scattering factors do not depend on temperature. However, with a change in temperature, the dominant scattering mechanism may change, and the value of the scattering factor may alter accordingly.

Strongly degenerated carriers

As we noted in §3.2.6, the transport properties of degenerate carriers are determined predominantly by the carriers with energies around the Fermi level, $E \sim E_F$ (3.122), (3.123). Therefore we obtain from (3.127)

$$\langle \tau^m \rangle \sim \tau^m|_{E=E_F}. \tag{3.261}$$

By substituting this expression into (3.258), we obtain

$$r_m \simeq 1 \tag{3.262}$$

for arbitrary m.

The Hall factor

The scattering factor r_2 given by (3.259) for $m = 2$,

$$r_2 = \langle \tau^2 \rangle / \langle \tau \rangle^2 \tag{3.263}$$

is of special importance. Denoted by r_{H0} (3.172), it appears as a proportionality factor in the formulae for the low magnetic field values of the Hall mobility (3.171) and the Hall coefficient (3.185). For this reason, the scattering factor r_2 is called the *Hall scattering factor*.

The notion of the Hall scattering factor, defined by (3.263), is reasonable and useful only for charge carriers with spherical constant energy surfaces and an energy-independent effective mass. If these conditions are not fulfilled, then it is not only the scattering process that gives rise to a difference between the Hall mobility and the drift mobility. Other effects, such as anisotropy and the mixing of carriers with different effective masses, may also take part in creating this difference. For this reason, it is useful to generalize the notion of the Hall scattering factor. We have already done so in discussing the Hall effect in n-type silicon (see §3.3.8). The new generalized factor is called the *Hall factor* (without the attribute 'scattering'). The Hall factor is defined either in analogy with (3.171) by

$$r'_H = |\mu_H|/\mu \tag{3.264}$$

or, according to (3.185), by

$$r_H = |R_H|qn. \tag{3.265}$$

This equation holds for a strongly extrinsic material, with n denoting the majority carrier concentration.

It is usually tacitly assumed that the above two definitions of the Hall factors are equivalent. However, we shall show later that this is true only at nearly zero magnetic inductions. For this reason, we shall be using different notations for the Hall factors defined in one or the other way: r'_H and r_H.

In the case of spherical constant energy surfaces and a constant effective mass of carriers, the Hall factor reduces to the Hall scattering factor: $r_H = r_{H0}$ (3.172) $= r_2$ (3.263). Otherwise, the Hall factor can usually be expressed as the product

$$r_H = r_{H0}a \tag{3.266}$$

where r_{H0} summarizes the influence of scattering, and a characterizes anisotropy and mixing effects. We have already applied (3.266) for electrons in silicon (see (3.212)).

Very often, the terms 'Hall factor' and 'Hall scattering factor' are used without making any distinction between them. The notations in use include: r_H, r', r, r^* and A.

According to the simple model, the Hall factor is expected to be $r_H \sim 1$ for strongly degenerated carriers (3.262), and $r_H = 1 \ldots 1.93$ if one or the other scattering mechanism prevails (see r_2 in table 3.1). In some important applications of Hall devices, more accurate values of the Hall factor are

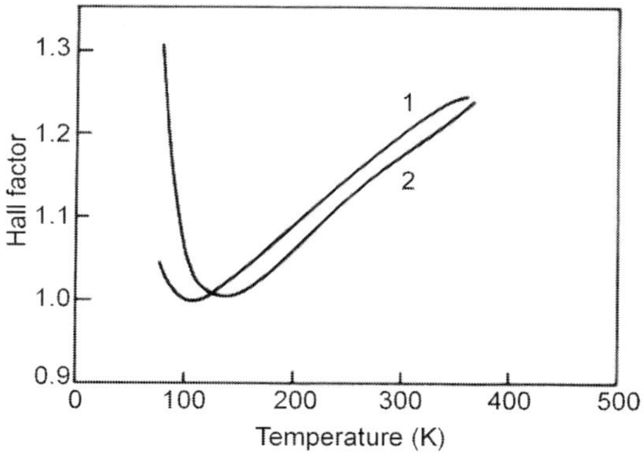

Figure 3.23. The Hall factor of electrons against temperature for n-type silicon, with $N_{DI} \approx 8 \times 10^{13}\,\text{cm}^{-3}$ (curve 1) and $N_{DI} \approx 9 \times 10^{14}\,\text{cm}^{-3}$ (curve 2). Experimental (after [23]).

needed. For this reason, a lot of work has been done to characterize precisely the Hall factor for various materials. Most of the results on r_H, together with other properties for many semiconductors, which were available by 1980, are summarized in [22]. We present below a few results on r_H that are not considered in [22].

The Hall factor of electrons in low-doped n-type silicon was studied in [23]. The theoretical model was based on the Herring–Vogt theory. To explain the experimentally observed temperature dependence of r_H in addition to intravalley scattering, intervalley phonon scattering was also included. A typical result is shown in figure 3.23. It was found that, at about room temperature, the function $r_H(T)$ is almost independent of the phosphorus concentration. However, the increase in $r_H(T)$ at low temperatures depends on the impurity concentration. At low temperatures, the impurity scattering prevails. For pure impurity scattering, it should be $r_H(T) = 1.68$: see (3.212)–(3.214), with $r_{H0} = 1.93$ and $a = 0.87$.

Similar results were obtained by careful experiments on various low-doped n-type silicon crystals [24] (see figure 3.24).

The Hall factor for electrons in n-type silicon is shown in figure 3.25 as a function of donor concentration [25]. As the doping level increases towards about $3 \times 10^{18}\,\text{cm}^{-3}$, the Hall scattering factor also increases. This was to be expected, since with the increase in impurity concentration, the ionized impurity scattering starts to dominate. With a further increase in doping concentration, the Hall factor decreases. This is because of degeneration. At $N > 10^{20}\,\text{cm}^{-3}$, the Hall factor approaches the theoretical value of $r_H \simeq 0.87$: see (3.212)–(3.214), with $r_{H0} \to 1$ (strongly degenerated carriers).

The Hall factor of holes in silicon was studied in [26]. In addition to scattering, the non-parabolicity of the $E(\boldsymbol{k})$ relation at the top of the

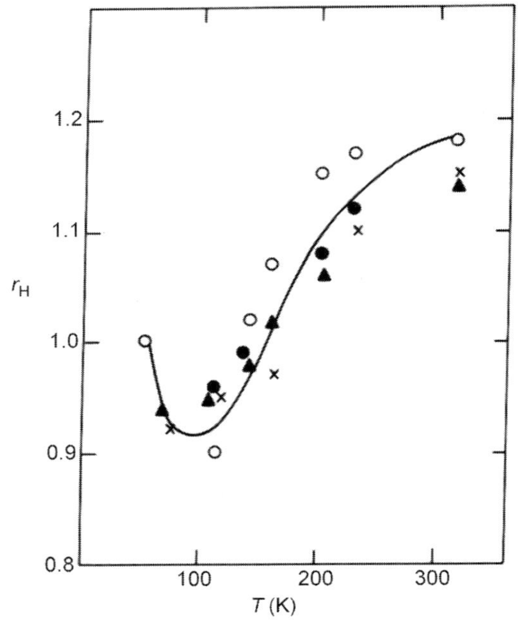

Figure 3.24. The Hall factor of electrons against temperature for various low-doped n-type silicon samples. ●, As-grown Czochralski (Cz); ○, Cz after 1 h annealing at 1300 °C; ×, as-grown float zone (FZ); ▲, FZ after 1 h annealing at 1300 °C. Experimental (after [24]).

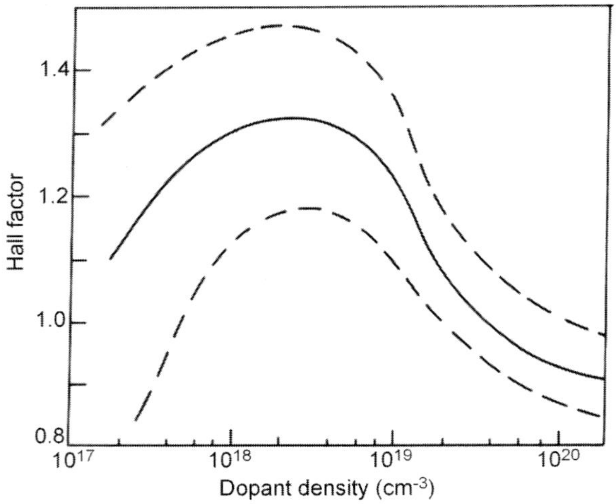

Figure 3.25. The Hall factor of electrons against phosphorous concentration in silicon at room temperature. The broken curves indicate the experimental error (after [25]).

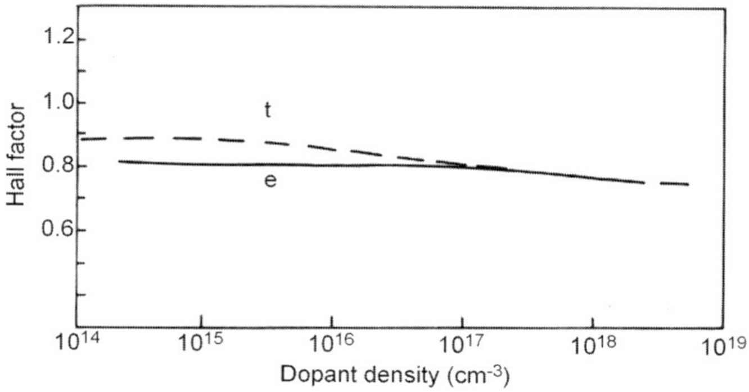

Figure 3.26. The Hall factor of holes against dopant density in p-type silicon: e, experimental; t, theoretical (after [26]).

valence band, its degeneracy (light and heavy holes), and anisotropy (constant energy surfaces have a form of warped spheres) were also taken into account. A result is presented in figure 3.26. The Hall factor of holes in silicon is about 0.8.

Figure 3.27 shows the Hall factor of electrons in InP. Such curves are rather typical for III–V compounds: at lower temperatures, r_H strongly depends on the total impurity concentration; with an increase in temperature, the Hall factor decreases and eventually levels off at a value of about 1.1.

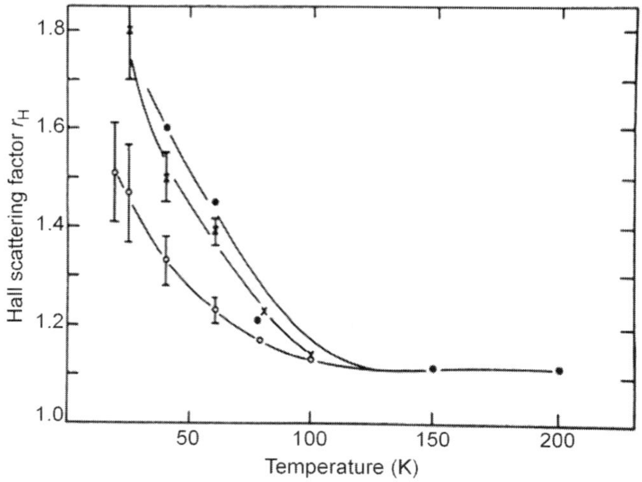

Figure 3.27. Measured variation of Hall factor with temperature for several InP samples (reprinted from [27]).

3.4.2 Corbino conductivity

The Corbino conductivity σ_B is a coefficient in the basic equation (3.256). This is the effective material conductivity due to a certain kind of carrier, in the direction of the total electric field, when the sample is exposed to a perpendicular magnetic field (see §3.3.5). It is the effective material conductivity of a Corbino disc exposed to a perpendicular magnetic field. The conductivity σ_B is given by (3.150), which we repeat:

$$\sigma_B = e^2 K_1. \tag{3.267}$$

At zero magnetic induction, the conductivity σ_B reduces to the usual conductivity (3.145). The dimensions of the Corbino conductivity are those of the usual conductivity, namely (ohm-cm)$^{-1}$.

To find the magnetic field dependence of σ_B we will make use of the approximations for the kinetic coefficient K_1 found in §3.2.7.

Non-degenerated carriers

For a weak magnetic field, by substituting (3.128) into (3.267), we obtain

$$\sigma_B \simeq e^2 \frac{n}{q} \langle\mu\rangle \left(1 - \frac{\langle\mu^3\rangle}{\langle\mu\rangle} B^2\right). \tag{3.268}$$

This can be rewritten as

$$\sigma_B \simeq \sigma_0 (1 - r_3 \mu^2 B^2) \tag{3.269}$$

where σ_0 is the conductivity at $\boldsymbol{B} = 0$ (3.146), $\mu = \langle\mu\rangle$ is the drift mobility, and r_3 is a scattering factor (3.258). Equation (3.269) can also be put in the form

$$\sigma_B \simeq \sigma_0 \left(1 - \frac{r_3}{r_2^2} \mu_H^2 B^2\right) \tag{3.270}$$

where r_2 is the Hall scattering factor and μ_H is the Hall mobility.

For a strong magnetic field, by substituting (3.134) into (3.267), and keeping only the first term, we have

$$\sigma_B \simeq e^2 \frac{n}{q} \langle\mu\rangle \frac{1}{B^2} \frac{\langle\mu^{-1}\rangle}{\langle\mu\rangle}. \tag{3.271}$$

This can be rewritten as

$$\sigma_B \simeq \sigma_0 \frac{r_{-1}}{\mu^2 B^2} \tag{3.272}$$

where r_{-1} is the scattering factor given by

$$r_{-1} = \langle\tau^{-1}\rangle\langle\tau\rangle. \tag{3.273}$$

Strongly degenerated carriers

By substituting (3.123) into (3.267), we readily obtain

$$\sigma_B \simeq \frac{\sigma_0}{1+\mu^2 B^2}. \tag{3.274}$$

An identical expression was obtained before by approximate analysis (see (3.37)). This is plausible: in §3.1.2, we did not account for the energy distribution of charge carriers, and by deriving (3.274), we considered only those carriers with energies around the Fermi level.

According to (3.272) and (3.274), at very strong magnetic fields the conductivity σ_B tends to be very small. This is in accordance with our discussion in §3.1.3: at a very high magnetic field, a carrier manages to make a few cycloids before it is scattered. During this time, the carrier does not drift along the direction of the electric field, but transverse to it (see figure 3.10(b)). Therefore, the conductivity in the direction of the electric field is very small.

The dependence of the conductivity σ_B on temperature and impurity concentration is essentially the same as that of the zero-field conductivity (see §2.6.1).

The magnetic field dependence of the conductivity $\bar{\sigma}_B$ of a mixed conductor can easily be found by substituting into (3.191) the appropriate formula found above for each kind of carrier.

3.4.3 Hall mobility

The Hall mobility μ_H is a coefficient in equations (3.256). The Hall mobility is an effective mobility of certain kinds of carriers for their transverse drift under the action of magnetic forces (see §3.3.5). Thus the Hall mobility determines the relative intensity of the transverse galvanomagnetic effect. In particular, the product $\mu_H B$ defines the Hall angle (see (3.152)).

At a low magnetic field, the Hall mobility of carriers with spherical constant energy surfaces is given by (3.171):

$$\mu_H(0) = \text{sign}[e] r_{H0} \mu \tag{3.275}$$

where r_{H0} denotes the Hall scattering factor, and μ is the drift mobility of the carriers under consideration. The Hall mobility bears the sign of the corresponding carriers, and its magnitude approximately equals the drift mobility (since $r_H \simeq 1$). The dimensions of the Hall mobility are those of the drift mobility, namely $m^2\,V^{-1}\,s^{-1}$.

Generally, the Hall mobility is defined by (3.169):

$$\mu_H = \frac{e}{m^*}\frac{K_2}{K_1}. \tag{3.276}$$

Since the kinetic coefficients depend on the magnetic field, so also does the Hall mobility. We shall now evaluate this dependence.

Non-degenerated carriers

For low magnetic inductions, by substituting (3.128), (3.129) and (3.84) into (3.276), we obtain

$$\mu_{\mathrm{H}} = \mathrm{sign}[e] r_{\mathrm{H}0} \frac{1 - (\langle\mu^4\rangle / \langle\mu^2\rangle) B^2}{1 - (\langle\mu^3\rangle / \langle\mu\rangle) B^2} \langle\mu\rangle \tag{3.277}$$

where $\langle\mu\rangle = q\langle\tau^m\rangle/m^*$. This expression can be simplified if the denominator is expanded into a series. After rearrangement, and keeping only terms up to B^2, the Hall mobility attains the form

$$\mu_{\mathrm{H}} \simeq \mu_{\mathrm{H}}(0)(1 - \beta\mu^2 B^2). \tag{3.278}$$

Here $\mu_{\mathrm{H}}(0)$ is given by (3.275), and β is a coefficient given by

$$\beta = \frac{\langle\mu^4\rangle}{\langle\mu^2\rangle\langle\mu\rangle^2} - \frac{\langle\mu^3\rangle}{\langle\mu\rangle^3} = \frac{r_4}{r_2} - r_3 \tag{3.279}$$

where r_2, r_3 and r_4 are the scattering factors (3.258).

According to (3.186), $E_{\mathrm{H}} \sim \mu_{\mathrm{H}} B$. At $B \simeq 0$, $\mu_{\mathrm{H}} \simeq \mu_{\mathrm{H}}(0) \neq f(B)$, and E_{H} is linearly proportional to B. But due to (3.278), at higher inductions this linear dependence is eventually disturbed. For this reason, the coefficient β is called the non-linearity coefficient of the Hall mobility. Some numerical values of β (3.279), calculated using the r values from table 3.1, are listed in table 3.2.

For *high magnetic inductions*, by substituting (3.134) and (3.135) into (3.276), we obtain

$$\mu_{\mathrm{H}} \simeq \mathrm{sign}[e] \frac{1}{\langle\mu^{-1}\rangle} \frac{1 - \langle\mu^{-2}\rangle B^{-2}}{1 - (\langle\mu^{-3}\rangle / \langle\mu^{-1}\rangle) B^{-2}}. \tag{3.280}$$

For very strong magnetic fields, this expression reduces to

$$\mu_{\mathrm{H}} \simeq \mathrm{sign}[e] \frac{1}{r_{-1}} \mu. \tag{3.281}$$

Table 3.2. The numerical values of the non-linearity coefficient β, figuring in equation (3.278), for a few scattering processes. β was calculated using (3.279) and the r_m values from table 3.1.

Scattering process	Acoustic phonon	Ionized impurity	Neutral impurity
p	−1/2	3/2	0
β	1.26	1.08	0

Thus at strong magnetic fields, in the expression for the Hall mobility the Hall scattering factor $r_{H0} = r_2$ is replaced by the factor $1/r_{-1}$.

Strongly degenerated carriers

By substituting (3.123), with $s = 1$ and $s = 2$, into (3.276), we readily obtain

$$\mu_H = \frac{e}{m^*}\tau(E_F) = \text{sign}[e]\mu. \tag{3.282}$$

Therefore, for strongly degenerate carriers, the Hall scattering factor equals unity. Then, except for the sign, the Hall mobility exactly equals the drift mobility.

Up to now, we have been assuming that the kinetic properties of carriers can be characterized by a parabolic $E(\boldsymbol{k})$ relation and spherical constant energy surfaces in the $\boldsymbol{k}$ space. But even if these assumptions are not fulfilled, it turns out that the results obtained above hold by and large. Then the Hall mobility μ_H can be related to the drift mobility μ by the phenomenological relation (3.210). This relation can be further generalized to be applicable to both carrier types:

$$\mu_H = \text{sign}[e]r'_H\mu \tag{3.283}$$

where r'_H denotes the Hall factor (3.264). In the case of spherical constant energy surfaces and low magnetic inductions, (3.283) reduces to (3.275), with $r'_H = r_{H0} = r_2$ given by (3.263).

The magnetic field dependence of the Hall mobility is usually expressed as

$$\mu_H(B) = \text{sign}[e]r'_H(B)\mu(0) \tag{3.284}$$

where $\mu(0) = \mu(\boldsymbol{B} = 0)$. Thus the $\mu_H(B)$ dependence is fully ascribed to a magnetic field dependence of the Hall factor. At low magnetic fields, (3.284) can be put in a form similar to (3.278):

$$\mu_H(B) = \text{sign}[e]r'_H(0)(1 - \beta_a\mu^2B^2)\mu(0) \tag{3.285}$$

where, with reference to (3.266),

$$r'_H(0) = r_H(0) = r_{H0}a(0). \tag{3.286}$$

Here $r_H(0) = r_H(B = 0)$, $r_{H0} = r_2$ is the Hall scattering factor for the isotropic model (3.263), $a(0)$ is the anisotropy factor for $B = 0$, and β_a is a non-linearity coefficient. For the isotropic model, β_a reduces to β (3.279).

At very strong magnetic fields, the anisotropy factor approaches unity, and r'_H in (3.284) approaches the value of $1/r_{-1}$.

For strongly degenerated carriers, $r_{H0} \simeq 1$ and $r'_H \simeq a$; if $B \to \infty$, r'_H approaches unity.

The dependence of the Hall mobility on temperature and impurity concentration is essentially determined by the corresponding dependences of the conductivity mobility (see §2.6.1). The temperature and impurity concentration dependences of the Hall factor are negligible in comparison with those of the mobility.

3.4.4 Planar Hall-current coefficient

The coefficient Q_H relates the current density component collinear with the magnetic field to the product $\boldsymbol{B}(\boldsymbol{E} \cdot \boldsymbol{B})$ (see (3.256)). It is given by (3.251), which we repeat:

$$Q_H = \frac{e^4}{m^{*2}} K_3. \tag{3.287}$$

For *non-degenerated carriers* in an isotropic energy band, we obtain the following approximations. For a *low magnetic field*, by substituting (3.130) into (3.287), we obtain

$$Q_H \simeq qn\langle\mu^3\rangle = qn\mu^3 r_3 = \sigma_0 \mu^2 r_3 \tag{3.288}$$

where we have also made use of (3.258) and (3.146). Some numerical values of the scattering factor r_3 are listed in table 3.1. For a *high magnetic field*, by making use of (3.136), we obtain

$$Q_H \simeq qn\mu \frac{1}{B^2} = \frac{\sigma_0}{B^2}. \tag{3.289}$$

For *strongly degenerate carriers*, with the aid of (3.123), we find

$$Q_H \simeq nq \frac{q^3}{m^{*3}} \tau^3 \frac{1}{1 + \mu^2 B^2} = \sigma_0 \frac{\mu^2}{1 + \mu^2 B^2} \tag{3.290}$$

where we have also made use of (2.86).

3.4.5 Magnetoresistivity

The coefficient ρ_b, figuring in the basic equation (3.257), denotes the effective material resistivity due to certain kinds of carriers, in the direction of the current density, when a sample is exposed to a perpendicular magnetic field. Therefore, this is also the effective material resistivity of an infinitely long Hall device exposed to a perpendicular magnetic field (see §3.3.6). The magnetoresistivity ρ_b is given by (3.176), which we repeat:

$$\rho_b = \frac{1}{1 + \left(\dfrac{e}{m^*} \dfrac{K_2}{K_1}\right)^2 B^2} \frac{1}{e^2 K_1}. \tag{3.291}$$

Let us analyse the magnetic field dependence of ρ_b, assuming the simple isotropic $E(\boldsymbol{k})$ relation.

Non-degenerated carriers

For low magnetic inductions, we may substitute (3.128) and (3.129) into (3.291). Keeping only the terms proportional to B^0 and B^2, we obtain

$$\rho_b \simeq \frac{1}{\sigma_0[1 - (r_3 - r_2^2)\mu^2 B^2]} \tag{3.292}$$

$$\rho_b \simeq \rho_0\{1 + [(r_3/r_2^2) - 1]\mu_H^2 B^2\} \tag{3.293}$$

where $\sigma_0 = \sigma_B(B = 0)$, $\rho_0 = 1/\sigma_0 = \rho_b(B = 0)$, and r_2 and r_3 are the scattering factors (3.258). Note that the approximation (3.181) that we have found before can be also rewritten in the form (3.293).

According to above equations, at zero magnetic induction, the magnetoresistivity ρ_b reduces to the usual resistivity. Also ρ_b does not depend on B if $r_3 = r_2 = 1$. This happens if the relaxation time of carriers does not depend on energy (see table 3.1). In this case, there is no spread in the deflections of micro-currents due to carriers with different energies. Hence, to a first approximation, the physical magnetoresistance effect does not appear.

At a *very strong magnetic field*, by substituting (3.134) and (3.135) into (3.291), the resistivity attains the form

$$\rho_b \simeq r_{-1}\rho_0 \tag{3.294}$$

where r_{-1} is the scattering factor given by (3.273). Thus at very strong magnetic fields, the magnetoresistivity ρ_b saturates at a finite value, and the physical magnetoresistance effect vanishes. The ultimate increase in the magnetoresistivity ρ_b due to a magnetic field strongly depends on the scattering mechanism involved (see the r_{-1} values in table 3.1).

Strongly degenerated carriers

By substituting (3.123) for $s = 1$ and $s = 2$ into (3.291), we obtain

$$\rho_b = \frac{1}{\sigma_0}. \tag{3.295}$$

Within applied approximations, the physical magnetoresistance effect does not exist in strongly degenerate materials.

3.4.6 Hall coefficient

The Hall coefficient is a material parameter that characterizes the efficiency of generating the Hall electric field in the material. The Hall coefficient figures in equation (3.257), where it relates the product $[\boldsymbol{J} \times \boldsymbol{B}]$ to the Hall electric field (see also (3.183)). The Hall coefficient of a strongly extrinsic semiconductor is given by (3.177):

$$R_H = \frac{1}{1 + \left(\frac{e}{m^*}\frac{K_2}{K_1}\right)^2 B^2} \frac{1}{em^*}\frac{K_2}{K_1^2} \tag{3.296}$$

where K_1 and K_2 denote the first two kinetic coefficients (3.116). By making use of (3.176) and (3.276), R_H (3.296) may be expressed in terms of Hall mobility and magnetoresistivity, as in (3.184):

$$R_H = \mu_H \rho_b. \tag{3.297}$$

At very low magnetic inductions, the Hall coefficient reduces to (3.185):

$$R_H(0) \simeq \text{sign}[e]\frac{r_H(0)}{qn} \tag{3.298}$$

where r_H denotes the Hall factor §3.4.1. Otherwise, the *Hall coefficient depends on the magnetic induction*. Let us consider this dependence, again assuming a simple model of carriers with spherical constant energy surfaces.

Non-degenerated carriers

Assume first a weak magnetic field. By substituting (3.128), (3.129) and (3.84) into (3.296), we have, after omitting the terms of higher orders,

$$R_H = \text{sign}[e]\frac{r_2}{qn}\frac{(1 - \langle\mu^4\rangle/\langle\mu^2\rangle)B^2}{[1 - 2(\langle\mu^3\rangle/\langle\mu\rangle)B^2](1 + \mu_H^2 B^2)}. \tag{3.299}$$

Expanding the expressions in the denominator into a series and keeping only the first two terms, we obtain

$$R_H = R_{H0}\left(1 - \frac{\langle\mu^4\rangle}{\langle\mu^2\rangle}B^2\right)\left(1 + 2\frac{\langle\mu^3\rangle}{\langle\mu\rangle}B^2\right)(1 - \mu_H^2 B^2). \tag{3.300}$$

Neglecting here the higher-order terms, we obtain

$$R_H = R_{H0}\left[1 - \left(\frac{\langle\mu^4\rangle}{\langle\mu^2\rangle} - 2\frac{\langle\mu^3\rangle}{\langle\mu\rangle} + \frac{\langle\mu^2\rangle^2}{\langle\mu\rangle^2}\right)B^2\right] \tag{3.301}$$

where $R_{H0} = R_H(0)$ is given by (3.298), and the terms $\langle\mu^m\rangle$ denote the energy-weighted averages of the mobility, $\langle\mu^m\rangle = q\langle\tau^m\rangle/m^*$. Equation (3.301) can be rewritten as

$$R_H = R_{H0}\left[1 - \left(r_2^2 - 2r_3 + \frac{r_4}{r_2}\right)\mu^2 B^2\right] \tag{3.302}$$

where r_2, r_3 and r_4 are the scattering factors (3.259); or as

$$R_H = R_{H0}[1 - \alpha\mu_H^2 B^2] \tag{3.303}$$

Table 3.3. The numerical values of the non-linearity coefficient α, figuring in equation (3.303), for some scattering processes. α was calculated using (3.304) and the r_m values from table 3.1.

Scattering process	Acoustic phonons	Ionized impurities	Neutral impurities
p	$-1/2$	$3/2$	0
α	0.99	0.49	0

with

$$\alpha = 1 - 2\frac{r_3}{r_2^2} + \frac{r_4}{r_2^3}. \tag{3.304}$$

A magnetic field dependence of the Hall coefficient brings about a non-linearity in the functions $\boldsymbol{E}_{\mathrm{H}} = f([\boldsymbol{J} \times \boldsymbol{B}])$ (3.183) or $V_{\mathrm{H}} = f(IB)$ (3.18). For this reason, the quantity α (3.304) is called a non-linearity coefficient. To distinguish it from other non-linearity coefficients of the Hall devices to be discussed later, we shall call α the material non-linearity coefficient. By substituting the numerical values of the scattering factors from table 3.1 into (3.304), we obtain the values of α given in table 3.3.

For high magnetic inductions, by substituting (3.134) and (3.135) into (3.296), we obtain

$$R_{\mathrm{H}} \simeq \mathrm{sign}[e]\frac{1}{r_{-1}}\mu \Bigg/ \sigma_0 \frac{r_{-1}}{\mu^2 B^2}\left[1 + \left(\frac{1}{r_{-1}}\mu\right)^2 B^2\right]. \tag{3.305}$$

At very high magnetic fields, this reduces to

$$R_{\mathrm{H}} \simeq \mathrm{sign}[e]\frac{1}{qn} \tag{3.306}$$

where we have made use of (3.146). Thus at very strong magnetic fields, the Hall coefficient ceases to depend on the scattering process.

Strongly degenerated carriers

By substituting (3.282) and (3.295) into (3.297), we readily obtain

$$R_{\mathrm{H}} \simeq \mathrm{sign}[e]\frac{1}{qn}. \tag{3.307}$$

This equals the value for the Hall coefficient found by approximate analysis (see (3.17)).

For semiconductors with non-spherical constant energy surfaces or non-parabolic $E(\boldsymbol{k})$ relations, the above results are only approximately valid. To account for the anisotropy and non-parabolicity, one usually makes use of the Hall factor, in a similar way as we did when discussing the Hall mobility (see (3.283)). Thus, the Hall coefficient of a strongly extrinsic semiconductor can be expressed as

$$R_H = \text{sign}[e]\frac{r_H}{qn} \tag{3.308}$$

where r_H is the Hall factor (3.265). Equation (3.308) is a generalization of (3.215) and (3.298). The magnetic field dependence of the Hall coefficient can now be interpreted as a magnetic field dependence of the Hall factor: $r_H = r_H(B)$.

At a low magnetic field, (3.308) may be put in the same form as (3.303):

$$R_H(B) = \text{sign}[e]\frac{r_H(0)(1 - \alpha_a \mu_H^2 B^2)}{qn} \tag{3.309}$$

where, as before (3.286),

$$r_H(0) = r_{H0} a(0) \tag{3.310}$$

and α_a is a non-linearity coefficient.

For a semiconductor with spherical constant energy surfaces, (3.309) reduces to

$$R_H(B) = \text{sign}[e]\frac{r_{H0}(1 - \alpha \mu_H^2 B^2)}{qn} \tag{3.311}$$

which is identical to (3.303). Here $r_{H0} = r_2$ is the Hall scattering factor (3.263), and α is given by (3.304).

At very strong magnetic fields, the anisotropy factor a in (3.310) approaches unity. By comparing (3.308) with (3.306), we conclude that the Hall factor now approaches unity.

For strongly degenerate semiconductors, $r_{H0} \simeq 1$ and $r_H \simeq a$; if $B \to \infty$, r_H approaches unity.

Therefore, if $\mu^2 B^2 \gg 1$ (3.131),

$$R_H = \text{sign}[e]\frac{1}{qn} \tag{3.312}$$

irrespective of the form of the constant energy surfaces.

It is interesting to note that the ascribing of the magnetic field dependence of the Hall mobility and that of the Hall coefficient to a magnetic field dependence of a common Hall factor is inconsistent. Actually, at a low magnetic induction, the factors β_a in (3.285) and $r_2^2\alpha$ in (3.311) are not identical (compare (3.279) and (3.304)); and at a very high induction, r'_H in (3.284) approaches the value of r_{-1}, and r_H in (3.308) approaches unity. Therefore, the two Hall factors r'_H (3.264) and r'_H (3.265) are identical only at very low magnetic inductions.

Experimental results

Experimentally, a quadratic dependence of the Hall coefficient on the magnetic induction can usually be found:

$$R_H = R_{H0}(1 - \alpha^* B^2). \tag{3.313}$$

An example of such a dependence for a silicon Hall device is shown in figure 3.28 [28]. There we measured the parameter

$$NL = -\alpha^* B^2 \tag{3.314}$$

as a function of the magnetic induction B. Obviously, (3.313) holds. By comparing (3.303) and (3.313), we found the material non-linearity coefficient α for n-type silicon as a function of temperature and doping concentration (figures 3.28–3.31).

The experimental results corroborate the predictions of the simple isotropic theory listed in table 3.3 surprisingly well. At low temperatures and higher impurity densities (figures 3.29 and 3.30) the measured coefficient α attains a value of approximately 0.3–0.4. This compares well with the theoretical value of $\alpha = 0.49$ for ionized impurity scattering. At a low doping density and higher temperatures (figure 3.31) α reaches a value of

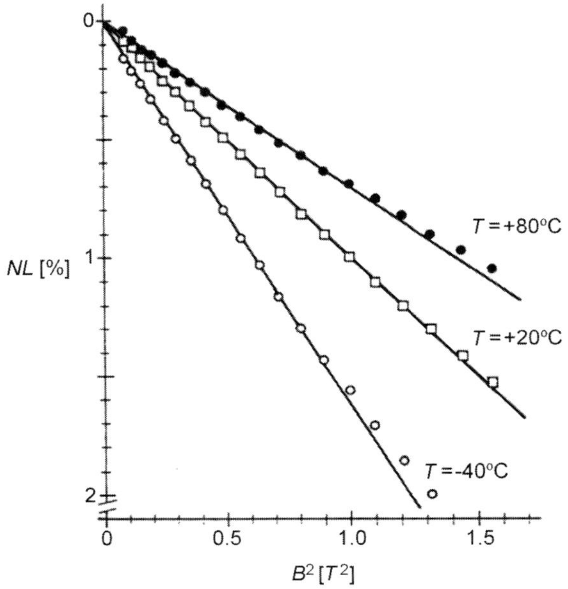

Figure 3.28. Non-linearity in the Hall coefficient of n-type silicon as a function of squared magnetic induction, at different temperatures. Experimental. The sample was phosphorus doped, with $N_D \simeq 10^{15}$ cm^{-3} (reprinted from [28]).

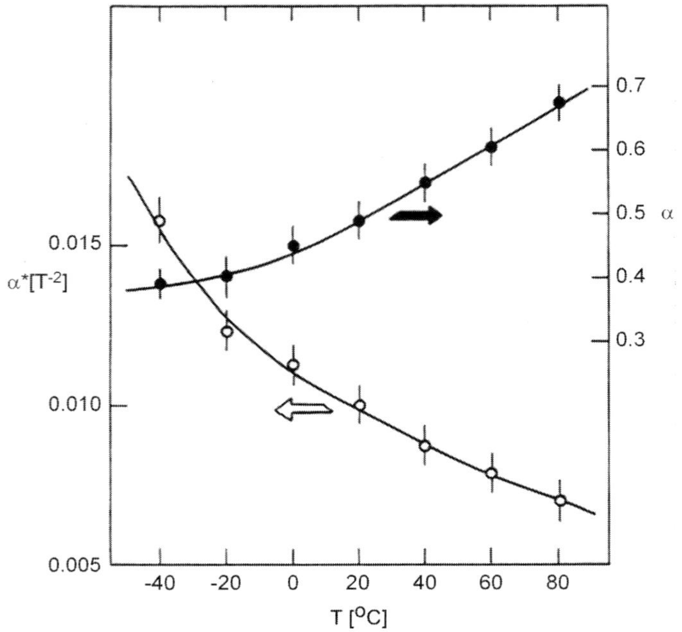

Figure 3.29. The non-linearity parameter α^* and the material non-linearity coefficient α as functions of temperature. Experimental, n-type silicon, donor concentration $N_D \simeq 10^{15}\,\text{cm}^{-3}$ (reprinted from [28]).

approximately 1.1. This agrees very well with the theoretical prediction of $\alpha \simeq 1$.

For n-type GaAs, a value of $\alpha \simeq 0.1$ was deduced [28] from the experimental data published in [29].

In the exhaustion range of an extrinsic semiconductor, the carrier concentration is almost temperature independent (see (2.45)). Thus the temperature dependence of the Hall coefficient is then determined by the temperature dependence of the Hall factor. For example, the temperature dependence of the Hall coefficient of n-type silicon follows the curves shown in figures 3.23 and 3.24: at room temperature, it varies with a change in temperature for about $0.1\%\,\text{K}^{-1}$.

Outside the exhaustion range, the temperature dependence of the Hall coefficient is predominantly determined by the temperature behaviour of the carrier concentration. Then the temperature coefficient of R_H is much higher than in the exhaustion range.

The magnetic field and temperature dependences of the average Hall coefficient R_H of a mixed conductor (3.202) can easily be found by making use of the appropriate coefficients μ_H and ρ_b for each kind of carrier involved.

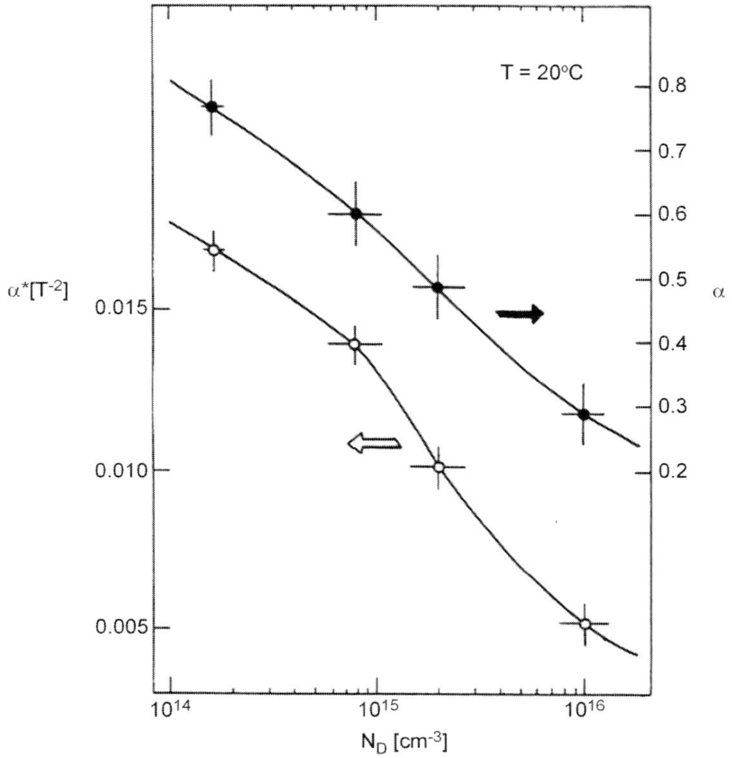

Figure 3.30. The non-linearity coefficients α^* and α of n-type silicon as functions of impurity concentrations at $T = 20\,^\circ\mathrm{C}$. Experimental (reprinted from [28]).

3.4.7 Planar Hall coefficient

The planar Hall coefficient is a material parameter that characterizes the efficiency of generating the planar Hall electric field in the material. The planar Hall coefficient relates the product $(\boldsymbol{J}\cdot\boldsymbol{B})\boldsymbol{B}$ to the planar Hall electric field (3.236). The planar Hall coefficient of a strongly extrinsic semiconductor is given by (3.237):

$$P_{\mathrm{H}} = \frac{1}{1+\left(\dfrac{e}{m^*}\dfrac{K_2}{K_1}\right)^2 B^2} \frac{\left(\dfrac{e}{m^*}\dfrac{K_2}{K_1}\right)^2 - \dfrac{e^2}{m^{*2}}\dfrac{K_3}{K_1}}{e^2 K_1\left(1+\dfrac{e^2}{m^{*2}}\dfrac{K_3}{K_1}B^2\right)}. \tag{3.315}$$

As we saw in §3.3.11, the planar Hall coefficient is strongly related with the intrinsic magnetoresistivity.

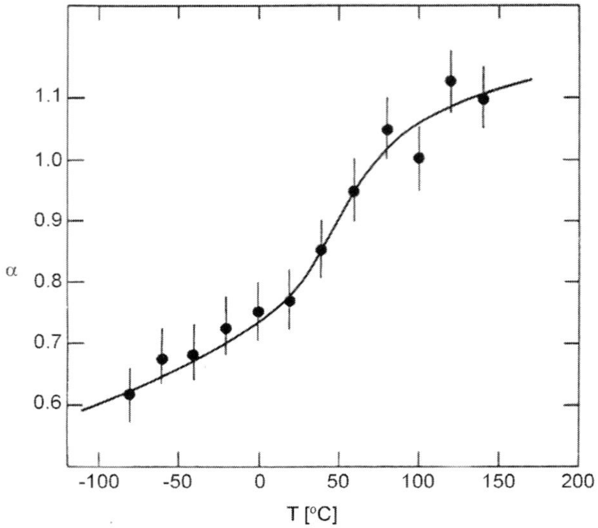

Figure 3.31. The material non-linearity coefficient α of low-doped n-type silicon as a function of temperature. The sample was phosphorus doped, with $N_D \simeq 1.7 \times 10^{14}\,\mathrm{cm}^{-3}$ (reprinted from [28]).

At low magnetic fields, the planar Hall coefficient is given by (3.240):

$$P_H \simeq \frac{\langle\tau\rangle}{nm^*}\left(\frac{\langle\tau^2\rangle^2}{\langle\tau\rangle^4} - \frac{\langle\tau^3\rangle}{\langle\tau\rangle^3}\right). \tag{3.316}$$

Comparing this expression with (3.293), we find

$$P_H = -\frac{(\rho_b - \rho_0)}{B^2} \tag{3.317}$$

which is in accordance with (3.252).

3.5 Related effects

Throughout this chapter, we have discussed the Hall effect and magnetoresistance effect in their classical forms: we presumed carriers at thermodynamic equilibrium, a weak electric field, and no other external influences. Now we shall briefly consider the effects that arise if one or other of these assumptions is not fulfilled.

3.5.1 Diffusion current. Complete galvanomagnetic equation

Owing to their random thermal motion, quasi-free charge carriers by and large tend to move in such a way as to even out any spatial non-uniformity in their distribution. The corresponding transport process is called diffusion (see §2.6.3). In the presence of a magnetic field, the magnetic part of the Lorentz force (3.1) acts on each moving carrier. Hence it is reasonable to expect that a magnetic field affects the diffusion process. It does indeed; and it even turns out that a current in a semiconductor sample is affected by a magnetic field in roughly the same manner, irrespective of whether the current is due to drift or the diffusion process.

The basic equation, related to pure drift/magnetic effects, is given by (3.107) or (3.139). We shall now extend this equation to account also for diffusion/magnetic phenomena.

Recall that we derived (3.107) by starting with the Boltzmann kinetic equation (3.58). By making use of the relaxation time approximation, and after neglecting the diffusion term, this equation reduced to (3.67). If we do not neglect the diffusion term, the Boltzmann kinetic equation in the relaxation time approximation attains the form

$$\frac{\boldsymbol{F}}{\hbar}\cdot\nabla_k F(\boldsymbol{k},\boldsymbol{r}) + \boldsymbol{v}\cdot\nabla_r F(\boldsymbol{k},\boldsymbol{r}) = -\frac{F(\boldsymbol{k},\boldsymbol{r}) - F_0(\boldsymbol{k},\boldsymbol{r})}{\tau}. \tag{3.318}$$

The first term on the left-hand side of this equation is related to the drift process (see (3.67)) and the second term accounts for the diffusion process. Now we assume that the carrier distribution function depends on the positions in the $\boldsymbol{k}$ space and in the $\boldsymbol{r}$ space: $F(\boldsymbol{k},\boldsymbol{r})$.

To solve equation (3.318), we can proceed in an analogous way as before (see equations (3.67)–(3.82)). In particular, we may again seek the non-equilibrium distribution function in the form (3.68), namely

$$F(\boldsymbol{k},\boldsymbol{r}) = F_0(\boldsymbol{k},\boldsymbol{r}) + F_1(\boldsymbol{k},\boldsymbol{r}) \tag{3.319}$$

where F_0 is the equilibrium part of the distribution function, and F_1 is a change in the distribution function caused by external fields. Now in (3.69) also the gradient of the distribution function in the $\boldsymbol{r}$ space must be taken into account. The major part of the latter is given by

$$\nabla_r F_0(\boldsymbol{k},\boldsymbol{r}) = -\frac{\partial F_0}{\partial E}\left(\nabla_r E_{\mathrm{F}} + (E - E_{\mathrm{F}})\frac{\nabla_r T}{T}\right) \tag{3.320}$$

where $\partial F_0/\partial E$ is given by (3.72). Equation (3.320) shows that the diffusion effect in a broader sense is due to two separate causes: (i) The gradient in the Fermi energy $\nabla_r E_{\mathrm{F}}$, which is related to the gradient in the carrier concentration; and (ii) the temperature gradient $\nabla_r T$. This latter 'driving force' is responsible for the thermoelectric effect. We shall keep the assumption of isothermal conditions and thus neglect the second term in (3.319).

Note that the notion of the Fermi level E_{F} was originally related to the carrier distribution function at thermodynamic equilibrium (see (2.17)). However, it also turns out that the distribution of non-equilibrium carriers can be approximated by a function of the same form as (2.17). Then the reference energy in such a function, E_F, is called the *quasi-Fermi level.* At equilibrium, the quasi-Fermi level coincides with the Fermi level. Since (3.320) describes a non-equilibrium phenomenon, E_F denotes the quasi-Fermi level.

The disturbance in the distribution function, F_1 in (3.319), can also be expressed in a form similar to (3.75):

$$F_1(\boldsymbol{k}, \boldsymbol{r}) = -\frac{\partial F_0}{\partial E} X(\boldsymbol{k}, \boldsymbol{r}) \cdot \boldsymbol{v}. \tag{3.321}$$

The vector function $\boldsymbol{X}$ now attains a more general form than before (3.80). It is given by

$$\boldsymbol{X}(\boldsymbol{k}, \boldsymbol{r}) \simeq \frac{Y(\boldsymbol{k}, \boldsymbol{r})}{1 + (e\tau/m^*)^2 B^2}$$

$$Y(\boldsymbol{k}, \boldsymbol{r}) = \tau(e\boldsymbol{E} - \nabla_r E_F) + \frac{e\tau^2}{m^*}[(e\boldsymbol{E} - \nabla_r E_F) \times \boldsymbol{B}] \tag{3.322}$$

$$+ \frac{e^2\tau^3}{m^{*2}} \boldsymbol{B}[(e\boldsymbol{E} - \nabla_r E_F) \cdot \boldsymbol{B}].$$

Therefore, an electric force and a gradient in the quasi-Fermi level are equivalent in creating transport of charge carriers. The quantity $(e\boldsymbol{E} - \nabla_r E_F)$ is a generalized driving force for the transport of carriers. Correspondingly, the quantity

$$\boldsymbol{E}_{ec} = \boldsymbol{E} - \frac{1}{e}\nabla_r E_F \tag{3.323}$$

is the equivalent electric field that accounts for both drift and diffusion transport. The equivalent field E_{ec} is also called the *electrochemical field,* since the Fermi level is also known as the *electrochemical potential.*

Using (3.322), we can now calculate the current density in the same way as we did in §3.2.5, or we may formally substitute the electric field in (3.107) by the electrochemical field (3.323). The result is given by

$$\boldsymbol{J} = e^2 K_1\left(\boldsymbol{E} - \frac{1}{e}\nabla E_F\right) + \frac{e^3}{m^*}K_2\left[\left(\boldsymbol{E} - \frac{1}{e}\nabla E_F\right) \times \boldsymbol{B}\right]$$

$$+ \frac{e^4}{m^{*2}} K_3 \boldsymbol{B}\left[\left(\boldsymbol{E} - \frac{1}{e}\nabla E_F\right) \cdot \boldsymbol{B}\right] \tag{3.324}$$

where K_1, K_2 and K_3 are the kinetic coefficients (see §3.2.6). This is the *complete basic equation*, describing the galvanomagnetic transport process of one kind of carriers under isothermal conditions. This equation should be used instead of (3.139) whenever the diffusion process cannot be neglected.

Let us express equation (3.324) in terms of conventional transport coefficients. Consider in particular non-degenerate electrons. With the aid of equations (2.24) and (2.102), we find the gradient of the Fermi level to be

$$\nabla E_{\mathrm{F}} = q\frac{D_{\mathrm{n}}}{\mu_{\mathrm{n}}}\frac{\nabla n}{n}. \tag{3.325}$$

Here D_{n} is the diffusion coefficient, μ_{n} is the drift mobility, and n is the density of quasi-free electrons. By making use of the relations (3.325), (3.150), (3.151) and (3.251), equation (3.324) attains the form

$$\begin{aligned} \boldsymbol{J}_{\mathrm{n}} = \sigma_{\mathrm{Bn}}\left(\boldsymbol{E} + \frac{D_{\mathrm{n}}}{\mu_{\mathrm{n}}}\frac{\nabla n}{n}\right) + \sigma_{\mathrm{Bn}}\mu_{\mathrm{Bn}}\left[\left(\boldsymbol{E} + \frac{D_{\mathrm{n}}}{\mu_{\mathrm{n}}}\frac{\nabla n}{n}\right) \times \boldsymbol{B}\right] \\ + Q_{\mathrm{H}}\boldsymbol{B}\left[\left(\boldsymbol{E} + \frac{D_{\mathrm{n}}}{\mu_{\mathrm{n}}}\frac{\nabla n}{n}\right) \cdot \boldsymbol{B}\right]. \end{aligned} \tag{3.326}$$

The coefficients σ_{B}, μ_{H} and Q_{H} have been defined in §3.4. Note the similarity of equations (3.250) and (3.326). In fact, the latter equation follows directly from the former if we substitute the electric field $\boldsymbol{E}$ in this equation by the electrochemical field $\boldsymbol{E}_{\mathrm{ec}}$. The electrochemical field (3.323) is then given by

$$\boldsymbol{E}_{\mathrm{ec}} = \boldsymbol{E} + \frac{D_{\mathrm{n}}}{\mu_{\mathrm{n}}}\frac{\nabla n}{n}. \tag{3.327}$$

At a weak magnetic induction, we may put $\sigma_{\mathrm{Bn}} \simeq qn\mu_{\mathrm{n}}$ (3.268) and $Q_{\mathrm{H}} \simeq qn\mu_{\mathrm{n}}^3 r_3$ (3.288). Then equation (3.326) becomes

$$\begin{aligned} \boldsymbol{J}_{\mathrm{n}} \simeq qn\mu_{\mathrm{n}}\boldsymbol{E} + qD_{\mathrm{n}}\nabla n + \mu_{\mathrm{Hn}}[qn\mu_{\mathrm{n}}\boldsymbol{E} \times \boldsymbol{B}] + \mu_{\mathrm{Hn}}[qD_{\mathrm{n}}\nabla n \times \boldsymbol{B}] \\ + r_3\mu_{\mathrm{n}}^2\boldsymbol{B}(qn\mu_{\mathrm{n}}\boldsymbol{E} \cdot \boldsymbol{B}) + r_3\mu_{\mathrm{n}}^2\boldsymbol{B}(qD_{\mathrm{n}}\nabla n \cdot \boldsymbol{B}) \end{aligned} \tag{3.328}$$

where r_3 is the scattering factor for $m = 3$ (see §3.4.1). At zero magnetic induction, equation (3.328) reduces to (2.109).

The complete equations describing the galvanomagnetic transport of holes can be obtained by replacing the subscripts n by p in equations (3.326) and (3.328), and changing the sign of the diffusion terms.

3.5.2 Magnetoconcentration effect

The magnetoconcentration effect is a variation in charge carrier concentration in a current-carrying sample due to a magnetic field. The magnetoconcentration effect is also known as the Suhl effect: it was discovered by Suhl and Shockley [30].

In its simplest form, the magnetoconcentration effect appears in a semiconductor slab of nearly intrinsic conduction (see figure 3.32) [31]. As a result of the 'magnetic pressure', both electrons and holes are pushed towards the same insulating boundary, and their concentrations build up there. This leads to the formation of a localized region of high conductivity and consequently to current crowding. The reverse happens on the other side,

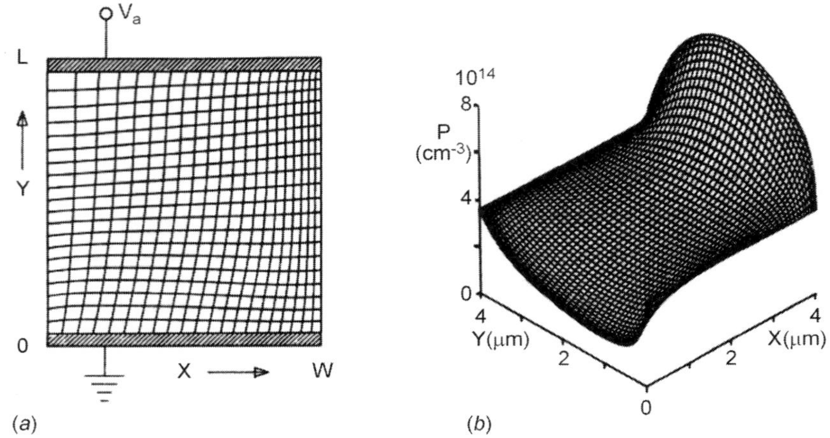

Figure 3.32. Modelling results of carrier transport in the presence of a magnetic field, demonstrating the magnetoconcentration effect. The sample was assumed to be of low-doped silicon. By putting $T = 500\,\mathrm{K}$, the material became nearly intrinsic. Parameters: $L = W$, $V = 0.1\,\mathrm{V}$, $\mu_n B = 0.21$, $\mu_p B = 0.007$. (a) Current lines (connecting the contacts) and equipotential lines (approximately parallel with the contacts). (b) Hole concentration (reprinted from [31]).

where there is a depletion of both mobile charge concentrations and current lines.

Figure 3.33 illustrates another form of the magnetoconcentration effect. A long semiconductor slab of intrinsic conduction is prepared such that the two faces of the sample, S_1 and S_2, have two very different surface

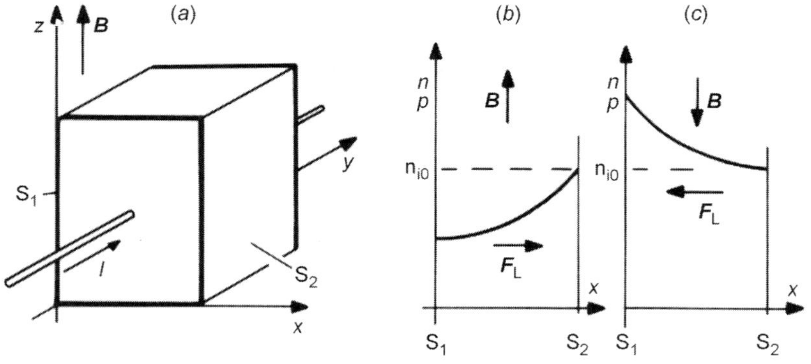

Figure 3.33. The magnetoconcentration effect in an intrinsic slab with a low recombination velocity at one surface (S_1) and a high recombination velocity at the other (S_2). (a) Sample configuration; (b) and (c), the carrier concentrations across the sample for the two opposite directions of a magnetic induction.

recombination velocities, s_1 and s_2, respectively (figure 3.33(a)). Suppose $s_1 \simeq 0$ (no recombination centres) and $s_2 \to \infty$. According to (2.113), we infer that $s_2 \to \infty$ means that the carrier concentration at the surface always stays at equilibrium. We also assume that the lifetime of carriers in the sample bulk is relatively large.

Figure 3.33(b) and (c) illustrates the carrier concentrations in the sample for various magnetic inductions. At zero magnetic field, the carrier concentrations are constant over the sample cross-section and equal to the equilibrium intrinsic concentration, $n = p = n_{i0}$. If a magnetic induction collinear with the z axis is established (figure 3.33(b)), the magnetic part of the Lorentz force pushes both electrons and holes towards the high-recombination surface S_2. The carriers recombine there, but their concentration at S_2 stays practically at equilibrium. Because the generation rate in the rest of the sample is small, it cannot compensate for the loss in carriers at the surface S_2. Therefore, the concentration of carriers in the largest part of the sample cross-section decreases. As a consequence, the resistance between the contacts increases. In the opposite direction of the magnetic field (figure 3.33(c)), everything happens analogously, but in the opposite direction, and the resistance of the sample decreases.

It is important to note that a magnetic field equally affects the concentrations of both carrier types. It must be so because of the charge neutrality condition (2.36). For this reason, the change in a sample resistance as a result of the magnetoconcentration effect is significant only if the concentrations of holes and electrons are comparable. This is why we assumed nearly intrinsic conduction in the samples discussed above.

The magnetoconcentration effect also appears as a parasitic effect in any Hall effect or magnetoresistance device. However, modern Hall devices and magnetoresistors are usually made of an extrinsic semiconductor with low-recombination surfaces. In such devices, the magnetoconcentration effect is negligible.

3.5.3 Galvanothermomagnetic effects

Galvanothermomagnetic effects are non-isothermal galvanomagnetic effects. They include the Ettingshausen effect, which is the appearance of a transverse temperature gradient in a sample as a consequence of a Hall effect taking place in the sample; the Nernst effect, which is the generation of a voltage in a Hall-device-like sample with the heat flow replacing the current; and the Righi–Leduc effect, which is a thermal analogue of the Hall effect [32].

Consider, for example, the *Ettingshausen effect*. Recall our discussion of the physical magnetoresistance effect in §3.3.6. In a long Hall device, two transverse forces act on each carrier: the magnetic force and the electric force due to the Hall electric field. Faster carriers are affected more by the magnetic force, and drift also in a transverse direction; slower carriers are affected more by the Hall field, and also move transversely but in the opposite direction. Therefore, the carriers in a Hall device are separated according to their velocities, that is, according to their kinetic energies. A result is the appearance of a transverse temperature gradient; this is the Ettingshausen effect. Thus the Ettingshausen effect and the physical magnetoresistance effect are the appearances of a single basic physical phenomenon.

Phenomenologically, the three galvanothermomagnetic effects can be characterized by the following expressions.

The temperature gradient due to the Ettingshausen effect is given by

$$\nabla T_{\mathrm{E}} = P_{\mathrm{E}}[\boldsymbol{J} \times \boldsymbol{B}] \tag{3.329}$$

where $\boldsymbol{J}$ is the current density in the device, $\boldsymbol{B}$ is the magnetic induction, and P_{E} is the Ettingshausen coefficient.

The electric field due to the Nernst effect is given by

$$\boldsymbol{E}_{\mathrm{N}} = Q_{\mathrm{N}}[\nabla T \times \boldsymbol{B}] \tag{3.330}$$

where ∇T is the driving temperature gradient and Q_{N} is the Nernst coefficient.

The temperature gradient due to the Righi–Leduc effect is given by

$$\nabla T_{\mathrm{RL}} = S_{\mathrm{RL}}[\nabla T \times \boldsymbol{B}] \tag{3.331}$$

where S_{RL} is the Righi–Leduc coefficient.

The Ettingshausen coefficient and the Nernst coefficient are related by the Bridgman relation

$$P_{\mathrm{E}} = Q_{\mathrm{N}} T/k \tag{3.332}$$

where $\boldsymbol{k}$ denotes the thermal conductivity.

The galvanothermomagnetic effects may, as parasitic effects, interfere with the Hall effect.

Example

Consider the influence of the *Ettingshausen effect* on the Hall voltage in the parallelepiped-shaped Hall device, shown in figure 3.3.

As a result of the Ettingshausen effect, a temperature difference appears between the device faces and the sense contacts. The temperature difference is given by

$$\Delta T = w P_{\mathrm{E}} J B_{\perp} \tag{3.333}$$

where we have made use of (3.329). Since the current density is given by $J = I/wt$, (3.333) becomes

$$\Delta T = \frac{SP_{\mathrm{E}}}{t} I B_{\perp}. \tag{3.334}$$

As a result of the thermoelectric effect, this temperature difference produces a voltage difference between the sense contacts:

$$\Delta V_{\mathrm{E}} = \frac{SP_{\mathrm{E}}}{t} IB_{\perp} \tag{3.335}$$

where S is the thermoelectric power coefficient. The voltage ΔV_{E} adds to the Hall voltage and produces an error in it. The relative error in the Hall voltage due to the combined Ettingshausen/thermoelectric effect is given by

$$\frac{\Delta V_{\mathrm{E}}}{V_{\mathrm{H}}} = \frac{SP_{\mathrm{E}}}{R_{\mathrm{H}}} \tag{3.336}$$

where we have made use of (3.335) and (3.18). With the aid of (3.332), the above equation yields

$$\frac{\Delta V_{\mathrm{E}}}{V_{\mathrm{H}}} = \frac{SQ_{\mathrm{N}}T}{kR_{\mathrm{H}}}. \tag{3.337}$$

Substituting here the numerical values for low-doped n-type silicon [22], $S = 1\,\mathrm{mV\,K^{-1}}$, $Q_{\mathrm{N}} = 0.5\,\mathrm{cm^2\,s^{-1}\,K^{-1}}$, $T = 300\,\mathrm{K}$, $k = 1.5\,\mathrm{W\,cm^{-1}\,K^{-1}}$ and $R_{\mathrm{H}} = 1.4 \times 10^{-3}\,\mathrm{m^3\,C^{-1}}$ (3.19), we obtain $\Delta V_{\mathrm{E}}/V_{\mathrm{H}} \simeq 0.7 \times 10^{-4}$.

3.5.4 The Hall effect at a high electric field

In discussing the Hall effect so far, we have assumed a weak electric field. Now we shall briefly see what happens if the applied electric field becomes very strong. We aim here at an intuitive understanding of the main phenomena involved. A more detailed treatment of the subject can be found in the review paper [33].

In §2.6.2, we discussed the drift of carriers in a strong electric field. Recall the main features of this transport phenomenon: the carriers get hot, phonon scattering dominates, and the drift velocity saturates. But the electron gas may still have a distribution function reminiscent of the Maxwell–Boltzmann function, in spite of the high electric field. This is because, at higher carrier energy, the relaxation time for acoustic phonon scattering decreases (see table 2.2). A short relaxation time means a little disturbance in symmetry of the distribution function (see (3.80)–(3.82)). Note, however, that the function $F_0(\boldsymbol{k})$ in (3.81) now denotes simply the even part of the distribution function, and not its equilibrium part.

At high electric fields, we must distinguish between two different relaxation processes: momentum relaxation and energy relaxation. The *momentum relaxation process* relies on collisions that scatter carriers efficiently into accidental directions. If frequent enough, such carrier scattering keeps the spatial central symmetry of the distribution function almost conserved, in spite of unidirectional forces that may act upon the carriers. But these scattering events may be almost elastic, and thus very inefficient in dissipating carrier energy. This is exactly the case with acoustic phonon scattering (see

(2.78)). The *energy relaxation process* relies on scattering events with a higher-energy transfer efficiency, such as the creation of optical phonons. The energy relaxation events limit the temperature of hot carriers, and play a dominant role during the process of cooling carriers after switching off the high electric field.

The efficiency of momentum relaxation by acoustic phonons increases with the crystal temperature (see table 2.2). Thus at a high-enough temperature, the disturbance in the central symmetry of the hot carrier distribution function stays small, in spite of a high electric field. It turns out that room temperature is already high enough for this. Therefore, with good reason, we may also apply the theory of the Hall effect, developed in §§3.2 and 3.3, to the case of a high electric field. In particular, the basic equations of the Hall effect, (3.139) and (3.141), also hold at a very strong electric field. However, some of the results of the application of these equations are quite different in the cases of weak and strong electric fields.

First of all, the hot carrier distribution function is less affected by a magnetic field than that corresponding to a weak electric field. This is obvious from (3.80): with a decrease in the relaxation time τ, the terms proportional to τ^2 and τ^3 become less important relative to the pure drift term, which is proportional to τ. A consequence of this fact is that the strong magnetic field condition (3.131) becomes hard to reach with a strong electric field. The same conclusion can be inferred by taking into account the decrease of the drift mobility at high electric fields (2.99).

Another general consequence of the decrease in the relaxation time and the drift mobility is a decrease in the magnetic field dependence of all transport coefficients. In particular, the magnetoresistance effect becomes rather weak at a high electric field.

Let us consider now some galvanomagnetic coefficients and quantities of hot carriers. We assume the simple model with the isotropic effective mass of carriers increase in electric field.

The Hall scattering factor $r_{H0} = r_2$ (3.263) does not change much with the increase in electric field. As the carrier temperature increases, the Hall scattering factor approaches the value characteristic for the acoustic phonon scattering (see table 3.1). This is because, at a high electric field, the dominant momentum scattering mechanism is the acoustic phonon scattering.

The Hall mobility (3.275) is strongly affected by a high electric field. Since drift mobility decreases with the increase in electric field (2.99), and the Hall factor stays close to unity, the Hall mobility also decreases at a high electric field. From (3.275) and (2.99), it follows that

$$\mu_H(|\boldsymbol{E}| \to \infty) \simeq \operatorname{sign}[e] r_H \frac{v_{d,sat}}{|\boldsymbol{E}|}. \tag{3.338}$$

Detailed numerical calculations for electrons in silicon corroborate this result [34].

A natural consequence of the decrease of Hall mobility is a reduction of the Hall angle (3.152) at a high electric field. This has been experimentally observed [35].

The Hall coefficient (3.298) of an extrinsic semiconductor varies little with the increase in the electric field, as long as the field does not alter the carrier concentration. Then the Hall coefficient may vary only as a result of a change in the Hall scattering factor, which is small.

A strong increase in the Hall coefficient of n-type germanium was observed at 200 K and with a high electric field [36]. The increase is due to electron transfer to a higher-energy valley, where the mobility of electrons is small. With the aid of (3.204), it can be shown that such a redistribution of carriers brings about an increase in the average Hall coefficient. Otherwise, a sudden and strong decrease in the Hall coefficient is expected when carrier generation by impact ionization begins.

The Hall electric field tends to saturate at a finite value when the driving electric field becomes very high. This is obvious from (3.8), if we bear in mind that the drift velocity saturates at high electric fields:

$$|\boldsymbol{E}_{\mathrm{H,sat}}| \simeq v_{\mathrm{d,sat}} B \tag{3.339}$$

where $v_{\mathrm{d,sat}}$ is the saturation drift velocity. Alternatively, by substituting (3.338) into (3.186), we obtain

$$|\boldsymbol{E}_{\mathrm{H,sat}}| \simeq r_{\mathrm{H}} v_{\mathrm{d,sat}} B. \tag{3.340}$$

Since $r_{\mathrm{H}} \simeq 1$, the last two equations are almost equal. By substituting here $v_{\mathrm{d,sat}} \simeq 10^7\,\mathrm{cm\,s^{-1}}$, we find the physical limit of the Hall electric field for a given magnetic induction [37]:

$$|\boldsymbol{E}_{\mathrm{H,sat}}|/B \lesssim 10^5\,\mathrm{V\,m^{-1}\,T^{-1}}. \tag{3.341}$$

This value is appropriate for many semiconductor materials of practical importance, for the saturation velocities of carriers in various semiconductors are roughly equal.

At very low temperatures, say below 77 K, other physical effects dominate. Then the acoustic phonons are only a little excited, and the corresponding scattering mechanism is of little importance. The carrier drift velocity soon becomes limited by the emission of optical phonons, much as we supposed in our model in §2.6.2. This type of carrier drift is called the streaming motion [38]. If we neglect acoustic phonons, the free transit time then equals the time elapsed until an electron acquires the energy of an optical phonon from the electric field. Thus at a very low temperature and high electric field, the transit times of all electrons become almost equal. Accordingly, the Hall scattering factor reduces to unity.

At very low temperatures, it is easy to achieve the conditions of combined strong electric and strong magnetic fields. Scattering events are then scarce, and the situation is similar to that we described in connection with figure 3.10(b). In particular, the Hall angle tends to saturate at $\Theta_H \simeq \pi/2$ [38].

3.5.5 The piezo-Hall effect

The piezo-Hall effect is the alteration of the Hall voltage of a Hall device upon the application of a mechanical force. The effect is best characterized by the mechanical stress dependence of the Hall coefficient. The relative change in the Hall coefficient is proportional to stress $\boldsymbol{X}$:

$$\frac{\Delta R_H}{R_H} = PX \tag{3.342}$$

where P denotes the piezo-Hall coefficient. Figure 3.34 shows an example of the experimental set-up and the measurement results of the piezo-Hall effect.

The piezo-Hall effect is closely related to the piezoresistance effect (see §2.7.4). Similarly as in the piezoresistance effect, the piezo-Hall effect is stronger in semiconductors with anisotropic and degenerated energy minima than in isotropic semiconductors.

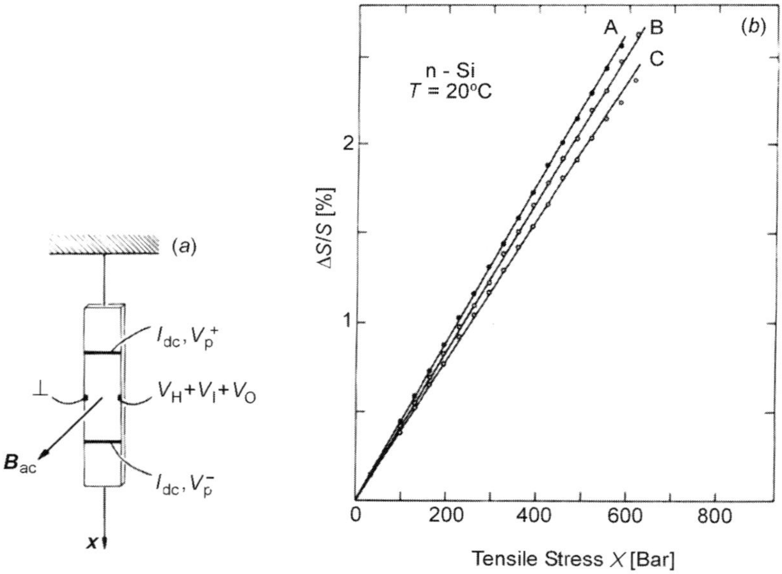

Figure 3.34. The sensitivity of a Hall device (which is proportional to the Hall coefficient R_H; see chapter 5) depends on a mechanical stress. (a) Experimental set-up: a Hall device on a long semiconductor strip. The stress $\boldsymbol{X}$ was realized by hanging weights on the string. (b) The relative change in sensitivity against stress of silicon Hall devices with $\boldsymbol{B} \parallel \langle 100 \rangle$ and $\boldsymbol{X} \parallel \boldsymbol{J} \parallel \langle 010 \rangle$. The devices were phosphorus doped, with A: $N_D \simeq 1.8 \times 10^{14}\,\mathrm{cm}^{-3}$; B: $N_D \simeq 1.5 \times 10^{15}\,\mathrm{cm}^{-3}$; C: $N_D \simeq 6 \times 10^{15}\,\mathrm{cm}^{-3}$ (adapted from [39]).

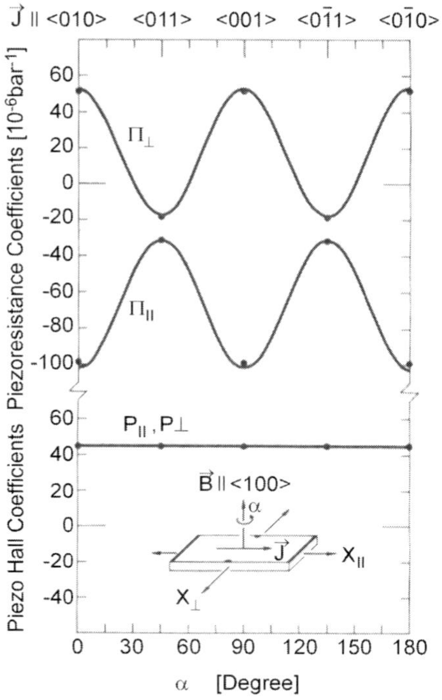

Figure 3.35. Piezoresistance (π) and piezo-Hall (P) coefficients of n-type silicon for the current flow in the (100) crystal plane: $\|$, longitudinal; $\perp$, transverse; $\alpha = 0$ for $\boldsymbol{J} \parallel \langle 010 \rangle$. Full line, theory; points, experimental; $T = 20\,^\circ\mathrm{C}$, $n = 1.8 \times 10^{14}\,\mathrm{cm}^{-3}$ (reprinted from [39]).

In a many-valley semiconductor, such as n-type silicon, the piezo-Hall effect is due to the repopulation of the valleys upon the application of a stress. With no stress, all six energy valleys of a silicon crystal are, on average, equally populated with electrons. If the crystal is uniaxially stressed, the conduction band edge at some energy minima sinks, and that at others rises. Owing to the intervalley scattering process, the electrons repopulate the valleys: the lower valleys are then more populated, the higher less.

According to the theory of the Hall effect in many-valley semiconductors (§3.3.8), a change in the concentrations of electrons associated with different valleys must bring about a change in the Hall factor (3.233). A change in the Hall factor shows up as a change in the Hall coefficient; hence relation (3.342).

The piezoresistance and the piezo-Hall effect in n-type silicon are roughly equally strong [39]. Figures 3.35 and 3.36 show the corresponding coefficients for two crystal planes and various current directions. It was

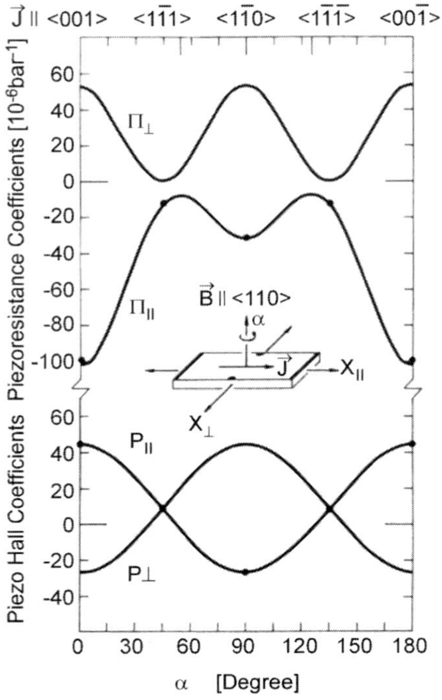

Figure 3.36. Piezoresistance (π) and piezo-Hall (P) coefficients of n-type silicon for the current flow in the (110) crystal plane. $\|$, longitudinal; $\perp$, transverse; $\alpha = 0$ for $\boldsymbol{J} \parallel \langle 001 \rangle$. Full line, theory; points, experimental. $T = 20\,^\circ\mathrm{C}$, $n = 1.8 \times 10^{14}\,\mathrm{cm}^{-3}$ (reprinted from [39]).

also found that the piezo-Hall coefficient decreases with an increase in temperature and at higher doping concentrations.

References

[1] Hall E H 1879 On a new action of the magnet on electric currents *Am. J. Math.* **1** 287–92

[2] Kireev P S 1974 *Semiconductor Physics* ch 4 (Moscow: Mir)

[3] Smith R A 1987 *Semiconductors* (Cambridge: Cambridge University Press)

[4] Putley E H 1960 *The Hall Effect and Semi-conductor Physics* (New York: Dover)

[5] Beer A C 1963 Galvanomagnetic effects in semiconductors *Solid-State Phys.* suppl. 4 (New York: Academic Press)

[6] Nag B R 1972 *Theory of Electrical Transport in Semiconductors* (Oxford: Pergamon)

[7] Seeger K 1989 *Semiconductor Physics* (Berlin: Springer)

[8] Askerov B M 1985 *Electronye Javljenija Perenosa v Poluprovodnikah* (Moscow: Nauka)

[9] Corbino O M 1911 Elektromagnetische Effekte, die von der Verzerrung herriihren, welche ein Feld an der Bahn der Ionen in Metallen hervorbringt *Phys. Z.* **12** 561

[10] Andor L, Baltes, H B, Nathan A and Schmidt-Weinmar H G 1985 Numerical modeling of magnetic-field sensitive semiconductor devices *IEEE Trans. Electron Devices* **ED-32** 1224–30

[11] Herring C and Vogt E 1956 Transport and deformation-potential theory for many-valley semiconductors with anisotropic scattering *Phys. Rev.* **101** 944–61

[12] Abramowitz M and Stegun I A (eds) 1972 *Handbook of Mathematical Functions* (New York: Dover) p 255

[13] Sze S M 1981 *Physics of Semiconductor Devices* (New York: Wiley) pp 50–1

[14] Herring C 1955 Transport properties of a many-valley semiconductor *Bell Syst. Tech. J.* **34** 237–90

[15] Long D 1960 Scattering of conduction electrons by lattice vibrations in silicon *Phys. Rev.* **120** 2024–32

[16] Eagles P M and Edwards D M 1965 Galvanomagnetic coefficients in semiconductors with spherical energy surfaces and strong ionized impurity scattering *Phys. Rev.* **138** A1706–17

[17] For example: Ref. 7, pp 216–54, and references therein

[18] Beer A C 1980 Hall effect and beauty and challenges of science *The Hall Effect and its Applications* ed. C L Chien and C R Westgate (New York: Plenum) pp 229–338

[19] Goldberg C and Davis R E 1954 New galvanomagnetic effect, *Phys. Rev.* **94** 1121–5

[20] Pierre-Andre Besse, *Private communication*, EPFL, October 2002

[21] See, for example, equation (15.18) in: O'Handley R C 2000 *Modern Magnetic Materials* (New York: Wiley)

[22] Landolt-Bornstein 1982 Numerical and functional relationships in science and technology *Semiconductors* vol. III/17a (Berlin: Springer)

[23] Okta E and Sakata M 1978 Temperature dependence of Hall factor in low compensated n-type silicon *Jpn. J. Appl. Phys.* **17** 1795–804

[24] Rijks H J, Giling L J and Bloem J 1981 Influence of charge carrier scattering on the exact form of the Hall curve *J. Appl. Phys.* **52** 472–5

[25] del Alamo J A and Swanson R M 1985 Measurement of Hall scattering factor in phosphorus-doped silicon *J. Appl. Phys.* **57** 2314–17

[26] Lin J F, Li S S, Linares L C and Teng K W 1981 Theoretical analysis of Hall factor and Hall mobility in p-type silicon *Solid-State Electron.* **24** 827–33

[27] Anderson D A and Apsley N 1986 The Hall effect in III–V semiconductor assessment *Semicond. Sci. Technol.* **1** 187–202

[28] Popovic R S and Halg B 1988 Nonlinearity in Hall devices and its compensation *Solid-State Electron.* **31** 681–8

[29] Ref. 22, p 535

[30] Suhl H and Shockley W 1949 Concentrating holes and electrons by magnetic fields *Phys. Rev.* **75** 1617–18

[31] Baltes H and Nathan A 1989 Integrated magnetic sensors *Sensor Handbook 1* ed. T Grandke and W H Ko (Weisheim: VCH) pp 195–215

[32] Ref. 7, pp 91–7

[33] Kachlishvili Z S 1976 Galvanomagnetic and recombination effects in semiconductors in strong electric fields *Phys. Status Solidi* (a) **33** 15–51

[34] Heinrich H and Kriechbaum M 1970 Galvanomagnetic effects of hot electrons in n-type silicon, *J. Phys. Chem. Solids* **31** 927–38

[35] Kordic S 1987 Offset reduction and three-dimensional field sensing with magneto-transistors *PhD Thesis* Delft University of Technology

[36] Heinrich H, Lischka K and Kreichbaum M. 1970 Magnetoresistance and Hall effect of hot electrons in germanium and carrier transfer to higher minima *Phys. Rev.* B **2** 2009–16

[37] Popovic R S 1984 A MOS *Hall device free from short-circuit effect Sens. Actuators* **5** 253–62

[38] Komiyama S, Kurosawa T and Masumi T 1985 Streaming motion of carriers in crossed electric and magnetic fields *Hot-Electron Transport in Semiconductors* ed. L Reggiani (Berlin: Springer)

[39] Halg B 1988 Piezo-Hall coefficients of n-type silicon *J. Appl. Phys.* **64** 276–82

Chapter 4

Hall plates and magnetoresistors. Basic properties

In the previous chapter, we were interested in the basic theory of the galvanomagnetic effects. The devices in which these effects take place were of secondary importance: we considered them just as means to study the effects. Now we shall move our attention closer to practical aspects of the galvanomagnetic effects: we shall study the basic properties of Hall plates and magnetoresistors. Recall that a Hall plate is a four-terminal solid-state electron device, similar to that with which Hall discovered his effect [1]; and in principle, a magnetoresistor is a two-terminal version of a Hall plate. We have already met Hall devices in previous chapters (see, for example, figure 3.3).

The discussion in this chapter shall be rather general and thus applicable to any Hall or magnetoresistance device, irrespective of the used fabrication technology or of the intended application of the device. Therefore we shall analyse the galvanomagnetic devices in their simplest and idealized form: they are plate-like, made of uniform and single-carrier-type material, and fitted at the sides with ideal ohmic contacts. Further application-, design-, and technology-specific properties of Hall devices will be considered in the later chapters.

Some of the issues discussed in this chapter have been treated in the books [2–4] and the review papers [5–7].

4.1 Galvanomagnetic effects: integral quantities

At the device level, it is convenient to describe the galvanomagnetic effects using easily-measurable, so-called integral or 'global' quantities. Let us consider a Hall plate shown in figure 4.1. The plate is biased via the input terminals C1, C2 by the input voltage V_{in} driving a current I. The input voltage and the current are related by the device input resistance: $V_{in} = R_{in}I$. If the plate is exposed to a magnetic induction $\boldsymbol{B}$, then R_{in}

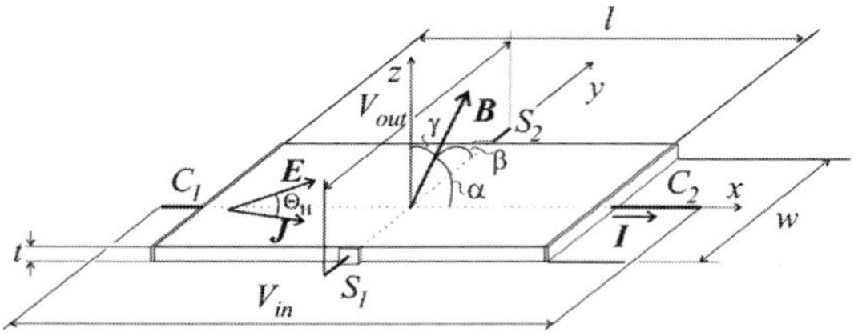

Figure 4.1. A rectangular Hall plate with local and global galvanomagnetic quantities. The local galvanomagnetic transport quantities are the electric field $\boldsymbol{E}$, the current density $\boldsymbol{J}$ and the Hall angle Θ_{H}. The global (observable) galvanomagnetic quantities are the input voltage V_{in}, the input current I and the output voltage V_{out}. The vector $\boldsymbol{B}$ encloses the angles α, β and γ with the axes x, y and z, respectively. Compare with figures 3.3 and 3.22.

changes, and an output voltage V_{out} appears across the terminals S1, S2. V_{in}, I, R_{in}, V_{out} are examples of integral (or 'global') quantities that are now of our interest. Obviously, the integral quantities depend on the local (differential) quantities, such as the electric field $\boldsymbol{E}$ and the current density $\boldsymbol{J}$, which we used to describe the galvanomagnetic effects in chapter 3. In this section we shall find the relationships between the local and the integral galvanomagnetic quantities.

4.1.1 From the basic equation to integral quantities

Let us assume that our Hall plate shown in figure 4.1 is very long, i.e. $l \gg w$, and that the sense contacts S1, S2 are very small. The corresponding boundary conditions (§3.3.4) impose that the current density vector $\boldsymbol{J}$ is collinear with the x axis. Then, if we know the device current I, we can find the current density as

$$\boldsymbol{J} \simeq \frac{I}{tw}\boldsymbol{i}. \tag{4.1}$$

Here t and w are the thickness and the width of the plate, and $\boldsymbol{i}$ is the unit vector collinear with the x axis (see figure 4.1).

The electric field in the plate is given by the basic equation (3.243). For easy reference, we repeat this equation:

$$\boldsymbol{E} = \rho_{\mathrm{b}}\boldsymbol{J} - R_{\mathrm{H}}|\boldsymbol{B}|[\boldsymbol{J} \times \boldsymbol{b}] + P_{\mathrm{H}}B^2(\boldsymbol{J} \cdot \boldsymbol{b})\boldsymbol{b}. \tag{4.2}$$

Now we can calculate the other integral quantities of interest as follows.

The *input voltage of the device*, i.e. the voltage drop between the two current contacts C1 and C2, is given by

$$V_{\mathrm{in}} = \int_{\mathrm{C1}}^{\mathrm{C2}} \boldsymbol{E}\,\mathrm{d}\boldsymbol{x} \simeq \boldsymbol{E} \cdot \boldsymbol{i}\,l. \tag{4.3}$$

By substituting here (4.2) and (4.1), we obtain

$$V_{\text{in}} \simeq (\rho_b + P_H B^2 \cos^2 \alpha) \frac{l}{tw} I. \tag{4.4}$$

Note that the current density vector $\boldsymbol{J}$ is collinear with the unit vector $\boldsymbol{i}$ and, therefore, the product $[\boldsymbol{J} \times \boldsymbol{B}] \cdot \boldsymbol{i}$ vanishes.

The term in the parentheses of (4.4) is obviously the *effective resistivity* of our very long device. We have discussed the effective resistivity before—see (3.254), which we repeat:

$$\rho(B) \simeq \rho_b + P_H B^2 \cos^2 \alpha. \tag{4.5}$$

From (4.4) and (4.5), we can deduce the expression for the resistance of a very long rectangular sample exposed to a magnetic field:

$$R(B) = \rho(B) \frac{l}{tw}. \tag{4.6}$$

This expression has the same form as the standard formula for the resistance of a rectangular sample. The only difference is that here the resistivity depends on the magnetic field due to the intrinsic magnetoresistance effect.

The output voltage of the Hall plate, i.e. the voltage between the two sense contacts S1 and S2, is given by

$$V_{\text{out}} = \int_{S1}^{S2} \boldsymbol{E}\, \mathrm{d}\boldsymbol{y} \simeq \boldsymbol{E} \cdot \boldsymbol{j} w. \tag{4.7}$$

Here $\boldsymbol{j}$ denotes the unit vector collinear with the y axis. By substituting here (4.2) and (4.1), we obtain

$$V_{\text{out}} = V_H + V_P \tag{4.8}$$

$$V_H \simeq -\frac{R_H}{t} IB \cos \gamma \tag{4.9}$$

$$V_P \simeq \frac{P_H}{t} IB^2 \cos \alpha \cos \beta. \tag{4.10}$$

We recognize the first term in (4.8), i.e. V_H (4.9), as the *Hall voltage* (see (3.18)). Here R_H is the Hall coefficient (§3.4.6), t is the plate thickness, I is the bias current, and $B \cos \gamma = B_\perp$ is the component of the magnetic induction perpendicular to the device plane.

The second term in (4.8), i.e. V_P (4.10), is called the *planar Hall voltage*. Here P_H is the planar Hall coefficient (§3.4.7), and $B \cos \alpha$ and $B \cos \beta$ are the components of the magnetic induction collinear with the x and the y axes, respectively. Recall that we discussed the planar Hall effect in §3.3.9. An expression equivalent with (4.10) can be obtained by integrating the planar Hall electric field (3.238) over the device width w. This gives

$$V_P \simeq \frac{P_H}{t} IB_P^2 \frac{1}{2} \sin 2\alpha. \tag{4.11}$$

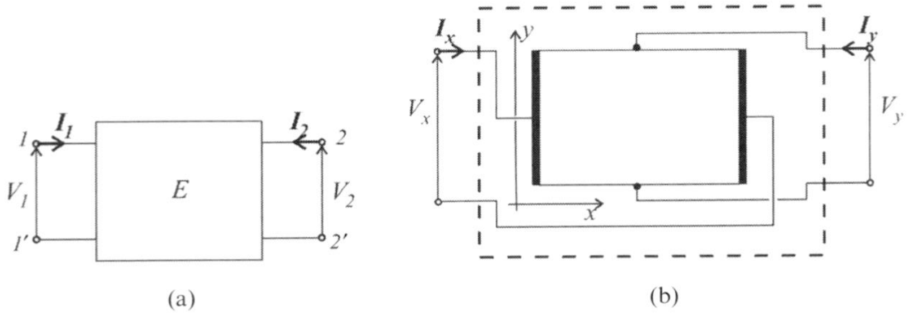

Figure 4.2. Two-port representation of a Hall plate. (a) Two-port black-box element. (b) A Hall plate imbedded into the black box.

Equation (4.10) reduces to this one for a planar magnetic field, $\boldsymbol{B} \rightarrow \boldsymbol{B}_{\mathrm{P}}$.

Note that we derived the above expressions for the integral quantities for the case of a very long Hall plate. If this condition is not valid, then the above expressions must be corrected (see §4.2).

4.1.2 Hall plate as a two-port device

We often find in the literature a description of the behaviour of Hall devices using the language of circuit theory. This is especially useful when the physical details of the Hall plate are not known and one has to determine the parameters of the plate experimentally. Then a Hall device is represented as a 'black box', a passive circuit element with two ports, as shown in figure 4.2(a). Usually, the Hall plate is imbedded in the black-box so that the input voltage produces in the plate an electric field and a current essentially collinear with the x axis (see figure 4.2(b)). Therefore, the (input) voltage and current of the port 1 are denoted with V_x and I_x, and the (output) voltage and current of the port 2 are denoted with V_y and I_y, respectively.

Then the function of the Hall plate can be described by the following *matrix equation*:

$$\begin{bmatrix} V_x \\ V_y \end{bmatrix} = \begin{bmatrix} R_{xx} & R_{xy} \\ R_{yx} & R_{yy} \end{bmatrix} \begin{bmatrix} I_x \\ I_y \end{bmatrix} \tag{4.12}$$

where the R matrix contains the parameters of the Hall plate. Note that the elements of the R matrix have the dimensions of resistance. We shall now find the relationships between the elements of the R matrix and the physical parameters of a Hall plate.

In most applications of a Hall plate, the output current is negligibly small, i.e. $I_y \cong 0$. Therefore, (4.12) reduces to

$$V_x \simeq R_{xx} I_x \qquad V_y \simeq R_{yx} I_x. \tag{4.13}$$

By comparing the first of these equations with (4.4)–(4.6), we readily find

$$R_{xx} = R(B) = \rho(B)\frac{l}{tw} \tag{4.14}$$

where $\rho(B)$ is the effective resistivity of the plate, and l, t, and w are the dimensions of the plate defined in figure 4.1.

By comparing the second equation (4.13) with (4.9)–(4.11), we find that the off-diagonal matrix element is given by

$$R_{yx} = -\frac{R_H}{t}B\cos\gamma + \frac{P_H}{t}B^2\cos\alpha\cos\beta. \tag{4.15}$$

For the case of a perpendicular magnetic field, this reduces to

$$R_{yx} \simeq -\frac{R_H}{t}B \tag{4.16}$$

which is often cited in the literature.

All variables and parameters of the matrix equation (4.12) are integral quantities. We can formally transform this equation into a 'local' form by expressing the currents in it by the current densities using (4.1) and multiplying (4.12) by the matrix $[1/l \quad 1/w]$. In this way we obtain

$$\begin{bmatrix} E_x \\ E_y \end{bmatrix} = \begin{bmatrix} \rho_{xx} & \rho_{xy} \\ \rho_{yx} & \rho_{yy} \end{bmatrix} \begin{bmatrix} J_x \\ J_y \end{bmatrix}. \tag{4.17}$$

This equation, which is identical with (3.248), relates the planar electrical field and the current density components in the plate. Note that the elements of the ρ matrix have the dimensions of resistivity. By comparing (4.17) with the standard basic equation (4.2), we can find the relations of the ρ matrix elements with the conventional parameters of a Hall plate. For the case of a *perpendicular magnetic field*, the relationships are as follows:

$$\rho_{xx} = \rho_b \tag{4.18}$$

$$\rho_{yx} = -R_H B. \tag{4.19}$$

See also (3.249).

4.2 Influence of the plate geometry

In our discussions up to now we always considered Hall plates of particularly simple shapes: the plates were rectangular and either very long or very short. This greatly simplified our theoretical analysis and gave us an understanding of the asymptotic limits of the galvanomagnetic effects. But in reality the geometry of galvanomagnetic devices is rarely close to these extremes. We shall now analyse the influence of the geometry of more realistic Hall plates on the observed galvanomagnetic effects.

4.2.1 Plates neither long nor short. Geometrical factors

Let us have a look at the distributions of the current density and the electrical potential in a 'realistic' galvanomagnetic sample. Figure 4.3 shows such a two-terminal rectangular plate, which is somewhere between 'very long' and 'very short'. By inspecting the current and the equipotential lines, we notice the following. Near the two isolating boundaries, current is forced to flow along the boundaries and the equipotential lines are strongly inclined: a Hall electric field is generated. We infer that in these regions also the physical magnetoresistance effect shows up. Near the two ohmic contacts, the Hall field is short-circuited, so the current deflection and the geometrical magnetoresistance effects prevail. In the regions far away from both the isolating boundaries and the ohmic contacts (in the middle of the device), both current and equipotential lines are moderately inclined. Therefore, we may imagine that in a galvanomagnetic device, which is neither very long nor very short, a combination of the effects inherent to the two extreme device shapes appears.

It is then to be expected that a device with an intermediate geometry has external characteristics that are somewhere between those of a very long and a very short device. Really, it turns out that the Hall voltage of a Hall plate with an arbitrary shape can be expressed as

$$V_H = G_H V_{H\infty} \tag{4.20}$$

where G_H is a parameter called the *geometrical correction factor of Hall voltage*, and $V_{H\infty}$ denotes the Hall voltage in a corresponding infinitely long strip, given by (4.9). 'Corresponding' means that the two devices have identical Hall coefficients and thickness, are biased by identical currents,

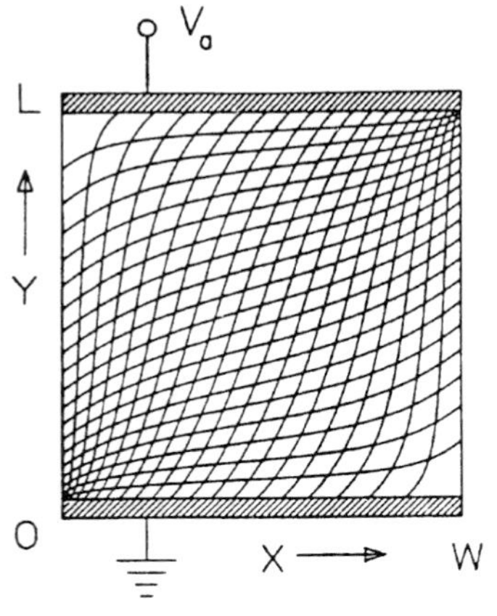

Figure 4.3. A result of the numerical modelling of a square Hall plate ($l = w$) with no sense contacts, exposed to a perpendicular magnetic field, with $\mu B = 0.21$. Shown are current lines (connecting the contacts) and equipotential lines (approximately parallel with the contacts) [8].

and are exposed to identical homogeneous magnetic inductions. If there is no danger of confusion, we shall further on often drop the precision 'correction of Hall voltage' and refer to G_H as the 'Hall geometrical factor' or simply as a 'geometrical factor'. The geometrical factor summarily represents the diminution of the Hall voltage due to a non-perfect current confinement in a finite length Hall device. The Hall geometrical factor is a number limited by $0 < G_H < 1$. For a very long Hall device, $G_H \approx 1$; for a very short Hall device, $G_H \approx 0$.

Similarly, for the resistance of a device exposed to a magnetic field we may write

$$R(B) = G_R R_\infty(B) \tag{4.21}$$

where G_H is a parameter called the *geometrical factor of magnetoresistance*, $R(B)$ is the resistance of a considered magnetoresistor at an induction B, and $R_\infty(B)$ denotes the resistance of a corresponding part of an infinitely long strip, given by (4.6).

The geometrical factors are important parameters of real Hall and magnetoresistance devices. To determine the value of G_H for a particular Hall plate shape according to (4.20), one must somehow calculate the Hall voltage for a plate of this shape V_H, and the Hall voltage of the corresponding infinitely long device $V_{H\infty}$ (4.9). An analogous procedure applies for calculating G_R according to (4.21).

The Hall voltages and resistances for plates of various shapes have been calculated using the following methods: conformal mapping techniques [9–13], boundary element methods [14], and finite difference [15, 16] or finite element [17] approximations. The latter two approximations have also been used to construct Hall device analogues in the form of electrical networks [18, 19] that may be solved using a circuit-analysis program [19]. Most of the works on the subject up to 1980 are reviewed in [14]. Below we shall briefly present two of these calculation methods.

4.2.2 Conformal mapping: method and results

We shall now sketch the main features and results of the conformal mapping method. This method is very convenient for developing an intuitive understanding of the influence of geometry on the characteristics of a galvanomagnetic device. Furthermore, the method yields simple and very useful analytical formulae for the geometrical factors. The method is equally applicable for both Hall devices and magnetoresistors. By way of example, we shall take the more complex device, which is a Hall plate.

The conformal mapping method for evaluating Hall devices was first proposed by Wick [9]. Generally, the conformal mapping method is used to solve the Laplace equation (3.137) in two dimensions, subject to some boundary conditions [20].

To be solvable by this method, a Hall device must fulfil the following conditions.

(i) No space charge in the active region: see condition (ii) of §3.3.1.
(ii) Two-dimensionality: the device must be planar and plate-like. For a plate of a finite thickness, no quantity, such as material resistivity, may vary in the direction perpendicular to the device plane.
(iii) Uniformity: the magnetic induction and the galvanomagnetic material properties must not vary over the device plane.

Condition (ii) practically excludes the application of the method to low-doped devices and devices with a badly prepared surface. In such devices, the magnetoconcentration effect may occur (see §3.5.2). Condition (iii) is related to an essential property of the conformal mapping, as explained below.

In the conformal mapping method, one region of, say, a complex t plane is mapped on to a region of another shape in the complex w plane. Thus the crossing angle of any two lines in the original t plane equals the crossing angle of the images of these two lines in the w plane. Suppose that these two lines are a current density line and an electric field line of a Hall device. Then the crossing angle of the two lines is the Hall angle. As we know by now, in a Hall plate with homogeneous properties, exposed to a uniform magnetic induction, the Hall angle is constant over the whole device area.

The main task in conformal mapping is to find a function that maps the active region of the analysed Hall device into an image, which is so simple that it can be solved by inspection. This is in principle possible only if the Hall angle is constant over the whole device area. Hence the above condition (iii).

Figure 4.4 illustrates the steps in a conformal mapping analysis of a rectangular Hall plate. We shall refer to the Hall device to be analysed as the original. The original is embedded into the complex t plane (figure 4.4(a)). Typically, the original is first mapped conformally on to an intermediate complex z plane. The active region of the original is mapped on to the upper half of the z plane such that the boundary of the original goes into the real x axis of the z plane (figure 4.4(b)). For polygon-shaped originals, this intermediate mapping is accomplished by using the Schwarz–Christoffel transformation [9, 11], and for a circle-shaped original, by a bilinear transformation [13]. This mapping could intuitively be imagined as the stretching of a rectangle, as illustrated in the inset of figure 4.4.

Finally, the upper half of the z plane is conformally mapped on to the complex w plane, such that the real x axis of the z plane goes into the boundary of a skewed parallelogram (see figure 4.4(c)). This is again done by the Schwarz–Christoffel transformation. The tilting angle of the parallelogram is chosen to be equal to the Hall angle Θ_H. The two fins are the folded images of the two side contacts 2-3-4 and 7-8-9.

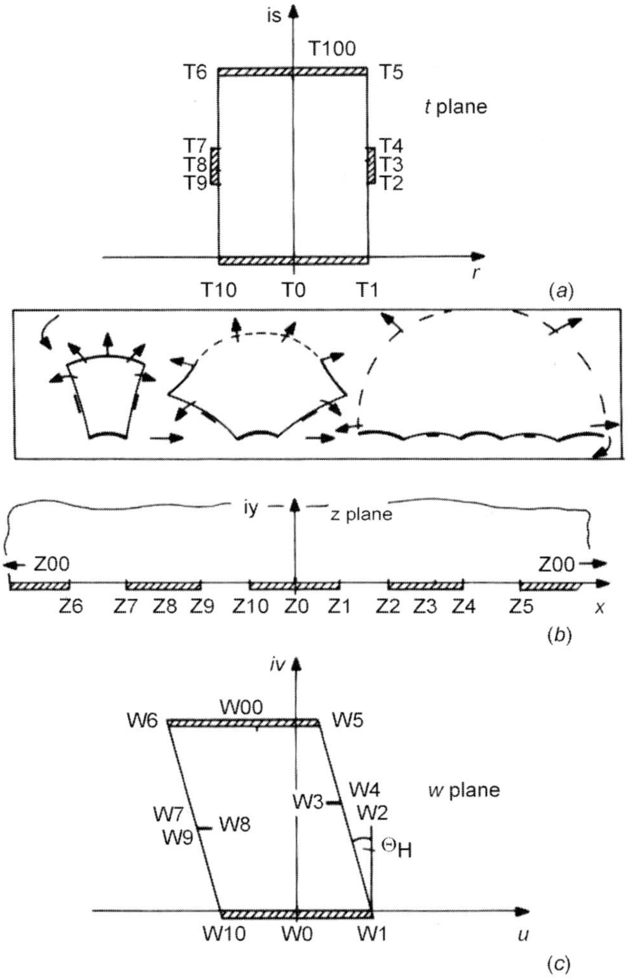

Figure 4.4. Transforming a rectangular Hall plate (a) into the half plane device (b), and then into a skewed Hall plate (c). Wi, Zi and Wi, with $i = 0, 1, 2, \ldots, 10$, are the corresponding points. The inset shows three steps in an intuitive transformation, that is, stretching of the rectangular into the half plane.

The Schwarz–Christoffel transformation, which maps conformally the upper half of the complex z plane on to an n-sided polygon in the t plane, is given by

$$t(z) = A \int_0^z \frac{\mathrm{d}u}{\prod_{k=1}^{n} (u - x_k)^{\alpha_k}}. \tag{4.22}$$

Here x_k ($k = 1, 2, \ldots, n$) are the points on the real x axis of the z plane, corresponding to the corners t_k of the polygon in the t plane; the external angles of the polygon are $\alpha_k \pi$, and A is an integration constant.

The bilinear transformation

$$z(t) = \frac{(1 - i)(t + i)}{(1 + i)(t - i)} \tag{4.23}$$

maps conformally a unit disc in the complex t plane on to the upper half of the complex z plane. The periphery of the disc, the unit circle, goes into the real x axis of the z plane.

Figure 4.5 illustrates the current density and electric field distribution in the three devices from figure 4.4. The devices are biased by a current I, the voltage drop between the current supply contacts is V, and the magnetic induction $\boldsymbol{B}$ is directed into the drawing plane. We assume that the devices are made of a p-type material. The correct positions of the vectors of electric field $\boldsymbol{E}$ and current density $\boldsymbol{J}$ for this case are shown in the inset. Note that the directions of the electric field and current density at boundaries are determined by the boundary conditions (see figure 3.18).

The sense of the whole story becomes evident by inspecting the final image of the Hall device shown in figure 4.5(c). There all current lines are parallel with the two skewed isolating boundaries, all electric field arrows are vertical and perpendicular to all four electrodes, and the equipotential lines are horizontal and equidistant. In comparison with (a) and (b), (c) is a rather boring picture, albeit very practical: the voltage drop V and the Hall voltage V_H can now be found by elementary calculations.

The main results of the analysis of Hall plates using the conformal mapping method can be summarized as follows.

(i) *As far as the geometrical correction factor is concerned, in principle, all shapes of Hall devices are equivalent*. This disposes of the idea, for example, that the efficiency of a Hall device can be increased by an ingeniously shaped boundary [9]: for an arbitrarily shaped single-connected original can be mapped on to the upper half of the z plane (figure 4.2(b)) and, conversely, the upper half of the z plane can be mapped back on to any other, arbitrarily shaped region in the t plane.

(ii) *A long Hall device*, the notion of which we have been using extensively so far, is equivalent to *a small-contact Hall device*. To understand why, consider the two rectangular devices in figure 4.6(a) and (b). The device (a) is a long device, and (b) is a small-contact device. It turns out that both devices, after mapping on to the upper half of the z plane, yield roughly the same image, such as that shown in figure 4.6(c). Then all three devices (a), (b) and (c) are approximately equivalent. Any long device and any small-contact device can be transformed

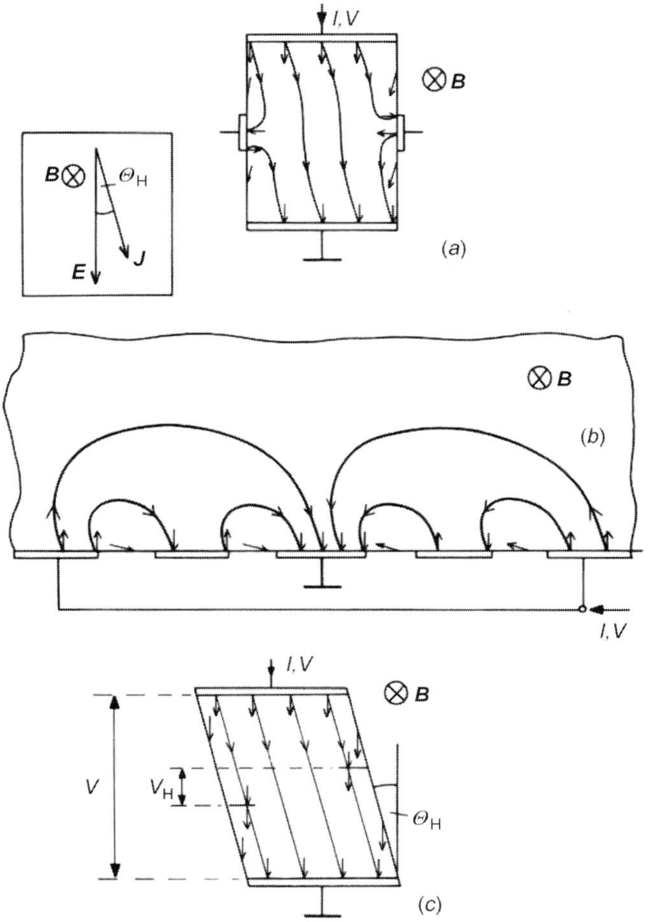

Figure 4.5. Current density and electric field in the three plates from figure 4.2. The lines connecting the contacts are the current lines, and the short arrows indicate the local direction of the total electric field. The inset shows the general relation among the vectors of electric field (***E***), current density (***J***) and magnetic induction (***B***).

into a half plane image (c) which satisfies the following conditions:

$$\frac{c}{b_1} \ll 1 \qquad \frac{s}{b_1} \ll 1 \qquad \frac{s}{b_2} \ll 1. \tag{4.24}$$

A very important particular case is that of an *infinitely long Hall device*. If in figure 4.6(a) $l/w \to \infty$, then in figure 4.6(c) $c/b_1 \to 0$ and $s/b_1 \to 0$. Consequently, an infinitely long Hall device is equivalent to a *point-contact Hall device*.

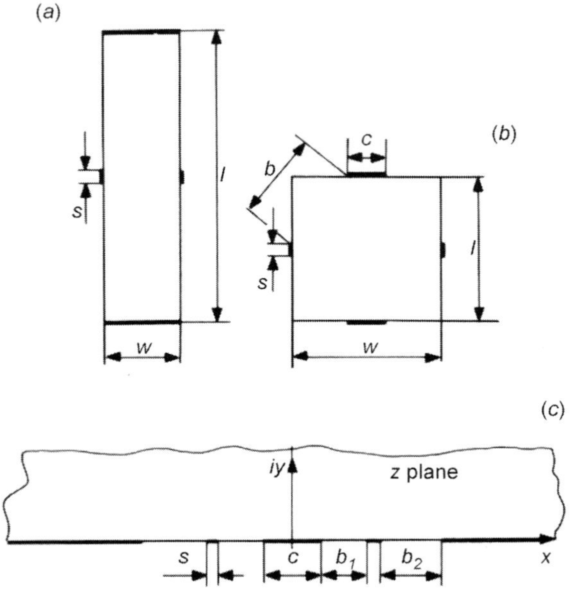

Figure 4.6. A long Hall device (a) and a small-contact Hall device (b) are galvanomagnetically equivalent: they yield similar half plane images (c).

For a device of a simple convex shape such as those in figure 4.6(a) and (b), it is easy to recognize intuitively if it is a small-contact one or not. Indeed, the condition (4.24) may be fulfilled only if the original satisfies the conditions:

Device (a):

$$\frac{s}{w} \ll 1 \qquad \frac{s}{\frac{1}{2}l - s} \ll 1, \ldots \tag{4.25}$$

Device (b):

$$\frac{c}{w} \ll 1 \qquad \frac{c}{b} \ll 1, \ldots \tag{4.26}$$

(iii) The Hall voltage varies with the distance from the finite-sized current contacts. Recall that in the vicinity of the current supply contacts there is no Hall field (see figure 3.18(cb)). In figure 4.7 the numerical results for a few rectangular Hall devices are shown [11]. By inspecting these results, we notice: (a) the largest Hall voltage appears at the middle between the two current contacts; this is why the voltage-sensing contacts are usually positioned there; (b) the largest Hall voltage develops in the longest device; as we inferred before, such a device has the greatest geometrical correction factor.

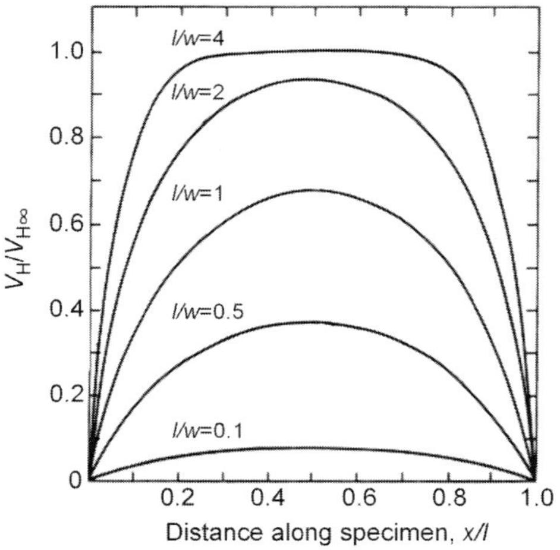

Figure 4.7. Normalized Hall voltage as a function of the position of sense electrodes along the length of a rectangular Hall plate. Magnetic field is weak. The labels denote the pertinent length-to-width ratio (adapted from [11]).

(iv) Finite-sized voltage-sensing contacts reduce the geometrical correction factor in a similar way as finite-sized current supply contacts. The sense contacts are responsible for short-circuiting a part of the supply current (see figure 4.5(a)). Since the two current lines that pass through the sense contacts are missed at the middle of the device, the current density at the middle is reduced and so also is the Hall electric field.

4.2.3 Conformal mapping: examples

We shall illustrate the application of the conformal mapping method by two examples.

Example 1

Let us determine the positions and dimensions of the contacts in a half plane Hall plate, such that it becomes equivalent to a given circular Hall plate.

Figure 4.8 illustrates the two Hall plates. The circular plate (a) is embedded in the complex t plane, and the half-plate device (b) coincides with the upper half of the complex z plane. We assume that all characteristic points T0, T1, ..., T10 of the circular plate are known: they are determined by the radius R and the central angle α. The problem is to find the positions of the corresponding characteristic points Z0, Z1, ..., Z10 of the half plane device.

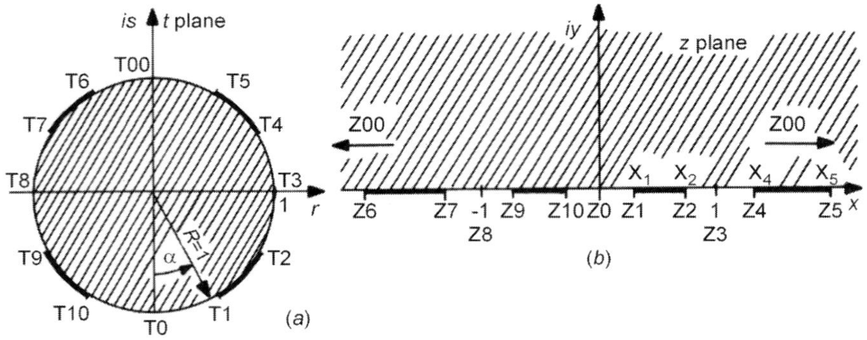

Figure 4.8. These two Hall plates are galvanomagnetically equivalent. (a) A circular Hall plate invariant for a rotation of 90°. (b) The corresponding half-plane Hall plate in which the current and sense contacts are also interchangeable. T_i and Z_i, with $i = 0, 1, 2, \ldots, 10$, are the corresponding points.

We may choose the scale in the two planes such that the radius of the circle R equals the unit of length. Then the circular plate represents a unit disc, and its periphery a unit circle.

To solve the problem, we have to map conformally the unit disc in the t plane on to the upper half of the complex z plane. Thereby the periphery of the disc, the unit circle, should go into the real x axis of the z plane—analogously to the mapping illustrated in figure 4.4(a) and (b). The mapping can be accomplished by the bilinear transformation [21], given by (4.23). The latter may be rearranged as

$$z(t) = -i\frac{t+i}{t-i}. \tag{4.27}$$

To see that the above function provides the required mapping, let us find the real and the imaginary parts of z:

$$\mathrm{Re}(z) = \tfrac{1}{2}(z + \bar{z}) \tag{4.28}$$

$$\mathrm{Im}(z) = \frac{1}{2i}(z - \bar{z}) \tag{4.29}$$

where $\bar{z}$ denotes the conjugate of z. By making use of (4.27), (4.28) yields

$$\mathrm{Re}(z) = \frac{1}{2}\frac{4\,\mathrm{Re}(t)}{t\bar{t} + 1 - 2\,\mathrm{Im}(t)}. \tag{4.30}$$

Since at the unit circle $t\bar{t} = |t|^2 = 1$, the latter equation reduces to

$$\mathrm{Re}(z) = x = \frac{r}{1-s}. \tag{4.31}$$

In a similar way, we find from (4.29) and (4.27) that

$$\mathrm{Im}(z) = y = 0 \tag{4.32}$$

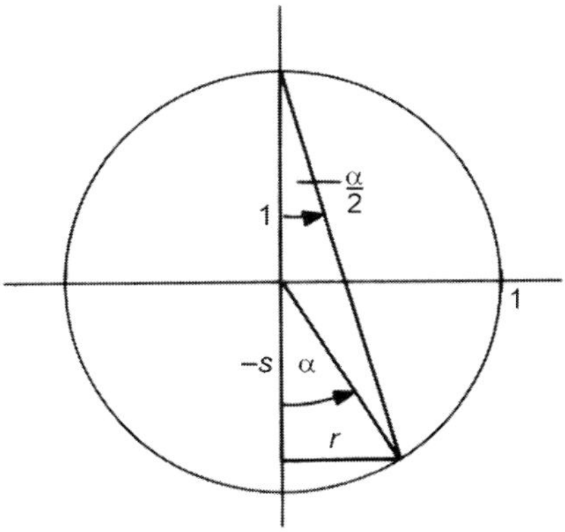

Figure 4.9. Illustrating the geometrical interpretation of equation (4.31).

if $t\bar{t} = 1$. Therefore, the unit circle indeed goes into the x axis. It can also be easily shown that if $t\bar{t} < 1$, $\mathrm{Im}(z) > 0$. Thus the inner part of the unit disc maps on to the upper half of the z plane.

With reference to figure 4.9 we see that (4.31) may be rewritten as

$$x = \tan\frac{\alpha}{2}. \tag{4.33}$$

This formula gives the required transformation of the points T at the periphery of the circular plate into the points Z at the periphery of the half plane plate. In particular, the point T0(0, −1) goes into Z0(0, 0), T3(1, 0) → Z3(1, 0), T00(0, i) → (∞, 0), and T8(−1, 0) → Z8(−1, 0). This is also obvious directly from (4.27).

For a rotation symmetrical Hall plate, shown in figure 4.10, it has been found [14] that

$$\begin{aligned} x_1 &= \tan\left(\frac{\pi}{8} - \frac{\theta}{2}\right) \qquad & x_2 &= \tan\left(\frac{\pi}{8} + \frac{\theta}{2}\right) \\ x_2 x_4 &= 1 & x_1 x_5 &= 1 \end{aligned} \tag{4.34}$$

where x_n denotes the value of $x = \mathrm{Re}(z)$ at the point Zn.

Example 2

A finite version of the hypothetical half plane Hall device analysed in example 1, figure 4.8(b), is of a considerable practical interest. Let us see how the symmetry of the half plane device changes owing to the introduction of the plate limits, i.e. additional insulating boundaries.

To attain the intuitive understanding of the influence of limiting the active area of the half plane device, let us map the latter conformally on a unit circle (see figure 4.11). First, we map a rotation-symmetrical circular Hall plate in the t plane (a) on to the upper half of the z plane (b). (We

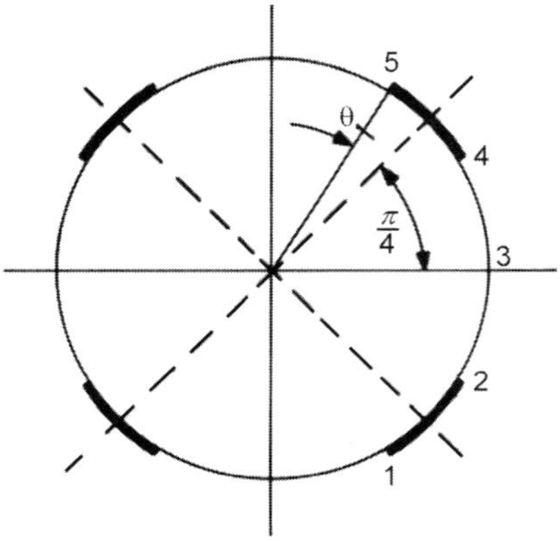

Figure 4.10. Defining the parameter θ for a circular Hall plate invariant for a rotation of $\pi/2$.

discussed this mapping in detail in the previous example.) Next, we cut a part of the z plane and then map the z plane back on to the t plane (see figure 4.11(c)–(h)). The area of the missing parts of the discs in (d), (f) and (h) will give us an indication of the importance of the cut made.

The mapping (a)–(b) is accomplished by (4.27). We assume again $R = 1$, and the latter transformation yields (4.33).

Suppose now we cut out a part of the z plane defined by

$$\mathrm{Im}(z) > y_1 \tag{4.35}$$

(see figure 4.11(c)). To see how this region maps on to the t plane, let us eliminate $\mathrm{Re}(z)$ from (4.27). Substituting (4.27) into (4.29), we obtain:

$$\frac{t+i}{t-i} + \frac{\bar{t}-i}{\bar{t}+i} = -2\,\mathrm{Im}(z). \tag{4.36}$$

After rearranging, and substituting $t\bar{t} = r^2 + s^2$ and $\mathrm{Im}(z) = y_1$, (4.36) yields

$$r^2 + s^2 - \frac{2y_1}{y_1+1}s + \frac{y_1 - 1}{y_1 + 1} = 0. \tag{4.37}$$

This is the equation of a circle, with centre at the point $(0, y_1/(y_1+1))$ and radius given by

$$r_1 = \frac{1}{(y_1+1)}. \tag{4.38}$$

This circle, denoted by C1, is shown in figure 4.11(d).

Imagine now we cut out a part of the z plane defined by

$$\mathrm{Re}(z) > x_2 \tag{4.39}$$

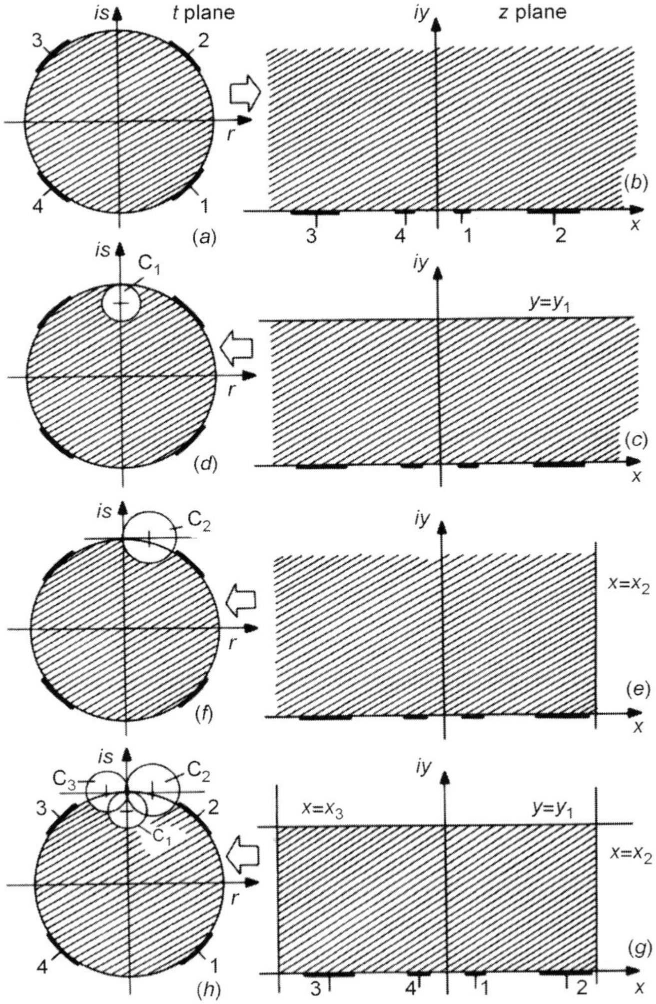

Figure 4.11. Illustrating the degradation in symmetry due to limiting the active area of a half plane symmetrical Hall plate. (a) A symmetrical circular Hall plate. (b) The half plane symmetrical Hall plate, equivalent to (a). (c) The plate (b), after cutting off the portion $y > y_1$. (d) The circular Hall plate equivalent to (c). (e) The plate (b), after cutting off the portion $x > x_2$. (f) The circular Hall plate equivalent to (e). (g) The plate (b), after cutting off the portions $y > y_1$, $x > x_2$ and $x < x_3$. (h) The circular Hall plate equivalent to the device (g).

(see figure 4.11(e)). By eliminating $\mathrm{Im}(z)$ from (4.34), and substituting $\mathrm{Re}(z) = x_2$, we find

$$r^2 + s^2 - \frac{2}{x_2} r - 2s + 1 = 0. \tag{4.40}$$

This is the equation of a circle, with centre at the point $(1/x_2, 1)$ and radius given by

$$r_2 = 1/x_2. \tag{4.41}$$

This circle, denoted by C2, is shown in figure 4.11(f). Analogously, everything similar happens if we make a cut by $x = x_3$: the circle C3 appears in the image.

Thus if we limit the active area of the half plane device in figure 4.11(b) in the way shown in (g), the rectangular device (g) will be equivalent to the truncated disc-shaped plate (h). If the Hall plate (a) is invariant for a rotation through $\pi/2$, the resistances between any pair of neighbouring contacts, such as 1-2 and 2-3 in the plates (a) and (b), are equal. This is not the case with the plates (g) and (h). The cuts bring about changes in the resistances, the largest change being that between the contacts 2 and 3.

We may get an idea about this change in resistance by speculating in the following way. Assume that the contacts cover a large part of the periphery of the disc-shaped device. If a current is forced through the device, the current density is approximately constant over the disc area. Then the resistance increases roughly proportionally with the area of the missing part of the disc. Accordingly, the relative change in the resistance R_{23} is given by

$$\frac{\Delta R_{23}}{R_{23}} \lessapprox \frac{A_1 + \frac{1}{2}(A_2 + A_3)}{A} \tag{4.42}$$

where A_1, A_2 and A_3 are the areas of the circles C1, C2 and C3, and A is the area of the whole unit disc. Assume $y_1 \gg 1$ and $y_1 \approx x_2 \approx x_3$. By making use of (4.38) and (4.41), (4.42) yields

$$\frac{\Delta R_{23}}{R_{23}} \lessapprox \frac{2}{y_1^2}. \tag{4.43}$$

For example, if $y_1 \approx x_2 \approx x_3 > 10$, the maximal change in the resistances between the contacts of the device (g), owing to the limits of its active area, should be less than about 1%.

The change in the resistances between the contacts can be neutralized by an appropriate rearrangement of the contacts. The proper contact arrangement can be found by mapping the symmetrical disc (a) on to the rectangle (g), as was shown in [22].

4.2.4 Numerical analysis using circuit simulation method

In the cases when the conditions of the strict uniformity of a galvanomagnetic plate (see §4.2.2, conditions (i)–(iii)) are not fulfilled, one cannot apply the conformal mapping method. Then the integral galvanomagnetic properties of the device can be predicted only by numerical analysis. One

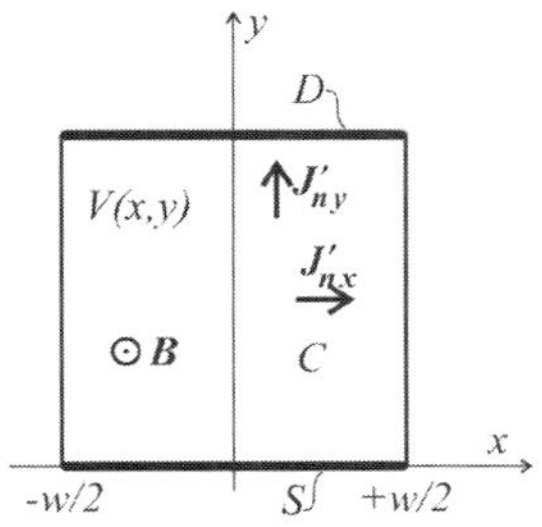

Figure 4.12. Geometry of the analysed MOS Hall plate. D: drain, C: channel, S: source, ***B***: magnetic induction.

efficient way of performing the numerical analysis of a galvanomagnetic device is to construct first a device analogue in the form of an electrical circuit [18, 19], and then to solve the circuit using a circuit-simulation program. We shall illustrate this approach for the case of an *MOS Hall plate* [19]. As we shall see in §7.1.1, the active region (channel) of a MOS Hall device is highly non-uniform and rather complex. In spite of this fact, using the circuit-simulation method, we can find a solution in a straightforward way.

Consider the two-dimensional current flow in the channel region of an N-type channel MOS device in the x–y plane subject to a perpendicular magnetic field (figure 4.12). The general equation describing the current flow is given by (3.326). For our case with the perpendicular magnetic field, we may drop the planar (last) term. Assuming a strong inversion in the channel, we may use expression (3.274) for the Corbino conductivity σ_B. So the current density in the channel is given by

$$\boldsymbol{J}_n = \frac{1}{1+(\mu_n B)^2}[(q\mu_n n\boldsymbol{E} + qD_n\nabla n) + \mu_n \boldsymbol{B} \times (q\mu_n n\boldsymbol{E} + qD_n\nabla n)]. \quad (4.44)$$

Here n denotes the electron density in the channel, D_n is the diffusion coefficient of electrons, and other notation is as usual.

The electrical field figuring in (4.44) is given by

$$\boldsymbol{E} = -\nabla V(x,y) \quad (4.45)$$

where $V(x,y)$ denotes the surface potential in the channel region. By substituting this expression into (4.44) and integrating (4.44) in the z direction over the channel depth, we obtain the following two-dimensional equations:

$$J'_{nx} = \frac{1}{1+(\mu_n^H B)^2}(J'_{n0x} + \mu_n^H B J'_{n0y})$$
$$J'_{ny} = \frac{1}{1+(\mu_n^H B)^2}(J'_{n0y} - \mu_n^H B J'_{n0x}). \quad (4.46)$$

Here

$$J'_{\rm nx} = \int_0^{z_i} J_{\rm nx}\,{\rm d}z \qquad J'_{\rm ny} = \int_0^{z_i} J_{\rm ny}\,{\rm d}z \tag{4.47}$$

are the total line current densities (per unit length) in the x and y direction, respectively;

$$\begin{aligned} J'_{\rm n0x} &= \left(-\mu_{\rm n}\frac{\partial V}{\partial x} + D_{\rm n}\frac{\partial}{\partial x}\right)Q_{\rm n} \\ J'_{\rm n0y} &= \left(-\mu_{\rm n}\frac{\partial V}{\partial y} + D_{\rm n}\frac{\partial}{\partial y}\right)Q_{\rm n} \end{aligned} \tag{4.48}$$

are the line current densities that would exist in the channel if there were no magnetic field; and

$$Q_{\rm n} = q\int_0^{z_{\rm i}} n\,{\rm d}z \tag{4.49}$$

is the surface charge density of conduction electrons in the channel. The integrals (4.47) and (4.49) are taken from the semiconductor–gate insulator interface, where $z = 0$, to the 'intrinsic' point below the channel $z_{\rm i}$, where $n(z_{\rm i}) \cong 0$.

The current components (4.46) are subject to the condition

$$\frac{\partial J'_{\rm nx}}{\partial x} + \frac{\partial J'_{\rm ny}}{\partial y} = 0 \tag{4.50}$$

which follows from the steady-state continuity equation. The current density (4.49) depends not only on the biasing conditions of the device, but also on the local value of the surface potential (4.45). However, with the present method, the function

$$Q_{\rm n}(x, y) = f(V(x, y)) \tag{4.51}$$

will be automatically taken into consideration through the appropriate model of the MOS transistor—see below.

We can convert the equations (4.46) and (4.50) into the finite-elements form by multiplying them by finite line elements ΔX and ΔY. So we obtain

$$\begin{aligned} \Delta I_{\rm x} &= \frac{1}{1 + (\mu_{\rm n}^{\rm H} B)^2}\left(\Delta I_{\rm 0x} + \mu_{\rm n}^{\rm H} B\frac{\Delta y}{\Delta x}\Delta I_{\rm 0y}\right) \\ \Delta I_{\rm y} &= \frac{1}{1 + (\mu_{\rm n}^{\rm H} B)^2}\left(\Delta I_{\rm 0y} + \mu_{\rm n}^{\rm H} B\frac{\Delta x}{\Delta y}\Delta I_{\rm 0x}\right) \end{aligned} \tag{4.52}$$

and

$$\Delta I_{\rm x} + \Delta I_{\rm y} = 0 \tag{4.53}$$

which is obviously Kirchhoff's first law.

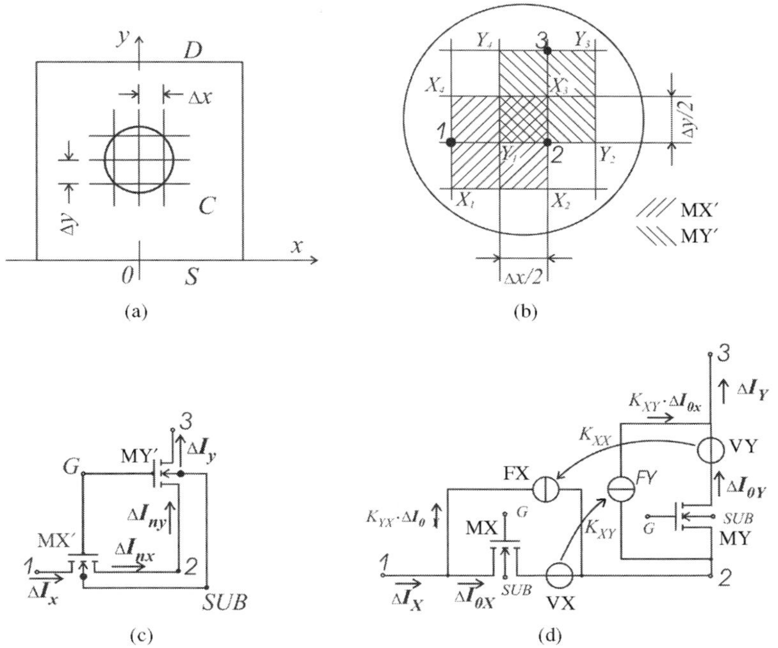

Figure 4.13. Development of the basic circuit cell for the simulation of an MOS galvanomagnetic device. (a) The channel region cut into rectangles. (b) Two partly overlapping rectangles X_1-X_2-X_3-X_4 and Y_1-Y_2-Y_3-Y_4 make up the basic MOS elements MX′ and MY′, respectively. (c) MX′ and MY′ yield an L-shaped basic cell. (d) The final basic circuit cell, consisting only of conventional circuit elements.

With reference to figure 4.13, we can construct a physical model that obeys the above equations as follows: figure 4.13(a) shows the channel region of the MOS device of figure 4.12 cut into small rectangles by two sets of lines parallel with the x and y axes, respectively. There are two current components in each rectangle, equation (4.52). But we may assume that our device consists of two sets of overlapping MOS elements, MX′ and MY′, with one-dimensional current in each of them (figure 4.13(b)). MX′ has source and drain connected to nodes 1 and 2; and MY′ has source and drain connected to nodes 2 and 3. The two devices make up an L-shaped basic circuit cell, shown in figure 4.13(c). We use a special symbol for MX′ and MY′ (with a dot) to point out that currents of these devices depend on a magnetic field.

The next step is to replace the special devices MX′ and MY′ by a circuit consisting of only conventional elements (figure 4.13(d)). Here MX is a conventional MOS transistor with a channel length $L_x = \Delta X$ and a channel width $W_x = \Delta Y$. MY is, analogously, a MOST with $L_y = \Delta Y$ and

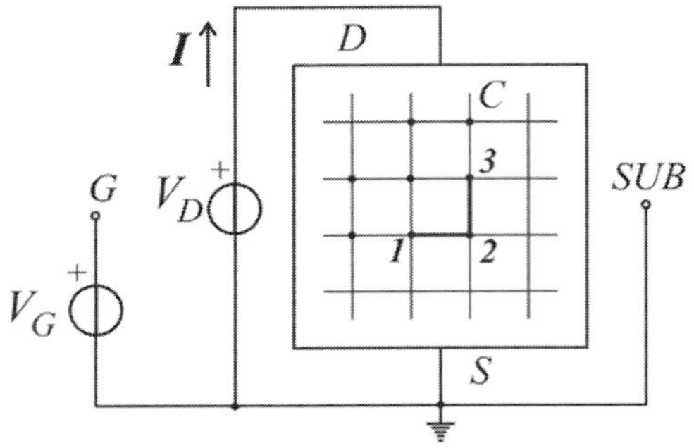

Figure 4.14. Equivalent circuit of the analysed MOS Hall device. Note the L-shape of the basic circuit cell 1-2-3, which is shown in detail in figure 4.13(d).

$W_y = \Delta X$. The currents in MX and MY depend only on the drift and diffusion phenomena, equation (4.48). The currents are only modulated by the magnetoresistance effect through the factor $1/(1 + \mu^2 B^2)$. VX and VY are zero-value voltage sources used to measure currents through MX and MY. The main influence of the magnetic field is simulated by the two current-controlled current sources FX and FY, the control currents being those measured by VX and VY, respectively. This is shown symbolically by the arrows in figure 4.13(d). The 'current gains' K_{xy} and K_{yx} are according to (4.52)

$$K_{xy} = \mu_n^{\mathrm{H}} B \frac{\Delta x}{\Delta y} \qquad K_{yx} = \mu_n^{\mathrm{H}} B \frac{\Delta y}{\Delta x}. \tag{4.54}$$

Finally, the whole device is simulated by a network of identical basic circuit elements of figure 4.13(d), as illustrated in figure 4.14. This is a lumped circuit model of the MOS plate of figure 4.12. The behaviour of the equivalent circuit can be now analysed using a conventional circuit analysis computer program, such as SPICE [23]. An example of the results obtained in this way is presented in §7.1.1.

So instead of solving directly equations (4.46) and (4.50) subject to appropriate boundary conditions, we can solve the corresponding lumped equivalent circuit, using a conventional circuit analysis program. This approach is rather convenient for the following reasons. A particular device form can be simply taken into account by an appropriate equivalent circuit topology. The fulfilment of the boundary conditions (see §3.3.4) is then a trivial task: the boundary conditions will be automatically met by the circuit topology and the supply conditions. For example, the condition at the insulating boundary (3.160) at the sides of the plate shown in figure 4.14 is met simply by the absence of the circuit elements outside of the device area. Moreover, the method can handle virtually any non-uniformity of the plate. For example, if the plate is uniform, we can replace the MOS transistors in figure 4.13(d) by resistors; but if the channel charge of a plate depends on the local channel potential (4.51), which can be also magnetic field dependent, no problem: this will be automatically resolved if we use an appropriate MOSFET or JFET model for the basic elements MX and MY.

4.2.5 Common shapes of Hall plates and magnetoresistors

Hall plates

Generally, any piece of electrically conductive material, fitted with four electrical contacts, could be used as a Hall device. However, to be reasonably efficient and convenient for practical applications, a Hall device should preferably be made in the form of a plate (but there are some important exceptions: see §5.4). The contacts should be positioned at the boundary of the plate, far away from each other. The two contacts for sensing the Hall voltage should be placed so as to attain approximately equal potentials at a zero magnetic field. Then the potential difference at the sense contacts equals the Hall voltage. The common potential at the sensing contacts should roughly equal the mid-potential of the current contacts. Then the largest Hall voltage can be measured (see figure 4.3). To simplify the design and fabrication of a Hall plate, a highly symmetrical shape and a uniform material and thickness of the plate are usually chosen.

Some commonly used shapes of Hall plates are shown in figure 4.15. We have seen earlier that, as far as the galvanomagnetic properties of Hall plates are concerned, all single-connected shapes can be made to be equivalent (§4.2.2). However, this is not so if additional properties are considered, such as the ease of fabricating a structure, Joule heating, required space, etc. Thus a particular Hall plate shape might be better suited to one application than the other shapes. This will become clear from the following comments on the shapes shown in figure 4.15.

The rectangular-shaped plate (a) is an approximation of the strip-like Hall device treated in §§3.1.1 and 4.1.1. This shape is very simple. However, if a high value of the geometrical correction factor is to be achieved, then very small sense contacts must be made.

The *bridge-shaped sample* (b) approximates very well an infinitely long Hall device, in spite of relatively large contacts. If the contacts are not ohmic or are generally bad, it is useful if they are large: this minimizes parasitic effects, such as contact resistance, heating, and noise. The Hall voltage is measured between the one or the other pair of the sense contacts across the bridge (for example, S1-S1). A pair of the sense contacts along the bridge (S1-S2) is used for the precise measurement of the voltage drop along the sample. The fact that there is no current through these contacts eliminates the error due to the contact resistance.

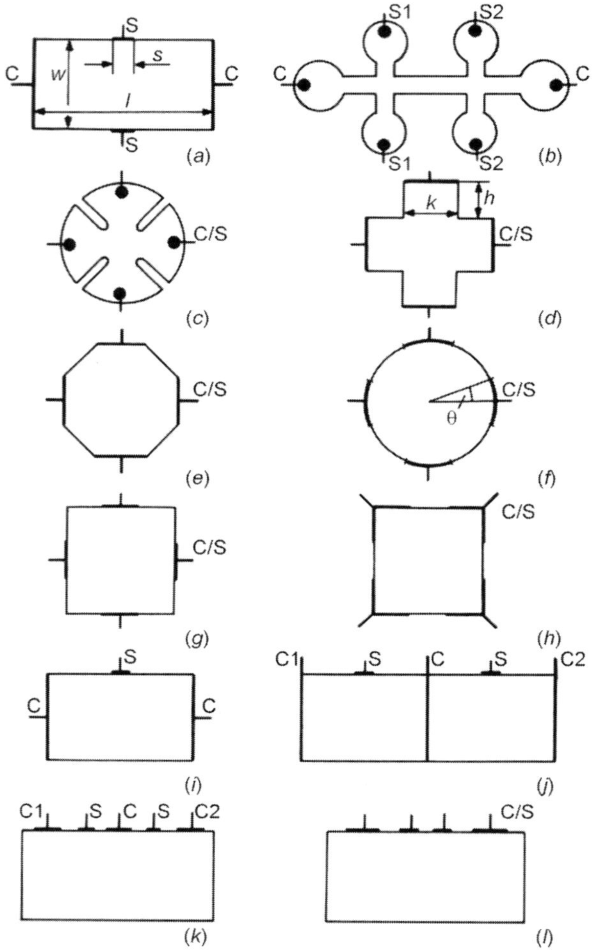

Figure 4.15. Various shapes of Hall plates. Current contacts are denoted by C, sense contacts by S, and C/S indicates that the current and sense contacts are interchangeable.

The *clover-leaf shape* (c), also known as the van der Pauw shape [24], is a four-contact version of the bridge shape (b). It possesses all the advantages of the latter (in particular, a geometrical correction factor of $G \approx 1$ can easily be achieved), but it is simpler and more compact than the bridge. Similar characteristics can also be obtained with a cross-shaped device (d).

The devices (c)–(h) are *invariant for a rotation through* $\pi/2$. All four contacts are equivalent. Thus the current and the sense contacts are interchangeable. As we shall see later, this fact can be used to cancel the offset voltage.

The devices (g) and (h) are examples of so-called *dual* or *complementary structures*. The shapes and dimensions of the active device regions are identical, but where there are contacts in one device, there are insulating boundaries in the dual device, and vice versa.

The plate (i) [25] has only one sense contact. The voltage at this contact can be measured relative to a reference voltage, which may be formed by a voltage divider [26]. In this device, one uses only one-half of the generated Hall voltage. The device (j) [27] is a combination of two one-arm devices (i). The differential voltage between the two sense contacts equals the sum of two half Hall voltages. The other two halves are lost. The shapes (i) and (j) are useful if only one side of a Hall plate is accessible for making a sense contact.

In the shapes (k) [28] and (l) [22] all contacts are consequently positioned at only one side of the plate. Such an arrangement makes possible the realization of a *vertical Hall device* [28] using integrated circuit technology. Note that the device (l) can be considered as a materialization of the truncated halfplane Hall device shown in figure 4.11(g). Recall that the standard procedure in the conformal mapping of one conventional Hall plate shape into the other includes the intermediate auxiliary step of mapping on to the complex half plane (§4.2.2). Therefore, in principle, a device of the shape (l) can be made equivalent to any other device of conventional shape. This also holds for the shape (k): it is dual to the shape (l). The shape (l) can be designed to be electrically equivalent to the rotation symmetrical shapes (c)–(h) [22].

Magnetoresistors

Similarly as in the case of Hall devices, any piece of electrically conductive material, fitted with at least two electrical contacts, could be used as a magnetoresistor. However, to be reasonably efficient and convenient for practical applications, also a magnetoresistor should preferably be made in the form of a plate. To simplify the design and fabrication of a magnetoresistor, a highly symmetrical shape and a uniform material and thickness of the plate are usually chosen. Some commonly used shapes of magnetoresistor plates are shown in figure 4.16.

The bridge-shape, figure 4.16(a), and the clover-leaf van der Pauw shape (b) [24] samples, that we described above as Hall devices (see figure 4.15(b) and (c)), are also used to study the magnetoresistance effect. As in the case of Hall devices, such samples approximate very well an infinitely long device, in spite of relatively large contacts. Therefore, they are very convenient to study the physical magnetoresistance effect. We shall discuss the techniques of resistance measurements in chapter 6.

In order to allow the full development of the geometrical magnetoresistance effect, we need an infinitely short sample. A virtually perfect

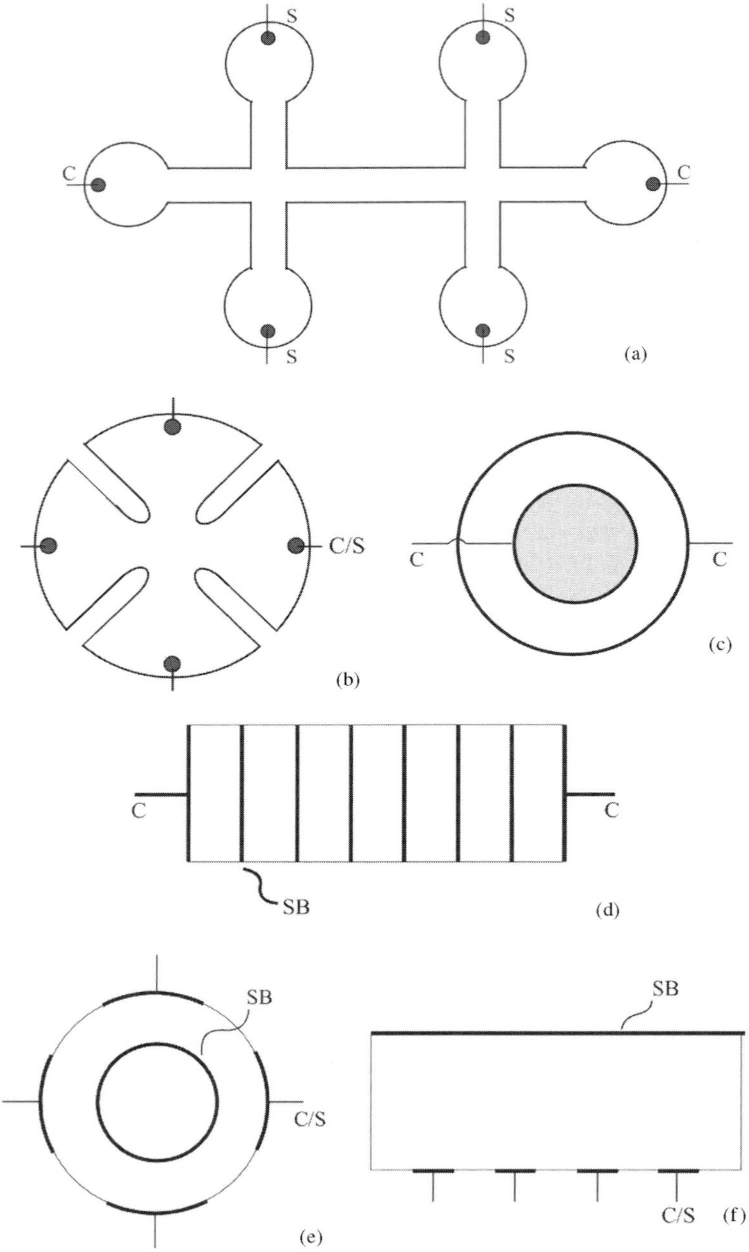

Figure 4.16. Various shapes of magnetoresistance samples. Current contacts are denoted by C, sense contacts by S, C/S are the interchangeable current and sense contacts, and SB indicates the short-circuiting bars.

materialization of an infinitely short sample is a *Corbino disc*, shown in figure 4.16(c) (see also figure 3.5). If a rectangular shape is preferred or imposed, one can use a rectangular very short sample. Since a very short sample often has a too small resistance, it is useful to make a series connection of several such samples. This is usually done by introducing shorting bars (SB) in a resistor layer (d). The shorting bars are highly-conductive stripes, in a good ohmic contact with the resistive layer, positioned perpendicularly to the current direction at $\boldsymbol{B} = 0$.

Figure 4.16(e) [29] has elements of a van der Pauw shape (b) and of a Corbino disc (c). The shorting bar (SB) is introduced in the centre in order to boost the geometrical magnetoresistance effect. Using the conformal mapping example of §4.2.3, it was shown that the structures (e) and (f) can be made to be galvanomagnetically almost equivalent [30].

4.2.6 Geometrical factor of Hall voltage

We have introduced the notion of the geometrical factor G_H in §4.2.1. Briefly, this factor describes the diminution of the Hall voltage in a finite-size Hall device compared to that of a corresponding infinitely long device. This diminution comes about due to a non-perfect current confinement in Hall devices of finite dimensions. According to (4.20) the Hall geometrical factor is defined by

$$G_H = V_H / V_{H\infty}. \tag{4.55}$$

Here V_H is the Hall voltage of an actual device, and $V_{H\infty}$ that of a corresponding infinitely long or point-contact device.

A part of the device geometry, the geometrical correction factor also depends on the Hall angle: G_H increases with the Hall angle. To understand the reason for this dependence, it is helpful to inspect the general shape of the current lines in figure 4.3. In figure 4.17, we have sketched two of the current lines for two different Hall angles. To stress the point, each line is represented by just three straight-line portions. Two peripheral line portions are skewed by the Hall angle Θ_H with respect to the normal on the contacts, and the middle portion is parallel to the isolating boundary. If the Hall angle becomes larger, the portion of the line, which is parallel to the isolating boundaries, grows longer. Recall that the Hall field develops only if the current is confined to flow parallel to an isolating boundary. Therefore, in device (b) the Hall voltage will be less affected by the contacts than that in device (a): device (b) appears effectively longer, and has a bigger geometrical factor.

According to this model, at $\Theta_H \to \pi/2$, the geometrical correction factor should tend to unity, irrespective of the device shape. Detailed calculations and experiments prove this speculation to be sound.

We shall now review the most important results on the geometrical correction factor available in the literature.

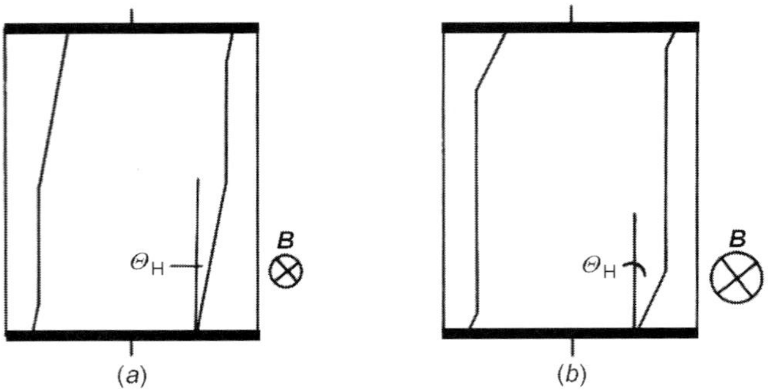

Figure 4.17. Simplified sketches of the current lines in a rectangular Hall device for two different Hall angles. Compare with figure 4.3! The larger the Hall angle, the longer the portion of the current lines running parallel to the insulating boundaries.

Lippmann and Kuhrt [31] and Haeusler [32] have calculated the geometrical correction factor of a *rectangular Hall plate* using the conformal mapping method. The results are represented graphically in figure 4.18. The shape of the plate is shown in figure 4.15(a). It was presumed that the sense contacts are infinitely small, $s \rightarrow 0$, and positioned at mid-length between the current contacts. The geometrical correction factor is a function of the length-to-width ratio l/w and the Hall angle Θ_H. At $l/w \geq 3$, a

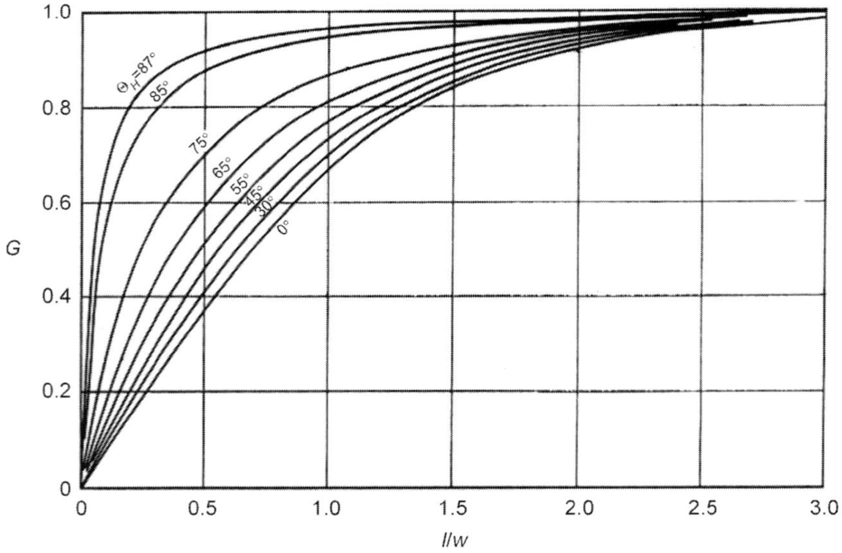

Figure 4.18. The geometrical correction factor (4.55) for a rectangular Hall plate with point sense contacts as a function of the length-to-width ratio (l/w) (figure 4.15(a), with $s \rightarrow 0$). The parameter is the Hall angle (adapted from [31]).

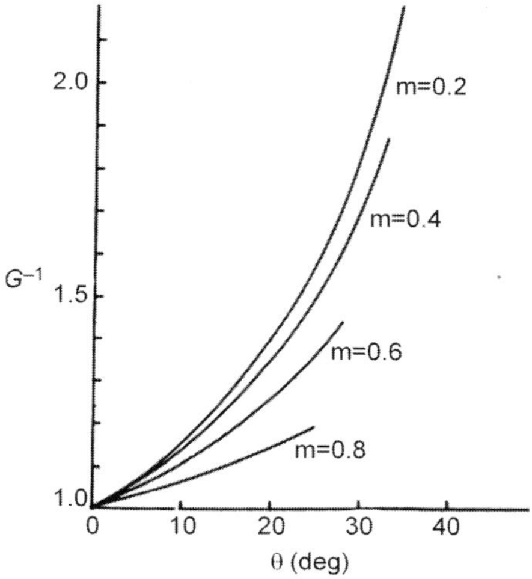

Figure 4.19. The reciprocal value of the geometrical correction factor for the circular Hall plate of figure 4.15(f) as the function of the angle θ. The parameter m is related to the Hall angle (4.56) (adapted from [21]).

rectangular device can be considered as an infinitely long strip. The geometrical correction factor for a rectangular Hall plate was also treated in [33].

Versnel has calculated the geometrical correction factor for a *circular plate* [21] and several other *rotation symmetrical Hall plates* [34]. He also applied the conformal transformation technique. The results are represented graphically in figures 4.19–4.21. The notation used in these figures is as follows. The parameter

$$m = \frac{\Theta_H}{\pi/2} \tag{4.56}$$

denotes the relative Hall angle (note that $\Theta_H \rightarrow \pi/2$ if $B \rightarrow \infty$), and λ denotes the ratio of the sum of the lengths of the contacts c and the total length of the plate boundary b:

$$\lambda = c/b. \tag{4.57}$$

For example, for a circular Hall plate (figure 4.15(f)), $\lambda = 4\theta/\pi$.

In [35], a comparison between equivalent cross-shaped and rectangular Hall plates was made. It was shown that a geometrical correction factor of $G \approx 1$ can easily be achieved in a cross-shaped device, in spite of its large contacts. On the other hand, in a rectangular device with $G \approx 1$, the sense contacts must be impractically small.

In a limited range of parameters, the geometrical correction factor can be expressed by an analytical function. We will comment below on some of these approximations.

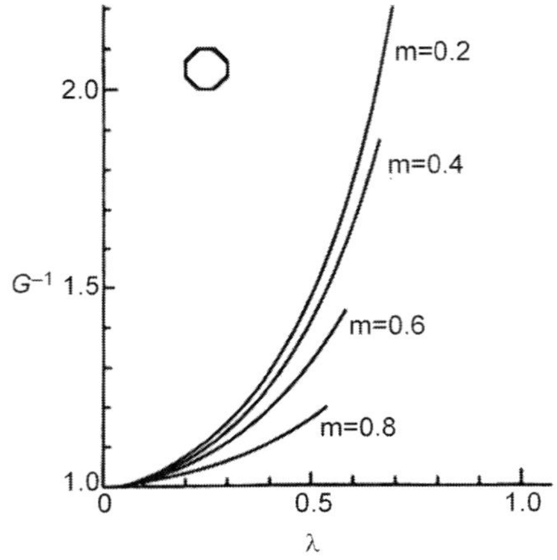

Figure 4.20. The reciprocal value of the geometrical correction factor for the octagon Hall plate of figure 4.15(e) as the function of the parameter λ (4.57). The parameter m is given by (4.56) (adapted from [34]).

For a rectangular Hall plate with point sense contacts, such as that shown in figure 4.15(a) with $s \to 0$, the geometrical correction factor can be approximated [31] as

$$G_H \simeq 1 - \frac{16}{\pi^2}\exp\left(-\frac{\pi}{2}\frac{l}{w}\right)\left[1 - \frac{8}{9}\exp\left(-\frac{\pi}{2}\frac{l}{w}\right)\right]\left(1 - \frac{\Theta_H^2}{3}\right). \tag{4.58}$$

This holds if $0.85 \leq l/w < \infty$ [2] and $0 \leq \Theta_H \leq 0.45$.

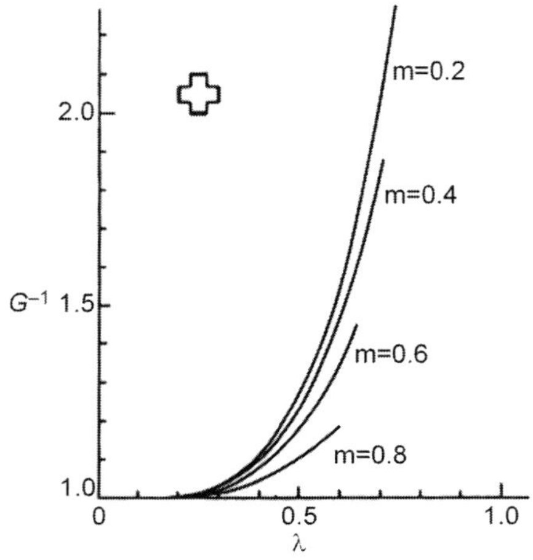

Figure 4.21. The reciprocal value of the geometrical correction factor for the cross-like Hall plates of figure 4.15(d) as the function of the parameter λ (4.57). The parameter m is given by (4.56) (adapted from [34]).

For a relatively long Hall plate, with $l/w > 1.5$ and small sense contacts $s/w < 0.18$ (figure 3.15(a)), it was found for small Hall angles [32] that

$$G_H \simeq \left[1 - \frac{16}{\pi^2}\exp\left(-\frac{\pi}{2}\frac{l}{w}\right)\frac{\Theta_H}{\tan\Theta_H}\right]\left(1 - \frac{2}{\pi}\frac{s}{w}\frac{\Theta_H}{\tan\Theta_H}\right). \tag{4.59}$$

The term in the square brackets represents the influence of the current contacts (compare with (4.58)), and the second term is due to the sense contacts. Within the above limits, the accuracy of (4.59) is better than 4%. G_H approaches unity if $l/w > 3$ and $s/w < 1/20$.

For a short rectangular Hall plate with point sense contacts, the geometrical correction factor can be approximated by (after [31])

$$G_H \simeq 0.742\frac{l}{w}\left[1 + \frac{\Theta_H^2}{6}\left(3.625 - 3.257\frac{l}{w}\right)\right]. \tag{4.60}$$

This expression is accurate to within 1% if $l/w < 0.35$ and $\Theta_H < 0.45$. If $\Theta_H \approx 0$, (4.60) reduces to

$$G_H(0) \simeq 0.742\, l/w \tag{4.61}$$

For Hall plate shapes which are invariant under a rotation through $\pi/2$, such as the configurations shown in figure 4.15(c)–(h), the geometrical correction factor can be approximated by a function of the following general form [34]:

$$G_H(\lambda, \Theta_H) \simeq 1 - g(\lambda)\frac{\Theta_H}{\tan\Theta_H}. \tag{4.62}$$

Here λ is the ratio defined above (4.57) and $g(\lambda)$ is a function of the plate geometry which is specific for each plate shape. The formula (4.52) gives a good approximation of the curves in figures 4.19–4.21 for small Hall angles and small λ. Simple approximate expressions of the function $g(\lambda)$ for some Hall plate shapes are listed in table 4.1.

If we expand the term $\Theta_H/\tan\Theta_H$ in (4.62) into a series and retain only the first two terms, we obtain the approximation [36]

$$G_H(\lambda, \Theta_H) \simeq 1 - g(\lambda)(1 - \tfrac{1}{3}\Theta_H^2). \tag{4.63}$$

Table 4.1. Functions $g(\lambda)$ (see (4.62) and (4.63)) for some rotation symmetrical Hall plate shapes [34]. The parameter λ is defined by (4.57), and h and k in figure 4.15(d).

Shape	Figure 4.15	$g(\lambda)$ or $g(h/k)$	Accurate within 0.5% if
Circle	(f)	0.636λ	$0.31 \geq \lambda$
Square	(g)	0.062λ	$0.3 \geq \lambda$
Square	(h)	$0.696\lambda^2$	$0.59 \geq \lambda$
Octagon	(e)	$1.940[\lambda/(1 + 0.414\lambda)]^2$	$0.73 \geq \lambda$
Cross	(d)	$1.045\exp(-\pi h/k)$	$0.38 \geq k/2h$

This can be rearranged to attain a form similar to that of (4.60):

$$G_H \simeq G_{H0}(1 - \beta\Theta_H^2). \tag{4.64}$$

Here G_{H0} is the value of the geometrical correction factor at $B = 0$, and β is a numerical coefficient. Comparing (4.63) and (4.64), we find that for Hall plates with relatively small contacts

$$G_{H0} \simeq 1 - g(\lambda) \tag{4.65}$$

$$\beta \simeq \frac{1 - G_{H0}}{3G_{H0}}. \tag{4.66}$$

These hold if $\lambda \leq 0.73$ (see table 4.1, octagon), and thus if $G_{H0} \geq 0.4$. On the other hand, from (4.64) and (4.60), we find that for Hall plates with relatively large contacts

$$G_{H0} \simeq 0.742\, l/w \tag{4.67}$$

$$\beta \simeq 0.604 - 0.534\, l/w. \tag{4.68}$$

The latter two equations yield

$$\beta \simeq 0.604 - 0.732 G_{H0}. \tag{4.69}$$

These are good approximations if $l/w < 0.35$, and thus if $G_{H0} < 0.260$.

Therefore, the geometrical correction factor of Hall plates exposed to a weak magnetic field can be expressed by a quadratic function of the Hall angle. The coefficient β depends only on the initial value of G_H, that is G_{H0}. Although we inferred this conclusion by inspecting the solutions for some particular shapes of Hall plates, it is valid for a Hall plate of arbitrary geometry, for any Hall plate shape can be transformed to one of the shapes analysed here.

4.2.7 Geometrical factor of magnetoresistance

While discussing the galvanomagnetic effects in samples of different shapes in §§3.3.5 and 3.3.6, we saw that the magnetoresistance effect also depends on the sample geometry. Briefly, in a very long sample exposed to a magnetic field, the separation of carriers according to their longitudinal velocities takes place. The corresponding change in resistance is called the physical (or intrinsic) magnetoresistance effect and is characterized by the resistivity ρ_b (see §3.4.5). In a very short sample, the change of resistance is additionally increased owing to the overall current deflection. This contribution to the change in resistivity is called the geometrical magnetoresistance effect. The magnetoresistance effect in an infinitely short sample is characterized by the Corbino conductivity σ_B (see §3.4.2).

In order to describe the magnetoresistance effect in an 'intermediate' sample, we introduced the geometrical factor of magnetoresistance (4.21).

This factor is given by

$$G_R = R(B)/R_\infty(B). \tag{4.70}$$

Here $R(B)$ is the resistance of a magnetoresistor of an arbitrary geometry, measured between two ohmic contacts at an induction B, and $R_\infty(B)$ denotes the resistance of a corresponding part of an infinitely long strip.

In order to relate the factor G_R to our previous analysis of the magnetoresistance effect and to the relevant literature, it is convenient to consider the ratio $R(B)/R(0)$ of devices of various geometries. Generally, according to (4.6),

$$\frac{R(B)}{R(0)} = \frac{\rho(B)}{\rho(0)} \tag{4.71}$$

where $\rho(B)$ is an effective average material resistivity in a sample. In an *infinitely long device* we have only the *physical magnetoresistance effect*, $\rho(B) = \rho_b(B)$, and (4.71) reduces to

$$\frac{R(B)}{R(0)} = \frac{\rho_b(B)}{\rho_b(0)}. \tag{4.72}$$

For an *infinitely short sample* or a *Corbino disc*, the appropriate resistivity is given by $\rho(B) = 1/\sigma_B(B)$. By substituting this into (4.71), and making use of (3.270), we have

$$\frac{R(B)}{R(0)} = \frac{\rho_b(B)}{\rho_b(0)}(1 + \mu_H^2 B^2). \tag{4.73}$$

For a device which is neither very long nor very short, the ratio $R(B)/R(0)$ is somewhere between the values given by (4.72) and (4.73). Using the notion of the geometrical correction factor of magnetoresistance (4.70), the resistance ratio for a device of arbitrary shape may be expressed by

$$\frac{R(B)}{R(0)} = \frac{\rho_b(B)}{\rho_b(0)} G_R. \tag{4.74}$$

For a rectangular plate-like resistor, $R(B) = (l/wt)\rho(B)$. Then, according to (4.72) and (4.73), if $l/w \to \infty$, $G_R \to 1$; if $l/w \to 0$, $G_R \to 1 + \mu_H^2 B^2$. So the geometrical factor of magnetoresistance is a number always larger than unity. It expresses the extent of the development of the geometrical magnetoresistance effect in a sample. $G_R = 1$ means that there is no geometrical magnetoresistance effect, and only the physical magnetoresistance effect takes place in the sample. If $1 < G_R < 1 + \mu_H^2 B^2$, then, in addition to the physical magnetoresistance effect, also some geometrical magnetoresistance effect shows up. And $G_R = 1 + \mu_H^2 B^2$ means that both the physical magnetoresistance effect and the full geometrical magnetoresistance effect are at work.

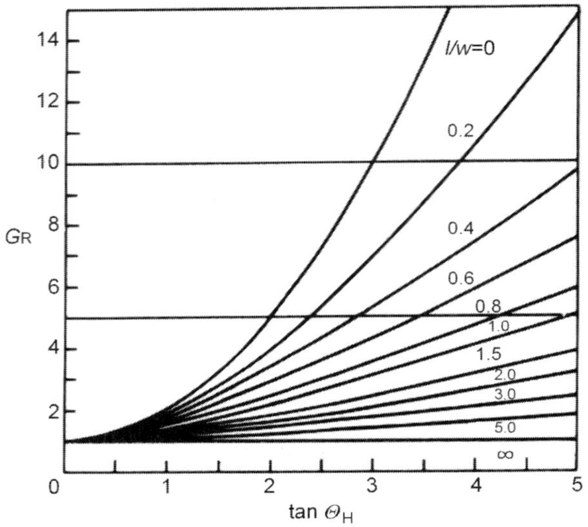

Figure 4.22. The geometrical factor of magnetoresistance for a rectangular sample as functions of the product $\mu_H B$. The labels show the length-to-width ratio l/w [38].

The geometrical factor of magnetoresistance for rectangular plate-like resistors of various l/w ratios was calculated using the conformal mapping technique [37, 38]. The results are shown in figure 4.22.

If $\mu_H^2 B^2 \ll 0.45$ and $l/w \ll 0.35$, the function $G_R(\mu_H B, l/w)$ can be approximated by

$$G_R \simeq 1 + \mu_H^2 B^2 \left(1 - 0.54 \frac{l}{w}\right) \tag{4.75}$$

with an accuracy better than 1%. Further approximate expressions for G_R are listed in [39].

4.3 The Hall voltage mode of operation

We shall consider in this section the electrical characteristics of a Hall plate operated in the conventional way: the plate is exposed to a perpendicular magnetic field, is biased via its two non-neighbouring leads, and the Hall voltage occurs between the other two non-neighbouring leads. The Hall voltage is considered as the output signal. We call this way of operation of a Hall device the Hall voltage mode. In §4.4 we shall see that this is not the only mode of operation of a Hall device.

We already started the analysis of the voltage mode of operation of a Hall plate in §4.1. But there we considered an ideal device. Now we shall

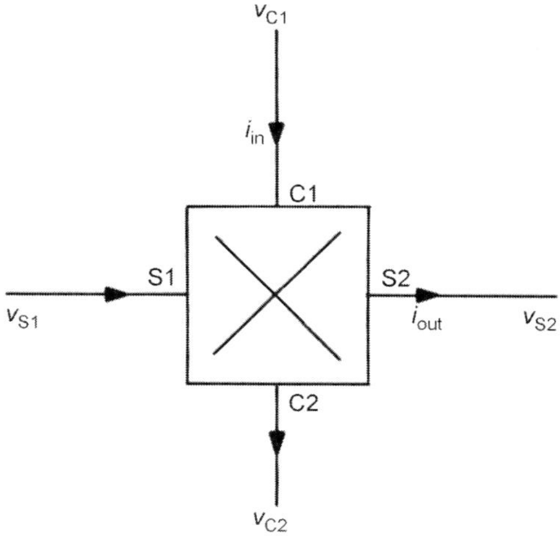

Figure 4.23. A symbolic representation of a Hall element. C1 and C2 are the current leads, and S1 and S2 are the sense leads.

take into account also the influence of device geometry, and of some basic parasitic effects, such as offset and noise.

Figure 4.23 is a symbolic representation of a Hall device as a circuit element: a rectangle with a big '×', which stands for a multiplier. The multiplier symbol indicates the fact that the Hall voltage is proportional to the product of current and magnetic induction. The terminals C1 and C2 denote the current leads. They are also-called the input terminals. The terminals S1 and S2 denote the sense leads for sensing the Hall voltage. They are also-called the output terminals. We use here the terms 'leads' or 'terminals' for the accessible external part of the metallic contacts to the device.

The bias current i_{in} is also-called the input current, and the bias voltage $v_{in} = v_{C1} - v_{C2}$ the input voltage. Analogously, the voltage between the output terminals $v_{out} = v_{S1} - v_{S2}$ is called the output voltage, and the current i_{out} through the output terminals, the output current. When there is no danger of misunderstanding, we shall use the simpler notation: $i_{in} = I$, $v_{in} = V$.

The resistance between the current leads of a Hall device is called the input resistance. The resistance between the sense leads is called the output resistance. In figure 4.24, the two resistances are denoted by R_{in} and R_{out}, respectively. In a rotation symmetrical device, $R_{in} = R_{out}$.

4.3.1 Input and output resistances

The resistance of a Hall device consists of the intrinsic resistance of the active device region and the resistances of the contacts. We shall assume here that the contacts are ohmic and their resistances are negligible.

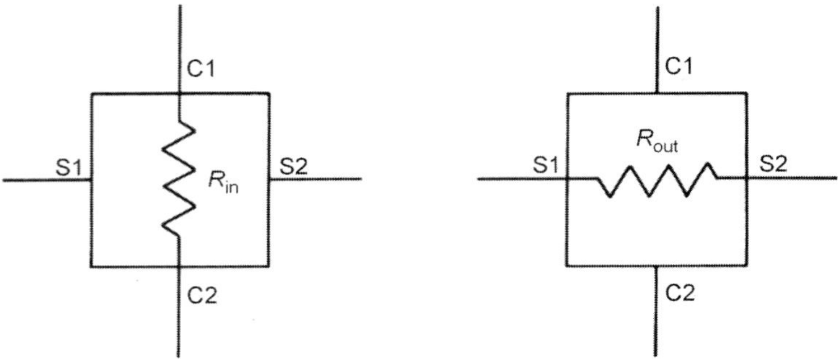

Figure 4.24. Defining the equivalent input and output resistances of a Hall device.

The input resistance of a rectangular Hall device with infinitely small sense contacts, such as that shown in figure 3.3, at $\boldsymbol{B} = 0$ is readily given by

$$R(0) = \rho \frac{l}{wt}. \tag{4.76}$$

Here ρ is the material resistivity, and l, w and t are the dimensions defined in figure 3.3. If $\boldsymbol{B} \neq 0$, by substituting (4.76) into (4.74), the resistance becomes

$$R(B) = \rho_b(B) \frac{l}{wt} G_R. \tag{4.77}$$

Here ρ is the magnetoresistivity (§3.4.5), and G_R is the geometrical factor of magnetoresistance, discussed in §4.2.7.

The input resistance of a Hall plate with an arbitrary shape may be found with the aid of the transformation of this plate into an equivalent rectangle. However, if the Hall geometrical factor G_H of the device for which the resistance is required is known, the transformation through conformal mapping can be avoided. Then the aspect ratio l/w of the equivalent rectangle can be determined simply by making use of the inverse function of $G_H(l/w)$: $l/w = f(G_H)$. For example, figure 4.18 may be used for this purpose. Alternatively, one may use the approximate formulae for $G_H(l/w)$, as illustrated below.

Suppose we seek the resistance of a Hall plate with a geometrical correction factor which equals G_{H0} at $B = 0$. Let G_{H0} be such as to correspond to a relatively long device. Then the appropriate equation giving the geometrical correction factor of an equivalent rectangular plate is (4.58). By substituting $\Theta_H = 0$ and $G_H = G_{H0}$ in this equation, and assuming $l/w \gg 1$, we find the length-to-width ratio of the required equivalent rectangular device:

$$\left.\frac{l}{w}\right|_{\text{eff}} \simeq -\frac{2}{\pi} \ln\left(\frac{\pi^2}{16}(1 - G_{H0})\right). \tag{4.78}$$

This expression is reasonably accurate if $l/w > 1$.

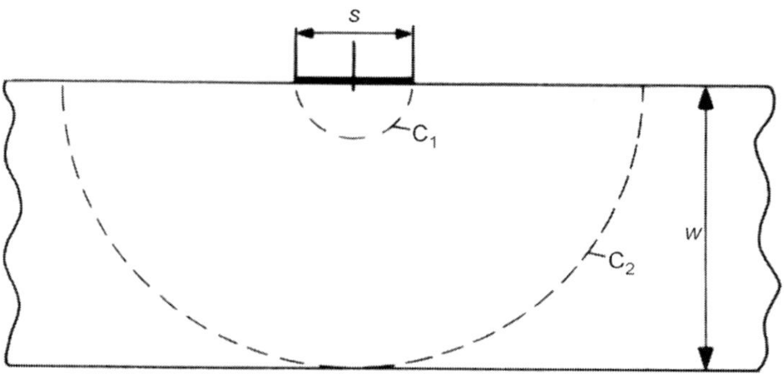

Figure 4.25. A small contact at a boundary of a long strip. Its contribution to the device resistance can be estimated by calculating the resistance of the plate between the half-cylinders C_1 and C_2.

If G_{H0} is small, say $G_{H0} < 0.3$, the appropriate function $G_H(l/w)$ is given by (4.60), so we find

$$\left.\frac{l}{w}\right|_{\text{eff}} \simeq \frac{1}{0.742} G_{H0}. \tag{4.79}$$

This is reasonably accurate if $l/w < 0.5$. The device resistance can now be obtained by substituting the value l/w into (4.77).

The above method may yield a good estimate of the resistance between a pair of not-too-small contacts. If a contact is very small in comparison with the other dimensions of a Hall device, a better estimation of its contribution to the resistance can be made by a direct calculation of the resistance. Figure 4.25 shows a small sense contact placed at a long insulating boundary. To simplify the calculation, we may approximate the flat contact surface by a cylindrical surface C_1, and substitute the next contact by another cylindrical surface C_2. Then we find the plate resistance between the two surfaces C_1 and C_2 to be

$$R = \frac{\rho}{\pi t} \ln\left(\frac{w}{s}\right). \tag{4.80}$$

Accordingly, the output resistance of a long rectangular Hall plate with small sense contacts is given by

$$R_{\text{out}} \simeq 2\frac{\rho_b}{\pi t} \ln\left(\frac{w}{s}\right) \tag{4.81}$$

provided $s \ll w < l$.

The resistances of a Hall device depend on temperature and mechanical stress: see §§2.6.1 and 2.7.4.

4.3.2 Output voltage

The output voltage of a Hall device is the differential voltage appearing between the sense terminals of the device. The output voltage can generally be expressed as the sum

$$v_{\text{out}} = v_{\text{s}}(I, B) + v_{\text{N}}(t) + v_{\text{ind}}(B(t)) \tag{4.82}$$

where $v_s(I, B)$ denotes the voltage between the sense contacts, which depends on the device current I and the magnetic induction $\boldsymbol{B}$ in a deterministic way; v_N is the noise voltage; and v_{ind} is the voltage induced in the circuit due to the time variations of the magnetic induction. The induced voltage can virtually be eliminated by a careful design of the Hall device and the circuit. Therefore we shall neglect it from now on.

The deterministic part of the output voltage of a Hall device is given by the integral of the electric field over a line connecting the two sense contacts. Consider a point-contact Hall plate, shown in figure 4.26. Let the sense contacts be at the points M and N.

Assume also that the output of the device is not loaded. We may choose the integration path MN such that MN = MN′ + N′N [40]. Then the line MN′ is normal on the current lines, and the line N′N goes along the insulating boundary. Thus the output voltage is given by

$$v_{\text{out}} = \Delta V_{\text{MN}} = \int_{\text{M}}^{\text{N}'} \boldsymbol{E}\,\text{d}\boldsymbol{l} + \int_{\text{N}'}^{\text{N}} \boldsymbol{E}\,\text{d}\boldsymbol{l}. \tag{4.83}$$

If we neglect noise, and assume a pure perpendicular magnetic field, the electric field $\boldsymbol{E}$ is given by (3.175). Substituting this expression into (4.83), we obtain

$$v_{\text{out}} \simeq v_{\text{s}}(I, B) = -\int_{\text{M}}^{\text{N}'} R_{\text{H}}[\boldsymbol{J} \times \boldsymbol{B}]\,\text{d}\boldsymbol{l} + \int_{\text{N}'}^{\text{N}} \rho_{\text{b}} \boldsymbol{J}\,\text{d}\boldsymbol{l}. \tag{4.84}$$

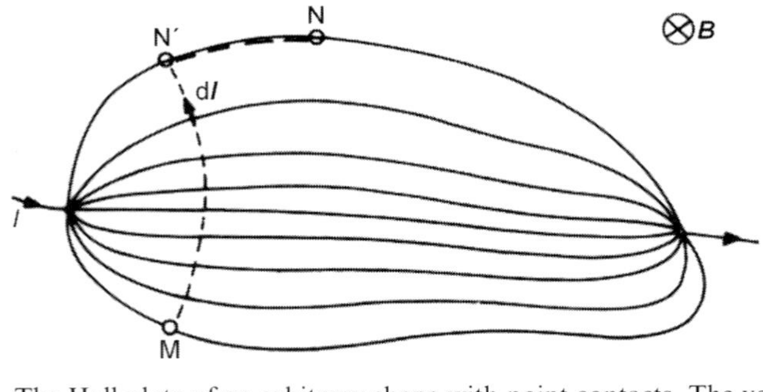

Figure 4.26. The Hall plate of an arbitrary shape with point contacts. The voltage difference between the output contacts M and N can be calculated by integrating the electric field over a line MN.

The integrand of the first term in this equation equals the Hall electric field (3.183). Therefore, the integral over the line MN′ yields the Hall voltage (see also (3.11) and (4.7)–(4.9)). The second integral yields the voltage drop between the points N′ and N. Therefore, (4.84) may be rewritten as

$$v_s(I, B) = V_H + V_{off} \tag{4.85}$$

where V_H denotes the Hall voltage and V_{off} is called the offset voltage.

If the magnetic induction varies, the current distribution in the considered point-contact device does not change. (Recall that a point-contact device is equivalent to an infinitely long device.) Thus the position of the point N′ does not depend on the magnetic field. But in a finite contact device, it does. Then equation (4.85) still holds, only both V_H and V_{off} depend on the magnetic field in a more complicated way.

Substituting (4.85) into (4.82), the latter becomes

$$v_{out} = V_H + V_{off} + v_N(t). \tag{4.86}$$

In the applications of Hall devices, the Hall voltage is considered as a useful signal, and the offset voltage and the noise voltage as disturbances; see also (2.123).

4.3.3 Hall voltage

According to (4.84), (4.85) and figure 4.26, the Hall voltage of a point contact Hall device is given by

$$V_H = -\int_M^{N'} R_H [\boldsymbol{J} \times \boldsymbol{B}]\, \mathrm{d}\boldsymbol{l}. \tag{4.87}$$

Recall the notation: R_H is the Hall coefficient, $\boldsymbol{J}$ is the current density, and $\boldsymbol{B}$ denotes the magnetic induction. If the Hall device is a homogeneous plate of thickness t, and the magnetic induction is uniform over the plate, equation (4.87) can be rearranged as

$$V_H = -\frac{R_H}{t} B_\perp \int_M^{N'} Jt\, \mathrm{d}l. \tag{4.88}$$

Here $B_\perp$ denotes the normal component of the magnetic induction. The integral in equation (4.88) yields the total bias current of the device I. Therefore

$$V_H = \frac{R_H}{t} I B_\perp \tag{4.89}$$

where we have omitted the minus sign. Equation (4.89) is identical to (3.18) and (4.9), which we derived while assuming that the points M and N lay on the same equipotential plane at $\boldsymbol{B} = 0$. Therefore, the Hall voltage does not depend on the exact position of the sense contacts: V_H does not change as long as the current flux crossing the vertical surface MN stays constant.

The Hall voltage of a Hall device with finite contacts is given by (4.20). In view of (4.89), this gives

$$V_H = G_H \frac{R_H}{t} I B_\perp \tag{4.90}$$

where G_H denotes the *geometrical correction factor of Hall voltage* (see §4.2.6). For an extrinsic semiconductor, the Hall coefficient is given by (3.308), and equation (4.90) becomes

$$V_H = G_H \frac{r_H}{qnt} I B_\perp \tag{4.91}$$

where we consider only the absolute value. Recall the notation: r_H is the Hall factor, q is the electron charge, and n is the carrier density in the plate material. Note that the product nt equals the number of charge carriers per unit plate area, and qnt equals the mobile charge per unit plate area.

By making use of (4.77), the Hall voltage (4.90) can be expressed in terms of the bias voltage $V = R_{in} I$:

$$V_H = \frac{G_H}{G_R} \frac{R_H}{\rho_b} \frac{w}{l} V B_\perp \tag{4.92}$$

where G_R denotes the geometrical factor of magnetoresistance, and w/l is the width-to-length ratio of an equivalent rectangular device. At $B \approx 0$, and for an extrinsic semiconductor, (4.92) reduces to

$$V_H \simeq \mu_H \frac{w}{l} G_H V B_\perp . \tag{4.93}$$

This equation clearly shows the importance of high carrier mobility materials for efficient Hall devices.

The Hall voltage of a Hall device biased by a constant voltage increases continuously with the increase in the width-to-length ratio of the device. For a very short device, G_H is given by (4.60). Thus at a weak magnetic field, (4.93) and (4.60) yield

$$V_H \simeq 0.742 \mu_H V B_\perp . \tag{4.94}$$

This is the *maximal Hall voltage* that can be obtained with a Hall device at a given product $VB_\perp$. For $V = 1\,\text{V}$ and $B_\perp = 1\,\text{T}$, equation (4.94) yields, at 300 K, $V_H \approx 0.126\,\text{V}$ and $V_H \approx 0.67\,\text{V}$ for low-doped n-type silicon ($\mu_{Hn} \approx 0.17\,\text{m}^2\,\text{V}^{-1}\,\text{s}^{-1}$) and GaAs ($\mu_{Hn} \approx 0.9\,\text{m}^2\,\text{V}^{-1}\,\text{s}^{-1}$), respectively. Put another way, the maximum Hall voltage at room temperature amounts to about 12% of V_{in} per tesla and 67% of V_{in} per tesla for silicon and GaAs, respectively.

From (4.91) and (4.93), we also obtain the Hall voltage as a function of the power $p = VI$ dissipated in the device:

$$V_H = G_H \left(\frac{w}{l}\right)^{1/2} r_H \left(\frac{\mu}{qnt}\right)^{1/2} P^{1/2} B_\perp . \tag{4.95}$$

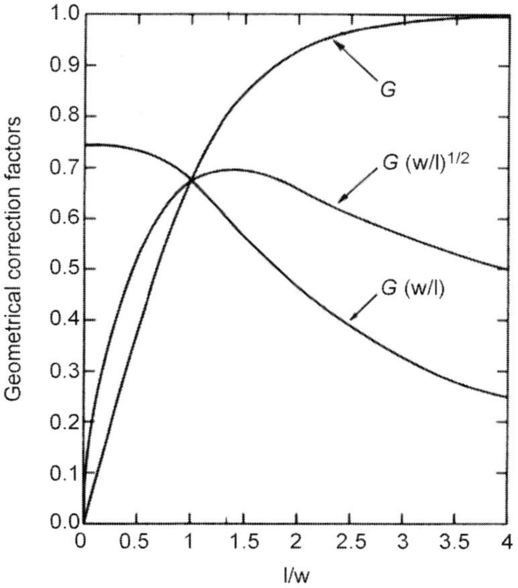

Figure 4.27. The geometrical factor G_H, and the quantities $(w/l)G_H$ (4.93) and $(w/l)^{1/2}G_H$ (4.95) for a rectangular Hall plate, as functions of the length-to-width ratio (w/l). It is assumed that the sense contacts are point-like ($s \to 0$ in figure 4.15(a)), and the magnetic induction is low.

Figure 4.27 shows the quantities G_H, $(w/l)G_H$ and $(w/l)^{1/2}G_H$, figuring in (4.91), (4.93) and (4.95), respectively, as functions of the effective length-to-width ratio.

According to this figure, the *highest Hall voltage* can be obtained: with a given bias current—if a Hall device is long, say with $l/w > 3$; with a given input voltage—if a Hall device is short, with $l/w < 0.1$; and with a given power dissipated in a Hall device—if it has an effective length-to-width ratio of $l/w \approx 1.3$–1.4.

The Hall voltage varies with a *change in temperature* in quite different ways, depending on the device biasing conditions. If a Hall device is biased by a constant current, $V_H(T) \sim R_H(T)$ (4.90). In the exhaustion range of an extrinsic semiconductor, the temperature dependence of R_H is determined by the temperature dependence of the Hall factor, which is rather weak. If a Hall device is biased by a constant voltage, $V_H(T) \sim \mu_H(T)$ (4.93). This dependence is usually much stronger than the previous one.

Example

Let us calculate more exactly the Hall voltage of the device analysed in the example of §3.1.1 (3.19)–(3.22).

Geometrical correction factor

The appropriate formula for calculating G_H is given by (4.58). We have

$$g_c = \frac{16}{\pi^2} \exp\left(-\frac{\pi}{2}\frac{l}{w}\right) \simeq 0.0146. \tag{4.96}$$

The second exponential term in (4.58) can be neglected. The term with the Hall angle can be approximated as:

$$g_b = 1 - \frac{\Theta_H^2}{3} \simeq 1 - \frac{\mu_H^2 B^2}{3}. \tag{4.97}$$

From figures 3.23, we find the Hall factor $r_{Hn} \approx 1.15$. So we correct (3.23) and obtain

$$\mu_H B = r_{Hn}\mu_n B \simeq 0.161. \tag{4.98}$$

Substituting this into (4.97), we find $g_b \approx 0.991$. The geometrical correction factor now becomes

$$G_H \simeq 1 - g_c g_b \simeq 0.986. \tag{4.99}$$

Consequently, this device can indeed be treated as an infinitely long device.

The Hall coefficient is given by (3.308):

$$|R_H| = r_H/qn. \tag{4.100}$$

Substituting here $r_{Hn} \approx 1.15$ and $n \approx N_D \approx 4.5\times10^{15}\ \mathrm{cm}^{-3}$, we obtain

$$|R_H| \simeq 1.60 \times 10^{-3}\ \mathrm{C}^{-1}\,\mathrm{m}^3. \tag{4.101}$$

The Hall voltage. Using (4.89), we find

$$V_H = G_H \frac{R_H}{t} I B_\perp \simeq 0.52\,\mathrm{V}. \tag{4.102}$$

Note that in this case the Hall voltage amounts to only 5.2% of the input voltage. The difference relative to (3.22) is mostly due to the influence of the Hall factor, which was neglected in the approximate analysis.

4.3.4 Offset voltage

The offset voltage is an unavoidable parasitic voltage, which adds to the Hall voltage (4.85). If not precisely known, which is in practice usually the case, the offset voltage limits the precision with which we can determine the Hall voltage.

With reference to (4.84), and figure 4.26, the offset voltage of a point-contact Hall device is given by

$$V_{\mathrm{off}} = \int_{N'}^{N} \rho_b \boldsymbol{J}\, d\boldsymbol{l}. \tag{4.103}$$

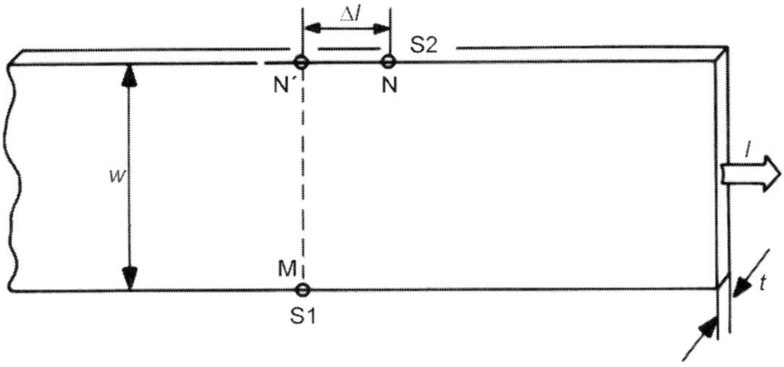

Figure 4.28. Part of a long strip-like Hall device in the absence of magnetic field. S1 and S2 are the sense contacts. A geometrical offset Δl of the sense contacts with respect to the equipotential line MN′ brings about an offset voltage between the sense contacts.

Let us evaluate this integral for the case of the strip-like device shown in figure 4.28. The current density in the strip is given by $J = I/wt$; substituting this into (4.103), we obtain

$$V_{\rm off} = \rho_{\rm b} \frac{I}{wt} \Delta l. \tag{4.104}$$

Here Δl denotes the geometrical offset of the sense contacts relative to an equipotential plane, such as MN′.

The major causes of offset in real Hall devices are imperfections in the device fabrication process, such as misalignment of contacts and non-uniformity of material resistivity and thickness. A mechanical stress in combination with the piezoresistance effect can also produce offset. All causes of offset can be represented using the simple *bridge circuit model* of a Hall plate, shown in figure 4.29 [41]. The offset voltage caused by the shown asymmetry of the bridge is given by

$$V_{\rm off} = \frac{\Delta R}{R} V_{\rm in}. \tag{4.105}$$

The offset voltage is sometimes also referred to as the *misalignment voltage* or the output *asymmetry voltage*.

The offset voltage may also depend on a magnetic field. The dependence comes about through the magnetoresistance effect. For example, in a point-contact Hall device biased by a constant current, $V_{\rm off}(B) \sim \rho_{\rm b}(B)$ (see (4.104)). In a Hall device biased by a constant voltage, $V_{\rm off}(B) \neq f(B)$ (see (4.105)).

The largest magnetic field dependence of the offset voltage appears in a large-contact Hall device biased by a constant current. By substituting $V_{\rm in} = R_{\rm in} I$ and $R_{\rm in} \approx R$ in (4.75), we obtain

$$V_{\rm off} = \Delta R I. \tag{4.106}$$

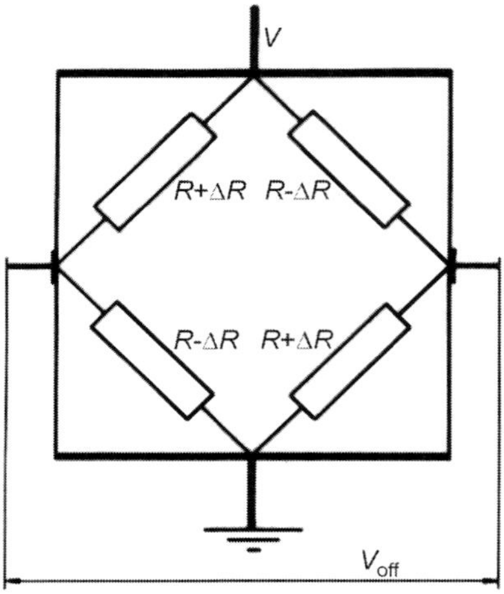

Figure 4.29. Bridge circuit model of a Hall plate. Ideally, the four resistors should be equal. Resistance variations ΔR lead to bridge asymmetry. The result is the offset voltage V_{off}.

If the R in figure 4.29 vary because of a magnetic field (see (4.74)), so also do the ΔR and V_{off}. At a low magnetic induction, the offset voltage can be expressed as

$$V_{\mathrm{off}}(B) \simeq \Delta R(0)(1 + \gamma\mu^2 B^2)I. \tag{4.107}$$

The coefficient γ may attain a value between zero and more than one (see (4.74)). The actual value of the coefficient γ depends on the type of carrier scattering and device geometry.

If $B \to 0$, $V_{\mathrm{H}} \to 0$ (4.90), but V_{off} tends to a constant value $V_{\mathrm{off}}(0)$ (4.107). Then, neglecting noise, the output voltage (4.86) reduces to

$$v_{\mathrm{out}}(B=0) \approx V_{\mathrm{off}}(0). \tag{4.108}$$

For this reason, the offset voltage is sometimes also-called the *zero-field output voltage*. In this definition the dependence $V_{\mathrm{off}} = f(B)$ is neglected.

4.3.5 Noise voltage

The noise voltage at the output terminals of a Hall device puts the ultimate limit on the precision with which we can measure the deterministic part of the Hall output voltage, i.e. both the Hall voltage and the offset voltage (4.86).

The noise in a Hall device is due to thermal noise, generation–recombination noise, and $1/f$ noise (see §2.7.3). Generation–recombination noise is often negligible in comparison with $1/f$ noise. Then the voltage noise spectral density across the sense terminals of a Hall device is given by [42]

$$S_{\mathrm{NV}}(f) = S_{\mathrm{V}\alpha}(f) + S_{\mathrm{VT}} \tag{4.109}$$

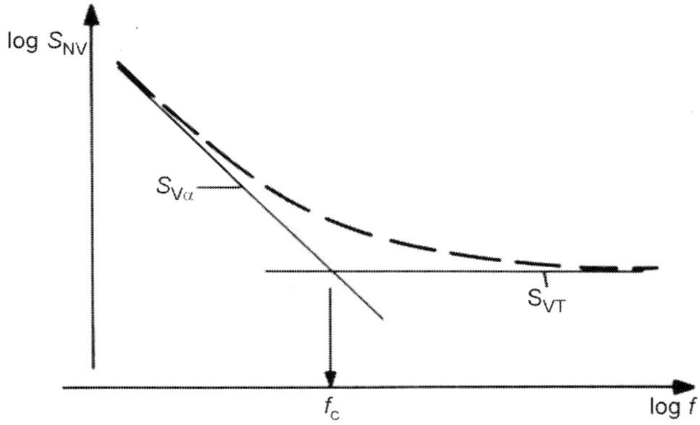

Figure 4.30. Sketch of a typical voltage noise spectral density S_{NV} at the output of a Hall device, as a function of frequency f. $S_{V\alpha}$ and S_{VT} are contributions of the l/f noise and the thermal noise, respectively. f_c is the corner frequency.

where $S_{V\alpha}$ and S_{VT} denote the *voltage noise spectral densities* due to $1/f$ and thermal noise, respectively, and f is the frequency (see figure 4.30).

For some simple device shapes $S_{V\alpha}$ can be found directly by making use of (2.129). More complicated structures can be cut into finite elements, the $1/f$ noise of each being given by (2.129). In this way, the geometry dependence of $S_{V\alpha}$ for various four-contact structures has been studied [43, 44]. For a rectangular Hall plate it was found [45] that (in our notation)

$$S_{V\alpha} \simeq \alpha \left(\frac{V}{l}\right)^{1/2} (2\pi n t f)^{-1} \ln\left(\frac{w}{s}\right) \tag{4.110}$$

where α denotes the Hooge $1/f$ noise parameter (see (2.129)).

S_{VT} can readily be obtained by substituting the output resistance of a Hall device into (2.126). For example, for a rectangular Hall plate with small sense contacts, by substituting (4.81) into (2.126) we obtain

$$S_{VT} \simeq 8kT \frac{\rho_b}{\pi t} \ln\left(\frac{w}{s}\right) \tag{4.111}$$

with T denoting the absolute temperature.

Given the noise voltage spectral density, the root-mean-square noise voltage can be calculated using (2.125).

4.3.6 Signal-to-noise ratio

In the applications of a Hall device, one measures the output voltage (4.86) in order to determine the Hall voltage. If the magnetic field is static, the Hall voltage cannot be distinguished from the offset voltage; besides, the noise voltage interferes with the Hall voltage too. Therefore, we define a generalized signal-to-noise ratio (SNR) by

$$\mathrm{SNR}(f=0) = V_{\mathrm{H}}/(V_{\mathrm{off}} + V_{\mathrm{NV}}) \tag{4.112}$$

where V_{off} is considered as a zero-frequency 'noise' voltage.

At a steady magnetic field, we can filter out the white part of the noise voltage, but not the $1/f$ part. At a higher frequency, we can filter out the offset voltage, and the signal-to-noise ratio reduces to the conventional form:

$$\mathrm{SNR}(f) = V_{\mathrm{H}}(f)/[S_{\mathrm{NV}}(f)\Delta f]^{1/2} \tag{4.113}$$

where Δf denotes a narrow bandwidth around a frequency f, and $S_{\mathrm{NV}}(f)$ denotes the voltage noise spectral density at the device output (4.109). Here are a few examples for the signal-to-noise ratio.

By substituting into (4.102) the expressions for the Hall voltage (4.89), the offset voltage (4.104), and (3.298), (3.295) and (3.146), and neglecting noise, we obtain

$$\mathrm{SNR}(f=0) = \mu_{\mathrm{H}} B \frac{w}{\Delta l}. \tag{4.114}$$

This holds for a very long Hall device with very small sense contacts and $\Delta l \cong 0$. For example, for a silicon Hall device with $\mu_{\mathrm{H}} \cong 0.15\,\mathrm{m}^2\,\mathrm{V}^{-1}\,\mathrm{s}^{-1}$, $B = 10^{-2}\,\mathrm{T}$ and $w/\Delta l \cong 1000$, (4.114) yields $\mathrm{SNR}(0) \cong 1.5$.

At low frequencies, $1/f$ noise dominates. By substituting (4.92) and (4.110) into (4.113), we obtain [42, 6]

$$\mathrm{SNR}(f) \simeq \frac{\mu_{\mathrm{Hn}}}{\alpha^{1/2}} \left(\frac{2\pi ntlw}{\ln(w/s)}\right)^{1/2} G_{\mathrm{H}} \left(\frac{w}{l}\right)^{1/2} \left(\frac{f}{\Delta f}\right)^{1/2} B_{\perp}. \tag{4.115}$$

This relation holds for $f < f_{\mathrm{c}}$, f_{c} being the corner frequency defined by $S_{\mathrm{V}\alpha}(f_{\mathrm{c}}) = S_{\mathrm{VT}}$ (see figure 4.29). The *corner frequency* depends on biasing conditions and temperature: from (4.110) and (4.111), we find

$$f_{\mathrm{c}} = \frac{q\mu\alpha}{16kT}\left(\frac{V}{l}\right)^2. \tag{4.116}$$

Note that in a very symmetrical Hall device, the signal-to-noise ratio at a static magnetic field might be determined by the $1/f$ noise, rather than by the offset: then the $1/f$ law on $S_{\mathrm{V}\alpha}(f)$ holds down to frequencies as low as 10^{-6} [46], and according to (4.110), if $f \to 0$, $S_{\mathrm{V}\alpha} \to \infty$.

At high frequencies, above the corner frequency, the thermal noise dominates. By substituting (4.89) and (4.111) into (4.113), we obtain

$$\mathrm{SNR}(f) = \left(\frac{\pi\mu}{8kTqnt\ln(w/s)\Delta f}\right)^{1/2} r_{\mathrm{H}} G_{\mathrm{H}} I B_{\perp}. \tag{4.117}$$

Therefore, at high frequencies, the SNR increases with the increase in device current. The maximal SNR is limited only by the maximal acceptable heat dissipation P_{max}. According to (4.95) and figure 4.26, the maximal Hall voltage for a given power dissipation is given by

$$V_{Hmax} \simeq 0.7 r_H \left(\frac{\mu}{qnt}\right)^{1/2} P^{1/2} B_\perp. \tag{4.118}$$

Inserting this expression and (4.111) into (4.113), we find the *maximal possible SNR* [42, 6]:

$$\mathrm{SNR}_{max} \simeq 0.44 \mu_H B_\perp \left(\frac{P_{max}}{kT\Delta f \ln(w/s)}\right)^{1/2}. \tag{4.119}$$

This relation holds for $f \gg f_c$. However, note that $f_c \sim V^2$ (4.116) and thus $f_c \sim P$. Therefore, the maximal SNR may be obtained only at very high frequencies.

4.3.7 Self magnetic fields

Self magnetic fields of a Hall device are the magnetic fields generated by the biasing current of the device. In this section we shall briefly review some influences of the self magnetic fields on a Hall device operation.

Let us first look at the values and the distribution of the self magnetic fields. With reference to figure 4.31, we distinguish the following fields: $\boldsymbol{B}_i$ and $\boldsymbol{B}_e$, which are the internal and the external magnetic fields of the bias current in the Hall plate itself; and $\boldsymbol{B}_c$, a field generated by the supply current in the connection wires of the device. The values of these fields can be calculated using the Biot–Savart law. Figure 4.31(b) shows a result of such a calculation.

An estimation of the magnetic field value at a distance y from the device symmetry axis (here x axis) can be found using the Ampere law for a long cylindrical conductor:

$$B \simeq \mu_0 \frac{I}{2\pi y}. \tag{4.120}$$

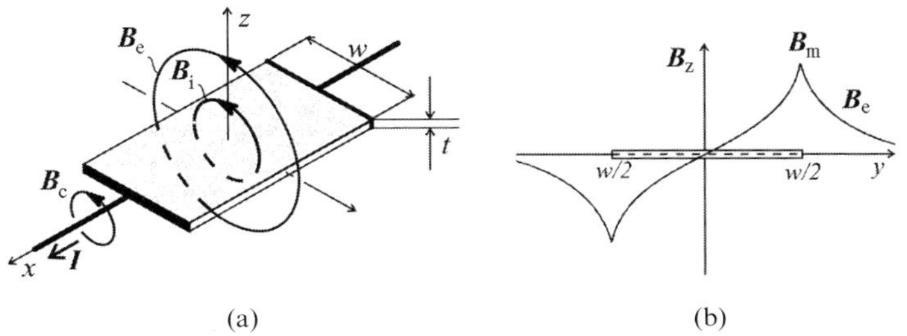

Figure 4.31. Self magnetic fields of a Hall plate. (a) Illustrating the internal field $\boldsymbol{B}_i$, the external field $\boldsymbol{B}_e$, and the field around the connections $\boldsymbol{B}_e$. (b) The calculated distribution of the z component of the self-fields $\boldsymbol{B}_i$ and $\boldsymbol{B}_e$ (adapted from [3]).

Here μ_0 denotes the vacuum permeability and I is the Hall device supply current. For $I = 1\,\text{mA}$ and $y = 10\,\mu\text{m}$, equation (4.120) gives $B = 20\,\mu\text{T}$. Equation (4.120) gives us also an estimation of the maximal value of the self magnetic field at the boundaries of the plate, where $B_{\text{m}} \approx B_{\text{i}}(w/2) = B_{\text{e}}(w/2)$ (see figure 4.30(b)).

The internal magnetic field becomes especially important if a Hall plate is placed in a narrow air gap of a magnetic core, since then the effective length of the magnetic field lines ($2\pi y$ in (4.120)) reduces to the double length of the air gap; and in small Hall plates operating with high current densities.

Self magnetic fields can have the following influences on the operation of a Hall plate.

(a) *Offset*. In a symmetrical structure as that shown in figure 4.31(a), the mean value of the self magnetic field over the Hall plate is zero. But if the Hall plate or the supply connection wires are not symmetrical with respect to the symmetry axis of the Hall plate, then the Hall plate will 'measure' a non-zero average self-field. This will produce a component of the output voltage proportional to I^2.
(b) *A.C. input–output cross talk*. If the supply current is alternating, and if the input and the output connections of a Hall plate are not properly placed, then the a.c. self-field may generate an inductive signal in the loop formed by the output circuit of the Hall plate.
(c) *Coupling with eddy currents*. This effect gives an additional Hall voltage if a Hall plate operates at a high-frequency magnetic field. We shall explain this effect in §4.6.

Theoretically, the first two of the above effects of self magnetic field, (a) and (b), can be eliminated by a good design of a Hall plate; but the third one (c) is an intrinsic effect and cannot be avoided.

4.4 The Hall current mode of operation

As we have seen in chapter 3, the generation of the Hall voltage is but one of the two complementary ways of the appearance of the Hall effect: the Hall effect may also show up as the current deflection effect. We shall now investigate the relationship of the current deflection in a sample to a change in the device terminal currents. We shall see that this change in the terminal currents is related to the so-called Hall current. The Hall current can also be regarded as the output signal of a Hall device. In this case, we say that the Hall device operates in the Hall current mode.

4.4.1 Large-contact Hall devices

Since the current deflection effect prevails in short Hall devices, let us first consider some general properties of such devices. Instead of using the term

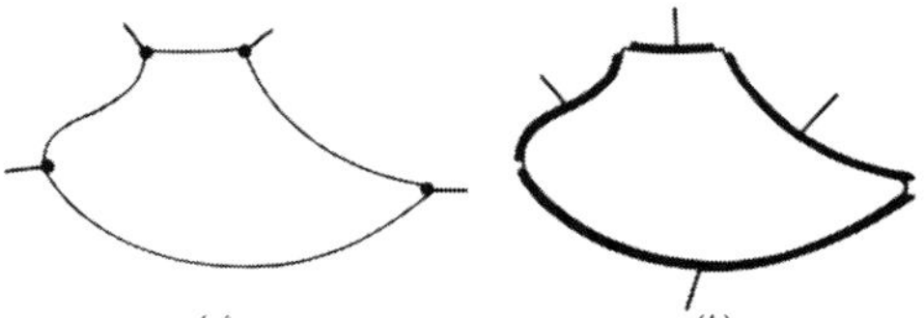

Figure 4.32. Dual point-contact (a) and point-isolation (b) Hall devices. The shapes and dimensions of the active regions of two dual devices are identical, but where there are contacts in one device, there are insulating boundaries in the other device, and vice versa.

short Hall device, it is here more adequate to introduce the term large-contact Hall device. Bearing in mind the discussion of §4.2.6, as far as the galvanomagnetic properties of Hall devices are concerned, the two terms are equivalent.

A large-contact Hall device is a Hall device with a boundary predominantly covered by highly conductive contacts. The short strip and the Corbino disc, shown in figures 3.4 and 3.5 respectively, are large-contact Hall devices. Another example is the Hall device dual to a point-contact Hall device (see figure 4.32). The boundary of the large-contact Hall device (b) is, except for infinitesimally narrow isolating slits, completely covered by highly conductive contacts. We shall call such a device a point-isolation Hall device.

Various Hall devices can be roughly categorized according to their relative contact size with the aid of the λ factor (4.57). For a point-contact device and an infinitely long device, $\lambda \to 0$. For an infinitely short Hall device, Corbino disc, and point-isolation Hall device, $\lambda \to \infty$. We shall call such a device an infinitely-large-contact device. The notions of a point-isolation device and an infinitely-large-contact device are equivalent.

In an infinitely-large-contact Hall device, the Hall electric field cannot develop (see §3.1.2). Consequently, no Hall voltage can appear at the output terminals. Instead, the Hall effect shows up through the current deflection effect. The current deflection effect brings about the geometric contribution to magnetoresistance, but this effect can be also observed more directly by measuring the so-called Hall current [47].

4.4.2 Dual Hall devices. Hall current

To understand the notion of the Hall current, we shall exploit the geometrical similarity between the equipotential lines and the current density lines in dual Hall devices (see figure 4.33). In this figure, (a) shows a very long rectangular Hall device with point sense contacts. The device is biased by the voltage $2V_0$. In the absence of a magnetic induction, the equipotential lines in this device are parallel to the current contacts. Its dual device (b) is a very short Hall device, all four terminals of which are used as current contacts. This device

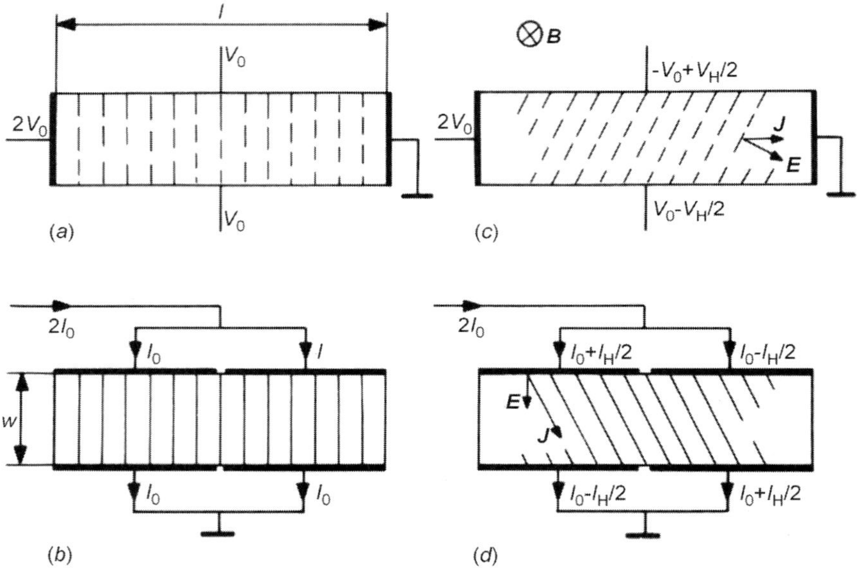

Figure 4.33. The analogy between a long Hall device operating in the Hall voltage mode ((a) and (c)) and its dual Hall device operating in the Hall current mode ((b) and (d)). The broken lines in (a) and (c) are the equipotential lines; the similarly distributed full lines in (b) and (d) are the current lines. A magnetic induction produces equal relative disturbances in the equipotential lines (c) and in the current lines (d). Therefore, $V_H/2V_0 = I_H/2I_0$.

is biased by the current $2I_0$. In the absence of a magnetic induction, the current density lines are perpendicular to the contacts. In the device (a), each sense contact attains an equal potential, V_0. In the device (b), in each terminal flows an equal current, I_0.

In the presence of a magnetic induction $\boldsymbol{B}$, in the point-contact device, the electric field and the equipotential lines rotate for the Hall angle (see figure 4.33(c)); and in the large-contact device (d) the current density lines also rotate for the same angle, but in the opposite direction (see also figure 3.12).

Owing to the rotation of the equipotential lines in the Hall device in figure 4.33(c), the potentials at its sense contacts alter. By pure geometrical and proportionality analysis of figure 4.33(c), we conclude that at each sense terminal the voltage changes for

$$\pm\frac{V_H}{2} = \frac{V_0}{l} w \tan\Theta_H \qquad (4.121)$$

with Θ_H being the Hall angle. The voltage difference appearing between the two sense contacts is V_H: this is, of course, the Hall voltage. Making use of

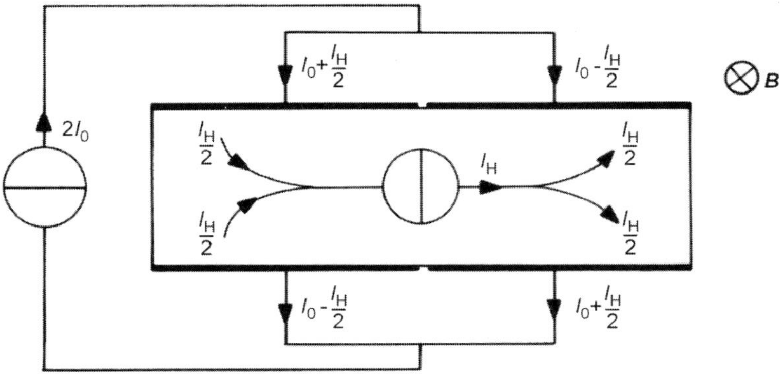

Figure 4.34. Equivalent circuit representation of a Hall device operating in the Hall current mode. As a consequence of the Hall effect, a current I_H appears in the direction transverse with respect to the externally forced current. The transverse current I_H is called the Hall current.

(3.152), (4.121) yields

$$V_H = \mu_H \frac{w}{l} 2V_0 B_\perp. \tag{4.122}$$

Note that this expression also follows from (4.93) when $G_H \to 1$ and $V = 2V_0$.

The rotation of the current density lines in the large-contact device (figure 4.33(d)) causes a change in its terminal currents. With a similar reasoning to the one above, we find the change of current in each terminal to be

$$\pm\frac{I_H}{2} = \frac{I_0}{l} w \tan\Theta_H. \tag{4.123}$$

These current changes could be produced by an equivalent current source of a transverse current I_H, as shown in figure 4.34. This transverse current is called the *Hall current*. From (4.123) and (3.152), we find the Hall current to be

$$I_H = \mu_H \frac{w}{l} 2I_0 B_\perp. \tag{4.124}$$

The Hall current characterizes the *current deflection effect*.

The analogy between the diagrams of electric potential and current density in figure 4.33, and between equations (4.122) and (4.124), is not a mere accident. It is a consequence of the analogy between the basic galvanomagnetic equations (3.165) and (3.175), and the boundary conditions (3.154) and (3.160), and (3.158) and (3.162) [47, 48].

Based on these facts, we infer that the analogy between the relations for V_H and I_H is not limited to a point-contact device and its dual device, the

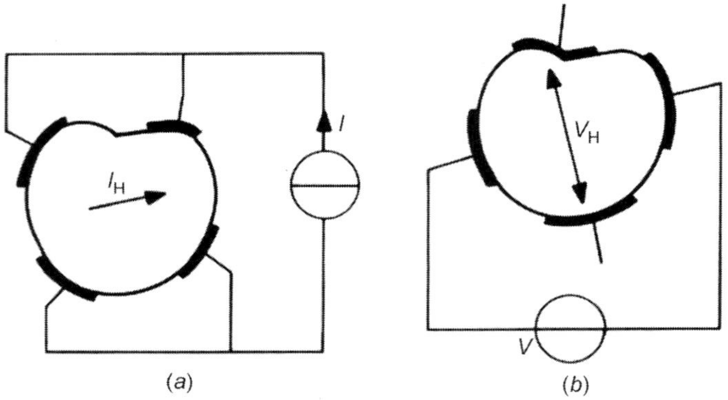

Figure 4.35. Analogous dual Hall devices of an arbitrary shape. (a) A Hall device operating in the Hall current mode. The output signal is the Hall current I_H, which produces a disturbance in the terminal currents. (b) The dual Hall device operating in the Hall voltage mode. The output signal is the Hall voltage V_H, which appears between the non-connected (sense) terminals.

infinitely-large-contact device. This analogy must hold for any pair of dual Hall devices as well. Therefore, by making use of (4.93), we may now generalize (4.124).

Consider two dual Hall devices of arbitrary shape (figure 4.35). Assume that the devices are made of an extrinsic material. Then in the Hall device operating in the Hall current mode, figure 4.35(a), the Hall current is given by

$$I_H = \mu_H \frac{w}{l} G_H I B_\perp. \tag{4.125}$$

Here, μ_H is the Hall mobility of majority carriers, I is the total bias current, $B_\perp$ is a normal magnetic induction, and w/l and G_H are the parameters pertinent to the dual Hall device operated in the Hall voltage mode, figure 4.34(b): G_H is the geometrical correction factor (see §4.2.6) and w/l is the width-to-length ratio of the equivalent rectangular device (see (4.78) and (4.79)).

It was estimated that the signal-to-noise ratio of a Hall plate does not depend on whether it is operated in the Hall voltage mode or in the Hall current mode [49].

4.5 The magnetoresistance mode of operation

In the magnetoresistance mode of operation, a plate, similar to one of those shown in figure 4.16, is incorporated into an electronic circuit in such a way that a change in device resistance can be measured. According to (4.74), the dependence of the device resistance on the magnetic field is given by

$$R(B) = R(0)\frac{\rho_b(B)}{\rho_b(0)}G_R(B). \tag{4.126}$$

Recall the notation: $R(0)$ is the device resistance at $B = 0$, ρ_b is the intrinsic magnetoresistivity (§3.4.5), and $G_R(B)$ is the *geometrical factor of magnetoresistance* (§4.2.7). The magnetic dependence of ρ_b is usually too weak for practical applications. Therefore, practical magnetoresistors rely on the geometrical magnetoresistance effect, i.e. on the magnetic dependence of the geometrical factor, which we stress here by writing $G_R(B)$.

4.5.1 Magnetoresistance ratio

In the magnetoresistance mode of operation, the useful signal of the device is usually proportional to the relative change of the device resistance under influence of a magnetic field:

$$\frac{\Delta R(B)}{R(0)} = \frac{R(B) - R(0)}{R(0)}. \tag{4.127}$$

This is called the magnetoresistance ratio, or, in short, magnetoresistance.

For a Corbino disc, from (4.73) we have

$$\frac{\Delta R(B)}{R(0)} \simeq \mu_H^2 B^2 \tag{4.128}$$

which is called a *Corbino magnetoresistance*. We obtain the same expression from (4.75) for a very short magnetoresistor at a low magnetic field. Otherwise, generally

$$\frac{\Delta R(B)}{R(0)} = G_R(B)\frac{\rho_b(B)}{\rho} - 1 \tag{4.129}$$

which follows from (4.126). By inspecting $G_R(B)$) in figure 4.22, we notice that G_R and, therefore, the magnetoresistance ratio (4.129) of reasonably short magnetoresistors at low magnetic fields follow a quadratic dependence on B, similar to (4.128); but at a higher magnetic field, the $G_R(B)$ eventually becomes linear. Only an infinitely short magnetoresistor, i.e. a Corbino disc, keeps the quadratic dependence of $G_R(B)$ without limit.

Due to the quadratic magnetic field dependence of the magnetoresistance ratio (4.128), at high magnetic bias, the signal-to-noise ratio of a very short device operating in the magnetoresistance mode might be better than that of a comparable device operating in any of the two Hall operating modes [50].

4.6 High-frequency effects

In our analysis of the galvanomagnetic effects and the basic galvanomagnetic devices up to now we have always assumed that all phenomena have enough time to settle down; and we completely neglected possible capacitive and inductive effects. This is quite reasonable for direct-current and low-frequency operation, whereby 'low frequency' may mean up to megahertz or gigahertz range. We shall now consider what happens if the excitation current and/or the measured magnetic field vary faster than that. Thereby we shall see that there are relevant effects that appear at any point of the material: we shall call them local high-frequency effects; and others, which appear at the global device level. In this second group belong the global capacitive and inductive effects.

4.6.1 Local high-frequency effects

The first question that we should ask ourselves in the present context is whether the basic theory of the galvanomagnetic effects, presented in chapter 3, is still valid if electric and magnetic fields vary rapidly. The answer is: not necessarily, at least for two reasons.

Relaxation time frequency limit

Recall that one of the key steps in the development of the theory of §3.2 was the so-called relaxation time approximation (3.62). According to this approximation, if conditions (i.e. electric and/or magnetic field) suddenly change, then the distribution function of charge carriers adapts itself with the time constant equal to the relaxation time τ (3.63). Accordingly, we conclude that the theory of the galvanomagnetic effects, as presented in chapter 3, is valid at any point of a material for frequencies of local electric and/or magnetic fields lower than

$$f_\tau = 1/(2\pi\tau). \tag{4.130}$$

This is the relaxation time frequency limit. The relaxation time of charge carriers in semiconductors is of the order of 10^{-13} s. Therefore, the relaxation time frequency limit f_τ is of the order of 1 THz (terahertz).

What happens at frequencies higher than f_τ? Briefly, then the drift velocity of carriers is no longer dependent on their free-transit time as in (2.84), but depends only on the effective mass of charge carriers (2.69). Moreover, electrons show up their inertia, and current density is in phase delay with respect to electric field. Under such conditions the Hall effect should still exist, but the basic parameters, such as Hall mobility, are probably quite different than those at low frequency.

Dielectric relaxation frequency limit

Another approximation pertinent to low-frequency operation is that we neglected the electrostatic polarization of the material in which the galvanomagnetic effects take place. A change of electrostatic polarization takes time, which brings about another frequency limit [51].

Suppose that at a point within the active volume of a galvanomagnetic device we have a non-equilibrium charge density q. Then from the third Maxwell's equation, i.e.

$$\operatorname{div}\boldsymbol{D} = \varepsilon \operatorname{div}\boldsymbol{E} = q \tag{4.131}$$

and the continuity equation

$$\operatorname{div}\boldsymbol{J} = \sigma \operatorname{div}\boldsymbol{E} = -\frac{\partial q}{\partial t} \tag{4.132}$$

we obtain

$$\frac{\partial q}{\partial t} + \frac{\sigma}{\varepsilon} = 0. \tag{4.133}$$

Here $\boldsymbol{D}$ is the electric flux density, $\boldsymbol{E}$ is the electric field, $\varepsilon = \varepsilon_r \varepsilon_0$ is the local permittivity of the material, and σ is the local conductivity of the material. Let at $t = 0$, $q = q_0$. Then the solution of (4.134) is

$$q = q_0 \exp\left(-\frac{t}{\tau_D}\right). \tag{4.134}$$

The quantity

$$\tau_D = \frac{\varepsilon}{\sigma} \tag{4.135}$$

is called the *dielectric relaxation time*. If for any reason there was a steady-state charge disequilibria in a material, and this reason suddenly disappears, then the charge disequilibria relaxes with the time constant τ_D. The dielectric relaxation time of good conductors is extremely small. But this is not necessarily the case with semiconductors relevant for galvanomagnetic devices. For example, for silicon of resistivity $1\,\Omega\,\text{cm}$ (Example of §3.1.1) with $\varepsilon_r = 11.9$, (4.135) gives $\tau_D \approx 10^{-8}$ s.

The corresponding frequency limit of the galvanomagnetic effect,

$$f_D = 1/(2\pi\tau_D) \tag{4.136}$$

is called the dielectric relaxation frequency limit. Depending on the density of charge carriers in the active region and on their mobility, f_D might be in the range from 1 MHz to 1 THz.

4.6.2 Global capacitive effects

Apart from the local dielectric relaxation of the device active material, we should take into account also the relaxation of the electrostatic polarization of the dielectric surrounding a galvanomagnetic device. Since we consider now the whole device, we use the term global capacitive effects. We can distinguish two slightly different global capacitive effects, which we shall call intrinsic and extrinsic, respectively.

Intrinsic capacitance is the capacitance related to the total area of the active part of a galvanomagnetic device. 'Active part' here means the part of the device in which the galvanomagnetic effect really takes place, excluding connecting plates and wires. During a normal operation of a galvanomagnetic device, the dielectric around the device is polarized. A change of electrostatic polarization takes time, which brings about yet another frequency limit.

Consider a model of a free-standing galvanomagnetic plate shown in figure 4.36. We assume that the plate is immersed in a homogeneous dielectric and is far away from any other conductive body. Yet in the steady state at the terminals of the plate there are potentials $+V/2$ and $-V/2$ (figure 4.36(a)). The potential difference may be due to any reason, including an external biasing voltage and (internal) Hall effect. Along the plate there is an internal electric field $\boldsymbol{E}_{\mathrm{i}}$,

$$E_{\mathrm{i}} = \frac{V}{L}. \tag{4.137}$$

Then, since a circulation of an electric field must be zero, there is also an external electric field $\boldsymbol{E}_{\mathrm{e}}$ all around the plate. Due to the symmetry of the plate, the electric field lines are symmetrical with respect to the yz plane. If we approximate the form of the electric field lines with a semi-circle, figure 4.36(b), then the external electric field is given by

$$E_{\mathrm{e}} \simeq \frac{2}{\pi} E_{\mathrm{i}}. \tag{4.138}$$

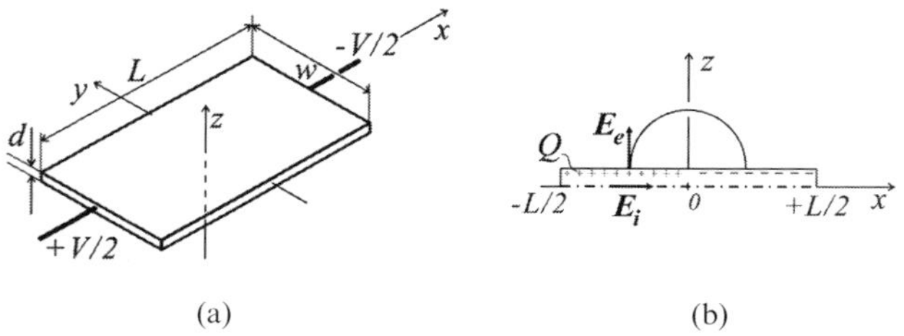

Figure 4.36. The model of a free-standing galvanomagnetic plate surrounded by a homogeneous dielectric. (a) Dimensions and polarization. (b) The cross-section of the upper half of the plate. The potential difference at the extremities of the plate comes along with the existence of both an internal ($\boldsymbol{E}_{\mathrm{i}}$) and external ($\boldsymbol{E}_{\mathrm{e}}$) electric fields. The external electric field is associated with a surface charge Q.

In accordance with the third Maxwell's law, at the issue of an electric field line in dielectric there exists an electric charge Q, as shown in figure 4.36(b). The surface density of this charge is given by

$$\frac{\mathrm{d}Q}{\mathrm{d}S} = D_\mathrm{e} = \varepsilon_\mathrm{e} E_\mathrm{e}. \tag{4.139}$$

Here $\boldsymbol{D}_\mathrm{e}$ is the external electric flux density and ε_e is the permittivity of the dielectric in which our plate is immersed. Note that within the present approximation, we obtain a constant surface charge density over the entire half plate, and not just a charge concentrated on the side surface, as illustrated in figure 3.3. Suppose that the plate is very thin, $d \ll L, W$. The total electric charge stored at the surface of the plate is then

$$Q = \int_{(1/2)} \boldsymbol{D}_\mathrm{e}\,\mathrm{d}\boldsymbol{S} \simeq \varepsilon_\mathrm{e} E_\mathrm{i} L W \frac{2}{\pi} \tag{4.140}$$

where we integrate the charge over the left half of the plate (the other half is negatively charged).

Let us assume now that the electromotive force that created the potential difference over the plate suddenly disappears. Then the stored charge will start to decay:

$$\frac{\partial Q}{\partial t} = -I = -\sigma E_\mathrm{i} W d. \tag{4.141}$$

This equation states the fact that all the current that eventually equilibrates the charges at the left and at the right half of the plate must pass the middle cross-section of the plate: I denotes the current at the middle of the plate ($x = 0$), σ is the electrical conductivity of the plate, and Wd is the cross-section of the plate.

From (4.140) and (4.141) we obtain

$$\frac{\partial Q}{\partial t} + \frac{\sigma}{\varepsilon_\mathrm{e}} \frac{d}{L} \frac{\pi}{2} Q = 0. \tag{4.142}$$

Let $Q = Q_0$ at $t = 0$. Then the solution of the differential equation (4.142) is

$$Q \simeq Q_0 \exp\left(-\frac{t}{\tau_\mathrm{i}}\right). \tag{4.143}$$

The quantity

$$\tau_\mathrm{i} \simeq \frac{\varepsilon_\mathrm{e}}{\sigma} \frac{L}{d} \frac{2}{\pi} \tag{4.144}$$

we call the *intrinsic relaxation time of the plate*. Note that the term $1/(\sigma d)$ figuring in (4.144) is the sheet resistance of the plate material—see §6.1.1. Note also that the numerical factor $2/\pi$ comes from our approximation of the forms of electric field lines by semi-circles. This approximation limits

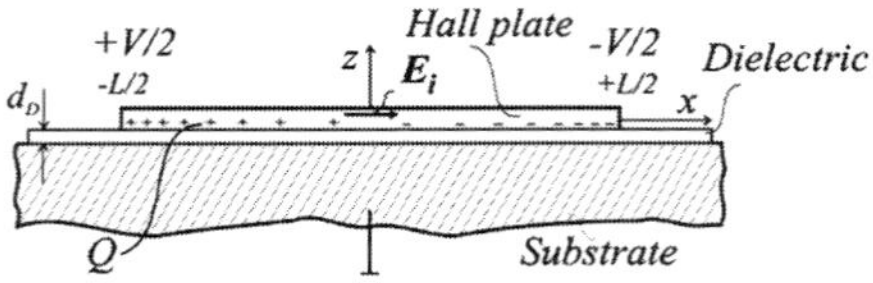

Figure 4.37. The cross-section of a Hall plate placed on a conductive substrate, with a dielectric film in between. The substrate is at ground potential, and the two ends of the plate are at the potentials $V/2$ and $-V/2$, as in figure 4.36(a). The charge density at the bottom surface of the plate Q increases with the distance from the plate centre (see (4.145)).

the volume of the polarized dielectric around the plate and brings about a sub-estimation of this effect. For very thin plates, numerical simulations give an about two-times higher value of the numerical factor [50].

In summary, if an electromotive force keeps a constant electric field in a plate, the surface of the plate is charged as illustrated in figure 4.36(b). If the electromotive force suddenly disappears, then the charge decays with the time constant τ_i. Thereby the still-existing charge ensures the continuity of the electric field in the plate, which decays along with the charge.

It is interesting to compare the local and the global dielectric relaxation times of a plate, equations (4.135) and (4.144). We see a striking similarity between the two expressions. For the case that the permittivity of the plate and that of the surrounding medium are equal, the global relaxation time is always greater than the local one of the plate material. The difference can be very important for big thin plates imbedded in a high permittivity dielectric. For example, for the silicon Hall plate of Example in §3.1.1, $\tau_D \approx 10^{-8}$ s; if the device is situated in vacuum, $\tau_i \approx 0.8 \times 10^{-8}$ s; but if it is imbedded in semi-insulating silicon, then $\tau_i \approx 10^{-7}$ s.

An important special case is when a plate is isolated with a thin dielectric layer of a thickness $d_D \ll (L, W)$, and of permittivity ε_e from a conductive substrate (figure 4.37). Suppose that along such a plate there is an internal electric field $\boldsymbol{E}_i$ as in the previous case (4.137). Above the plate, the dielectric will be polarized similarly as described before. For the sake of simplicity, we shall neglect the corresponding charges at the top surface of the plate. This is reasonable because now the bottom surface of the plate shall be charged much more. The surface charge density at the bottom surface of the plate is given by

$$\frac{dQ}{dS} \simeq -\varepsilon_e \frac{V}{L} \frac{1}{d_D} x. \tag{4.145}$$

By integrating the charge density over the left half of the bottom surface of the plate, we obtain the total stored electric charge,

$$Q \simeq \varepsilon_e E_i L W \frac{L}{8 d_D}. \tag{4.146}$$

Proceeding now as above, equations (4.141)–(4.143), we get the intrinsic relaxation time of a plate on a substrate,

$$\tau_{\text{is}} \simeq \frac{\varepsilon_{\text{e}}}{\sigma} \frac{L}{d} \frac{L}{8d_{\text{D}}}. \tag{4.147}$$

Compared with the case of a free-standing plate, now the surface charge density is not constant—see (4.145) and figure 4.37; and the numerical factor $2/\pi$ in (4.144) is replaced by the factor $L/(8d_{\text{D}})$. This factor can increase the relaxation time constant considerably. For example, for the silicon Hall plate of Example in §3.1.1 situated in vacuum, $\tau_{\text{i}} \approx 0.8 \times 10^{-8}$ s; but if it is isolated from a conducting silicon substrate with a depletion layer of thickness $d_{\text{D}} = 1\ \mu\text{m}$, then $\tau_{\text{is}} \approx 5.6 \times 10^{-6}$ s.

It is useful to define an equivalent intrinsic capacitance of a galvanomagnetic plate that gives the time constant τ_{i} (or t_{is}) estimated above. If we choose as the equivalent resistance the real resistance between the terminals of the device R, then the equivalent intrinsic capacitance is simply given by

$$C_{\text{i}} = \frac{\tau_{\text{i}}}{R}. \tag{4.148}$$

In order to estimate the total relaxation time of a galvanomagnetic plate, we should also take into account extrinsic capacitances. The *extrinsic capacitance* of a galvanomagnetic plate is the capacitance that exists in the input or the output circuit of the device apart from the intrinsic capacitance. For example, the extrinsic output capacitance of a Hall plate consists of the capacitances of the contacts, the connection lines Hall plate-amplifier, and the input capacitance of the amplifier.

Using (4.148) and neglecting the dielectric relaxation time of the plate material, the *input relaxation time* of a plate is given by

$$\tau_{\text{inp}} \simeq R_{\text{in}}(C_{\text{i}} + C_{\text{e}}) = \tau_{\text{i,in}} + R_{\text{in}} C_{\text{e,in}}. \tag{4.149}$$

We define also the *output relaxation time* of a Hall plate:

$$\tau_{\text{out}} \simeq \tau_{\text{i,out}} + R_{\text{out}} C_{\text{e,out}}. \tag{4.150}$$

For the operation of a Hall plate with rapidly varying biasing conditions, relevant is the *combined relaxation time* of a Hall plate, which is given by

$$\tau_{\text{tot}} \simeq \sqrt{\tau_{\text{inp}}^2 + \tau_{\text{out}}^2}. \tag{4.151}$$

This is so because the relaxation processes in the input and output circuits of a Hall plate behave much like those in two circuit blocks connected in series.

4.6.3 Inductive effects

Let us now consider secondary effects that may be produced in a galvanomagnetic device by alternative excitation current and/or alternative magnetic field.

Conventional inductive effects

As we saw in §4.3.7, an a.c. excitation current of a Hall plate produces an a.c. self-field, which may generate an inductive signal in the output circuit of the Hall plate. Also an external a.c. magnetic field can produce inductive currents or voltages in both input and output circuits of a galvanomagnetic device. These effects can, at least in principle, be avoided by a proper layout of the connection wires of the galvanomagnetic device.

Skin effect

An a.c. excitation current in a galvanomagnetic device is subject to the skin effect much as in any other device: the a.c. self magnetic field forces the current to avoid the centre and to concentrate at the extremities of the cross-section of the device. This produces an increase in the device resistance at high frequencies. When measuring a homogeneous magnetic field, the skin effect produces no change in the Hall voltage [2]. But for a very non-homogeneous magnetic field, this might not be the case. The skin effect was noticed in large-size galvanomagnetic device (plate width ~1 mm) placed in a narrow ferrite air-gap at a frequency higher than 10 MHz [2].

Eddy currents

If a galvanomagnetic device is exposed to a fast-varying magnetic field, then eddy currents will be induced in it much as in any other conductive material. Figure 4.38 illustrates the eddy currents in a rectangular Hall plate exposed to an a.c. perpendicular magnetic field. Eddy currents have the following two effects relevant to the measurement of high-frequency magnetic fields: first,

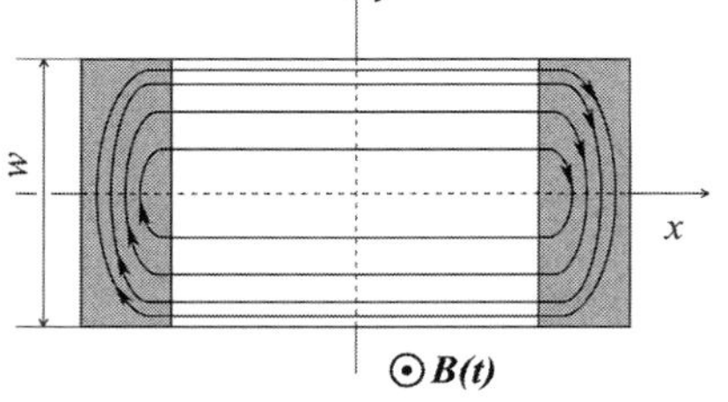

Figure 4.38. Illustrating eddy currents in a rectangular galvanomagnetic plate with highly-conductive current-supply contacts. The eddy current density increases linearly with the distance from the plate longitudinal axis.

in accordance with Lenz's law, the eddy currents tend to oppose the change in field inducing them, which results in a phase shift and a diminution of the magnetic field acting on the galvanomagnetic device; and second, in co-operation with the self magnetic field of a Hall plate, eddy currents produce an additional Hall effect. Let us briefly consider these two phenomena.

Suppose that a thin long rectangular Hall plate, like that in figure 4.38, is exposed to a perpendicular a.c. magnetic field

$$B(t) = B_0 \exp(\mathrm{i}\omega t). \tag{4.152}$$

We shall consider the case of a low-enough angular frequency ω so that the penetration depth of the magnetic field into the plate material is much higher than the plate thickness, and the reaction magnetic field produced by eddy currents is much smaller than the excitation field (4.152). Then we can calculate the density of the eddy currents as

$$J_x(y) \simeq -\frac{\sigma}{2L}\frac{\mathrm{d}\Phi}{\mathrm{d}t} = \cdots = -\sigma\frac{\mathrm{d}B}{\mathrm{d}t}y = -\mathrm{i}\omega\sigma y B(t). \tag{4.153}$$

So to the first approximation, the density of eddy currents increases linearly as we approach the plate periphery.

The *magnetic field produced by the eddy currents* has a maximum along the long symmetry axis of the plate, which is the x axis in figure 4.38. This maximum reaction field is given by

$$B_{\mathrm{m}}(t) \simeq 2\int_0^{W/2} \mu_0 \frac{J_x(y) d\, \mathrm{d}y}{2\pi y} = -\mathrm{i}B(t)\frac{\mu_0}{2\pi}\omega\sigma W d \tag{4.154}$$

where d denotes the thickness of the plate. The ratio of the maximum reaction field and the excitation magnetic field can be then put into the following form:

$$\left|\frac{B_{\mathrm{m}}}{B}\right| \simeq \mu_0 f \frac{W}{R_{\mathrm{S}}} \tag{4.155}$$

where f denotes the frequency of the excitation magnetic field and R_{S} is the sheet resistance of the plate,

$$R_{\mathrm{S}} = \frac{\rho}{d} = \frac{1}{\sigma d}. \tag{4.156}$$

Another consequence of the eddy currents is its participation in the creation of the Hall voltage. Consider again a long rectangular Hall plate as that shown in figure 4.38. At an a.c. excitation magnetic field, the total current density in the plate is given by

$$J = J_{\mathrm{i}} + J_{\mathrm{e}}. \tag{4.157}$$

Here J_i denotes the current density in the Hall plate produced by the input current,

$$J_i \simeq \frac{I}{Wd} \tag{4.158}$$

where we neglected a possible skin effect; and J_e denotes the eddy current density (4.153). The total magnetic field acting on the plate is given by

$$B \simeq B_i + B_s \tag{4.159}$$

where, neglecting the magnetic field of eddy currents, B_i denotes the input magnetic field (4.152); and B_s denotes the self-field, produced in the plate by the excitation current (see §4.3.7). To estimate the influence of the self magnetic field, with the aid of (4.120), we shall approximate the function shown in figure 4.31(b) by the linear function

$$B_s \simeq \mu_0 \frac{I}{2\pi(w/2)} \frac{y}{(w/2)}. \tag{4.160}$$

Now the total Hall voltage is given by the integral (4.87) over the plate width. The integrals of the products of even and odd terms of (4.157) and (4.159) vanish, and we obtain

$$V_H \simeq V_{Hi} + V_{He}. \tag{4.161}$$

Here V_{Hi} denotes the ordinary Hall voltage,

$$V_{Hi} = \frac{R_H}{d} IB(t) \tag{4.162}$$

and V_{He} denotes the Hall voltage produced by the coupling of the self magnetic field and eddy currents:

$$V_{He} = -\int_{-W/2}^{W/2} R_H [\boldsymbol{J}_e \times \boldsymbol{B}_s] \, dy. \tag{4.163}$$

After substituting here (4.153) and (4.160), we obtain

$$V_{He} \simeq -i\frac{\mu_0}{3} f\sigma W R_H IB(t). \tag{4.164}$$

To estimate the importance of this effect, let us find the ratio of (4.164) and (4.162). By neglecting the numerical factor 3, we may put this ratio into the following form:

$$\left|\frac{V_{He}}{V_{Hi}}\right| \simeq \mu_0 f \frac{W}{R_S} \tag{4.165}$$

which is similar to (4.155). Therefore, both effects of eddy currents, namely the creation of a reaction magnetic field and the creation of the new Hall voltage, become noticeable at about the same frequency. If we choose as a criterion that the high-frequency effects do not become noticeable in the form $B_m/B < 1/100$ (4.155) and $V_{He}/V_{Hi} < 1/100$ (4.165), we obtain

$$f\frac{W}{R_S} < \frac{1}{100\mu_0} \simeq 10^4\,\mathrm{m/H}. \tag{4.166}$$

For example, for $W = 1\,\mathrm{mm}$ and $R_S = 100\,\Omega$, this gives for the maximal acceptable frequency of the excitation field $f < 10^9\,\mathrm{Hz}$.

The magnetic high-frequency effects may become more important if a galvanomagnetic device is incorporated in an air-gap of a magnetic material [3].

References

[1] Hall E H 1879 On a new action of the magnet on electric currents *Am. J. Math.* 2287–92

[2] Kuhrt R and Lippmann H J 1968 *Hallgeneratoren* (Berlin: Springer)

[3] Weiss H 1969 *Physik und Anwendung galvanomagnetischer Bauelemente* (Braunschweig: Vieweg); 1969 *Structure and Application of Galvanomagnetic Devices* (Oxford: Pergamon)

[4] Wieder H H 1971 *Hall Generators and Magnetoresistors* (London: Pion)

[5] Popovic R S and Heidenreich W 1989 Galvanomagnetic semiconductor sensors *Sensors* vol. 5 *Magnetic Sensors* ed. R Boll and K J Oveshott (Weinheim: VCH) pp 43–96

[6] Baltes H P and Popovic R S 1986 Integrated semiconductor magnetic field sensors *Proc. IEEE* **74** 1107–32

[7] Popovic R S 1989 Hall-effect devices *Sens. Actuators* **17** 39–53

[8] Andor L, Baltes H B, Nathan A and Schmidt-Weinmar H G 1985 Numerical modeling of magnetic-field sensitive semiconductor devices *IEEE Trans. Electron Devices* **ED-32** 1224–30

[9] Wick R F 1954 Solution of the field problem of the germanium gyrator *J. Appl. Phys.* **25** 741–56

[10] Haeusler J 1966 Exakte Losung von Potentialproblemen beim Halleffekt durch konforme Abbildung *Solid-State Electron.* **9** 417–41

[11] Haeusler J 1968 Die Geometriefunktion vierelektrodiger Hallgeneratoren *Arch. Elektrotech.* **52** 11–19

[12] Kuhrt F and Lippmann H J 1968 *Hallgeneratoren* (Berlin: Springer) pp 72–82

[13] Versnel W 1981 Analysis of a circular Hall plate with equal finite contacts *Solid-State Electron.* **24** 63–8

[14] De Mey G 1983 Potential calculations in Hall plates *Advances in Electronics and Electron Physics* vol. **61** ed. P W Hawkes (New York: Academic Press)

[15] Newsome J P 1983 Determination of the electrical characteristics of Hall plates *Proc. IEE* **110** 653–9

[16] Mimizuka T 1978 The accuracy of the relaxation solution for the potential problem of a Hall plate with finite Hall electrodes *Solid-State Electron.* **21** 1195–7

[17] Nathan A, Alegreto W, Baltes H P and Sugiyama T 1987 Carrier transport in semiconductor magnetic domain detectors *IEEE Trans. Electron Devices* **ED-34** 2077–85

[18] Newsome J P 1967 Hall-effect analogues *Solid-State Electron.* **10** 183–91

[19] Popovic R S 1985 Numerical analysis of MOS magnetic field sensors *Solid-State Electron.* **28** 711–16

[20] For example: Bewley L V 1948 *Two-dimensional Fields in Electrical Engineering* (New York: Macmillan)

[21] Versnel W 1981 Analysis of a circular Hall plate with equal finite contacts *Solid-State Electron.* **24** 63–8

[22] Falk U 1990 A symmetrical vertical Hall-effect device *Transducers '89* vol. 2 ed. S Middelhoek (Amsterdam: Elsevier)

[23] For example: P Tuinenga 1988 *SPICE: A guide to circuit simulation using PSPICE* (Englewood Cliffs: Prentice Hall)

[24] van der Pauw L J 1958 A method of measuring specific resistivity and Hall effect of discs of arbitrary shape *Philips Res. Rep.* **13** 1–9

[25] Kataoka S, Yamada H and Fujisada H 1971 New galvanomagnetic device with directional sensitivity *Proc. IEEE* **59** 1349

[26] Putley E H 1960 *The Hall Effect and Semiconductor Physics* (New York: Dover)

[27] Fluitman J H J 1980 Hall-effect device with both voltage leads on one side of the conductor *J. Phys. E: Sci. Instrum.* **13** 783–5

[28] Popovic R S 1984 The vertical Hall-effect device *IEEE Electron Device Lett.* **EDL-5** 357–8

[29] Solin S A, Tineke Thio, Hines D R and Heremans J J 2000 Enhanced room-temperature geometric magnetoresistance in inhomogeneous narrow-gap semiconductors *Science* **289** 1530–2

[30] Zhou T, Hines D R and Solin S A 2001 Extraordinary magnetoresistance in externally shunted van der Pauw plates *Appl. Phys. Lett.* **78** 667–9

[31] Lippmann H J and Kuhrt F 1958 Der Geometrieeinfluss auf den Hall-Effekt bei rechteckigen Halbleiterplatten *Z. Naturforsch.* **13a** 474–83

[32] Haeusler J 1968 Die Geometriefunktion vierelektrodiger Hallgeneratoren *Arch. Elektrotech.* **52** 11–19

[33] Versnel W 1982 The geometrical correction factor for a rectangular Hall plate *J. Appl. Phys.* **53** 4980–6

[34] Versnel W 1981 Analysis of symmetrical Hall plates with finite contacts *J. Appl. Phys.* **52** 4659–66

[35] Haeusler J and Lippmann H J 1968 Hallgeneratoren mit kleinem Linearisierungsfehler *Solid-State Electron.* **11** 173–82

[36] Popovic R S and Halg B 1988 Nonlinearity in Hall devices and its compensation *Solid-State Electron.* **31** 681–8

[37] Lippmann H J and Kuhrt F 1958 Der Geometrieeinfluss auf den transverslilen magnetischen Widerstandseffekt bei rechteckformigen Halbleiterplatten *Z. Naturforsch.* **13a** 462–74

[38] Weiss H 1969 *Physik und Anwendung galvanomagnetischer Bauelemente* (Braunschweig: Vieweg)

[39] Heremans J 1993 Solid state magnetic field sensors and applications *J. Phys. D: Appl. Phys.* **26** 1149–68

[40] van der Pauw L J 1958/9 Messung des spezifischen Widerstandes und Hall-Koeffizienten an Scheibchen beliebiger Form *Philips Tech. Rundsch.* **20** 230–4

[41] Kanda Y and Migitaka M 1976 Effect of mechanical stress on the offset voltage of Hall devices in Si IC *Phys. Status Solidi* (a) **35 K** 115–18

[42] Kleinpenning T G M 1983 Design of an ac micro-gauss sensor *Sens. Actuators* **4** 3–9

[43] Vandamme L K J and van Bokhoven W M G 1977 Conductance noise investigations with four arbitrarily shaped and placed electrodes *Appl. Phys.* **14** 205–15

[44] Vandamme L K J and de Kuiper A H 1979 Conductance noise investigations on symmetrical planar resistors with finite contacts *Solid-State Electron.* **22** 981–6

[45] Kleinpenning T G M and Vandamme L K J 1979 Comment on 'Transverse $1/f$ noise in InSb thin films and the SNR of related Hall elements' *J. Appl. Phys.* **50** 5547

[46] Hooge F N, Kleinpenning T G M and Vandamme L K J 1981 Experimental studies on $1/f$ noise *Rep. Prog. Phys.* **44** 479–532

[47] Dobrovol'skii V N and Krolevets A N 1983 Hall current and its use in investigations of semiconductors (review) *Sov. Phys.-Semicond.* **17** 1–7

[48] Dobrovol'skii V N and Krolovets A N 1979 Determination of the electrical resistivity and Hall coefficient of samples dual to van der Pauw samples *Sov. Phys.-Semicond.* **13** 227–8

[49] Boero G, Demierre M, Besse P-A, Popovic RS 2003 Micro-Hall devices: performance, technologies and applications *Sensors and Actuators A*

[50] Boero G, private communication, EPFL January 2003

[51] See for example: Popovic B 1990 *Elektromagnetika* (Beograd: Gradjevinska knjiga), pp 185–6

Chapter 5

Hall magnetic sensors

In this chapter we come still another step closer to practical galvanomagnetic devices: we shall study Hall effect devices made to be used as magnetic field sensors. Moreover, we now take seriously into consideration real material properties and fabrication technology of the devices. This will force us to look into the consequences of non-idealities in device shape and of parasitic effects; on the other side, this will also give us the chance to discuss some ingenious realizations of practical Hall magnetic sensors.

A magnetic sensor is a device capable of sensing a magnetic field and transmitting the corresponding information. A magnetic sensor is therefore a transducer that converts a magnetic field into information. For example, a compass needle is a magnetic sensor. A galvanomagnetic device can also be applied as a magnetic sensor: for example, a Hall plate converts a magnetic field into a Hall voltage. Then the Hall voltage is the electronic signal carrying information on the magnetic field. A detailed account of various magnetic sensors can be found in [1].

A vast majority of magnetic sensors based on galvanomagnetic effects in non-magnetic materials that we find in practice are Hall plates operating in the Hall voltage mode. Therefore, in this chapter we will concentrate on such Hall devices and technologies that are compatible with them.

A Hall device may be a good magnetic sensor only if it is made of a material having a low concentration of high-mobility charge carriers. Since this criterion is best met in some semiconductors, only semiconductor Hall magnetic sensors are of practical importance. This is why the intensive work towards the magnetic sensor applications of Hall devices began with the development of semiconductor technology. In 1948, Pearson [2] proposed a germanium Hall device as a magnetic sensor. A few years after the discovery of high-mobility compound semiconductors, in the early 1950s, Hall magnetic sensors became commercially available. The early work on Hall magnetic sensors was summarized in [3–5]. Several books on Hall magnetic sensors were also published in Russian: see, for example, [6] and references therein.

In 1968, Bosch proposed the incorporation of a Hall device into a standard silicon bipolar integrated circuit [7]. A number of commercially successful products have been based on this idea [8, 9]. Currently (2003), more than two billion Hall magnetic sensors are manufactured and put to use each year [10], mostly as position sensors. Work on integrated Hall devices and some similar and related magnetic field sensors is reviewed in [10–13].

5.1 Basic characteristics

We shall now define and discuss the basic coefficients of a Hall device, which characterize its performance as a magnetic field to voltage transducer. We assume that the device is working in the Hall voltage mode of operation.

5.1.1 Sensitivity

Sensitivity is the most important figure of merit of a sensor. In a modulating transducer [14] such as a Hall device, one may define absolute sensitivity and relative sensitivities.

The *absolute sensitivity* of a Hall magnetic sensor is its transduction ratio for large signals:

$$S_A = \left| \frac{V_H}{B_\perp} \right|_C . \tag{5.1}$$

Here V_H is the Hall voltage, $B_\perp$ is the normal component of the magnetic induction, and C denotes a set of operating conditions, such as temperature, frequency and bias current. We shall no longer explicitly note the operating conditions.

The ratio of the absolute sensitivity of a modulating transducer and a bias quantity yields a relative sensitivity. For a Hall magnetic sensor, current-related sensitivity and voltage-related sensitivity are of interest.

Current-related sensitivity of a Hall device is defined by

$$S_I = \frac{S_A}{I} = \left| \frac{1}{I} \frac{V_H}{B_\perp} \right| \qquad V_H = S_I I B_\perp \tag{5.2}$$

where I denotes the bias current of the Hall device. The units of S_I are $\mathrm{V\,A^{-1}\,T^{-1}}$ (volts per ampere per tesla). Using (4.90), we find

$$S_I = G_H \frac{|R_H|}{t} . \tag{5.3}$$

If the plate is strongly extrinsic, the Hall coefficient R_H is given by (3.308), and (5.3) reduces to

$$S_I = G_H \frac{r_H}{qnt} . \tag{5.4}$$

Recall the notation: G_H is the geometrical correction factor (see §4.2.6), r_H is the Hall factor of the majority carriers (§3.4.1), n is the electron concentration in the Hall plate, and t is the thickness of the plate.

Note that the nt product, figuring in (5.4), denotes the surface charge carrier density in a homogeneous plate: $nt = N_s$. When the doping density in the plate depends on the depth z, as in the case of ion-implanted layers, the surface charge carrier density is given by

$$N_s = \int_0^t n(z)\, dz. \tag{5.5}$$

Then (5.4) attains the form

$$S_I = G_H \frac{r_H}{qN_s}. \tag{5.6}$$

The product qN_s equals the charge of free electrons per unit area of the plate. Note that the direct replacement of the nt product by the quantity N_s is correct only if the mobility and the Hall factor of carriers do not vary across the active layer of the device. This condition is approximately fulfilled in low-doped, small doping gradient layers. Such layers are usually used in Hall magnetic sensors. Otherwise, one has to operate with average quantities, as was done in §6.1.2.

Current-related sensitivity hardly depends on the plate material, since $r_H \simeq 1$, irrespective of the material. Typical values of S_I range between 50 and $500\,\mathrm{V\,A^{-1}\,T^{-1}}$. One of the highest values reported so far is $S_I \simeq 3100\,\mathrm{V\,A^{-1}\,T^{-1}}$ [15], which corresponds to $N_s \simeq 2 \times 10^{11}\,\mathrm{cm^{-2}}$. Such a low value of charge density in the active layer of a Hall device is impractical because of an extremely strong influence of surface charge and/or junction field effect—see §§5.1.5 and 5.1.7.

Voltage-related sensitivity is defined analogously to (5.2):

$$S_V = \frac{S_A}{V} = \left| \frac{1}{V} \frac{V_H}{B_\perp} \right| = \frac{S_I}{R_{in}} \qquad V_H = S_V V B_\perp. \tag{5.7}$$

The unit of S_V is 'per tesla' ($\mathrm{V\,V^{-1}\,T^{-1} = T^{-1}}$). For a strongly extrinsic semiconductor, by making use of (4.93) we obtain

$$S_V = \mu_H \frac{w}{l} G_H. \tag{5.8}$$

Recall the notation: μ_H is the Hall mobility of the majority carriers (§3.4.3), w/l is the width-to-length ratio of the equivalent rectangle (§4.3.3) and G_H is the geometrical correction factor of Hall voltage.

Typical values of voltage-related sensitivities are: $S_V \simeq 0.07\,\mathrm{T^{-1}}$ for silicon Hall plates [9], $S_V \simeq 0.2\,\mathrm{T^{-1}}$ for GaAs plates [16], and $S_V \simeq 3\,\mathrm{T^{-1}}$ for thin-film InSb Hall elements [10]. Voltage-related sensitivity increases with a decrease in the effective device length, yet there is an upper physical limit: with the aid of (4.94), we find

$$S_{V_{\max}} = 0.742\mu_H. \tag{5.9}$$

This yields, at 300 K, $S_{V_{\max}} \simeq 0.126\,\mathrm{T}^{-1}$ and $S_{V_{\max}} \simeq 0.67\,\mathrm{T}^{-1}$ for low-doped n-type silicon and GaAs, respectively.

5.1.2 Offset-equivalent magnetic field

The offset voltage V_{off} at the output of a Hall magnetic sensor (see §4.3.2) cannot be distinguished from a quasi-static useful signal. To characterize the error in measuring magnetic induction caused by offset, one can calculate the magnetic induction, which would yield a Hall voltage equal to the offset voltage. Using (5.1), we have

$$B_{\mathrm{off}} = V_{\mathrm{off}}/S_A. \tag{5.10}$$

This quantity is the offset-equivalent magnetic field.

By substituting in the above equation the offset voltage by (4.105), and the absolute sensitivity by (5.7) and (5.8), we find

$$B_{\mathrm{off}} = \frac{\Delta R}{R}\frac{1}{\mu_H}\frac{1}{wG_H}. \tag{5.11}$$

Thus the higher the mobility, the lower the offset-equivalent field for a given misalignment. For a silicon Hall plate with $\Delta R/R = 10^{-3}$, $l/w \simeq 1$ and $\mu_H \simeq 0.15\,\mathrm{m^2\,V^{-1}\,s^{-1}}$ (5.11) yields $B_{\mathrm{off}} \simeq 10\,\mathrm{mT}$. This is a typical value of offset found in good silicon Hall plates if no special offset-reduction measures are taken.

A study of the tolerances pertinent to the silicon bipolar integrated circuit fabrication process has given the following results [17]. The major causes of offset in Hall plates fabricated with this process are errors in geometry. The errors are due to etching randomness and rotation alignment. For a plate of dimensions $l \simeq w \simeq 500\,\mu\mathrm{m}$, the average etching tolerances of $0.1\,\mu\mathrm{m}$ cause an average offset induction of 6 mT.

Offset of a silicon Hall plate can be seriously affected by mechanical stress via the *piezoresistance* effect (§2.7.4). Kanda and Migitaka [18, 19] determined the maximum offset voltage due to a mechanical stress X:

$$V_{\mathrm{off}} = \frac{(\pi_l - \pi_t)X}{2[3 - 2(w/l)]}V. \tag{5.12}$$

Here π_l and π_t are the longitudinal and transverse piezoresistance coefficients for the bridge resistors in figure 4.29, and V is the bias voltage. By making use of (5.8), the corresponding equivalent magnetic field is given by

$$B_{\mathrm{off}} = \frac{(\pi_l - \pi_t)X}{\mu_H 2[3 - 2(w/l)]G_H(w/l)}. \tag{5.13}$$

Since the piezoresistive coefficients of silicon depend on crystal orientation (see table 2.3), stress-induced offset also depends on the Hall device placement in a crystal. The best choice for a silicon Hall plate is to make the plate parallel to the (110) crystal plane so that current flows in the $\langle 100 \rangle$ direction. For such a Hall plate, the encapsulation-induced strain of 10^{-4} produces the offset equivalent induction of $B_{\text{off}} \simeq 8.4\,\text{mT}$.

There are very efficient circuit techniques for reducing offset of Hall devices. We shall consider them in §5.6.3.

5.1.3 Noise-equivalent magnetic field

The noise voltage at the output of a Hall magnetic sensor, see §4.3.5, may be interpreted as a result of an equivalent magnetic field, acting on a noiseless Hall device. Replacing the Hall voltage in (5.1) by the mean square noise voltage (2.125), we obtain

$$\langle B_{\text{N}}^2 \rangle = \int_{f_1}^{f_2} S_{\text{NV}}(f)\,\text{d}f / S_A^2. \tag{5.14}$$

Here $\langle B_{\text{N}}^2 \rangle$ is the mean square noise-equivalent magnetic field in a frequency range (f_1, f_2), and $S_{\text{NV}}(f)$ is the noise voltage spectral density at the sensor output. From (5.14) we obtain

$$S_{\text{NB}}(f) = \frac{\partial \langle B_{\text{N}}^2 \rangle}{\partial f} = \frac{S_{\text{NV}}(f)}{S_A^2}. \tag{5.15}$$

This quantity is called the *noise-equivalent magnetic field spectral density* (NEMF spectral density).

The absolute sensitivity of a Hall device is frequency independent up to very high frequencies. Therefore, in most cases of practical interest, the NEMF spectral density follows exactly the frequency dependence of S_{NV}. In particular, we may talk about the $1/f$ range and the white-noise range of NEMF (see figure 4.30). By way of example, here are a few numerical values: in the silicon vertical Hall device such as that shown in figure 5.12, the quantities $S_{\text{NB}} \simeq 3 \times 10^{-13}\,\text{T}^2\,\text{Hz}^{-1}$ at $f = 100\,\text{Hz}$ ($1/f$ range) and $S_{\text{NB}} \simeq 1 \times 10^{-15}\,\text{T}^2\,\text{Hz}^{-1}$ at $f = 100\,\text{kHz}$ (white-noise range) have been measured [20]. The device was biased by $I = 0.5\,\text{mA}$ and $V = 2\,\text{V}$.

A convenient way of describing the noise properties of a sensor is in terms of *resolution*, which is also-called *detection limit*. The resolution for an AC measurand is the value of the measurand corresponding to a signal-to-noise ratio of one. From (4.113), (5.1) and (5.15), we find the resolution of a magnetic sensor as

$$B_{\text{DL}} = [S_{\text{NB}}(f)\Delta f]^{1/2} \tag{5.16}$$

in a frequency range (f_1, f_2). A high resolution can be achieved in a large Hall device made of a high-mobility material with a low Hooge $1/f$ noise parameter α (see (2.129)) when it operates at a high power level (see §4.3.6) and in a narrow frequency range. A resolution at a high frequency and $\Delta f = 1\,\text{Hz}$ as high as $B_N \simeq 4 \times 10^{-11}\,\text{T}$ has been theoretically assessed for a Hall device made of a material with $\mu_H = 6\,\text{m}^2\,\text{V}^{-1}\,\text{s}^{-1}$ and operated at $P = 0.5\,\text{W}$ [21].

5.1.4 Cross-sensitivity

The cross-sensitivity of a magnetic sensor is its undesirable sensitivity to other environmental parameters, such as temperature and pressure. A cross-sensitivity to a parameter P can be characterized by the quantity

$$PC = \frac{1}{S}\frac{\partial S}{\partial P} \tag{5.17}$$

which we shall call the P coefficient; for example, if P is temperature, then PC becomes TC, that is the temperature coefficient. In the above equation S denotes a magnetic sensitivity of the Hall device: S_A, S_I or S_V.

Let us consider, for example, the *temperature coefficient of the current-related sensitivity*:

$$TC_I = \frac{1}{S_I}\frac{\partial S_I}{\partial T}. \tag{5.18}$$

We shall assume that the Hall plate material is a strongly extrinsic n-type semiconductor. Therefore, S_I is given by (5.4). At a low magnetic field, the geometrical correction factor G_H depends only on the device shape (see §4.2.6). Thus $G_H \neq f(T)$. In the exhaustion temperature range, $n = N_D$ (see §2.3.5). Thus also $n \neq f(T)$. In most cases also $t \neq f(T)$. Therefore, (5.18) reduces to

$$TC_I = \frac{1}{r_{Hn}}\frac{\partial r_{Hn}}{\partial T} \tag{5.19}$$

where r_{Hn} is the Hall factor. Accordingly, under the above conditions, the variations of S_I due to a variation in temperature follows that of $r_{Hn}(T)$. The temperature dependence of the Hall factor is well documented for some important semiconductors (see §3.4.1). For example, for low-doped n-type silicon, around room temperature, $(1/r_{Hn})(\partial r_{Hn}/\partial T) \simeq 10^{-3}\,\text{K}^{-1}$ (see figures 3.23, 3.24).

In a similar way, we find the temperature coefficient of the voltage-related sensitivity for a strongly extrinsic n-type Hall plate:

$$TC_V = \frac{1}{S_V}\frac{\partial S_V}{\partial T} = \frac{1}{\mu_{Hn}}\frac{\partial \mu_{Hn}}{\partial T}. \tag{5.20}$$

Recall that $\mu_{Hn} = r_{Hn}\mu_n$. The active region of a Hall plate is usually low doped, and the $\mu_{Hn}(T)$ function is dominated by the strong temperature dependence of the drift mobility. In low-doped silicon, $\mu_n \sim T^{-2.4}$ (see figure 2.9). The corresponding temperature coefficient at room temperature is $(1/\mu_n)(\partial\mu_n/\partial T) \simeq -8 \times 10^{-3}\,\mathrm{K}^{-1}$.

For a silicon integrated Hall plate the following values have been found experimentally [8, 9]: $TC_I \simeq 0.8 \times 10^{-3}\,\mathrm{K}^{-1}$ and $TC_V \simeq -4.5 \times 10^{-3}\,\mathrm{K}^{-1}$ in the temperature range $-20\,^\circ\mathrm{C}$ to $+120\,^\circ\mathrm{C}$. For a GaAs Hall plate, $TC_I \simeq 0.3 \times 10^{-3}\,\mathrm{K}^{-1}$ [16].

The influence of the temperature sensitivity of S_I on the performance of a sensor system can be efficiently reduced with the aid of the circuit techniques described in §5.6.2.

5.1.5 Non-linearity

In some applications of Hall devices as magnetic sensors, it is particularly important that the proportionality relation $V_H \sim IB_\perp$ (5.2), or $V_H \sim VB_\perp$ (5.7), holds to a high degree of accuracy. Since this is not always the case, it is useful to define a sensor figure of merit called non-linearity [15]:

$$NL = \frac{V_H(I,B) - V_{H0}}{V_{H0}} = \frac{\Delta V_H}{V_{H0}}. \tag{5.21}$$

Here V_H denotes the measured Hall voltage at a bias current I and magnetic field B, and V_{H0} is the best linear fit to the measured values. If a Hall device works as a four-quadrant magnetic field-current multiplier around the point $B = 0$ and $I = 0$, the appropriate definition of V_{H0} is

$$V_{H0} = \left.\frac{\partial^2 V_H}{\partial B\,\partial I}\right|_{I=0,B_\perp=0} \times IB_\perp. \tag{5.22}$$

This is the first non-zero term in a MacLaurin series expansion of $V_H(I, B)$. The second derivative in (5.22) equals the current-related sensitivity at $I \simeq 0$ and $B \simeq 0$: S_{I0} (see (5.2)). Thus

$$V_{H0} = S_{I0} IB_\perp. \tag{5.23}$$

Substituting (5.2) and (5.23) into (4.21), we obtain

$$NL = \frac{S_I(I,B) - S_{I0}}{S_{I0}} = \frac{\Delta S_I}{S_{I0}}. \tag{5.24}$$

Therefore, a magnetic sensor will exhibit a non-linearity if its sensitivity depends on magnetic field. Figure 5.1 illustrates the two representations of non-linearity.

The current-related sensitivity S_I is given by (5.3). It may vary with a change in current or magnetic field if one of the quantities R_H, G_H or t does so. To be definite, let us consider an integrated junction-isolated Hall

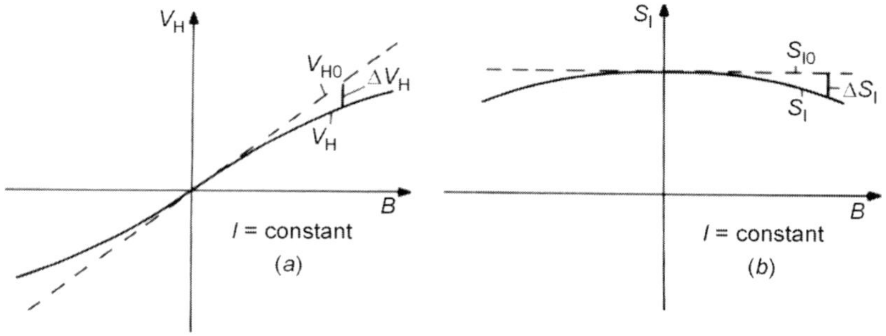

Figure 5.1. Two representations of the non-linearity of a Hall device. To a non-linear dependence $V_H = f(B)$ corresponds a non-constant sensitivity: $S_I = f(B)$. (a) Hall voltage V_H against magnetic field B at a constant bias current I. V_H denotes the actual Hall voltage, and V_{H0} is the linear extrapolation of the function $V_H = f(B)$ around the point $B \simeq 0$. (b) Relative sensitivity against magnetic field. S_I denotes the actual sensitivity, and $S_{I0} = S_I(B \simeq 0)$.

plate. Let the active region of the plate be made of a low-doped strongly extrinsic n-type semiconductor. We consider only a device biased by an ideal current source, so we neglect the magnetoresistance effect.

In §3.4.6, we found that, at a low magnetic field, the Hall coefficient depends on a magnetic field according to (3.303). Substituting this equation into (5.24), and assuming G_H/t = constant, we obtain

$$NL_M \simeq -\alpha\mu_H^2 B^2 \tag{5.25}$$

with α being the material non-linearity coefficient (see (3.304) and figures 3.29–3.31). The quantity NL_M is called the *material non-linearity of a Hall device.*

In §4.2.6, we found that, at a low magnetic field, the geometrical correction factor depends on a magnetic field according to (4.64). Substituting this equation into (5.24), and assuming R_H/t = constant, we obtain

$$NL_G \simeq \beta\mu_H^2 B^2 \tag{5.26}$$

where β is a numerical coefficient. It may take a value between $\beta_{min} \to 0$ if $G_0 \to 1$ (see (4.66)) and $\beta_{max} \to 0.604$ if $G_0 \to 0$ (see (4.69)). The quantity NL_G is called the *geometrical non-linearity of a Hall device.*

The material non-linearity (5.25) and the geometrical non-linearity (5.26) exhibit the same quadratic magnetic induction dependence, but have opposite signs. Moreover, the values of the coefficients α and β are of the same order of magnitude. These facts may be exploited to devise a Hall device in which the two non-linearity effects cancel each other. To this end, a suitable material has to be chosen, with an α coefficient matching the accessible β values [22]. Besides, geometrical non-linearity may be compensated by loading the Hall device output with a proper-value resistor [5–7].

In junction-isolated Hall devices, an additional source of non-linearity has been identified [22]. It comes about through a kind of feedback modulation of the plate thickness t by the Hall voltage. The modulation of the device thickness is due to the junction field effect.

Non-linearity due to the junction field effect depends on device structure and biasing conditions. By way of example, we shall show how the effect comes about in a buried homogeneously doped n-type Hall plate, such as that shown in figure 5.8. Consider the vertical cross section of such a Hall plate through the sense contacts as in figure 5.2(a) and (b). To simplify the analysis, we assume that the p-type jacket surrounding the active device region is heavily doped (c). We also assume that the jacket is biased by a constant negative voltage V_j, and that one sense contact is virtually at earth, so that the full Hall voltage appears at the other sense contact against earth.

Consider the electric potential difference between the Hall plate region along a line connecting the two sense contacts and the jacket. At the earthed sense contact, this potential difference is always equal to V_j. However, the potential difference varies linearly with the distance from that contact, reaching $V_j + V_H$ at the second sense contact. This potential difference is applied across the junction between the Hall plate and the jacket. Its variation causes a variation in the junction depletion-layer width and, consequently, a variation in the effective plate thickness. Therefore, the mean plate thickness t depends on the Hall voltage, and so also does the sensitivity. According to this simplified model, the variation in the sensitivity (5.3) is given by

$$\Delta S_I = \frac{\partial S_I}{\partial \bar{t}} \frac{\partial \bar{t}}{\partial V_H} V_H \tag{5.27}$$

where we assume $V_H \ll V_j$ and take V_H as a small variation in V_j, that is a $\delta V_j = V_H$. The mean plate thickness $\bar{t}$ is given by

$$\bar{t} = t_j - 2\bar{W} \tag{5.28}$$

where t_j is the plate thickness between the metallurgical junctions and W the depletion-layer width at the centre of the plate (see figure 5.2(c)):

$$\bar{W} = [2\varepsilon_s (V_{bi} - V_j + \tfrac{1}{2} V_H)/qn]^{1/2} \tag{5.29}$$

where ε_s is the permittivity of the material and V_{bi} the built-in voltage. Finally, (5.4), (5.24) and (5.27)–(5.29) yield

$$NL_{JFE} \simeq \frac{S_{I0}}{G_H r_H} \left(\frac{\varepsilon_s qn}{2(V_{bi} - V_j)} \right)^{1/2} IB_{\perp}. \tag{5.30}$$

Thus the non-linearity due to the junction field effect is proportional to both the magnetic field and the device bias current. This is in contrast to the

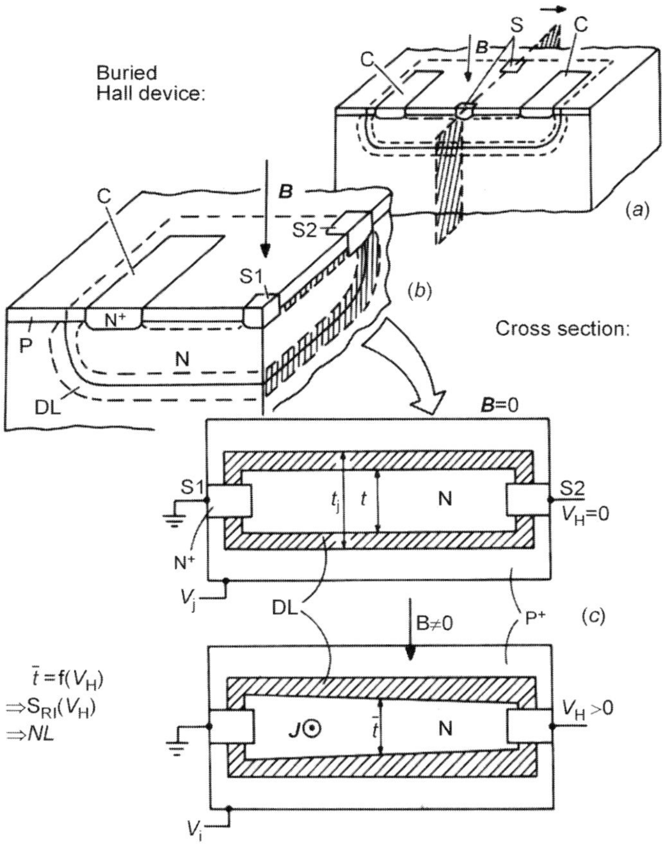

Figure 5.2. This is how the junction field effect may bring about a non-linearity of a junction-isolated Hall device. (a) A buried Hall device (see also figure 5.8). (b) The cross section through the sense contacts. (c) A model. If $\boldsymbol{B} = 0$, $V_H = 0$, and the thickness of the plate t is constant; but if $\boldsymbol{B} \neq 0$, $V_H \neq 0$, the depletion layer (DL) width is not constant over the cross section, and neither is the plate thickness. Consequently, the average plate thickness $\bar{t}$ depends on the Hall voltage, and so does the sensitivity.

geometrical and material non-linearities, which are proportional to the square of the magnetic field and are independent of the bias current.

The junction field-effect non-linearity may be the dominant non-linear effect in high-sensitivity junction-isolated devices. For example, for a silicon sensor with $S_{I0} \simeq 500\,\mathrm{V\,A^{-1}\,T^{-1}}$, $n = 10^{15}\,\mathrm{cm^{-3}}$, $V_j = 5\,\mathrm{V}$ and $G_H r_H \simeq 1$, at $B = 1\,\mathrm{T}$ and $I = 1\,\mathrm{mA}$, equation (5.30) gives $NL_{\mathrm{JFE}} \simeq 10^{-2}$.

Figure 5.3 demonstrates non-linearity in a high-sensitivity buried silicon Hall device. The area carrier concentration in the active device layer was $N_s \simeq 2.1 \times 10^{11}\,\mathrm{cm^{-2}}$. The plate was cross-shaped, with $h = k = 80\,\mu\mathrm{m}$ (see

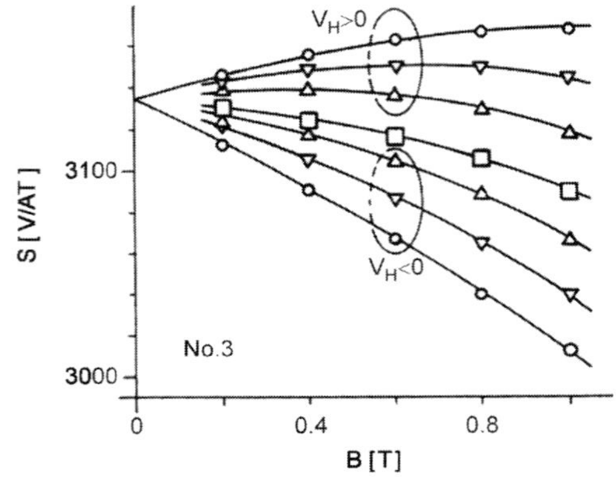

Figure 5.3. Sensitivity of a silicon buried Hall device as a function of the magnetic field B and the bias current. The active device region is phosphorus implanted, with the average impurity concentration $N_D \simeq 10^{15}\,\mathrm{cm}^{-3}$, and the area density $N_s \simeq 2.1 \times 10^{11}\,\mathrm{cm}^{-3}$. The average junction bias voltage is $V_j = 1.8\,\mathrm{V}$, and $T = 20\,^\circ\mathrm{C}$. The curves correspond to the following bias currents: ○, 60 μA; ▽, 40 μA; △, 20 μA; □, any of these currents, but with application of a non-linearity compensation method (see §5.6.2) (reprinted from [15]).

figure 4.15(d)). According to (4.62) and table 4.1(d), the geometrical correction factor at $B \simeq 0$ is $G_0 \simeq 0.955$.

The curves reveal both of the above non-linearity types: one proportional to B^2, and the other proportional to the BI product. Owing to the high value of G_0, the geometrical non-linearity is negligible in this case (see (4.66)).

Consequently, the B^2 dependence of the curves is predominantly related to the material non-linearity. We have discussed details of the material non-linearity coefficient α in §3.4.6 (see figures 3.29–3.31).

The BI-product non-linearity, which is a consequence of the junction field effect, is easily recognized by the fact that the sensitivity increases for positive Hall voltages and decreases for negative Hall voltages. For such a high-sensitivity device, the non-linearity due to the junction field effect is clearly the dominant one in the magnetic induction region shown. Using a feedback compensation technique (see §5.6.2), the non-linearity due to the junction field effect can be virtually eliminated. This is demonstrated in figure 5.3 by the curve with squares.

5.1.6 Frequency response

A Hall device has a flat frequency response to a magnetic field until one of the frequency-dependent limiting or parasitic effects, described in §4.6, becomes active enough to be noticeable. Let us now review these frequency-limiting effects.

The ultimate frequency limit of the Hall effect comes about because of the *relaxation time limit* (4.130). The functioning of the Hall effect at about 1 THz, which is close at this limit, was experimentally demonstrated [23]. This was done while using the current-mode Hall effect for the characterization of GaAs wafers. It is interesting to note that in the *Hall current mode of operation* (§4.4), we have equipotential conditions, so we do not expect any capacitive effect, described by (4.135) and in §4.6.2. In some other particular applications, such as electromagnetic power measurements in wave guides, Hall plates operating in the voltage mode were used at the 10 GHz frequency range [24]. But there are many other important applications of Hall plates where the practical frequency limit is much lower.

In discrete Hall probes for precise magnetic field measurements, the dominant frequency response limitation usually comes from the conventional *inductive effects*, mentioned in §4.6.3. By a very careful wiring of the device, the surface of the inductive loops can be minimized or almost compensated for a chosen direction. In the case of high-mobility and high-current Hall devices, this might work well until a frequency above 1 MHz. But in low-mobility Hall devices, such as silicon devices, the practical frequency limit is 10 kHz or so: at higher frequencies of the measured magnetic field, even a very tightly twisted cable picks up an inductive voltage, which interferes with the Hall voltage.

In integrated Hall sensors one can avoid inductive signals until much higher frequencies: IC technology allows a dramatic miniaturization of the inductive loops and their precise compensation. But then the frequency response is usually limited by the *global capacitive effects* (§4.6.2). Here the relevant parameter is the combined relaxation time of the Hall plate, given by (4.151). By way of example, a cross-shaped silicon integrated Hall plate (figure 4.15(d)) with $h = k = 50\,\mu m$, may have a resistance of about $3\,k\Omega$, and a total capacitance of about 3 pF. This gives a basic electrical frequency limit of about 10 MHz. If we also take into account frequency response of interface electronics, we come to a practical frequency limit which might be about 10 times lower than that.

5.1.7 Stability

When a Hall device is used as the input transducer in a magnetic-field-measuring instrument, its stability is a crucial issue, for the product of the transducer sensitivity and the amplifier gain makes up the sensor system transduction constant. Generally, very careful device design and processing may yield a sensor that could be regarded as a reference element in a system. Most present-day sensors rely on this concept and suffer from its unreliability [12].

The relative sensitivity of a Hall sensor is directly proportional to the Hall coefficient (5.3). The Hall coefficient, however, is not an absolutely stable quantity. For example, as a result of the *piezo-Hall effect* (§3.5.5), the Hall coefficient of low-doped n-type silicon may vary by as much as

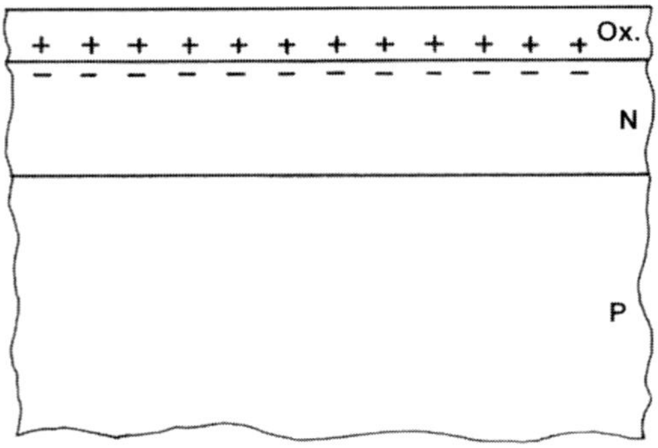

Figure 5.4. The influence of the surface charges on a Hall device. N is the active layer of the device, P the substrate, and Ox is the oxide layer. The positive charges in the oxide layer or at the Ox/N interface induce precisely the same density of charges in the active layer [12].

2.5% under the influence of a mechanical stress (see figure 3.34). A mechanical stress may be caused by the encapsulation of the device (see §5.7).

The relative sensitivity of a Hall sensor also depends directly on the surface density of charge carriers in the plate, that is the nt product in (5.4), or N_s in (5.6). Therefore, any physical effect leading to a variation in the carrier density may cause an instability in sensor sensitivity. *Surface effects* fall into this class.

As is well known, a solid-state surface always contains localized electron states, called traps. If the surface is covered with an oxide layer, the oxide may contain some additional trapped charges and mobile ionic charges. In the case of a thermally oxidized silicon surface, which is the most stable semiconductor surface now known, the surface densities of these charge carriers may be in the range 10^9–10^{13} cm^{-2} [24].

A charge at the interface induces a charge of opposite polarity in the underlying substrate (see figure 5.4). This induced charge adds to the existing charge in the substrate, altering its characteristics. In particular, a variation in the interface charge density δQ_s will cause a variation in the surface charge density in the Hall plate of $-\delta Q_s$. Assuming no inversion layer at the surface, the corresponding relative variation in the relative sensitivity of the sensor (see (5.6)) will be

$$\frac{\delta S_I}{S_I} = \frac{(\partial S_I/\partial Q_{\rm H})\delta Q_s}{S_I} = -\frac{\delta Q_s}{Q_{\rm H}}. \tag{5.31}$$

Here $Q_{\rm H} = qN_s$ denotes the total surface charge density in the Hall plate. Using (5.6), equation (5.31) can be rewritten as

$$\frac{\delta S_I}{S_I} \simeq S_I \delta Q_s \tag{5.32}$$

where we have put $r_{\rm Hn} G_{\rm H} \simeq 1$ (since $r_{\rm Hn} \geq 1$ and $G_{\rm H} \leq 1$).

The influence of surface instability on the instability of the sensor sensitivity is proportional to the sensitivity itself. In other words, high-sensitivity devices may tend to be unstable. As an example, let us assess the relative instability of a Hall sensor with $S_I \simeq 500\,\mathrm{V\,A^{-1}\,T^{-1}}$ ($N_s \simeq 10^{12}\,\mathrm{cm^{-2}}$), if $\delta Q_s/q \simeq 10^{10}\,\mathrm{cm^{-2}}$. From (5.31) we obtain $\delta S_I/S_I \simeq 10^{-2}$, that is 1%.

Uncorrelated fluctuations in interface charge may also produce asymmetry in a Hall plate resistivity and thus *offset*. The sheet resistance of the active device layer is given by

$$R_s = \frac{\rho}{t} = \frac{1}{\mu q n t} = \frac{1}{\mu Q_{\mathrm{H}}}. \tag{5.33}$$

The relative change in the layer resistance is then

$$\frac{\Delta R}{R} = -\frac{\delta Q_s}{Q_{\mathrm{H}}} \tag{5.34}$$

with the same notation as above. By substituting (5.34) into (5.11), we obtain the *offset-equivalent magnetic field* due to the fluctuation in surface charges:

$$B_{\mathrm{off}} \simeq \frac{S_I}{\mu_{\mathrm{H}}} \delta Q_s. \tag{5.35}$$

Here we have also made use of (5.6) and put $r_{\mathrm{H}} l/w \simeq 1$.

Therefore, the higher the sensitivity of a Hall device, the more pronounced will be the influence of the surface effects on its offset. As a numerical example, consider a silicon Hall device with $S_I \simeq 100\,\mathrm{V\,A^{-1}\,T^{-1}}$, $\mu_{\mathrm{H}} \simeq 0.15\,\mathrm{m^2\,V^{-1}\,s^{-1}}$ and $\delta Q_s/q \simeq 10^9\,\mathrm{cm^{-2}}$. From (5.35), we obtain $B_{\mathrm{off}} \simeq 10^{-3}\,\mathrm{T}$.

Fortunately, a very efficient method exists to cope with unstable surfaces: get away from them. In order to isolate the Hall device active region from the surroundings, one can use a reverse-biased p–n junction rather than a bare or oxidized surface. Such Hall devices are called *buried Hall devices* (see figure 5.8).

Hall sensors may incorporate ferromagnetic components, usually used as magnetic flux concentrators. If this is the case, the sensor suffers from another form of instability, caused *perming*. Perming is offset change after a magnetic shock. It is due to the remanent magnetization of the ferromagnetic parts.

5.2 Material and technology

Before starting the description of the successful realizations of Hall devices in §§5.3–5.7, we shall here first consider some general criteria for an optimal choice of appropriate material and technology for the purpose.

5.2.1 Choice of material

The main question we have to pose when selecting material for a sensor is: does the material show an efficient sensor effect? The efficiency of the Hall effect may be assessed by inspecting the relations of the Hall voltage with the bias quantities, namely (4.90), (4.93) and (4.95).

The first of these equations shows that the Hall voltage is proportional to the ratio Hall coefficient/plate thickness. In other words, the current-related sensitivity of Hall plate is proportional to this ratio, equation (5.3). This reduces to (5.4), which shows that the lower the carrier concentration and the thinner the plate, the larger sensitivity. Therefore, a good material for a Hall magnetic sensor must exhibit a low carrier concentration. This criterion is best met by low-doped semiconductors.

Although the very first Hall plate was made of a metal film, this is rarely done any more. To comprehend why it is so, consider the physical limit of the current-related sensitivity of a metal-plate Hall device. The thinnest metal 'plate' is a mono-atomic metal film. Let us assume the atomic diameter is 10^{-8} cm and that there is one free electron per atom. Then the electron density amounts to $N_s \sim 10^{16}\,\mathrm{cm}^{-2}$. According to (5.6) this would correspond to a sensitivity of $S_I \sim 0.1\,\mathrm{V\,A^{-1}\,T^{-1}}$. This is about 1000 times smaller than the value found in modern semiconductor Hall devices.

According to (4.93), the ratio of the Hall voltage and the input voltage is proportional to the Hall mobility. Put otherwise, the voltage-related sensitivity (5.8) of a Hall plate is proportional to the Hall mobility. Therefore, a good material for a Hall magnetic sensor must also exhibit high carrier mobility.

Both the above criteria are important if the efficiency of a Hall device with respect to the dissipated power P is considered. According to (4.95)

$$\frac{V_{\mathrm{H}}}{P^{1/2}} \sim \left(\frac{\mu}{nt}\right)^{1/2} B_{\perp} \tag{5.36}$$

where μ is the conductivity mobility. Recall that $\mu_{\mathrm{H}} \simeq \mu$ (3.275).

The semiconductors with the highest electron mobilities are InSb ($\mu n \simeq 8\,\mathrm{m^2\,V^{-1}\,s^{-1}}$) and InAs ($\mu n \simeq 3.3\,\mathrm{m^2\,V^{-1}\,s^{-1}}$) (see Appendix III). However, these materials have small band gaps: $E_{\mathrm{g}} = 0.18\,\mathrm{eV}$ and $E_{\mathrm{g}} = 0.36\,\mathrm{eV}$ for InSb and InAs, respectively. A small band gap means a high intrinsic carrier density (see (2.32)). Therefore, the criterion of low enough carrier concentration may be difficult to fulfil with these semiconductors, at least above room temperature.

A serious problem occurring in Hall devices made of small-band-gap materials is their high temperature sensitivity. According to (5.4), $S_I \sim 1/n$; and if the material is intrinsic, $n = n_i \sim T^{3/2} \exp(-Eg/2kT)$ (2.32). Much lower temperature sensitivity can be obtained if the semiconductor is extrinsic and operates in the exhaustion temperature range

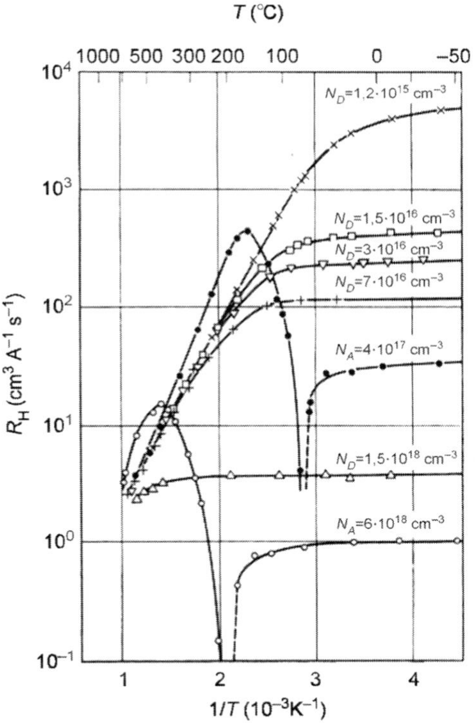

Figure 5.5. The Hall coefficient R_H of InAs as a function of temperature T for various doping densities. At room temperature, the material is in the exhaustion range, and R_H tends to saturate at a constant value. At high temperatures, the intrinsic range is approached, and $\ln(R_H) \sim T^{-1}$. The curves corresponding to the p-type samples have a range where $R_H \to 0$: see equation (3.207) (adapted from [25]).

(see §2.3.5). But then a small-band-gap semiconductor must be rather highly doped. Figure 5.5 illustrates the temperature dependence of the Hall coefficients of InAs specimens with different impurity concentrations [25]. In the intrinsic range, $\ln(R_H) \sim T^{-1}$; in the exhaustion range, $R_H \neq f(T)$. In p-type materials, $R_H = 0$ at a temperature at which the condition (3.207) is fulfilled.

Preferably, n-type semiconductors should be used to fabricate Hall devices, because the electron mobility is larger than the hole mobility in all known semiconductors. For example, μ_n/μ_p equals about 2.5 for silicon, 20 for GaAs and 70 for InAs. This is why in an n-type semiconductor the condition for $R_H = 0$ (3.207) is never fulfilled.

A semiconductor with a larger band gap stays extrinsic up to a high temperature in spite of a low doping (see (2.48)). For example, silicon

($E_g = 1.12\,\text{eV}$) with a donor concentration $N_D \geq 6 \times 10^{13}\,\text{cm}^{-3}$ stays strongly extrinsic up to $T = 125\,^\circ\text{C}$ (see the example of §2.3.5), while GaAs ($E_g = 1.42\,\text{eV}$) stays extrinsic even at much higher temperatures.

5.2.2 Technology

The next crucial issue in selecting the appropriate material for any sensor is its technological manageability. Is a material available, or can it be fabricated to meet the high-quality standards generally required in the fabrication of sensors? Is the process reproducible and economically efficient? Are the sensors made of this particular material reliable and stable?

All these questions can be answered positively if a sensor material is compatible with a mainstream technology, such as microelectronics technology. Silicon and GaAs are outstanding materials in this respect.

A compromise of all above criteria determines the following 'population density' in the land map of Hall magnetic sensors:

Early Hall devices for magnetic sensor applications were made in the form of a discrete plate, similar to that shown in figure 1.3. Usually high electron mobility semiconductors were used, such as InSb. Since the efficiency of a Hall magnetic sensor is higher if the plate thickness t is small, good Hall plates had to be very thin and thus difficult to fabricate.

With the advent of planar technology, much thinner semiconductor films have been fabricated by evaporation, sputtering or chemical vapor deposition, and precisely structured with the aid of photolithography [3–5]. A great majority of contemporary discrete Hall magnetic sensors are made in this way.

The first integrated Hall device was made using the MOS process [26]. Currently an ever-increasing proportion of all Hall sensors are integrated with electronics. A vast majority of such integrated Hall magnetic sensors are made of silicon, in spite of the fact that silicon has rather low electron mobility. This is because the structure of a Hall device can be made compatible with structures commonly used in silicon and GaAs integrated circuits. The possibility of using microelectronics technology outweighs the drawback of small carrier mobility. Integrated sensors, with on-chip signal processing, are called smart sensors [27].

5.3 Examples of Hall plates

In this section we shall present a few examples of industrial realizations of Hall plates. The main criterion for the selection of the examples is if the corresponding device has found a practical application in a commercially available product.

5.3.1 High mobility thin-film Hall plates

Hall elements made of thin-film high-electron-mobility semiconductors have been commercially available for a few decades [28]. The devices are manufactured applying the very efficient batch fabrication method of microelectronics industry. Briefly, the thin semiconductor film is prepared on large-area flat substrates, using vacuum deposition technologies, such as evaporation, sputtering, and molecular beam epitaxy; then, thousands of Hall elements are simultaneously structured using a photolithography process; and finally, the substrate is cut, and individual Hall elements are mounted and encapsulated in transistor-type plastic cases. A big advantage of this technology is that it allows the choice of the best material for the purpose—i.e. of a highest-mobility material. The resulting Hall sensors feature a very good performance–price ratio and are convenient for many practical applications. The negative feature of these highest-mobility materials is that they are not the standard materials used in microelectronics. Therefore, it is difficult to integrate electronics with high-mobility thin-film Hall elements, so that such magnetic sensors are usually discrete components.

Here a representative example of a Hall plate made of thin-film high-electron-mobility semiconductor material.

Thin-film indium-antimonide Hall plates

Mono-crystalline indium-antimonide (InSb) is a compound semiconductor with the highest known electron mobility (see Appendix C) of $8\,m^2/V\,s$. It is much easier to prepare a polycrystalline InSb thin film than a so thin mono-crystal of this material. But a polycrystalline InSb thin film has much lower electron mobility. A polycrystalline InSb thin film deposited on a mica substrate is a good compromise: it has a still very high electron mobility of about $2\,m^2/V\,s$ (see figure 5.6). The mobility degradation is due to a high concentration of scattering centres associated with the interface substrate–semiconductor film and with grain boundaries. These scattering centres are of the ionized impurity type, with a dominant influence at low temperatures; and of neutral impurity type, with a dominant influence at medium (room) temperatures (see table 2.2). A combination of these two scattering types produce a rather weak temperature dependence of the resulting mobility (see figure 5.6(a)).

According to figure 5.6(b), the Hall coefficient of the InSb thin film strongly decreases with the increase in temperature, with a temperature coefficient of about $-2\%\,{}^{\circ}C$. This is because of a very narrow band gap of InSb: low-doped InSb is intrinsic already well below room temperature. Therefore, the carrier concentration increases exponentially with temperature, and the Hall coefficient decreases accordingly.

The big difference in the temperature dependences of the electron mobility and of the Hall coefficient, shown in figure 5.6, determines the

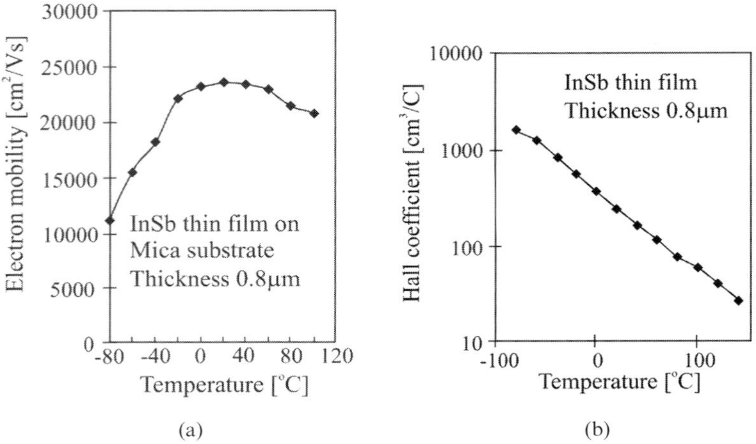

Figure 5.6. Temperature dependence of properties of an InSb thin film, 0.8 μm thick, formed by vacuum deposition on a mica substrate. (a) Electron mobility. (b) Hall coefficient (reprinted from [28]).

preferable mode of the biasing of InSb Hall elements in practical applications. Since the current-related magnetic sensitivity of a Hall sensor (5.3) is proportional to the Hall coefficient, and the voltage-related sensitivity (5.8) is proportional to the Hall mobility, then biasing with a constant voltage gives here much more stable sensor performance.

Typical characteristics of a commercially available InSb thin-film Hall element at room temperature are as follows: voltage-related sensitivity $S_V \simeq 1\,\mathrm{V/V\,T}$, offset voltage $V_{off} \leq 7\,\mathrm{mV}$, and input and output resistance between 240 and 550 Ω. If such a Hall element contains a magnetic flux concentrator (see §5.6), then its sensitivity is for a factor 3 to 6 higher [28].

5.3.2 Silicon and GaAs Hall plates

Silicon Hall plates

A great majority of integrated circuits (IC) have been made of silicon. Several books on integrated circuit technology are available: see, for example, [29, 30]. Silicon IC technology is very mature, easily accessible and low-cost. This makes this technology very attractive for the realization of Hall plates, in spite of moderate mobility of electrons in silicon. We shall now describe a few structures of Hall devices fabricated with the aid of silicon integrated circuit technology. We concentrate here just on Hall plates, leaving the system aspects of integrated Hall magnetic sensors for §5.6.

Figure 5.7 shows a Hall plate realized as a part of the n-type epitaxial layer in the *silicon bipolar integrated circuit process* [29]. The planar geometry

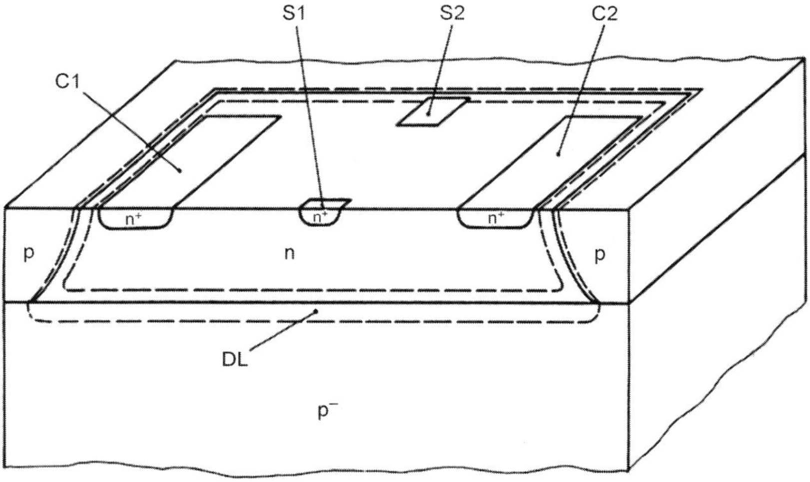

Figure 5.7. Rectangular Hall plate in silicon bipolar integrated circuit technology. The active device region is part of the n-type epitaxial layer. It is isolated from surrounding p-type material by the depletion layer DL. The heavily doped n^+ regions are to enable low-resistivity metal-semiconductor contacts (not shown). The notation of the contacts is the same as in figure 3.3: C1 and C2 are the current contacts, and S1 and S2 are the sense contacts. The device senses the magnetic field component perpendicular to the chip surface.

of the plate is defined by the deep p diffusion ('isolation') and the n^+ diffusion (emitter) regions. The n^+ layers are used to provide good ohmic contacts between the low-doped n-type active region and the metal layer (the contacts). The reverse-biased p–n junction surrounding the plate isolates it from the rest of the chip. Typical planar dimensions of integrated Hall plates are 100–400 μm.

Incidentally, the typical doping ($N_D \simeq 10^{15}$–$10^{16}\,\mathrm{cm}^{-3}$) and thickness ($t \simeq 5$–$10\,\mu\mathrm{m}$) of the epitaxial layers found in bipolar integrated circuits are almost optimally suited to be used as a Hall plate. The above values yield nt products (see (5.4)) in the range from 5×10^{11} to $10^{13}\,\mathrm{cm}^{-2}$, which corresponds to a current-related sensitivity of the order of 100 V/A T.

If the doping density in the epitaxial layer is less than required for the Hall plate, it may be increased by ion implantation. For example, phosphorus ion implantation with a dose of $5\times10^{12}\,\mathrm{cm}^{-2}$ has been used in one case [31]. If the implantation is done into an epitaxial layer with doping density N_E, and all dopants are ionized, equation (5.5) shall be replaced by

$$N_s = N_E t + \int_0^t N_I(z)\,\mathrm{d}z \tag{5.37}$$

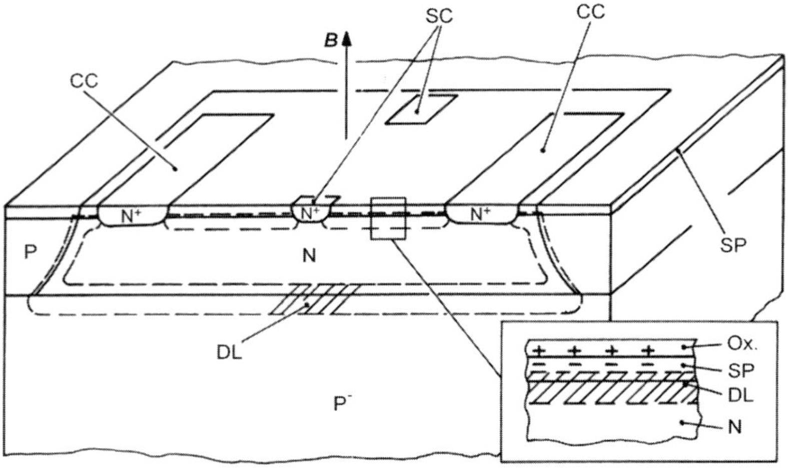

Figure 5.8. Buried Hall plate realized using silicon bipolar integrated circuit technology. N is the n-type active region, P^- the substrate, P the deep-diffused isolation walls, SP the shallow p layer, DL the depletion layer, CC the current contacts, and SC the sense contacts. The inset shows a detail at the device surface. In order to be effective, the shallow p layer must not be completely depleted; compare with figure 5.4.

with N_I denoting the density of implanted ions. If the penetration depth of implanted ions is smaller than t, the integral in (5.5) equals the implanted dose. Implantation is also useful because an implanted layer is usually more uniform than an epitaxial layer.

Figure 5.8 [32] illustrates a refinement of the structure from figure 5.7. This is the so-called *buried Hall device*. It has a structure reminiscent of a pinch collector-layer resistor in bipolar integrated circuits. The shallow p-type layer is added in order to push the active device region away from the upper surface. The shallow p-layer plays the role of an electrostatic shield between the silicon–silicon dioxide interface and the active region of the Hall device, improving its stability [12] (see §5.1.7).

In bulk *CMOS IC technology*, Hall plates are usually realized using the n-well layer. The standard doping of this layer corresponds to a current-related sensitivity of the Hall plate of about 100 V/A T and an input resistance of about 3 kΩ. Figures 5.2(a) and (b) show the structure of a rectangular buried Hall plate realized in n-well bulk CMOS technology. Another example of CMOS Hall device is given in figure 5.22.

The shape of all modern integrated Hall plates is chosen to be 90° rotation-symmetrical, such as those shown in figure 4.15(d)–(h) and in figure 5.22. This allows the application of one of the offset-cancellation techniques described in §5.6.3. The residual offset of a silicon integrated Hall sensor is about 1 mT.

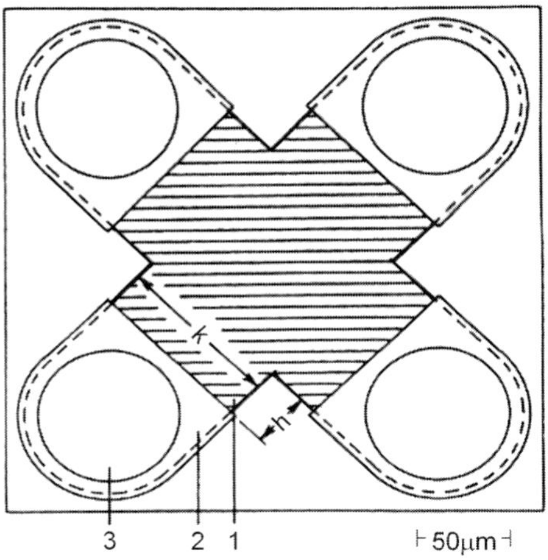

Figure 5.9. Structure of a GaAs Hall device. 1, active device area; 2, metallized region; 3, contact region. The cross shape is similar to that of figure 4.15(d) (adapted from [34]).

GaAs Hall plates

Electron mobility in gallium arsenide (GaAs) is about 5.5 times greater than in silicon. In addition, GaAs integrated circuits are capable of operating up to a temperature in excess of 200 °C. For these reasons, many attempts have been made to fabricate Hall devices using GaAs integrated circuit technology. The active region was made either in an epitaxial layer [33] or by ion implantation [16]. An example of the structure of a commercially available GaAs Hall plate is shown in figure 5.9 [34].

5.3.3 2DEG Hall plates

A very promising way of making high-performance Hall plates is to use the so-called two-dimensional electron gas (2DEG) layers. The term two-dimensional (2D) is used because a semiconductor layer with the 2D 'electron gas' (EG) is very thin, of the order of 10 nm. The electrons are confined in the 2D layer by two adjacent potential barriers, provided by two parallel layers of high-band-gap semiconductors. The distance between the barriers is much smaller than the mean free path of electrons. Therefore, for the movement of the electrons perpendicular to the layer, their energy is quantized. For this reason, we say that the 2D electron gas is confined in a *quantum well*. The charge of the electrons confined in the quantum well is neutralized by the charge of the ionized dopant atoms that are situated outside of the well. Such structures are called modulation-doped structures. The mobility of the electrons in the well of a modulation-doped structure is limited mostly by phonon scattering, and is very high, particularly at temperatures below 100 K.

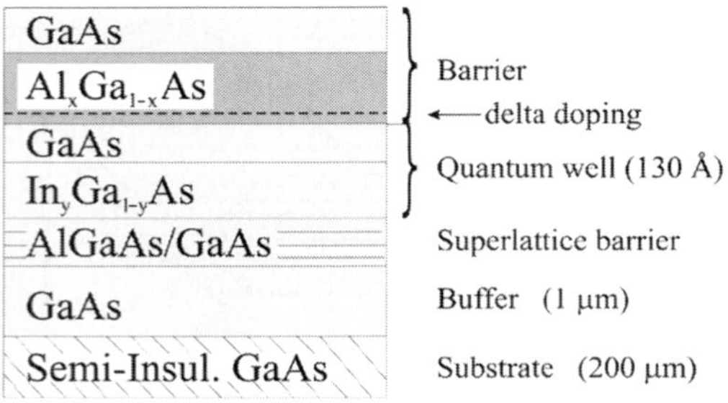

Figure 5.10. Cross-sectional view of the GaAs-based hetero-epitaxial structure applied for the realization of a 2DEG Hall plate used for electrical power metering (reprinted from [37]).

The concept of 2DEG was developed for high electron mobility transistors (HEMT), used at very high frequencies. The technology is rather complex and requires formation of high-quality III–V compound semiconductor hetero-junctions using sophisticated techniques, such as molecular beam epitaxy (MBE) [35]. Thanks to the major application for very high frequency transistors, 2DEG technology is now available also for niche applications, such as Hall magnetic sensors. Here two examples.

Figure 5.10 shows the structure of a Hall plate realized with a GaAs-based process similar to that used for HEMTs [36], [37]. A 13 nm thick InGaAs quantum well confines 2DEG with a sheet electron density of about $8.5 \times 10^{-11}\,cm^{-2}$ and mobility of $0.7\,m^2/V\,s$. The current-related sensitivity is about 700 V/A T, with the temperature coefficient of −140 ppm/°C, and the equivalent magnetic offset is less than 1 mT. The characteristics of the Hall element are stable enough to make it adequate for the application in electric power metering.

The Hall plate shown in figure 5.11 uses a 15 nm thick deep quantum well realized as an InAs layer sandwiched between two wide-band-gap AlGaAsSb layers. This 2DEG has electron mobility at room temperature as high as $2–3.2\,m^2/V\,s$. This is similar to the value of the mobility of the

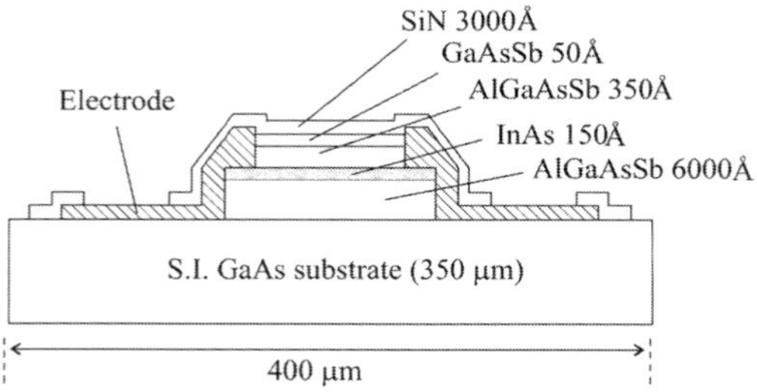

Figure 5.11. Cross-sectional view of a high-mobility Hall plate with 2DEG confined in an InAs quantum well (reprinted from [28]).

thin film InSb layer described above in §5.3.1. Consequently, the basic performance of the two Hall plates is similar. However, the big advantage of the quantum well device is that it stays stable and applicable up to higher temperatures, at least up to 150 °C [28].

5.4 Non-plate-like Hall magnetic sensors

Conventional Hall devices, such as those described in the previous section, are essentially 2D plate-like structures similar to that with which Hall discovered his effect. A plate-like Hall device is optimal in the case when we deal with a homogeneous magnetic field, and when we can position correctly the plate with respect to the measured magnetic field. But there are particular cases when we cannot fulfil one or the other of these conditions. For example, when the magnetic field to be measured is inhomogeneous over the volume of a Hall device, the plate-like structure of the Hall device might not be adequate. In this case, we should better adapt the structure of the Hall device to the shape of the magnetic field lines. Or, if due to technological constrains, we cannot realize a plate-like Hall device at all, we should consider if we can manage with a rough approximation of a plate. Still another example is when we have to measure two or even three components of a magnetic field in a very small volume, where we simply cannot place two or three properly oriented Hall plates. In all these cases we may find it preferable to use a 3-dimensional Hall device. This section is devoted to such devices, which we also call non-plate-like Hall devices [38].

5.4.1 Generalizing the criteria for a good Hall device

While living the solid ground of conventional Hall plates, it is useful to formulate a set of general criteria that a good Hall device must fulfil, independently if it is plate-like or not. In the present context, the essential are the following three criteria (in usual notation):

(a) The vectors of the current density and of the magnetic field must be mutually (approximately) orthogonal over the active region of the device. This is important since the Hall electric field is given by (3.183)

$$\boldsymbol{E}_{\mathrm{H}} = -R_{\mathrm{H}}[\boldsymbol{J} \times \boldsymbol{B}]. \tag{5.38}$$

(b) This active region must be accessible in order to retrieve the Hall voltage, which is given by (4.87)

$$V_{\mathrm{H}} = \int_{\mathrm{S1}}^{\mathrm{S2}} \boldsymbol{E}_{\mathrm{H}}\, \mathrm{d}\boldsymbol{S}. \tag{5.39}$$

(S1 and S2 indicate the positions of the sense contacts).

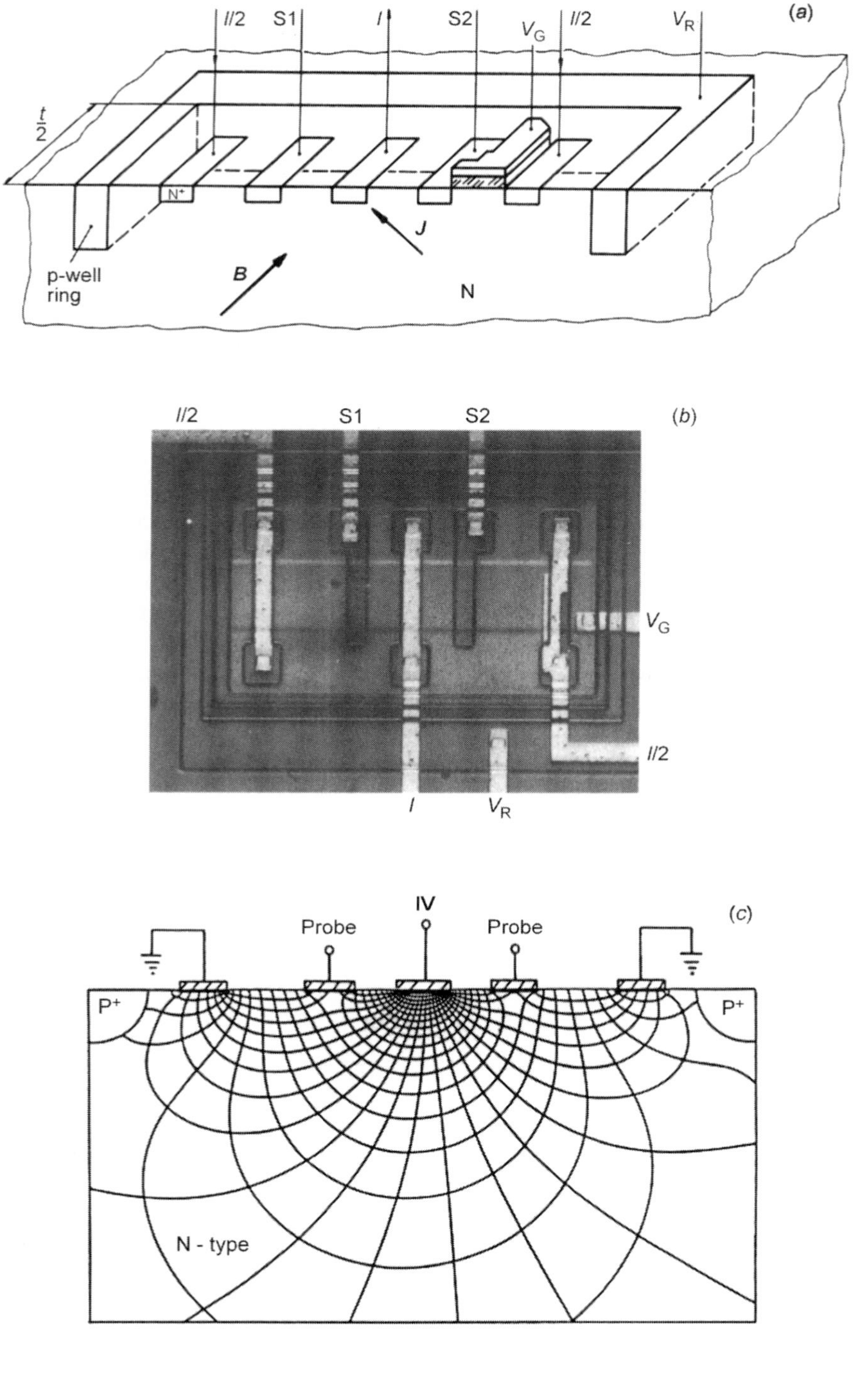
(a)
I/2
S1
I
S2
V_G
I/2
V_R
t/2
N⁺
J
p-well ring
B
N
(b)
I/2
S1
S2
V_G
I/2
I
V_R
(c)
IV
Probe
Probe
P⁺
P⁺
N - type

(c) Moreover, in the absence of a magnetic field, the voltage difference between the points S1 and S2 (offset voltage) should be zero. This means that along a suitable integration path S1, S2 (5.39), the biasing electrical field $\boldsymbol{E}$ and the Hall electric field $\boldsymbol{E}_\mathrm{H}$ should be mutually orthogonal, i.e.

$$\boldsymbol{E} \cdot \boldsymbol{E}_\mathrm{H} = 0. \qquad (5.40)$$

Now a possible way of thinking about Hall magnetic sensors is as follows: for a given orientation or distribution of a magnetic field, how should we structure a semiconductor device in order to best fulfil the above three criteria? And: can we influence the distribution of a magnetic field in the device in order to better achieve our objective? Such a way of thinking has led us to develop the non-plate-like Hall devices presented below.

5.4.2 Vertical Hall devices

The conventional Hall plates are parallel with the chip surface; so considering a chip as an ocean in which floats a Hall plate, we say that such a Hall plate is 'horizontal'. Horizontal Hall plates are sensitive to a magnetic field perpendicular to the chip plane. For applications where sensitivity to a magnetic field parallel to the device surface is preferred, I devised a vertical Hall device [39]. In a vertical Hall device the region that plays the role of the Hall plate is made to be perpendicular to the chip plane (therefore the attribute 'vertical'). To realize this, the general plate shape is chosen so that all contacts become available on the top chip surface. Thereby the device may have more than four terminals.

Figure 5.12 shows a five-terminal bottomless version of the vertical Hall device. The device shape is reminiscent of that shown in figure 4.15(k). Recall that we analysed a similar structure obtained by conformal mapping (see figure 4.5(b)). The active zone of the device in figure 5.12 is part of the n-type substrate material, partially confined into an approximately plate-like structure by a deep p-type ring and the depletion layer between the active zone and the ring. Note than the lower part of the sensor volume is not

---

Figure 5.12. Five-terminal vertical Hall device with the homogeneously doped active region without bottom. (a) Structure. (b) Photomicrograph. The vertical cut plane is a symmetry plane of the sensitive volume (therefore 'vertical'). The device senses the magnetic field parallel with the chip surface. (c) A cut through the 5-terminal vertical Hall device, showing the current and equipotential lines. The lines on the cut are the result of the numerical modelling of the device. The current lines connect the central current contact to the two side contacts. The equipotential lines are approximately normal to the current lines. Asymmetry of the lines is due to the Hall effect, which is characterized here by $\mu B = 0.2$ (adapted from [40]).

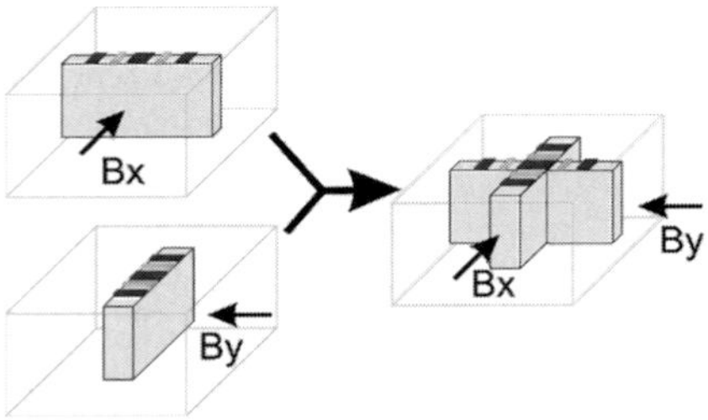

Figure 5.13. Merging two orthogonally placed vertical Hall devices. The two merged devices share the central current contact. The result is a compact two-axis vertical Hall device, which measures simultaneously the two horizontal components of a magnetic field B_x and B_y (reprinted from [10]).

clearly localized at all (therefore 'bottomless'); it is defined only by an increased current density. The symmetry plane of this quasi-plate is vertical with respect to the chip surface, but the whole sensitive volume is a 3D structure. By inspection we conclude that all three criteria given in §3.4.1 are well fulfilled. Notably, the output signal is the Hall voltage V_H, given by the integral (5.39) of the Hall field over a line following an equi-potential line (at $\boldsymbol{B} = 0$) connecting the two sense contacts.

An early version of this device [39] was fabricated using p-well CMOS technology. The active region doping and the effective plate thickness were $N_D \simeq 10^{15}\,\mathrm{cm}^{-3}$ and $t \simeq 12\,\mu\mathrm{m}$, respectively. These values yield the nt product of $1.2\times10^{12}\,\mathrm{cm}^{-2}$.

A later version of the five-terminal bottomless vertical Hall device became a commercially available product. It is fabricated using a dedicated silicon process. The device is used as a key component in a meter of electrical energy and in high-accuracy magnetic field transducers [41]. Its essential features are high current-related sensitivity (400 V/A T), low noise, and an unprecedented long-term stability: its magnetic sensitivity does not change more than 0.01% over a year.

It has been shown that a vertical Hall device, fabricated using the same process, can be made to be equivalent to a circular rotation symmetrical Hall device [42]. Such a vertical device lends itself to the application of the offset-reduction technique called 'spinning current' (see §5.6). The rotation symmetrical vertical Hall device is a materialization of the structure shown in figure 4.15(l); see also Example 2 of §4.2.3.

A merged combination of two five-terminal bottomless vertical Hall devices gives a good magnetic sensor for the simultaneous sensing of the two in plane components of a magnetic field [43]. Figure 5.13 illustrates the genesis of such a *two-axis magnetic sensor*. The major application of a two-axis vertical Hall device is in contactless angular position sensors [41] (see §5.9.2).

Vertical Hall devices have also been realized in conventional bipolar [44, 45] and n-well CMOS IC technologies [46]. These vertical Hall devices are situated in their own junction-isolated n-type islands: they are not bottomless as are the above devices. This brings about two important features. The good feature is that the device is isolated from the substrate and can be readily integrated with electronic circuits. But the bad news is that modern integrated circuits have only shallow and rather strongly doped layers. This is not compatible with the requirements for a high-sensitivity and low-offset vertical Hall device. The problem of the small depth of the available layers is particularly difficult: as we have seen in Example 2 of §4.2.3, limiting the depth of a vertical Hall device is equivalent to cutting a hole in the active region of a conventional Hall plate (see figure 4.11(c) and (d)). This brings about degradation in voltage-related sensitivity.

To soothe this difficulty, one can use a CMOS process with an additional deep n-well. Such CMOS process is available in some foundries and is used for high-voltage integrated circuits. However, a part of the problem persists, because the n-well is not uniformly doped: the doping follows a Gaussian distribution, with the maximum at the surface. Nevertheless, using this technology, vertical Hall devices with a reasonable voltage-related sensitivity of about 0.025 V/V T were demonstrated [46]. Figure 5.14 shows three versions

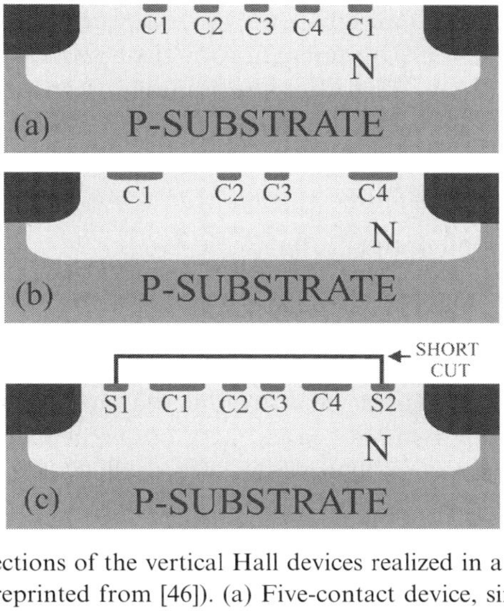

Figure 5.14. Cross-sections of the vertical Hall devices realized in a deep n-well island in CMOS technology (reprinted from [46]). (a) Five-contact device, similar to those shown in figures 4.15(k) and 5.12. (b) Four-contact device, similar to that shown in figure 4.15(l). If contacts C1 and C3 are used to inject a current, then the Hall voltage appears at the contacts C2 and C4, and vice versa. The device is suitable for the application of the spinning-current offset cancellation technique. (c) Six-contact device, which is an improved version of the four-contact device. The two short-circuited contacts help to decrease the offset voltage.

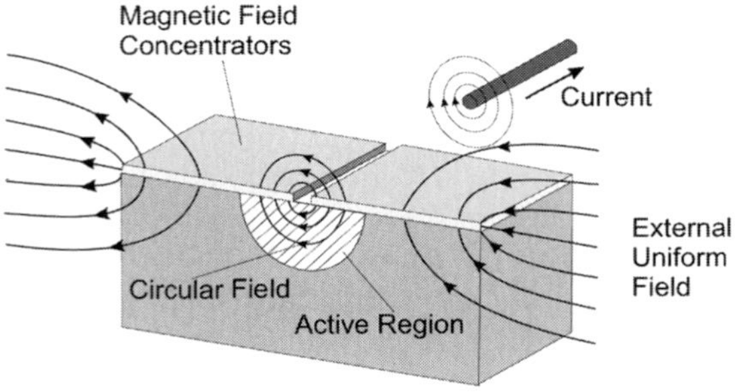

Figure 5.15. The magnetic fields around a current-carrying wire or around an air-gap of ferromagnetic flux concentrators have circular forms (reprinted from [38]).

of such vertical Hall devices. The five-contact device (a) is similar to the one shown in figure 5.12. When biased in the same symmetrical way, this device provides a relatively small offset voltage; however, it is difficult to further decrease this offset, because this device does not lend itself well for the application of the spinning-current technique. The four-contact version is better suited for the spinning current technique, but the original offset of the device cannot be made small enough: it is almost impossible to fulfil condition (c) of §5.4.1. The additional two short-circuited contacts in the six-contact device help to minimize the basic device offset, without compromising its suitability for the application of the spinning-current technique. The equivalent residual offset after the spinning-current offset cancellation is lower than 0.2 mT.

5.4.3 Cylindrical Hall device

In some cases, the magnetic field to be measured has the form, or could be made to have the form, of a circular field (figure 5.15). A well-known example of a circular field is the magnetic field around a current-carrying wire. Another example is the field around the air-gap between two ferromagnetic plates, called magnetoconcentrators, placed in a homogeneous magnetic field as shown in figure 5.15. We shall discuss this case in more detail in §5.7. In the present context it is only important to note that the stray magnetic field around the air gap has an almost circular form. In both examples, the intensity of the circular magnetic field is inversely proportional to the radius of the circles.

An elegant way to measure a circular magnetic field with a small radius is to adapt the form of a Hall device to the form of the field lines. This is possible using a so-called cylindrical Hall device [47] shown in figure 5.16. The magnetic field sensitive volume of a cylindrical Hall device has the form of a virtual half-cylinder, imbedded in the sensor chip. The generated

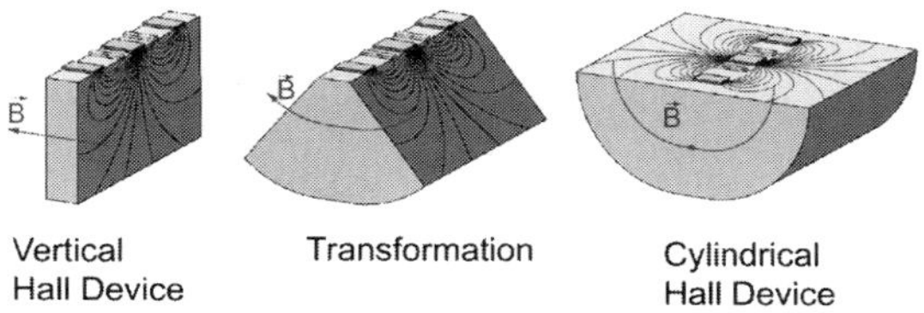

Figure 5.16. Genesis of a cylindrical Hall device by a conformal deformation of a vertical Hall device. While stretching the lower part of the vertical Hall device, including the magnetic field lines, like an accordion, the Hall device takes the form of a half-cylinder, and the magnetic field become half-circles. Within the whole active region, the current lines are approximately perpendicular to the magnetic field lines (reprinted from [38]).

Hall voltage is retrieved using two sense contacts at the surface of the device in the same way as in the vertical Hall device.

Evidently, the structure of a cylindrical Hall device is not even similar to a plate. Nevertheless, it fulfils well the criteria for a good Hall device (a)–(c) given in §5.4.1. The difference between a cylindrical device and a plate-like device is that the former uses the maximal possible volume for generating a Hall voltage. This might be an important advantage in the case when the volume with a high-enough circular magnetic field is very small.

Combined with the integrated planar magnetic flux concentrators, a cylindrical Hall device becomes a very sensitive magnetic field sensor (figure 5.17).

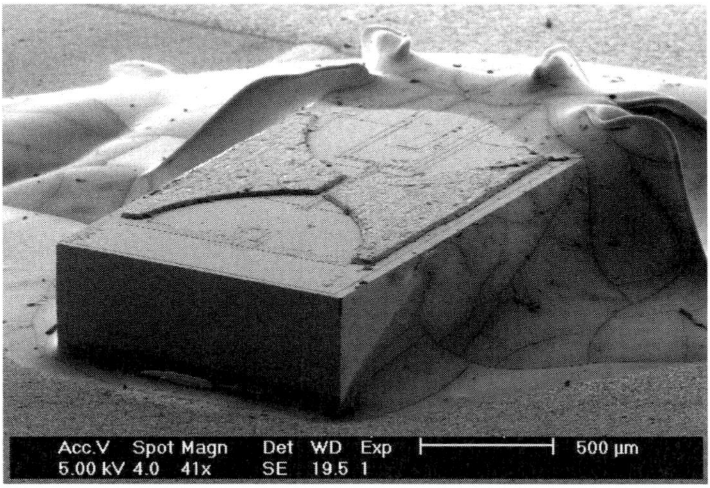

Figure 5.17. Photo of a cylindrical Hall device integrated with magnetic flux concentrators. Thanks to the concentrators, the sensitivity of this magnetic sensor is as high as 2000 V/A T. The detectivity threshold for a quasi-static magnetic field (in the bandwidth from 10^{-3} to 10 Hz) is 70 nT, which is a world record for a commercial Hall magnetic sensor (reprinted from [38]).

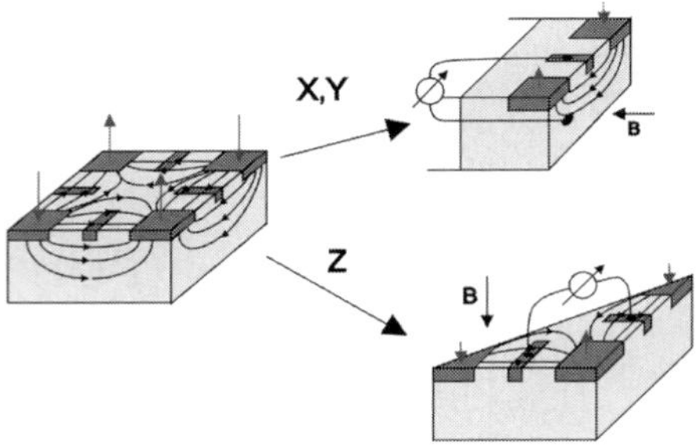

Figure 5.18. The three-axis Hall magnetic sensor. The eight dark regions at the chip surface represent the contacts, the vertical arrows are the bias currents, and the curved arrows are the current density lines. The voltmeter symbols indicate how the Hall voltages proportional to various magnetic field components are retrieved. For the X and Y components of a magnetic field, the sensitive regions are reminiscent of that of a three-terminal Hall device (figure 4.15(j)). For the Z component of a magnetic field, the sensitive region is similar to that of a Hall plate with the open bottom [48].

A cylindrical Hall sensor is applicable in many cases requiring high detectivity magnetic sensing, the domain used to be reserved exclusively for magnetoresistance sensors based on ferromagnetic thin films.

5.4.4 Three-axis Hall magnetic sensor

A conventional way of measuring all three components of a magnetic field is somehow to place three Hall magnetic sensors along the three-coordinate axis. Schott devised a Hall device capable of measuring all three components of a magnetic field using a single small volume of a semiconductor crystal [48]. Compatible with the vertical Hall technology, this magnetic sensor has ohmic contacts only at the chip surface (figure 5.18). Only six external contacts are needed to bias the device and to retrieve the three Hall voltages proportional to the three components of the measured magnetic field. Nevertheless, a simple inspection shows that the above three basic criteria for a good Hall device are well fulfilled.

Such a three-axis Hall chip is used as the input sensor for a three-axis teslameter (see §5.9.1).

5.5 Very small Hall devices

The dimensions of the Hall devices described in §§5.3 and 5.4 are generally chosen according to the following two criteria: on the one hand, a Hall device should be small enough as to not essentially influence the overall production costs of the Hall sensor chips; on the other hand, the dimensions of the Hall device should be big enough as to be not sensitive to production tolerances. A compromise of these two criteria gives the dimensions of conventional Hall devices of the order of 100 μm. The spatial resolution of the measured magnetic field is also of this order, which is more than enough for most practical applications.

But there are some magnetic measurement problems in engineering and science that cannot be solved by conventional Hall devices. For example, with conventional Hall devices we cannot read a modern magnetic memory and we cannot map the magnetic field around a micrometre-size magnetic particle. For such applications we need a Hall device of dimensions comparable with the dimensions of the source of the magnetic field to be measured. The question then arises, what happens with the characteristics of a Hall device if we start decreasing its dimensions? And how far can we go with the miniaturization? In this section we shall try to answer these and related questions.

5.5.1 Scaling down a Hall plate

While decreasing the dimensions of a Hall plate, we shall see that different physical effects predominantly determine its characteristics at different size ranges. We shall name these size ranges by the usual type of electron transport in a Hall plate of this size range.

For the following discussion it is useful to note that the absolute sensitivity of a Hall plate (5.1) is related to the width w of the plate and the drift velocity of the free carriers $\boldsymbol{v}$ by the following relation:

$$S_A = \frac{|V_{\mathrm{H}}|}{|B|} \simeq w|v|. \tag{5.41}$$

This follows from (5.7) and (5.8) if we substitute $G_{\mathrm{H}} \simeq 1$ and $\mu_{\mathrm{H}}(V/l) \simeq v$.

Linear drift range

This size range corresponds to conventional Hall devices analysed up to now. The minimal dimensions of a Hall plate belonging to this size range might be a few micrometres for a low-mobility material, or several tens of micrometres for a high-mobility material. The essential feature of the charge carrier transport in this size range is that the carrier drift velocity is linearly proportional to the applied electric field (2.84). Accordingly, Hall voltage is proportional to input voltage (4.92). In practice, the maximal input voltage is limited. For example, in certain CMOS technologies the voltage is kept constant at 5 V. Therefore, we consider the scaling down in the linear drift region under constant voltage. Then applied electric field as well as the drift velocity of carriers is inversely proportional to the linear dimension of the sensor. If we reduce all planar dimensions of a Hall plate working in the linear drift range by a scaling factor k (>1) while keeping the input voltage constant, we shall have the following consequences on its main characteristics:

(L.1) Sensitivity: does not change, see (5.41), (5.3) and (5.8).
(L.2) Input current, input and output resistances: do not change (4.77), (4.81).
(L.3) Offset-equivalent magnetic field: increases proportionally with k, if we assume that the main cause of offset is tolerances in dimensions (5.11).
(L.4) Noise-equivalent magnetic field (5.14): (i) in the white (resistive) noise range, it does not change since the resistance remains constant; but (ii) in the $1/f$ noise range, noise-equivalent magnetic field increases proportionally with k, because the $1/f$ noise spectral density is inversely proportional with the device volume (2.129), which scales as $1/k^2$. Therefore, $1/f$ noise may stay dominant until very high frequencies, in the hundreds MHz range.
(L.5) Frequency limit (at which we see the dependence of Hall voltage on frequency): increases at least proportionally with k (see §5.1.6).

If we keep increasing the scaling factor k, we shall eventually come to a limit of the linear drift range. Depending on the details of the Hall plate under consideration, we may reach first the limit due to overheating or due to velocity saturation.

Overheating

In the above-described scaling approach, we keep the total power dissipated in the Hall plate constant, whereas we decrease the plate surface by the factor k^2. If we do nothing to improve the heat removal from the Hall device, its temperature shall increase roughly proportionally with k^2. An increase in temperature brings about a redaction of carrier mobility, and, eventually, onset of intrinsic conditions of the active semiconductor layer. Overheating can be avoided by increasing the resistance of a Hall plate and/or by operating a Hall plate in a low-duty cycle regime.

Velocity saturation

By scaling down a Hall plate while keeping the input voltage constant, we increase the excitation electric field in it. Then the electric field shall eventually reach a value at which the drift velocity of charge carriers reaches saturation—see figure 2.13 in §2.6.2. Now if we continue reducing the Hall plate dimensions, the functional relationships between the dimensions and the sensor characteristics become quite different than those valid in the linear drift range.

Saturated velocity range

Electron drift velocity reaches a maximal value in GaAs at an electric field of about 3 kV/cm and in Si at about 30 kV/cm (see figure 2.13). At a bias voltage of 5 V, this corresponds to the lengths of the Hall plate of about 15 and 1.5 μm, respectively. If we continue scaling down a Hall plate below these dimensions while keeping the input voltage constant, the carrier drift velocity will not increase further. Consequently, the Hall electric field shall not increase either, since it is proportional to the carrier velocity (3.339). So it is not reasonable any more to keep the bias voltage constant. Instead, in this size range, we should better adopt the constant-field scaling rules, developed for MOS transistors [49]. This means: while reducing all planar dimensions of a Hall plate by a scaling factor k (>1), we also reduce the input voltage by the same factor. Then both excitation electric field and Hall electric field stay independent of k. We expect that such scaling down shall have the following consequences on the main characteristics of a Hall plate working in the velocity saturation range:

(S.1) Sensitivity: decreases proportionally with $1/k$. This is so because now the Hall electric field does not depend on k (see (3.340)), whereas the integration path MN′ for Hall voltage in figure 4.25 scales down as $1/k$. In other words, in (5.41), the velocity is constant (the saturation velocity) and the sensitivity becomes proportional to the sensor's width.

(S.2) Input current: decreases proportionally with $1/k$. This is so because we assume a constant carrier density, the carrier velocity is saturated, and the cross-section of the plate scales down as $1/k$.

(S.3) Input dynamic resistance: tends to infinity. If drift velocity is saturated, current no longer depends on the applied voltage. So $\mathrm{d}I/\mathrm{d}V \to 0$, and consequently, $R \to \infty$.

(S.4) Output resistance: slightly increases, as explained below.

(S.5) Noise: (i) the white (resistive) noise slightly increases because of S.4; but (ii) the $1/f$ noise further increases proportionally with k, much as above in the case L.4(ii). The noise is to be translated into the noise-equivalent magnetic field by taking into account that the sensitivity decreases proportionally with $1/k$. Therefore the noise-equivalent magnetic field increases slightly more than linearly with k in the case of white noise and quadratically with k for the $1/f$ noise.

(S.6) Frequency limit: increases, as above in the case L.5, as long as we do not reach a limit imposed by the local high-frequency effects (see §4.6.1).

At a first glance, the conclusions S.4 and S.5(i) about only slight increases in output resistance and in white noise seem to be unjustified. One would rather expect that, in view of S.3, the output resistance also tends to infinity. We can resolve this apparent paradox with the help of figure 5.19. By inspecting figure 5.19(a) we notice that the current density near the two current contacts is higher than that at the middle of the plate, $J_1 > J_2$. In view of (2.88), this implies $E_1 > E_2$, E_1 and E_2 being the electric fields in the regions with J_1 and

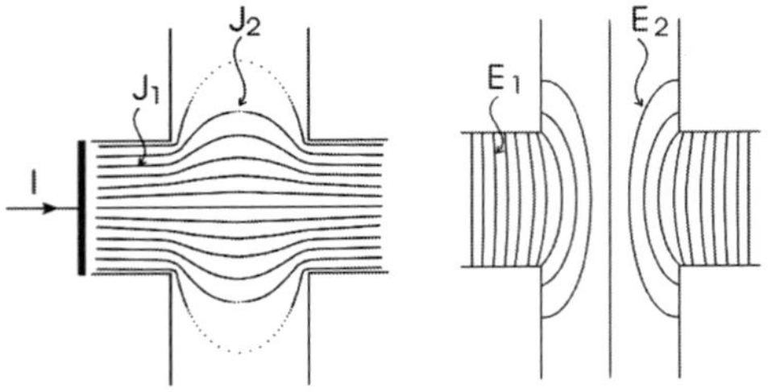

Figure 5.19. Current and voltage distributions in a cross-shaped Hall plate without magnetic field. Analytical calculation. (a) Current lines. A current I is injected into the plate via the current contacts, of which only the left one is shown. Near the current contacts, the current in the plate is more confined than at the middle of the plate: $J_1 > J_2$. (b) Equipotential lines. From the density of the equipotential lines we infer $E_1 > E_2$ (adapted from [50].

J_2, respectively. This is also evident from figure 5.19(b). Therefore, if the electrons making up the current density J_1 drift just at the onset of the saturation velocity, then the electrons making up the current density J_2 shall be still more or less in the linear range. It is interesting to note that, due to the increase in the material dynamic resistance near the current contacts, the *geometry factor of the Hall voltage* should increase with respect to the value of a similar plate working in the linear drift range.

In spite of the just described 'mitigating circumstance', the scaling consequences S.1 and S.5(ii) imply that scaling a Hall plate beyond the point of drift velocity saturation brings about a rapid degradation of its magnetic field resolution.

If we still keep scaling down a Hall plate, the number of successive collisions that an electron makes on its way between the two current contacts shall also decrease. Eventually, more and more electrons shall make the trip between the current contacts with only a few collisions. Then we enter the non-stationary transport range. Recall the two most important phenomena of non-stationary transport: velocity overshoot and ballistic transport—see §2.4.4 and [51].

Velocity overshoot range

Briefly, velocity overshoot means a swift increase in the drift velocity of charge carriers well above the value of stationary saturation drift velocity. This occurs when a Hall device is so small, and the applied voltage is so high, that electrons experience just a few collisions while passing the distance between the current contacts. For example, this may be the case in a high mobility 2DEG Hall plate, about 5 μm long, working in the cryogenic temperature range and biased by a voltage of about 5 V (see (2.112)).

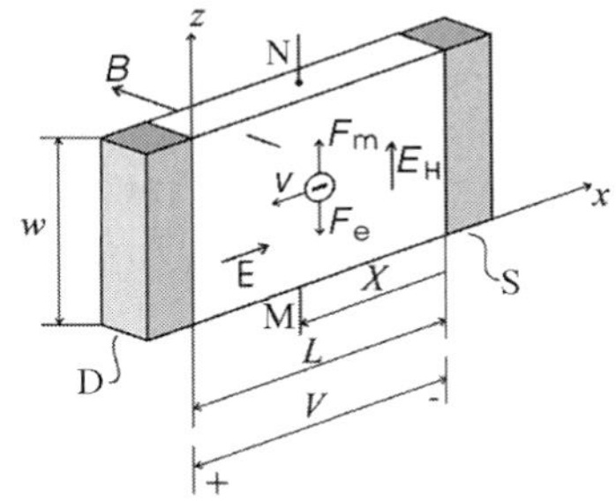

Figure 5.20. A model of a ballistic Hall plate. Ballistic electrons are emitted the source S and collected by the drain D; M and N are the points for the retrieval of the Hall voltage.

Assume that we have a Hall plate working just between the saturated velocity range and the velocity overshoot range. Then a further scaling down of the plate planar dimensions results in an increase of the velocity, which may compensate the decrease of the sensor width (see (5.41)).

Ballistic transport range

Finally, let us assume that the dimensions of a Hall plate are so small that we can treat electrons as little charged billiard balls that can 'fly' through it without any collision, as described in §2.4.4. To be definite, let us consider a *ballistic Hall bar* shown in figure 5.20. We assume that the material of the bar and the operation conditions are such that the length of the plate L is much smaller than the mean path of electrons between collisions (2.112). Now the notion of the plate resistance loses its sense and we assume that the device is designed in such a way that the source S limits the device current to a reasonable value. We also assume that there is no space charge effect in the plate so that the external biasing voltage V produces a constant electric field $\boldsymbol{E}$ over the length of the plate.

Generally speaking, the mechanism of the generation of a Hall voltage in the device shown in figure 5.20 is the same as that acting in the device shown in figure 3.1(n). Briefly, in both cases, the electrons move due to the biasing electric field along the x axis with a velocity $\boldsymbol{v}$. If the device is exposed to a magnetic field $\boldsymbol{B}$ along the y axis, then it appears a Lorentz magnetic force $\boldsymbol{F}_m$, which tends to 'press' the electrons against the upper border of the bar. The 'accumulated' charges (i.e. flying electrons with a slightly higher density) at the borders create a Hall electric field $\boldsymbol{E}_H$. The Hall electric field acts on the electrons by a Lorentz electric force $\boldsymbol{F}_e$, which exactly compensates the action of the magnetic force $\boldsymbol{F}_m$. The result is that the electrons move parallel to the x axis, as in the case without magnetic field.

The fundamental difference between the classical Hall effect (figure 3.1(n)) and the ballistic Hall effect (figure 5.20) is the following detail: whereas in the classical (drift) case, the carrier drift velocity $\boldsymbol{v}_d$ is constant along the bar, in the ballistic case the ballistic electron velocity $\boldsymbol{v}$ increases with the distance from the source X. As a consequence, whereas the classical Hall electric field is essentially constant along the length of the bar, the *ballistic Hall electric field* increases with the distance from the source electrode.

A ballistic Hall device normally operates at a low temperature. Accordingly, it is reasonable to assume that electrons leave the source with zero initial velocity. Then they are accelerated by the electric field (2.68) and after a fly-time t achieve a velocity given by (2.69). We can estimate the ballistic Hall electric field by replacing in (3.8) the drift velocity $\boldsymbol{v}_\mathrm{d}$ by the ballistic velocity (2.69). The result is

$$\boldsymbol{E}_\mathrm{H} \simeq \frac{q}{m^*} t[\boldsymbol{E} \times \boldsymbol{B}] \tag{5.42}$$

where we omit the negative sign. The fly-time can be expressed in terms of the fly distance X:

$$t = \sqrt{\frac{2X}{E}\frac{m^*}{q}}. \tag{5.43}$$

Now (5.42) becomes

$$\boldsymbol{E}_\mathrm{H} \simeq \sqrt{\frac{q}{m^*}\frac{2X}{E}}[\boldsymbol{E} \times \boldsymbol{B}]. \tag{5.44}$$

Suppose that the magnetic field $\boldsymbol{B}$ is normal to the plate and the probe points M and N are situated at the middle of the bar, i.e. $X = L/2$, Then, by integrating the Hall electric field (5.44) between the probe points (3.11), we calculate the ballistic Hall voltage as

$$V_\mathrm{H} \simeq \sqrt{\frac{q}{m^*}V} \cdot W \cdot B. \tag{5.45}$$

Here V denotes the bias voltage, $V = EL$.

Note that we cannot increase the bias voltage at will: the limit here is given by the occurrence of the impact ionization effect. When the carrier kinetic energy is greater than the band gap energy, electron–hole pairs can be produced by ionization. This considerably reduces the mean path between collisions. Therefore, we consider the case where the electrons reach at the drain D a kinetic energy just below the gap energy E_g. This implies that the applied voltage is constant just below the value $V = E_\mathrm{g}/q$. In this case, the magnetic sensitivity of a ballistic Hall device is proportional to the device width w.

Some good news about a Hall plate working in the ballistic regime is that it may have a small noise. Due to the nature of the carrier transport and to the fact that ballistic Hall device normally operates at a low temperature, we now expect to see predominantly *shot noise*.

The bias current I can be decomposed in single electrons flowing through the device at the rate I/q. Each electron produces a voltage difference between the two points M and N for a period corresponding to the transit time T_{trans}. From (5.43) we obtain the transit time as:

$$T_{\text{trans}} = \sqrt{\frac{2m^*}{Vq}} L. \tag{5.46}$$

In the time domain, this basic voltage pulses can be approximated by

$$\delta V(t) = \begin{cases} \Delta(q/\varepsilon W) & \text{for } t \in [0, T_{\text{trans}}] \\ 0 & \text{elsewhere} \end{cases}. \tag{5.47}$$

The factor Δ accounts for the electron trace position, averaged over all the electron's traces. Namely, the electrons flowing in the centre of the device do not produce a voltage imbalance, in contrast to the electrons flowing near the borders. ε is the high-frequency dielectric constant of the sensor material.

In the frequency domain, the voltage noise spectral density should be given by:

$$S_{\text{NV,out}} = 2\left(\frac{I}{q}\right)\left(\Delta \frac{q}{\varepsilon W} T_{\text{trans}}\right)^2 = 4\Delta^2 I \frac{m^*}{\varepsilon^2 V}\left(\frac{L}{W}\right)^2. \tag{5.48}$$

The first term represents the rate of the events, the second is the square of the Fourier transform of each single event.

Therefore, the voltage shot noise spectral density (5.48) at the output of a ballistic Hall plate is proportional to the device current. On the other hand, according to the above simple model, the Hall voltage does not depend on the device current at all (5.45). Accordingly, it should be possible to achieve a reasonable magnetic resolution of a ballistic Hall device just by reducing the emissivity of the source S.

We may *summarize* the results of our analysis of scaling down of Hall plates as follows:

- Depending on the predominant nature of carrier transport, the magnetic sensitivity of a Hall plate either stays more or less constant, or decreases proportionally with the plate dimensions, as shown in figure 5.21.
- In the stationary transport range, both the resistive noise and the $1/f$ noise increase (or at best stay constant) with the scaling factor k. The corner frequency of the $1/f$ noise shifts towards higher frequencies by scaling down.
- Therefore, magnetic resolution of a Hall plate in the stationary transport range generally quickly deteriorates with scaling down.
- In the ballistic transport range, we expect that shot noise dominates and the final outcome for magnetic resolution might not be so bad.

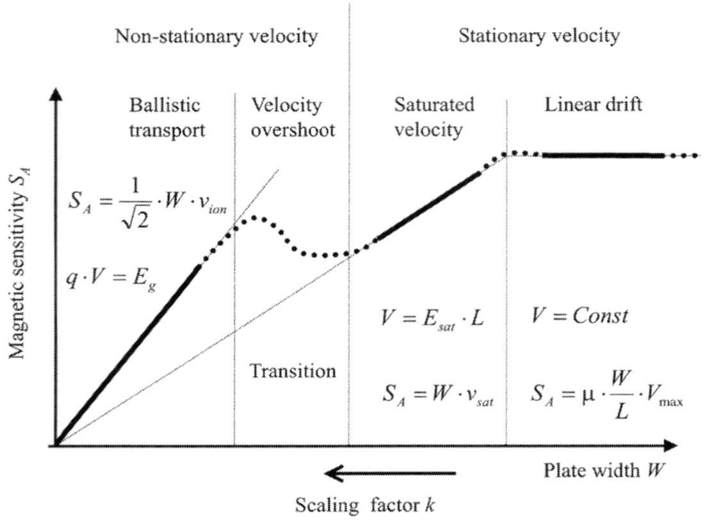

Figure 5.21. A qualitative piecewise linear approximation of the dependence of Hall plate magnetic field sensitivity on the plate planar dimensions and the scaling factor k. Plate dimensions are represented by the plate width W; $k \propto 1/W$.

5.5.2 Examples of micrometre and submicrometre Hall devices

A review of very small Hall devices based on the published results by summer 2002 is given in reference [52]. Here a brief summary of this review.

Most of the publications on Hall devices with dimensions smaller than 10 μm describe Hall plates of cross shape. They have been realized in Si [53], GaAs [54, 55], InSb [55], as 2D electron gas (2DEG) in various III–V heterostructures [50, 56–59], and using Bi thin films [60]. Table 5.1 gives the characteristics of these devices at room temperature estimated from the published data. The magnetic field resolution is calculated using (5.16) for the bandwidth of 1 Hz at the current specified in the table. This is the maximal Hall plate bias current quoted in the corresponding reference, not necessarily equal to the maximum possible current. Since both the Hall voltage and the

Table 5.1. Characteristics of some micro-Hall devices at 300 K

Ref.	[60]	[54]	[56]	[57]	[58]	[50]	[59]	[53]	[55]	[55]
Type	Bi	GaAs	2DEG	2DEG	2DEG	2DEG	2DEG	Si	GaAs	InSb
Area (μm^2)	$(0.1)^2$	$(0.3)^2$	$(0.4)^2$	$(0.4)^2$	$(0.8)^2$	$(1.5)^2$	$(2.0)^2$	$(2.4)^2$	$(4.0)^2$	$(4.5)^2$
S_I (V/AT)	3.3	30	230	230	300	700	350	87	3100	140
B_{min} (nT/$\sqrt{\text{Hz}}$)	70 000	130	10 000	180	4000	300	5	60	10	4
@I (mA)	0.04	2	0.001	0.3	0.003	0.1	2	2	0.3	3

$1/f$ noise are proportional to the bias current and the resistive noise is constant, the bias current is often set to the value at which the $1/f$ noise becomes larger than the thermal noise at the frequency of interest. This current is usually well below the limits imposed by the carrier velocity saturation and over-heating.

Sub-micron Hall devices based on high-mobility 2DEG structures, operating at temperatures below 77 K, are essentially ballistic. They show a magnetic field resolution better than $10\,\mathrm{nT}/\sqrt{\mathrm{Hz}}$ [61].

5.6 Incorporating a Hall device into a circuit

To operate, a Hall device requires a source of electrical energy, and its relatively low output signal must be amplified. Therefore, the Hall device must be incorporated into an appropriate electronic circuit. In addition of these basic functions, the circuit may fulfil more complex functions, such as compensation of temperature cross-sensitivity, and reduction of offset and noise. We shall now consider the interaction of a Hall device with its direct electronic circuit environment. As an example, we take a Hall device operating in the Hall voltage mode.

5.6.1 Circuit model of a Hall device

In order to facilitate the analysis of an electrical circuit incorporating a Hall device, it is convenient to represent the Hall device by an appropriate circuit model. A good circuit model of a Hall device should fulfil the following criteria.

(a) The model should be compatible with conventional circuit analysis techniques, including the compatibility with available computer circuit analysis programs, such as SPICE.
(b) The model should provide a clear relationship with the important physical effects in the device and with the material and technological parameters.
(c) The model should be exact but simplifiable. This means that, if we are not interested in all details of the behaviour of a Hall device in a circuit, we should be able to simplify its model, and so facilitate a rough estimation of a value of interest.

The representation of a Hall device by a lumped circuit model, described in §4.2.4, corresponds well with the above criteria (a) and (b): such a model contains only the conventional circuit components and it is traceable to basic equations describing the Hall effect. However, criterion (c) is not well fulfilled: if we start reducing the number of the L-shaped cells shown in figure 4.14, we shall eventually reach the stage in which we try to describe

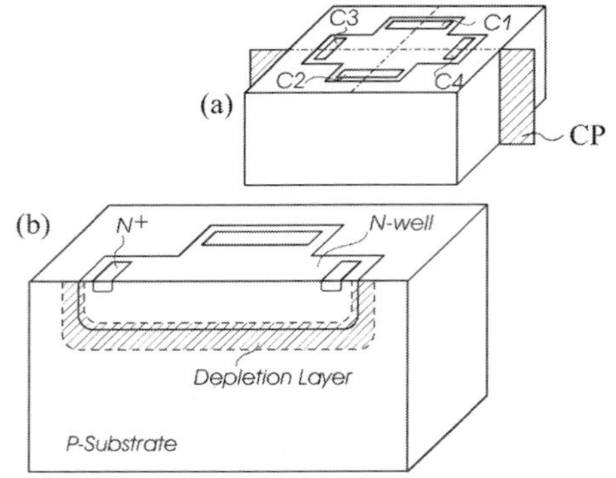

Figure 5.22. Cross-like Hall device in bulk CMOS technology. (a) General view: compare with figure 4.15(d). CP denotes a crossing plane. (b) View with the cross-section along CP. The Hall plate is realized using the N-well layer, which is normally used as a substrate for p-channel MOS transistors. The contact regions of the plate (C1–C4) are heavily doped—the layer N^+, which is normally used for source and drain regions of n-channel transistors. The Hall plate is insulated from the p-type substrate by the depletion layer of the p–n junction well/substrate. For clarity, only the essential semiconductor regions are shown, and the Hall device is not biased. When the Hall device is biased, the depletion layer width is not uniform.

a symmetrical device with a few (asymmetrical) L-cells, which is obviously inadequate. In order to avoid this difficulty while keeping the basic idea that meets the criteria (a) and (b), we shall now slightly modify the model introduced in §4.2.4 and make it symmetrical.

The idea of the symmetrical circuit model of a Hall device may be best understood by way of an example, so let us develop a circuit model of a *CMOS integrated cross-shaped Hall device* shown in figure 5.22. The active part of the device is realized using the so-called N-well layer. In bulk CMOS technology, the N-well is insulated from the p-type substrate by the depletion layer, which develops around the reversed-biased well/substrate p–n junction.

The width of the depletion layer depends on the potential applied over the junction (§2.5.2). Since in a normal operation, the electrical potential of the active region of the Hall device is position-dependent, so shall be also the width of the depletion layer. This means that the effective thickness (precisely: the surface charge carrier density, equation (5.5)) of the Hall plate will be also position-dependent. Recall that we had a similar situation in the case of the MOS Hall device in §4.2.4. There we found that the device can be conveniently modelled by a lumped circuit, with MOS transistors as the basic circuit elements. By analogy, we can model the Hall device of figure 5.22 by applying junction field-effect transistors (JFETs) as the basic elements of the *lumped circuit model*. Obviously, the active N-well region

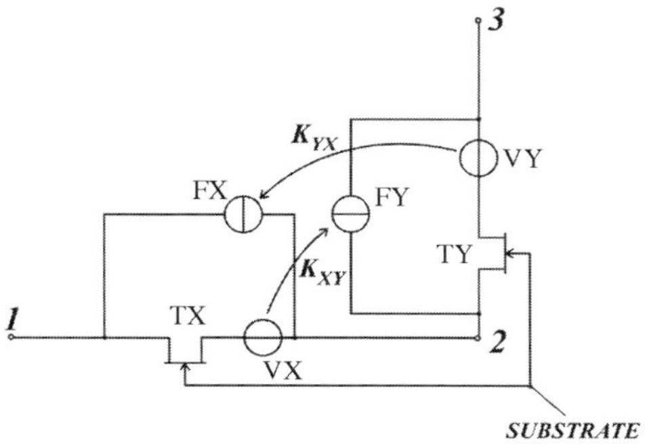

Figure 5.23. The basic circuit cell for the lumped circuit model of the junction-isolated Hall plate shown in figure 5.22. Compare with figure 4.13.

represents the channel of these JFETs and the substrate corresponds to their gates. Taking into account also the Hall effect, we conclude that we may model the present Hall device by the basic cells shown in figure 5.23.

Let us denote the basic circuit cell of figure 5.23 by the symbol ⌟, which symbolizes the general shape of the cell. Now we should note that the ⌟ cell is not the only circuit that is described by the equations (4.52). Indeed, also the circuits that can be obtained from the ⌟ cell by a rotation of 90° can be described by the same equations. So we can broaden our modelling approach of §4.2.4, and state that we may model a Hall plate using the basic cells of any of the following four equivalent forms: ⌟, ⌞, ⌝ and ⌜. We shall call these basic cells BC1, BC2, BC3 and BC4, respectively. In device simulation tasks, we usually represent a Hall device with a lot of basic cells, and the result of the simulation does not depend on the choice of the basic cell form. But in circuit simulation tasks, we usually want to minimize the number of elements in a device model. Then, in view of the need to keep the symmetry of the circuit model even if the number of the cells becomes small, it is convenient to model a Hall device using all four basic cells in a symmetrical way, as shown in figure 5.24.

The circuit model of the cross-shaped Hall device shown in figure 5.24 consists of an equal number N of the basic circuit cells BC1, BC2, BC3 and BC4. Therefore, the circuit model consists of $4N$ basic circuit cells. The larger the number of the group of four basic cells N, the more precise the description of the Hall device. But the smaller N is the easier and the faster analyses of a circuit with the Hall device will be. The minimum value of N is $N = 1$. Then the model of the Hall device reduces to the circuit shown in figure 5.25. This is the simplest circuit model of a rotation-by 90° symmetrical Hall device, which is still directly 'traceable' to the full lamped circuit model (figure 5.24) and to the physics of the Hall effect (§4.2.4).

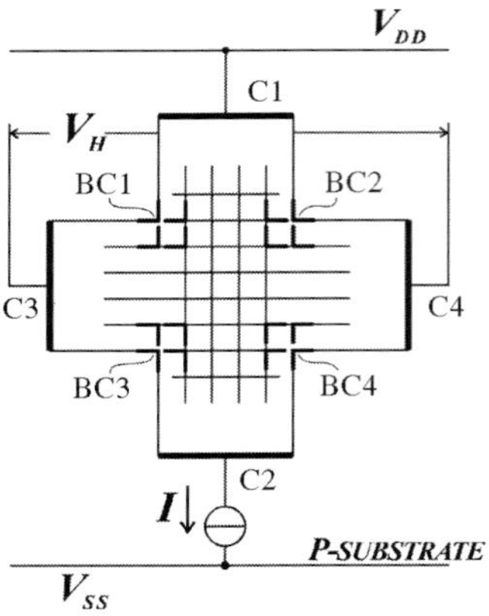

Figure 5.24. Lumped circuit model of the integrated Hall plate of figure 5.22. The model circuit consists of the basic circuit cells of four equivalent forms: ┘, └, ┐ and ┌, denoted by BC1–BC4, respectively. The cell ┘ (BC1) is shown in detail in figure 5.23. Compare with figure 4.14. If a current I is forced between the contacts C1 and C2, a Hall voltage V_H appears between the contacts C3 and C4. V_{DD} and V_{SS} are the notations of the supply voltage commonly used in CMOS ICs.

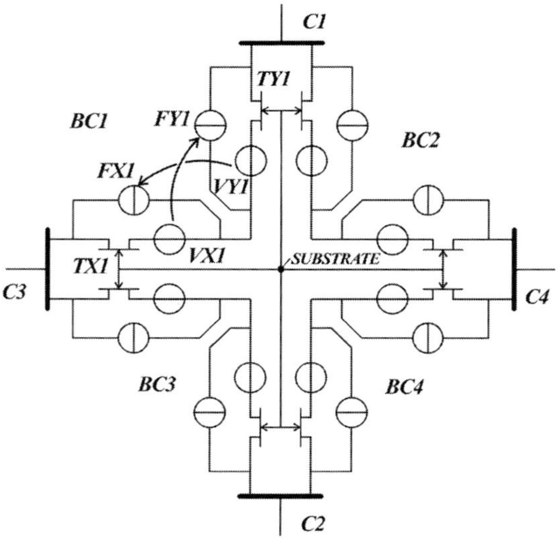

Figure 5.25. The simplest full circuit model of the integrated Hall plate of figure 5.22. The model consists of four basic circuit cells: BC1, which is the one of figure 5.23, and BC2, BC3 and BC4, which are obtained by a 90° rotation of BC1.

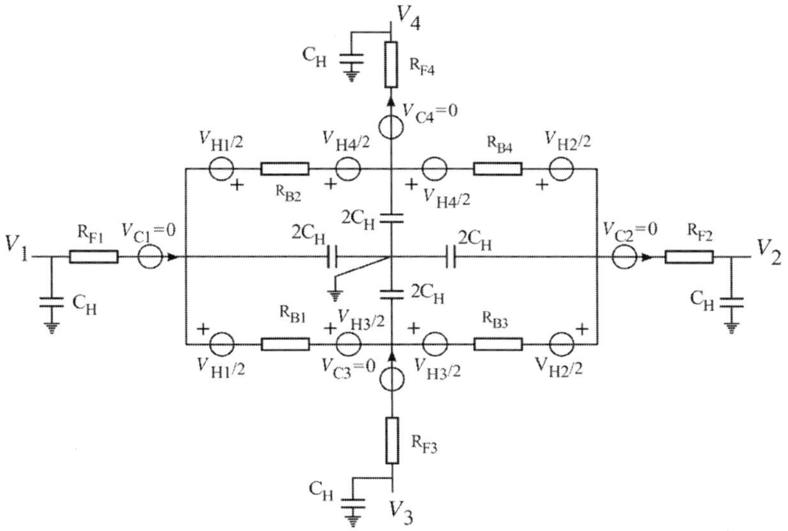

Figure 5.26. A heuristic circuit model of an integrated Hall plate that includes a.c. behaviour. Every Hall voltage generator $V_{Hi}/2$ is controlled by the current 'measured' by the zero-voltage source V_{ci}. The resistance bridge R_{B1}–R_{B4} represents the central part of the Hall cross, and the resistors R_{F1}–R_{F4} represent the resistances of the branches of the cross, near the contacts C1–C4. If the junction field-effect is important, the resistors should be replaced by JFETs (reprinted from [62]).

A circuit model like that shown in figure 5.24 (and to some extent, even that shown in figure 5.25) is suitable to simulate most of the effect in a Hall device of practical interest. First of all, the circuits model the Hall effect and the device resistances, including the magnetoresistance effect. Through the JFET part of the model, also the back-biasing and the associated non-linearity effects are taken into account. Offset can be easily simulated by introducing a non-symmetry into the circuit: for example, a difference between the JFETs in the cells BC1 and BC2 of figure 5.25. Moreover, if the basic model of the JFETs includes the junction capacitances, then the dominant high-frequency effects (§4.6.2) can also be considered.

If in a particular case we are not interested in all aspects of a Hall device, we may still further simplify the model of figure 5.25. Alternatively, we may apply a *heuristic model*, such as the one shown in figure 5.26 [62]. In this model, the JFETs are approximated by simple resistors, and the Hall voltage is simulated by controlled voltage sources. This is approximately equivalent with the model of figure 5.25. Note that the heuristic bridge circuit model shown in figure 4.29, which we used to describe the offset voltage, can be deduced either from figure 5.25 or from figure 5.26.

5.6.2 Biasing a Hall device. Temperature cross-sensitivity and non-linearity

A Hall device can function when biased either by a voltage source or a current source. However, the choice of biasing conditions affects the characteristics of the set-up with the Hall device, such as its temperature cross-sensitivity and non-linearity. See also §5.6.4.

The most important is the influence of the biasing conditions on the magnetic sensitivity of a Hall device (§5.1.1). Briefly, the (absolute) magnetic sensitivity of a voltage-biased Hall device is proportional to the product (Hall mobility) × (bias voltage), and that of the current-biased is proportional to the product (Hall coefficient/thickness) × (bias current). Depending on the details of the device design, these products may be extremely temperature dependent, which brings about a big temperature cross-sensitivity. Let us see how the temperature cross-sensitivity of magnetic sensitivity can be reduced by choosing an appropriate biasing method. As examples we take some of the conventional Hall plates described in §5.3.

High-mobility thin-film Hall plates (§5.3.1) are usually made of small band-gap semiconductor materials. Such a semiconductor is intrinsic at room temperature or at a slightly higher temperature. Consequently, its average Hall coefficient (§3.3.7) is strongly temperature dependent, as shown in figure 5.6(b). Obviously, biasing such a Hall plate by a current source would be not a good idea. On the other hand, the usual fabrication method of an InSb thin film brings about the interplay of a few effects that result in a rather temperature-independent electron mobility (figure 5.6(a)). Therefore, magnetic sensitivity of such Hall plate, biased by a voltage source, is virtually temperature-independent around room temperature. So in practice, high-mobility thin-film Hall devices are usually biased by a voltage source. But there is a big drawback of the voltage-biasing method in this case: the device current rapidly increases with temperature, which can cause a thermal runaway (breakdown) of the device.

On the other hand, Hall devices made of relatively large band-gap materials, such as silicon and GaAs (§5.3.2), are strongly extrinsic at the usual operating temperatures. Then, according to our analysis in §5.1.4, the temperature cross-sensitivity of magnetic sensitivity of a current-biased Hall device is much smaller than that of a voltage-biased Hall device.

Generally, compensation for the temperature variation of the Hall device parameters can be realized by measuring the temperature of the Hall device and using this information to adjust one or the other circuit parameter correspondingly. At low magnetic fields, when the magnetoresistance effect is negligible, a Hall device lends itself to the measurement of temperature: its input resistance is temperature dependent. At a constant bias current, the change in the input resistance produces a proportional variation in the input voltage. Thus the input voltage of a constant-current-biased Hall device can be used as a temperature indicator. Then we can use this

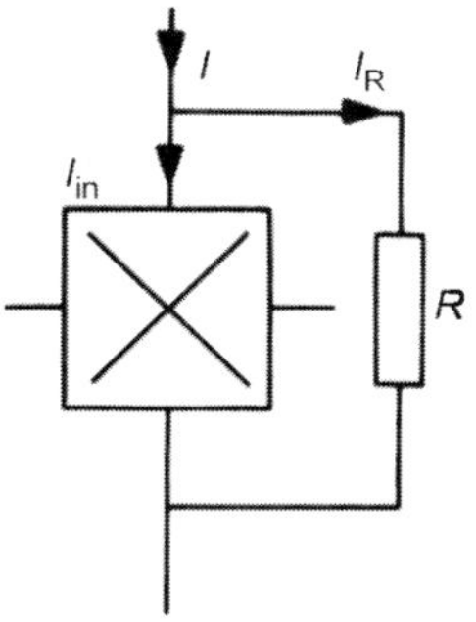

Figure 5.27. A circuit for compensating the variation of the Hall voltage due to temperature fluctuations. A change in temperature causes a change in the device input resistance, and thus a redistribution of the supply current among I_{in} and I_R. The change in the input current I_{in} of the Hall device compensates a change in the Hall coefficient due to the change in temperature.

temperature indicator to compensate the variation in magnetic sensitivity (see below) and/or to correct offset [63].

Figure 5.27 shows the simplest method for the temperature compensation of a Hall device by adjusting its bias current [64]. A resistor is connected in parallel with the input terminals of the Hall device, and a constant current I is forced through the parallel combination of the two devices. The bias current of the Hall device is given by

$$I_{in} = \frac{R}{R + R_{in}} I. \tag{5.49}$$

Suppose that the temperature variation of each of the resistances R and R_{in} can be described by a function of the type (2.1), namely

$$\begin{aligned} R(T) &\simeq R_0(1 + \alpha_R \Delta T) \\ R_{in}(T) &\simeq R_{in0}(1 + \alpha_{RH} \Delta T). \end{aligned} \tag{5.50}$$

with $\Delta T = T - T_0$, T_0 being a reference temperature. After substituting (5.50) into (5.49), and assuming $\alpha_R \Delta T$ is much less than unity, we obtain

$$I_{in} \simeq \frac{R_0 I}{R_0 + R_{in0}} \left(1 + \frac{R_{in0}}{R_0 + R_{in0}} (\alpha_{RH} - \alpha_R) \Delta T\right)^{-1}. \tag{5.51}$$

Suppose that the Hall voltage of our Hall device varies with temperature according to

$$V_H \simeq \left(G_H \frac{R_H}{t}\right)_{T = T_0} (1 + \alpha_s \Delta T) I_{in} B_\perp \tag{5.52}$$

where α_s is the temperature coefficient of the quantity $G_H(R_H/t)$ in (4.90). By inspecting (5.51) and (5.52), we see that, if we choose the parameters so as to make

$$\frac{R_{\text{in}0}}{R_0 + R_{\text{in}0}}(\alpha_{\text{RH}} - \alpha_{\text{R}}) = \alpha_s \tag{5.53}$$

the Hall voltage becomes

$$V_H \simeq \left(G_H \frac{R_H}{t}\right)_{T=T_0} \frac{R_0 I}{R_0 + R_{\text{in}0}} IB_\perp . \tag{5.54}$$

Therefore, within applied approximation, the Hall voltage becomes independent of temperature.

Note that the magnetic sensitivity of a Hall device biased by a current source depends also on the thickness of the Hall device. In junction-isolated Hall devices, the device thickness is not a constant quantity: it depends on the reverse voltage applied over the isolating junction. We call this a *junction field effect*. If the junction field effect exists, we should enlarge the meaning of the term 'biasing conditions' to include also the voltage applied on the isolating p–n junction of a Hall device. Obviously, this is important in the case of integrated Hall devices, and in particular when an integrated Hall device is supplied by a current source. For example, in this case the junction field effect participates in the temperature cross-sensitivity of magnetic sensitivity. This comes about because then the voltage drop over the Hall device depends on temperature, and so also does the distribution of the reverse voltage across the isolating junction.

In §5.1.5 we saw that in high-sheet-resistivity junction-isolated Hall devices, the dominant cause of non-linearity may be the junction field effect. But the non-linearity caused by the junction field effect can be successfully compensated by the use of just the same effect, namely the variation in the sensitivity due to the variation in the bias voltage of the junction surrounding the Hall plate. To this end, one should make voltage V_j of the p-type jacket dependent on the Hall voltage V_H in such a way that the effective thickness of the plate stays essentially constant. A simple circuit to perform this operation is shown in figure 5.28. If $R_1 = R_2$ is chosen, $V_j = (V_{\text{Ref}} + V_H)/2$ (V_{Ref} being a negative reference voltage), and the mean plate thickness remains, to a first approximation, independent of V_H.

5.6.3 Reducing offset and $1/f$ noise

The fabrication tolerances of Hall devices usually result in offsets (§§4.3.4 and 5.1.2) that are too big for intended applications. We shall present here two basic methods for offset reduction that can be implemented on the interface circuit level.

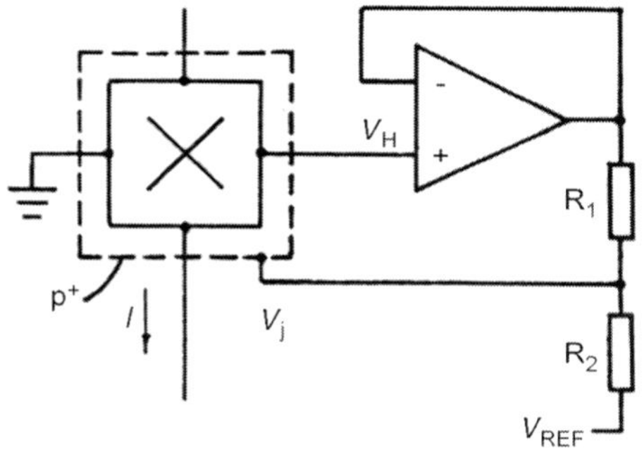

Figure 5.28. A circuit for the compensation of non-linearity due to the junction field effect. The broken square represents the p^+ jacket that surrounds the active area of the Hall device (reprinted from [15]).

Figure 5.29 shows a method suitable for compensating offset of a Hall device, invented by Thomson in 1856 [65]. The procedure is the following: while keeping the Hall device at a zero magnetic field, one slides the contact along the resistive wire and tries to find a position where the offset voltage disappears. Several variations on the theme are possible [66]. The advantage of the Thomson method is that the input current remains essentially unaffected by the shunt circuit. A variation of this method, called *trimming*, is also used on industrial level. A big disadvantage of the offset reduction by trimming is that the resulting zero-offset is usually not

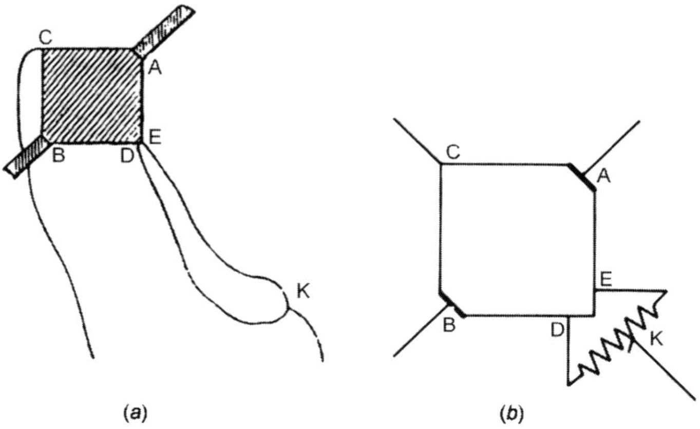

Figure 5.29. A circuit for compensating the offset voltage in a Hall device. (a) The arrangement applied by Thomson in 1856 in his experiments with the magnetoresistance effect. He used to eliminate the offset voltage of his four-terminal device reminiscent of a Hall plate by sliding the contact K along the resistive wire R (adapted from [65]). (b) A schematic representation of Thomson's arrangement.

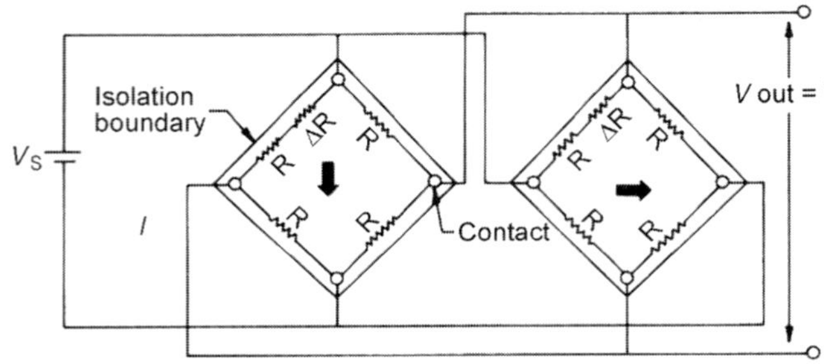

Figure 5.30. Offset cancellation by pairing of integrated Hall devices. Two equal Hall devices processed side by side in an integrated circuit exhibit matching properties, much as two transistors or resistors do. The figure shows two equal rotation symmetrical integrated Hall devices. ΔR denotes an incremental resistance in one of the bridge elements due to resistivity gradient, strain or misalignment. Thanks to the matching, the resistance increment appears at the same geometrical position in both bridges. When the two devices are interconnected as shown (electrically parallel, but with orthogonal current directions), the offset terms cancel in the output voltage (reprinted from [8]).

very stable: due to later thermal or mechanical shocks and natural ageing, offset may re-appear. Therefore, for precise magnetic measurements, it is necessary to re-zero offset at a zero-field each time before the measurement.

One of the basic concepts in integrated circuit engineering is the exploitation of the matching properties of devices. In integrated Hall magnetic sensors, offset can be substantially reduced by the *mutual compensation of asymmetries* in closely spaced Hall plates. If two or four identical integrated Hall plates, biased orthogonally, are properly connected, their asymmetries tend to cancel out, leaving the output signal virtually free of offset [8] (see figure 5.30). We call this technique *offset reduction by pairing*. The increase in the number of Hall devices connected together has a positive side-effect of also decreasing the equivalent noise of the sensor. Therefore, the overall accuracy of magnetic field measurement can be increased. The costs of this are a larger chip area and higher bias current. A possible problem with the offset reduction by pairing of Hall devices exists if their matching is not stable in time.

The drawback of instability of both trimming and pairing are avoided in the dynamic offset cancellation method called the *connection-commutation*, *switched Hall plate*, or *spinning-current technique*. The connection-commutation technique is particularly efficient when applied on a Hall device with a shape which is invariant under a rotation of $\pi/2$ (see §4.2.5). The simplest variation of the method is illustrated in figure 5.31 [67–69]. The diagonally situated contacts of the Hall devices are periodically commutated and alternately used as the current and the sense contacts. The commutation is performed using a group of switches (therefore the term 'switched Hall

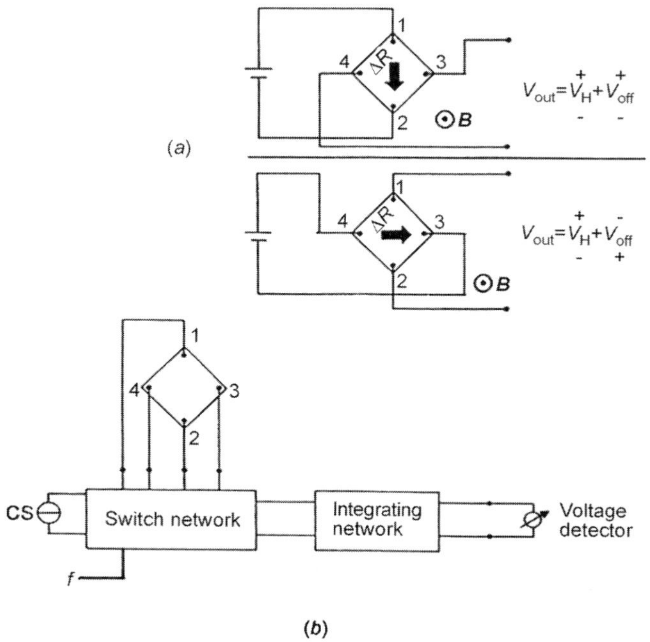

Figure 5.31. The connection-commutation (or spinning current) method for the cancellation of the offset voltage of a Hall device. (a) A rotation symmetrical Hall device is alternately biased in two orthogonal directions. Then the offset voltage due to the resistance asymmetry ΔR changes the sign, and the Hall voltage does not. (b) The circuit. The switch network connects periodically the terminals 1–4 of the Hall device to the current source CS as shown in (a). The integrating network is intended to even out the a.c. offset signal.

plate'). Thus the bias current virtually rotates in the device for 90° back and forth (therefore 'spinning-current'). Then the Hall voltage rotates with the bias current and the offset voltage does not. In the language of the circuit theory (§4.1.2), we say that the resistive matrix of the two-port Hall device (4.12) is *reciprocal* for the offset voltage, and *anti-reciprocal* for the Hall voltage. The offset voltage then appears as an a.c. signal with the switching frequency as the first harmonic and can be cancelled from the output voltage by filtering. The extraction of the Hall voltage from the overall measurement signal is possible only after several switching cycles. Thus the measurement must be extended in time. But in a measurement with a slowly varying magnetic induction (offset is a problem only in this case!), this is generally acceptable.

Note the similarity of the concepts for offset reduction shown in figures 5.30 and 5.31. In both cases, we have two orthogonal biasing currents. In the first case, the orthogonal currents are in two separate devices; in the second case, the orthogonal currents are in two separate time slots. In the first case, the measurement is extended in space; in the second case, it is extended in time. Both extensions lead to a reduction in uncertainty of the measured quantity. Roughly speaking, the uncertainty of the measured quantity appears to be proportional to the product (size $\times$ time).

In a way, the connection-commutation method is reminiscent of the *chopping technique* (see below §5.6.4), although the former does not require a 'de-connection' of the measurand from the sensing device. Similarly to chopping, the connection-commutation also reduces the $1/f$ noise of the measurement system. This can be best understood by imagining the $1/f$ noise voltage at the output terminals of a Hall device as a fluctuating offset voltage. Then we can model the $1/f$ noise source by a fluctuating asymmetry resistance ΔR in figure 5.31. If the biasing current of the Hall device spins fast enough so that the fluctuating voltage does not change essentially during one cycle, then the system shall not 'see' the difference between the static and the fluctuating offset and shall cancel both of them. In order to completely eliminate the $1/f$ noise of a Hall device, we have to set the spinning current frequency well above the $1/f$ noise corner frequency (see figure 4.30). For a very small Hall device, this may be a very high frequency (see §5.5).

5.6.4 Amplifying the output signal

A Hall voltage is usually very small and in most applications requires amplification. Thereby it is often convenient or even necessary to integrate into the system biasing and various corrective functions, discussed in the previous two sections. We shall present here a few ideas on how the system Hall device/amplifier/other interface circuits can be optimized for achieving certain performance.

One of the first attempts in the optimization concerns the choice of the biasing currents and the gain of the amplifier. The *bias current of the Hall device should* be large enough so as to ensure, if possible, that the Hall voltage for a magnetic field of interest is superior to the equivalent input noise voltage of the amplifier. The first amplifying stage should boost the output signal of the Hall device above the noise level of the subsequent circuitry. Then the *signal-to-noise ratio* of the whole system stays essentially equal to that of the Hall device itself. The notion of 'noise' here comprises both fluctuations and interfering signals.

The choice of the biasing method may affect the structure and performance of the first signal-amplifying stage. Figure 5.32 illustrates two examples of the current biasing of a Hall device. In figure 5.32(a), one current contact is held at a reference potential. Since the voltage drop over the device varies with a change in the bias current and temperature, the

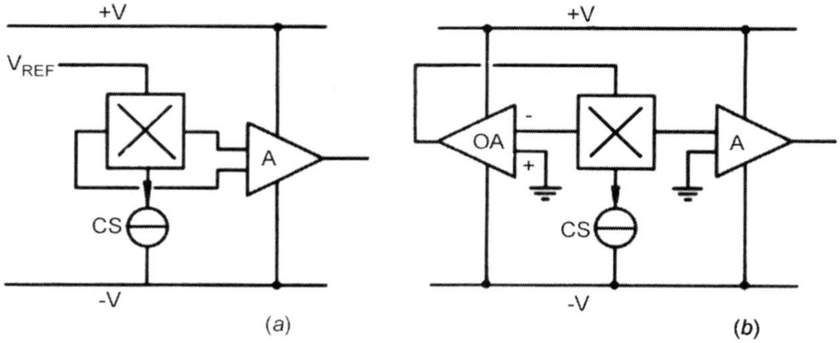

Figure 5.32. Circuits for biasing a Hall device and amplifying its output voltage. (a) In such a circuit a large common-mode voltage appears at the amplifier input. (b) With the aid of an operational amplifier (OA), the left sense contact is kept virtually at ground. Thus the full Hall voltage appears at the right sense contact against ground [70].

potential at the other current contact exhibits large excursions. Half of these potential variations occur as the variations in the *common-mode voltage* at the output terminals of the Hall device. Since only the differential signal at the output is of interest, this common-mode signal must be strongly rejected. This requires a complicated structure of the instrumentation amplifier A. The problem can be circumvented by the biasing method (figure 5.32(b)) [70]. Here, using the operational amplifier OA, the left sense contact is held at the virtual ground. Now the whole output voltage of the Hall device occurs at the other sense contact relative to ground. Thus a simple amplifier A suffices to amplify the output voltage.

After reducing the offset and $1/f$ noise of the Hall device using the connection-commutation technique (§5.5.3), the offset and $1/f$ noise of the amplifier itself may become dominant in the system. Then one may apply also the dynamic offset and $1/f$ cancellation techniques for the amplifier, such as auto-zeroing and chopper-stabilization [71].

An elegant and efficient way, to combine the two chopper-like *offset and $1/f$ noise cancellations* (for the Hall device and for the amplifier) into one was described by Bilotti *et al* [72]. They modified the first phase of the switching procedure of figure 5.31(a) so that the Hall voltage polarization reverses at the output of the circuit. With reference to figure 5.33, now at each change of state, the Hall voltage V_H changes polarity. In other words, the Hall voltage becomes modulated (multiplied) by the values $+1$ and -1 at the clock frequency. On the other side, the offset voltage of the Hall plate V_{op} remains quasi-constant. The consequence is that V_{oA}, the input-referred offset of amplifier A_1, will become indistinguishable from the offset of the Hall plate V_{op}. At the output of the amplifier A_1 we have the voltage

$$V_1 = A_1(V_H + V_{op} + V_{oA}) \tag{5.55}$$

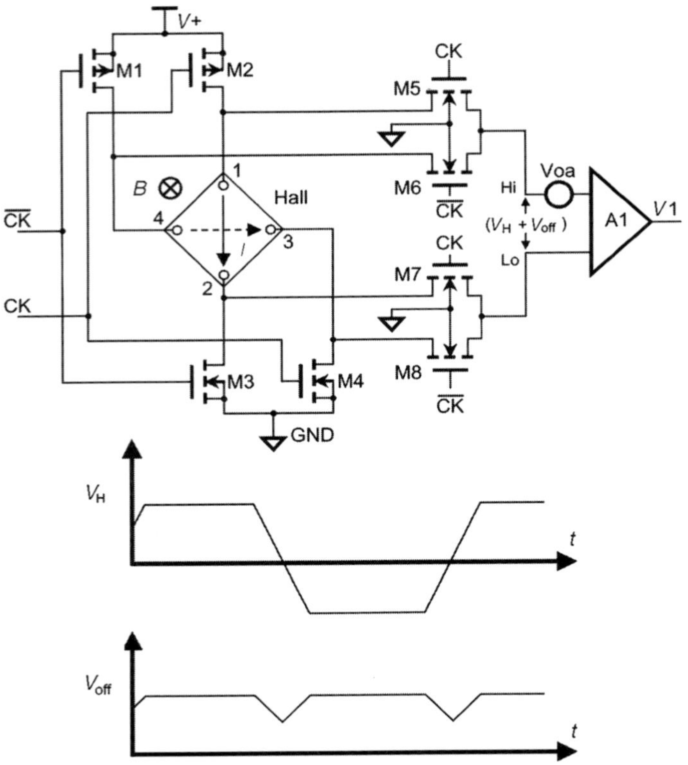

Figure 5.33. The front-end of a switched Hall plate system for the cancellation of the offsets and $1/f$ noises of both the Hall plate (Hall) and the amplifier (A_1). CK and $\overline{CK}$ are the clock signals, and M1–M8 are the switches implemented as complementary MOS transistors. When $\overline{CK}$ is 'High', the current I flows between 1 and 2, and when CK is 'High', the current flows between 4 and 3 (compare with figure 5.31). The lower part of the figure shows Hall voltage and the Hall plate offset signals versus time at the input of the amplifier (adapted from [72]).

of which the part $A_1 \times V_H$ is modulated (a.c.) and the part $A_1(V_{op} + V_{oA})$ is not modulated (quasi-d.c.).

The *demodulation* of the amplified Hall voltage and the cancelling of the sum of the two offset voltages $A_1(V_{op} + V_{oA})$ can be performed in one of the ways used in conventional chopper amplifier circuits. The principle is depicted in figure 5.34. The voltage V_1 at the output of the amplifier A_1 is now multiplied again by the values $+1$ and -1, synchronously with the first modulation. After a low-pass filtering, we have the amplified Hall voltage $A_1 \times V_H$, and virtually zero offset.

The maximum allowable Hall plate switching frequency is determined by the plate voltage settling time after each commutation transition. For

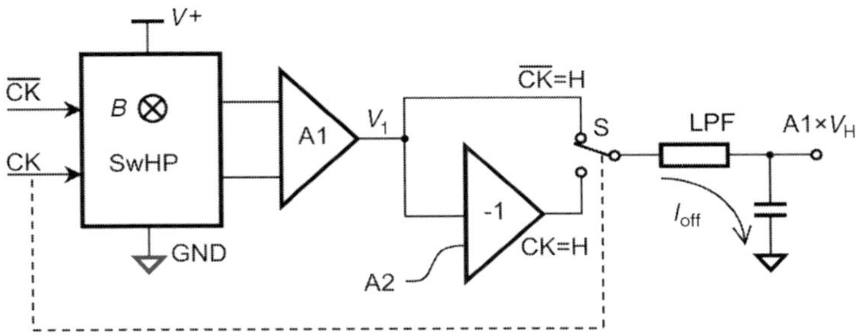

Figure 5.34. Schematic of the whole switched Hall plate system. The box SwHP represents the front-end of the Switched Hall Plate system shown in figure 5.33, without the amplifier A_1. The switch S takes the upper position when $\overline{\mathrm{CK}}$ is 'High', and the lower position when CK is 'High'. Since the switch S works synchronously with the switches in the box SwHP (M1–M8) in figure 5.33, the Hall voltage is demodulated, and the offset voltage is converted into an a.c. signal. In the low-pass filter LPF, the a.c. offset is cancelled (I_{off}), and the amplified Hall voltage is completely recovered.

example, with reference to figure 5.33, after the transition from the state 0° (current 4 → 3) to the state 90° (current 1 → 2), the voltage at the contact 4 has to decay from $V+$ to $V+/2$, and the voltage at the contact 3 has to increase from zero to $V+/2$. The corresponding time constant can be found experimentally or with the help of a model, such as the one shown in figure 5.26. For a cross-shaped integrated Hall plate of about 200 μm in length and with a resistance of 3 kΩ, the authors of [72] have found the maximum frequency of 200 kHz.

Because of the transition phenomena, charge injection of the switches, and other reasons, the cancelling of offset in a switched Hall plate system is not ideal: usually, a *residual offset* corresponding to a fraction of millitesla persists. One way to virtually cancel this residual offset is to apply the so-called nested chopper amplifier scheme [73].

5.6.5 Integrated Hall magnetic sensors

An ever-increasing proportion of commercially available Hall magnetic sensors comes in the form of integrated circuits. With a few exceptions, these are silicon monolithic integrated circuits. The integration allows the system approach to improve the sensor performance in spite of the mediocre characteristics of the basic silicon Hall cells. In other words, integrated Hall magnetic sensors are 'smart': they incorporate means for biasing, offset reduction, temperature compensation, signal amplification, and so on. The most important of this means were described in the preceding sections. In addition, practical integrated Hall sensors often incorporate additional functions, such as signal level discrimination, differential field measurement, various programming capabilities, and so on.

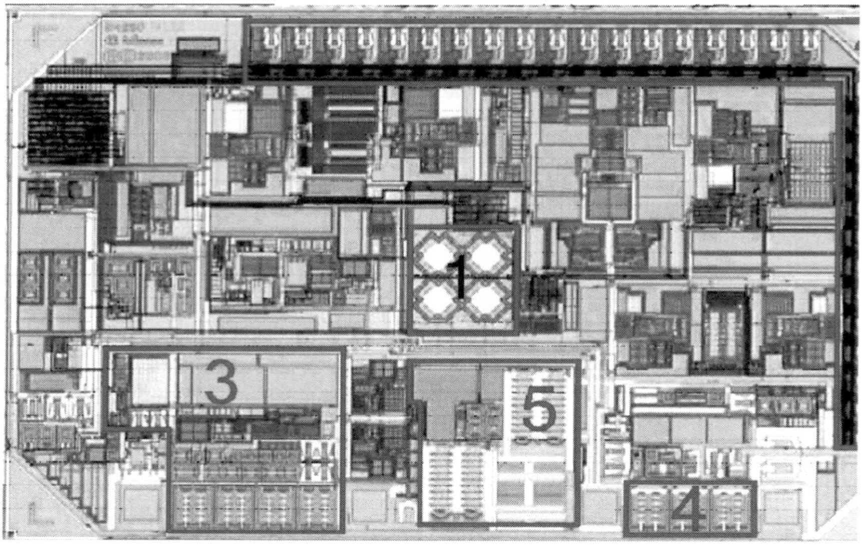

Figure 5.35. Photograph of a commercially available integrated Hall magnetic sensor chip (courtesy of Infineon Technologies). Legend: 1. Hall plate quadruple. 2. OTP (One Time Programmable) memory (30 bits). 3. Output amplifier. 4. Reverse polarity diode. 5. OBD circuit (On Board Diagnostic, detection of wire breakage). Four pads in the front (from left to right): TST (only for test purposes); OUT (linear output voltage and input port for calibration data); GND (ground); VDD (supply).

By way of *example*, figure 5.35 shows a photograph of an integrated Hall magnetic sensor chip.

In most of modern integrated Hall magnetic sensors, offset and $1/f$ noise are suppressed using the switching techniques described in §§5.6.3 and 5.6.4. This results in a residual offset (before eventual trimming) equivalent to about 1 mT and practically no $1/f$ noise. However, the offset and the $1/f$ noise sometimes reappear at the internal switching frequency (see figure 5.36). Such a '*switching noise*' in an integrated Hall sensor is usually equivalent to an input magnetic ripple of several 100 μT. This corresponds usually to an equivalent magnetic noise density of up to $1\,\mu T/\sqrt{Hz}$. One solution of the problem is to use an array of several Hall elements; then a noise density as low as $50\,nT/\sqrt{Hz}$ is achievable [74].

In table 5.2 we compare some characteristics of a few representative commercially-available integrated Hall linear magnetic field sensors [75]. Note the information on noise and *instability*, which are usually not given in the data sheets of vendors. We shall discuss the instability problems related to the packaging stress in §5.8.

One promising way to cope with the problem of the sensitivity dependence on stress, temperature and ageing is *self-calibration* using an

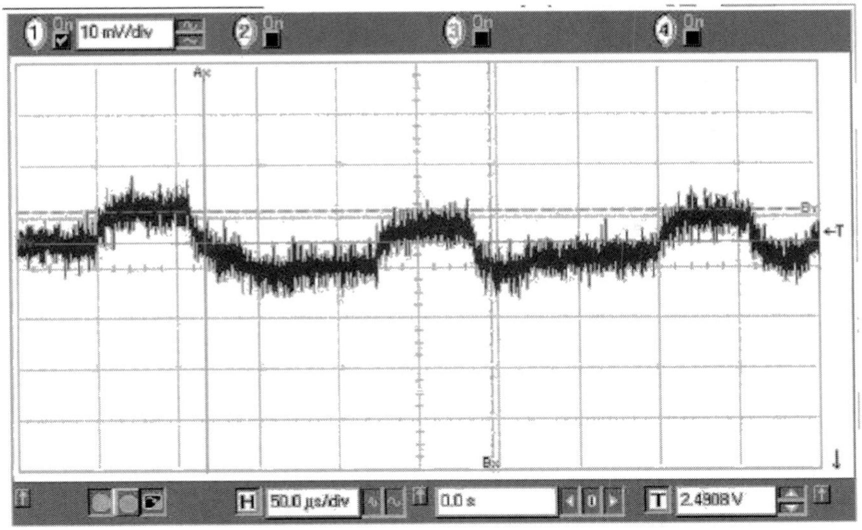

Figure 5.36. The output voltage of a commercially available CMOS integrated Hall magnetic field sensor based on the spinning-current and chopping principles. The peak-to-peak output voltage noise corresponds to an input magnetic field noise of about 170 µT [75].

Table 5.2. Characteristics of some commercially available integrated Hall magnetic field sensors

	Technology			
Characteristics	A CMOS: spinning current; programmable offset; gain; temperature coefficient	B Bipolar: quad Hall sensing element; laser trimmed thin film resistors	C Bi CMOS	Units
Bandwith	1.3	NA	23	kHz
Sensitivity (S)	52	31.25	13	V/T
$\Delta S/S$ [a]	−1.2 [b]	−1.66	−0.4	%
Offset (B_{offset})	−1.26 [b]	−0.73	−3.1	mT
ΔB_{offset} [c]	0.01 [b]	0.13	−0.72	mT
White noise	1540	64	30.15	$nT\sqrt{Hz}$
Ripple noise	0.460	1.29	1.0	mT_{pp}

[a] Thermal hysteresis: the relative change in sensitivity at room temperature, after performing a slow temperature cycle up to 125 °C and down to room temperature.
[b] At sensitivity set by programming to 10 V/T.
[c] Change in equivalent offset at room temperature, after performing a slow temperature cycle up to 125 °C and down to room temperature.

integrated coil. Proposed many years ago, this concept has become feasible only with the advent of the submicrometre IC technology. This is because the self-calibration is efficient only if the calibrating coil produces a sufficiently-high test magnetic field; and the smaller the coil, the higher the test field for a given coil current. In an example, a three-turn integrated coil provides a calibrating magnetic field of 0.15 mT per 1 mA of the coil current [74]. In a miniaturized combination integrated coil–Hall plate we obtained 0.39 mT/mA [76]. By using this device for the auto-calibration of a Hall magnetic sensor system, a temperature cross-sensitivity as low as 30 ppm/°C was obtained [77].

5.7 Hall sensors with magnetic concentrators

If a long ferromagnetic object is placed in a magnetic field parallel with the long axis of the object, the ferromagnetic object tends to collect the magnetic field lines in it (see figure 5.37). We can easily understand this fact with the help of the following analogy: the behaviour of a high-permeability soft ferromagnetic object placed in low-permeability milieu and exposed to a magnetization field $\boldsymbol{H}_0$ is analogous to the behaviour of a high-conductivity metallic object of the same shape placed in a high-resistivity milieu and exposed to an electrical field $\boldsymbol{E}_0$. As the current uses the lowest resistance paths and concentrates in the high-conductivity object, so also the magnetic flux uses the lowest reluctance paths and concentrates in the high-permeability object.

We say that the ferromagnetic object shown in figure 5.37 operates as a magnetic flux concentrator. We shall refer to such a device in short as a magnetic concentrator (MC). A Hall device placed close to an end of a magnetic concentrator shall 'see' a stronger magnetic field than elsewhere. This means that a magnetic concentrator can provide magnetic gain to a

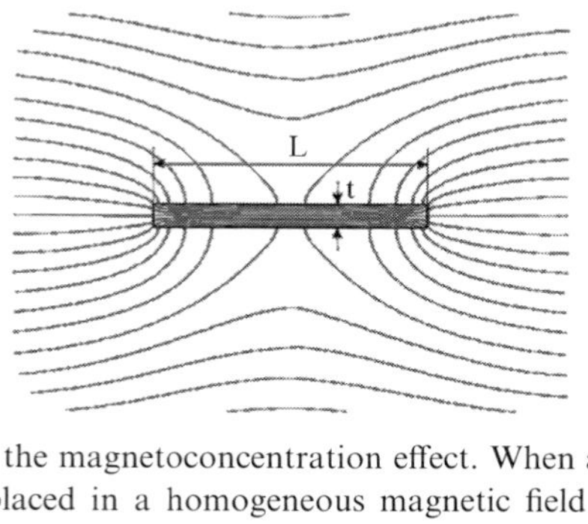

Figure 5.37. Illustrating the magnetoconcentration effect. When a high-permeability soft ferromagnetic plate is placed in a homogeneous magnetic field parallel with the plate plane, the plate concentrates the magnetic flux in itself. The distribution shown of the magnetic field ($\boldsymbol{B}$) lines is found by 2D numerical simulation.

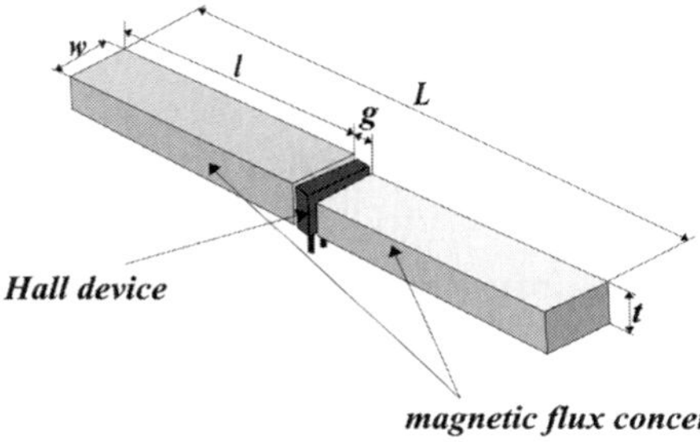

Figure 5.38. Classical combination of a Hall plate and a magnetic flux concentrator: the Hall plate is fitted in the narrow air gap between the two ferromagnetic rods. Since this magnetic concentrator (MC) consists of two equal parts, we call it a twin MC [78].

Hall sensor. A *magnetic gain* is a very useful property of a magnetic sensor: it improves its magnetic resolution. But a magnetic concentrator may also add offset and deteriorate linearity of a magnetic sensor. We shall now review the basic properties of magnetic concentrators and look at a few examples of their practical applications.

5.7.1 Basic characteristics of magnetic concentrators

Magnetic gain

We define the magnetic gain of an MC as the ratio

$$G_{\mathrm{M}} = B_{\mathrm{HE}}/B_0. \tag{5.56}$$

where B_{HE} is the concentrated ('amplified') magnetic induction 'seen' by the adjoined Hall element (HE) and B_0 is the magnetic induction component parallel with the axis of the structure, far away from the MC.

The magnetic gain of a given MC structure can be found using numerical simulations. As an example, let us consider the magnetic gain of the so-called twin MC shown in figure 5.38. Figure 5.39 shows the result of the numerical simulation of the magnetic field in and around a planar twin MC.

The magnetic gain is about 10: the density of the magnetic field lines is 10 times higher in the middle of the gap than that far away from the magnetic concentrator.

A rough estimation of the magnetic gain of a twin magnetic concentrator can be done semi-intuitively in the following way. Let us consider the asymptotic case of infinitely narrow gap, $g/t \to 0$. The distribution of the magnetic field lines in such a twin MC approaches that of a single magnetic concentrator, shown in figure 5.37. By inspecting figure 5.37, we find that the concentrator 'soaks' in itself the flux from a flux tube (far away from the MC) of a width approximately equal to the length of the

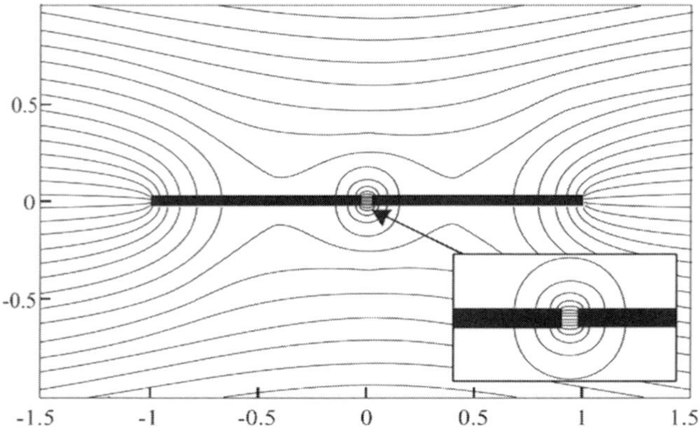

Figure 5.39. The result of a two-dimensional simulation of a twin planar MC, such as that from figure 5.38, with $l = 16t$, $g = t$ and $W \gg t$. The MC is introduced into a homogenous magnetic field parallel to the MC plane. The lines represent the ***B*** field (reprinted from [78]).

concentrator. Accordingly, we can write

$$G_{\mathrm{M}} \simeq L/t. \tag{5.57}$$

for the magnetic gain which is valid for a planar twin MC made of a high-permeability soft ferromagnetic material ($\mu_{\mathrm{r}} \to \infty$) with $L \gg t$, $W \gg t$ and $g/t \to 0$.

For larger gaps the gain becomes smaller than that predicted by (5.57). The reason is that then a good portion of the magnetic flux spreads out of the gap, as seen in figure 5.38. We shall look at a particular example of an MC with relatively large gap in §5.7.3.

Our notion of magnetic gain is, of course, related to the classical notions of the *demagnetization factor* and *apparent permeability*, used in textbooks on magnetic materials. By definition, the magnetic field (flux density) in an open (or shared) magnetic circuit is given by [79]

$$B_{\mathrm{i}} = \mu^{*}\mu_0 H_0. \tag{5.58}$$

where B_{i} means the magnetic flux density in the open magnetic circuit (B_{i} is supposedly constant), μ^{*} is the apparent relative permeability of the circuit, and H_0 is the average excitation magnetic field in the region. In our case, μ^{*} is the apparent relative permeability of MC, the notion of H_0 is the same, and B_{i} means the magnetic flux density in the infinitively narrow air gap in the middle of the magnetic concentrator. The apparent permeability is, in turn, related to the demagnetization factor N:

$$\mu^{*} \simeq 1/N. \tag{5.59}$$

This approximation follows from equation (2.16) of [80] if $\mu_r \gg \mu^*$. Therefore, in the above case of a flat twin high-permeability MC with a very narrow air gap,

$$G_M \simeq \mu^* \simeq 1/N \simeq L/t. \tag{5.60}$$

The standard approximation for the demagnetization factor of a thin plate such as that shown in figure 5.37, with $\mu_r \to \infty$ and $L \gg t$, is in our notation given by

$$N \simeq 2t/\pi L. \tag{5.61}$$

This corroborates our estimation for the magnetic gain (5.60).

Saturation field

By inspecting figure 3.37 we notice that the magnetic field lines have the highest density in the magnetic concentrator in the middle of the concentrator. If we keep increasing the external magnetic field, the flux density at the middle of the concentrator shall eventually reach the saturation induction B_{satM} of the MC material. Then our magnetic sensor shows a strong decrease in its magnetic sensitivity, i.e. the sensor response becomes non-linear. The external magnetic field at the onset of the saturation is given by

$$B_{0sat} \simeq B_{satM}/G_M \simeq B_{satM}t/L. \tag{5.62}$$

which is valid for a flat twin MC with $\mu_r \gg L/t$, $L \gg t$, $W \gg t$ and $g/t \to 0$. As the gap increases, the saturation field B_{0sat} first also increases, but then saturates at the value

$$B_{0sat} \simeq B_{satM}t/l. \tag{5.63}$$

This is valid for a planar twin MC with $\mu_r \gg l/t$, $l \gg t$, $W \gg t$ and $g/t \gg 1$. Note that in this case each part of a twin MC behaves much like a single MC shown in figure 5.37.

Remanent field

After exposure to a strong magnetic field, a magnetic concentrator may stay quasi-permanently magnetized. The maximal remanent magnetic field of a strongly shared magnetic circuit is given by [79] and [80]

$$B_r^* \simeq \mu^* \mu_0 H_c. \tag{5.64}$$

where H_c denotes the coercivity of the MC material and μ^* is given by (5.59). The remanent field produces an output signal of the associated Hall device, which is perceived as an additional offset, dependent on the magnetic excitation history of the sensor. This effect is called *perming* (see §5.1.7). In order to keep perming low, we have to use a very soft (low H_c) ferromagnetic material for the fabrication of MCs.

Magnetic noise

In principle, a magnetic concentrator adds both thermal and Barkhausen noise [81] to an adjacent Hall sensor. But this contribution is negligible even in the case of a much higher-resolution magnetic sensor, such as a flux-gate sensor [82].

5.7.2 Discrete and hybrid magnetic concentrators

Probably the first report on magnetic concentrators was published back in 1955 [83]. In order to amplify the magnetic field 'seen' by an InSb Hall plate, the authors put the Hall plate in the air gap between two long ferromagnetic rods. In this way they considerably increased the effective sensitivity and the resolution of the Hall element and could measure quasi-static magnetic fields down to milli-gauss range. Similar macroscopic systems Hall sensor–magnetic concentrators, made of discrete components, were investigated also by other researchers [84]–[86]. The best reported detectivity limit of a combination Hall sensor–magnetic concentrators is as low as 10 pT [85].

In [85] we find also a report on an attempt to incorporate ferromagnetic material into the package of a Hall device. This idea is also presented in [87], and is now used in several commercially available Hall magnetic sensors.

Figure 5.40 shows a *thin-film InSb Hall element*, of the kind described in §5.3.1, sandwiched between two ferromagnetic pieces [88]. The device is made by depositing a 0.8 μm thick layer of polycrystalline InSb on mica, then peeling the InSb layer and transferring it to a ferrite substrate. The upper ferrite chip is placed during the assembling phase of the production. The absolute magnetic sensitivity of the device is about 4 V/T at the supply voltage of 1 V. The magnetic concentrators saturates at a magnetic field of about 0.1 T.

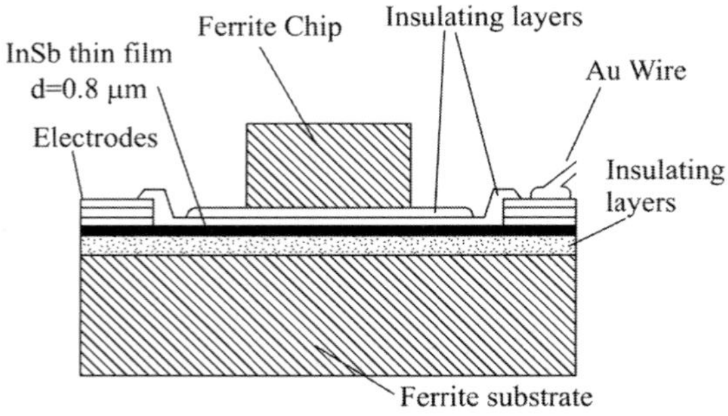

Figure 5.40. Structure of a commercially available high-mobility thin-film Hall element combined with a magnetic concentrator [88, 89].

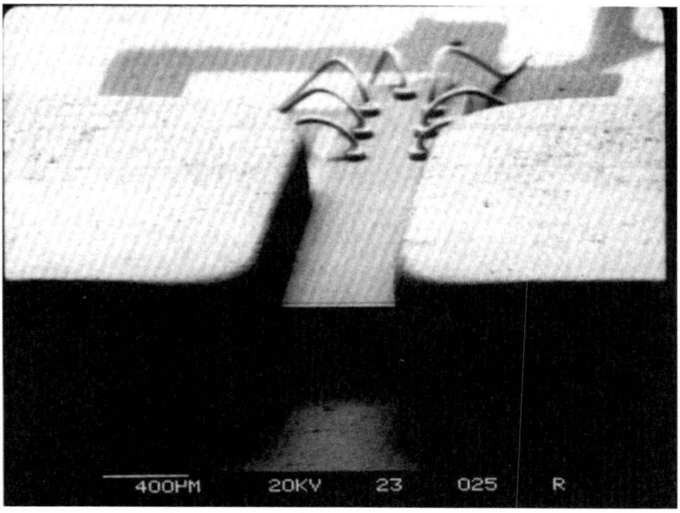

Figure 5.41. The vertical Hall device (the chip in the centre of the picture, with the bonds) assembled together with a twin magnetic concentrator (the two thicker chips at the sides). The device senses the horizontal magnetic field component, that is the magnetic field collinear with the axes of the magnetic concentrator [90].

Figure 5.41 shows another hybrid micro-system Hall sensor–magnetic concentrator [90], which became a part of a commercially available product. The Hall sensor is the discrete vertical Hall device, described in §5.4.2. Since it has the feature of being sensitive to a magnetic field parallel to the chip surface, the *vertical Hall device* lends itself to the application of a convenient magnetic flux concentrator: the sensor chip and both ferromagnetic pieces can be assembled in a plane, on a common substrate.

5.7.3 Integrated magnetic concentrators

The magnetic concentrators (MC) described so far are not compatible with integrated circuit technologies. In this section we describe another structure of the combination MC/Hall device, which is suitable for an integration. We shall refer to such Integrated Magnetic Concentrators in short as IMC. Like a conventional magnetic concentrator, an IMC works as a passive magnetic amplifier. In addition, an IMC changes the sensitivity direction of the sensor: whereas a conventional Hall element responds to a magnetic field perpendicular to the chip surface, a Hall element combined with an IMC responds to a magnetic field parallel to the chip surface. This opens the way for a number of interesting applications (see §5.9).

We shall describe two basic IMC structures: one with a twin magnetic flux concentrator and the other with a single magnetic flux concentrator.

Twin integrated magnetic concentrator

The idea of the integrated twin magnetic flux concentrator [91] is illustrated in figure 5.42. Obviously, the idea is inspired by the conventional combination Hall element/rod-like MC (see figure 5.38). The key difference is, however, that now the Hall element is placed not inside but near the air gap and under the concentrator. In this way, one can use conventional planar integrated Hall elements; and one can define by photolithography the shape of IMCs and the mutual positions in the system IMC–Hall elements.

By inspecting figures 5.39 and 5.42, we notice that the stray magnetic field near the gap has a vertical component. Here the IMC converts the external magnetic field, which is parallel to the chip surface, into a magnetic field perpendicular to the chip surface. The perpendicular component of the magnetic field is the strongest near the gap. There we place the Hall elements. We can use two Hall elements because two equivalent places, one under each part of the concentrator, are available. So we can increase the signal to noise ratio. The Hall plates below the different concentrator halves see the useful magnetic field in opposite direction. So the system is insensitive to an external field component perpendicular to the chip surface.

We can estimate the *magnetic gain of a twin IMC* as follows [78].

We assume that IMC material is of a very high permeability, $\mu_r \to \infty$. Then, with the aid of the above analogy with electricity (see the beginning of §5.7), we see that each part on an IMC is equipotential. Let us put the magnetic scalar potential difference (i.e. the equivalent excitation current I_{ex}) acting between the two adjacent parts of a twin MC in the form

$$I_{ex} = KH_0(2l + g). \tag{5.65}$$

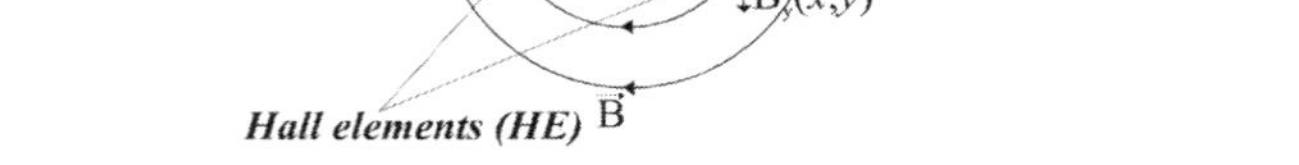

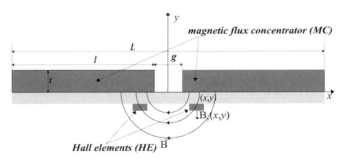

Figure 5.42. Schema of an integrated combination Hall magnetic sensor/twin magnetic flux concentrator. Whereas in the case of a conventional twin MC (figure 5.38) the Hall device is placed in the air gap between the two ferromagnetic rods, here two Hall elements are placed under the peripheries of two ferromagnetic plates. The semi-circular lines below the gap approximate the corresponding magnetic field lines shown in figure 5.39 [92].

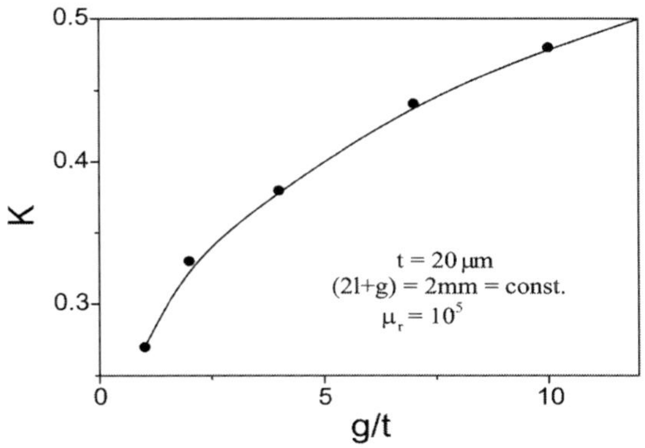

Figure 5.43. The values of the factor K (5.67) deduced from two-dimensional simulation of twin planar magnetic concentrator, such as that of figure 5.39 (adapted from [78]).

Here l and g are the dimensions defined in figure 5.42, H_0 is the magnitude of the excitation magnetic field in the region before introducing the MC,

$$H_0 = B_0/\mu_0 \tag{5.66}$$

and K is a numerical factor defined by

$$K = \frac{\int \boldsymbol{H}_g \, d\boldsymbol{l}}{H_0(2l + g)}. \tag{5.67}$$

Here $\boldsymbol{H}_g$ is the excitation magnetic field vector in the gap and the integration is done over the gap. The factor K tells us which part of the theoretically available equivalent excitation current along the whole length of the MC (which is equal to $H_0(2l + g)$) we can really use around the air gap. Figure 5.43 gives the numerical values of the factor K.

By inspecting figure 5.39 we see that the magnetic field lines below the MC and close to the gap have approximately a circular form. This is a consequence of the boundary conditions. With reference to figure 5.42 we can calculate the perpendicular component of the magnetic field $B_y(x, y)$ seen by a planar Hall element placed at a position (X, Y). Substituting this and (5.66) into (5.56), we find now the magnetic gain of a twin MC for a Hall element placed at (X, Y):

$$G_{XY} = \frac{K(2l + g)X}{\pi(X^2 + Y^2)} \tag{5.68}$$

for $X > g/2$, $g \simeq t$ and $\mu_r \gg 1$.

With an optimal layout of IMC, for given t, L and g, one can increase K and so the magnetic gain G_{XY} up to a factor of 2 [92]. The realistic values of G_{XY} for an integrated sensor are between 5 and 10. But adding external MCs, this can be easily increased to about 100.

The *saturation field of a twin IMC* has a value between those given by equations (5.62) and (5.63):

$$B_{\mathrm{satM}} t/L < B_{0\mathrm{sat}} < B_{\mathrm{satM}} t/L. \tag{5.69}$$

More precisely, the external magnetic field induction at the onset of the saturation is given by [78]

$$B_{0\mathrm{sat}} \simeq B_{\mathrm{Msat}} g[KL(1 + (2/\pi)(g/t)\ln(g))]^{-1} \tag{5.70}$$

for $g \simeq t$ and $\mu_r \gg L/t$.

For a sensor with the IMC of the total length $L = 2l + g = 2\,\mathrm{mm}$, $t = g = 20\,\mu\mathrm{m}$ and $B_{\mathrm{satM}} = 1\,\mathrm{T}$, we obtain $B_{0\mathrm{sat}} \simeq 10\,\mathrm{mT}$. By choosing an optimal form of the MC for given t, L and g, one can increase the saturation field up to a factor of 2 [92].

Application of twin IMC: a high-sensitivity single axis integrated Hall magnetic sensor

By integrating twin magnetic concentrators with Hall elements, we obtain a magnetic sensor system with the following combination of features: existence of a magnetic gain, which brings a higher magnetic sensitivity, lower equivalent magnetic offset, and lower equivalent magnetic noise than those in conventional integrated Hall sensors; and sensitivity to a magnetic field parallel with the chip surface, much as in the case of ferromagnetic magneto-resistance sensors (AMR and GMR [80]). Figure 5.44 shows one *example* of such an integrated magnetic sensor [93].

The electronic part of the magnetic sensor shown in figure 5.44 is realized in CMOS technology as a smart ASIC. The circuit includes biasing, amplification, cancelling offset, and temperature stabilization functions. The magnetic sensitivity, temperature coefficient of sensitivity, and residual offset of the sensor are programmable using the Zener-zapping technique.

Single integrated magnetic concentrator

A single integrated magnetic flux concentrator is shown in figure 5.45 [94, 95]. Here again, the concentrator converts locally a magnetic field parallel to the chip surface into a magnetic field perpendicular to the chip surface. The strongest perpendicular component of the magnetic field appears under the concentrator extremities. The Hall elements are placed under the concentrator extremities and 'see' this perpendicular component of the magnetic field much as in the case of the twin magnetic concentrator.

In order to estimate the *magnetic gain* of a single IMC, we model a disk-like magnetic concentrator with an oblate ellipsoid. For a very flat oblate ellipsoid magnetized parallel to a long axis, the demagnetization factor [80]

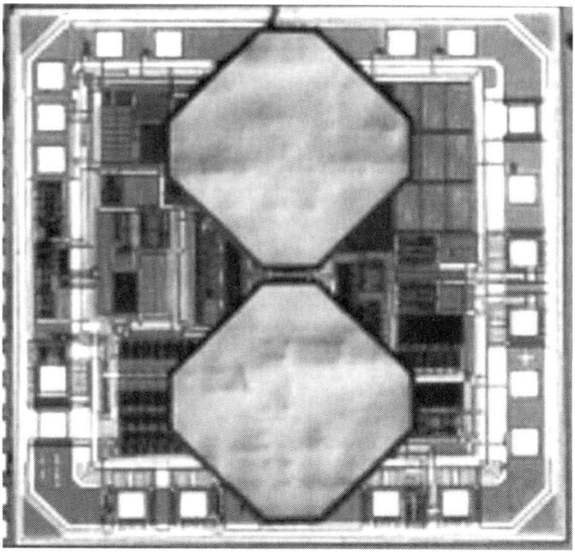

Figure 5.44. Photograph of a twin IMC Hall magnetic sensor chip (Sentron type 1SA-1M). The semiconductor part is a CMOS Hall ASIC. The twin magnetic concentrator has the form of the two octagons. The sensor responds to a magnetic field parallel to the in-plane symmetry axis through the two octagons. The magnetic sensitivity is programmable and can be as high as 300 V/T and the equivalent offset field is 0.03 mT. (Courtesy of SENTRON AG, Zug, Switzerland.)

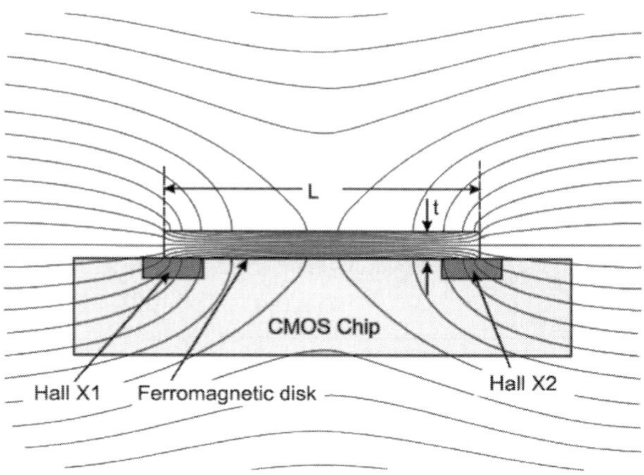

Figure 5.45. Cross-section of a Hall chip combined with a single magnetic flux concentrator. The magnetic concentrator usually has the form of a disc [95]. Compare with figure 5.37.

in our notation is

$$N \simeq \frac{\pi t}{4L}\left(1 - \frac{4t}{\pi L}\right) \quad \text{for } L/t \gg 1. \tag{5.71}$$

In a very high-permeability and not very elongated structure, the internal magnetic induction is

$$B_i \simeq \frac{\mu_0}{N} H_0. \tag{5.72}$$

Assuming that the external magnetic induction 'seen' by the Hall elements $B_{\mathrm{HE}} \simeq B_i$, we obtain for the magnetic gain

$$G_{\mathrm{M}} \simeq \frac{1}{N} \simeq \frac{4}{\pi(t/L) - 4(t^2/L^2)} \quad \text{for } \mu_{\mathrm{r}} \to \infty. \tag{5.73}$$

For a disc with $L/t = 10$, we calculate $G_{\mathrm{M}} \simeq 14.5$. Numerical simulations gave [95] a similar value for the total magnetic gain, but only about seven for the perpendicular field component.

The *saturation field of a single IMC* is given by equation (5.62). For a disk-like IMC with $L/t = 10$ and $B_{\mathrm{sat}} = 1\,\mathrm{T}$, this gives $B_{0\mathrm{sat}} \simeq 69\,\mathrm{mT}$.

Application of the single IMC: two-axis Hall magnetic sensor

An important application of the single integrated magnetic concentrator (IMC) described above is to build a two-axis magnetic sensor using conventional planar integrated Hall plates. Figure 5.46 shows the structure of a Hall sensor for two in plane components of a magnetic field parallel to the x and y axes [95]. The concentrator has the form of a thin disk. The Hall elements are positioned under the periphery of the disk. At least one

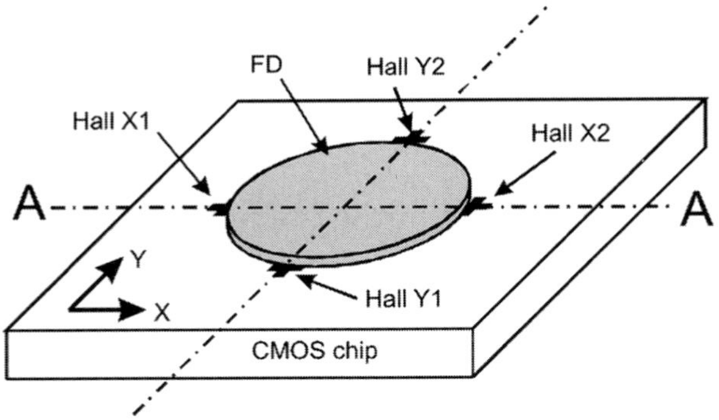

Figure 5.46. A schematic view of a two-axis Hall magnetic sensor. It consists of an integrated combination of a soft ferromagnetic disk (FD) and a few conventional Hall elements (Hall X1, X2, Y1, Y2) placed under the periphery of the disk. The role of the ferromagnetic disk is to convert an in plane magnetic field into a perpendicular magnetic field, as illustrated in figure 5.45 [95].

Hall element per sensitive axis is needed; but preferably, for each sensitive axis, we use two Hall elements placed at the two opposite ends of the disk diameter parallel to the corresponding axis. For example, for the component of a magnetic field collinear with axis x in figure 5.46, we use the Hall elements Hall X1 and Hall X2.

With reference to figure 5.45, the Hall element X1 'sees' a positive Z magnetic field component, whereas the Hall element X2 'sees' a negative Z magnetic field component. If we subtract the output voltages of the two Hall elements, we shall obtain a signal proportional to the mean value of the perpendicular components of the magnetic field at the two extremities of the ferromagnetic disk. Assuming a constant permeability of the disk and no saturation, the sensor output signal is proportional to the input (parallel) magnetic field.

According to (5.72) and (5.71), if the thickness t of a magnetic flux concentrator disk is much smaller than its diameter, then the magnetic field 'seen' by the Hall elements is given by

$$B_{\mathrm{H}} = CB_0L. \tag{5.74}$$

Here C denotes a numerical coefficient; B_0 is the input magnetic field induction, parallel to the chip plane; and L is the length of the magnetic flux concentrator collinear with the vector $\boldsymbol{B}_0$. In other words, for a given external planar magnetic field B_0, the Hall voltage is proportional to the length of the magnetic flux concentrator.

Consider now what happens if the magnetic field acting on the sensor shown in figure 5.45 rotates in the chip plane. Note first that, due to the rotational symmetry of the device, the form of the device 'seen' by the magnetic field does not change with the rotation. Therefore, if the disk is made of a very 'soft' ferromagnetic material, then the whole distribution of the magnetic field within and around the disk shall stay fixed and shall simply rotate synchronously with the rotation of the input magnetic field.

We can estimate the magnitude of the magnetic field 'seen' by a Hall element as the magnetic field rotates in the following way. Let us approximate again the ferromagnetic disk by an oblate ellipsoid of infinitely high permeability. We know that the magnetic field is homogeneous inside such an ellipsoid; and that the magnetic field outside of the ellipsoid is perpendicular to its surface – see figure 5.47. The density of the magnetic field lines at the periphery of the disk is then

$$B_\alpha = B_0 \cos\alpha. \tag{5.75}$$

By inserting here (5.75), we obtain

$$B_{\mathrm{H}} = CDB_0 \cos\alpha. \tag{5.76}$$

Now the difference of the Hall voltages of the two Hall elements Hall X1 and Hall X2 is given by

$$V_{\mathrm{H}X} = 2SCDB_0 \cos\alpha \tag{5.77}$$

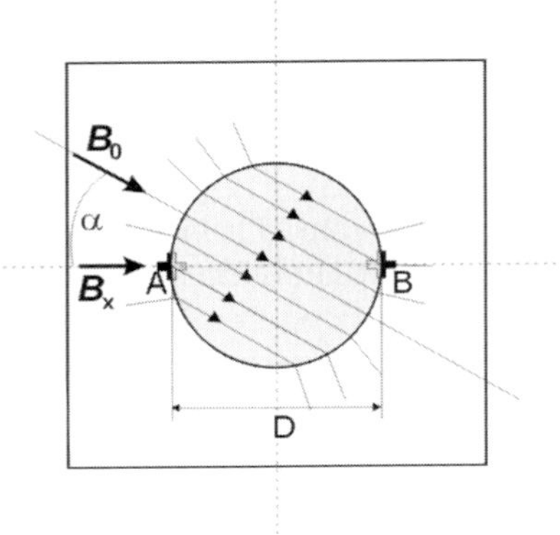

Figure 5.47. For a magnetic field vector rotated in the plane by an angle α with respect to the axis of the two Hall elements A and B, the magnetic field 'seen' by the Hall elements becomes $B_X = B_0 \cos\alpha$. Consequently the Hall voltages are also proportional with $\cos\alpha$.

and the difference of the Hall voltages of the two Hall elements Hall Y1 and Hall Y2 is given by

$$V_{HY} = 2SCDB_0 \sin\alpha. \tag{5.78}$$

Here S denotes the magnetic field sensitivity of the Hall elements.

From the two harmonic signals (5.77) and (5.78) phase-shifted by $\pi/2$, we can easily retrieve the information about the angular position of the magnetic field vector component parallel to the chip surface.

Figure 5.48 shows an integrated two-axis magnetic sensor based on this principle. The electronic part of the sensor is realized in CMOS technology as

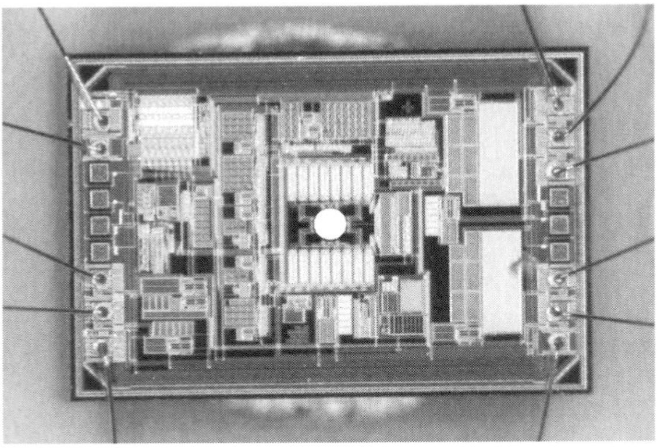

Figure 5.48. Photograph of the two-axis single IMC Hall sensor. The dimensions of the silicon chip are 2.7 mm × 1.9 mm. The white circle in the centre of the chip is the disk-shaped IMC of 200 μm in diameter. (Courtesy of SENTRON AG, Zug, Switzerland.)

a smart ASIC. The circuit includes biasing, amplification, offset cancellation and temperature stabilization functions. The residual offset of the sensor is programmable using the Zener-zapping technique [93].

When used as a magnetic angular position sensor, this device provides, without any calibration, accuracy better than 0.5° over the measurement range of 360°. Some applications of this two-axis Hall sensor are described in §5.9.2.

5.8 Packaging Hall magnetic sensors

Like other semiconductor devices, a Hall magnetic sensor needs a package. The package protects the chip and enables electrical contacts to be made between the chip and external circuits. Figure 5.49 illustrates a packaged Hall sensor chip.

The requirements of a good package for a magnetic sensor are more difficult to fulfil than those for conventional non-magnetic devices. Indeed, the conditions concerning protection (mechanical strength, hermeticity), electrical contacts and isolation are identical, irrespective of whether the device is 'magnetic' or not. In addition, unlike in figure 5.49, a magnetic sensor often requires a special form of case [96], such as very small outline [97]; the application of some commonly used materials is excluded because of their ferromagnetic properties [98]; and silicon Hall devices are stress sensitive and require special assembling techniques [96, 31].

5.8.1 Packaging stress

When a semiconductor chip is bonded on to a substrate or encapsulated in a plastic case, a mismatch in the thermal expansion coefficients of the chip and of its surroundings produces a mechanical stress in the chip. This stress is difficult to control, and it leads to undesirable piezo-effects [99, 100]. In a Hall device, a stress affects both the resistance and the Hall coefficient.

The packaging stress is the cause of a major problem with the state-of-the-art Hall magnetic sensors integrated in silicon: a *drift in their characteristics*. The instability is the consequence of a variable stress in the sensor chip

Figure 5.49. The two-axis Hall sensor chip, shown in figure 5.48, is packaged in a standard plastic SOIC-8 housing. The sensor responds to the two components of a magnetic sensor parallel to the chip surface.

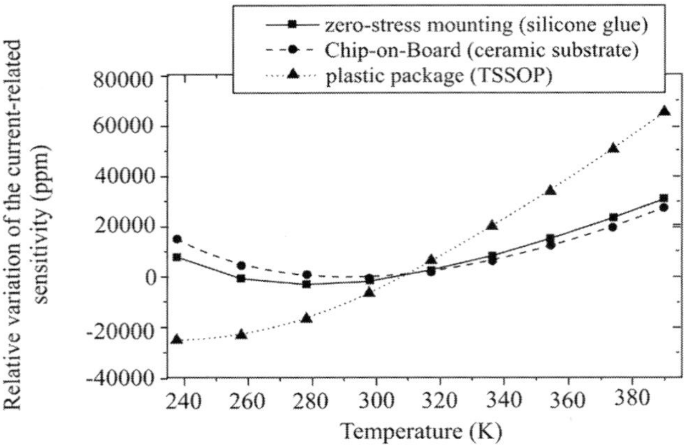

Figure 5.50. Measured variation of the current-related sensitivity of a CMOS Hall element as a function of temperature for three assembling and encapsulation cases. The variation is expressed in percentages relative to the value at 308 K.

created by the package. A variable stress in the sensor chip transforms itself, via the piezo-resistance and the piezo-Hall effects, into a drift in the sensor parameters. Whereas the disastrous influence of these effects on the offset is strongly reduced by the current spinning and the chopping [101], the influence on the sensitivity and its temperature dependence stays. The effect is particularly serious in plastic-encapsulated integrated Hall magnetic sensors. This is due to the visco-elastic phenomenon in the moulding compounds. Figures 5.50 and 5.51 illustrate the problem [75].

Briefly, we found that the encapsulation of an integrated Hall device in a plastic package may drastically change the *temperature coefficient* of its

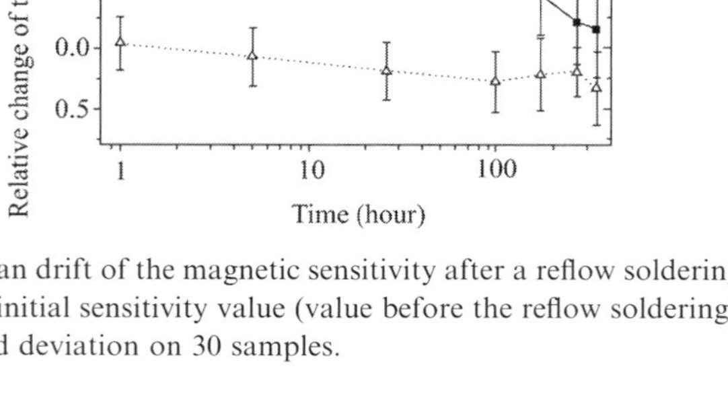

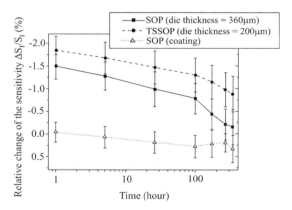

Figure 5.51. Mean drift of the magnetic sensitivity after a reflow soldering cycle. The drift is related to the initial sensitivity value (value before the reflow soldering). The error bars give the standard deviation on 30 samples.

magnetic sensitivity [102] (see figure 5.50). This makes difficult the thermal compensation. We also found that a thermal shock may produce a non-permanent change in sensitivity of an integrated Hall sensor as high as 2% [103] (see figure 5.51). The lines $\Delta S/S$ and ΔB_{offset} in table 5.2 (§5.6.5) illustrate the same phenomena. See also §6.3. The importance of these problems becomes particularly clear in view of the current need of the automotive industry for magnetic sensors working well in the temperature domain between −40 °C and 150 °C (and perhaps up to 175 °C).

5.8.2 Reducing packaging stress

The influence of encapsulation stress on a Hall device can be minimized by choosing the proper orientation of the device with respect to the crystal lattice of the chip material [31, 104].

Obviously, the smaller the difference between the thermal expansion coefficient (TEC) of the Hall chip and those of the substrate and the encapsulation material, the smaller the packaging stress will be.

We developed an assembling method, which allows the virtual elimination of the substrate–TEC-mismatch stress in a part of a sensor chip [105]. The idea is illustrated in figure 5.52. When the chip is bonded on to the substrate, the section of the chip that contains stress-sensitive elements is left free. Thus the stress-sensitive section is mechanically isolated from the substrate. Moreover, with the aid of micromachining, the sensitive portion of the chip can also be mechanically isolated from the rest of the chip. A reduction in the encapsulation stress of more than an order of magnitude has been achieved in this way.

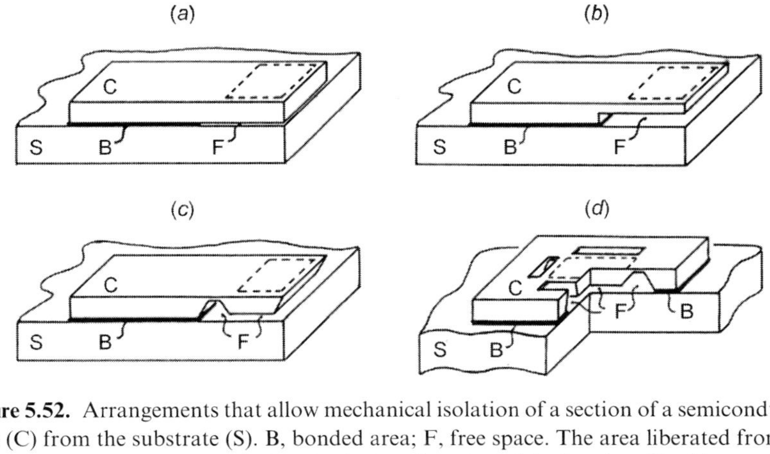

Figure 5.52. Arrangements that allow mechanical isolation of a section of a semiconductor chip (C) from the substrate (S). B, bonded area; F, free space. The area liberated from the substrate-related stress is marked by the broken line. A Hall device placed in this area is not affected by the substrate stress (reprinted from [105]).

Application of this technique allows the fabrication of silicon Hall probes with a long-term sensitivity-instability as low as 0.01% (see §5.9.1).

5.9 Applications of Hall magnetic sensors

A Hall magnetic sensor can be directly used to measure a magnetic field, for example, in order to characterize a current-carrying coil. Alternatively, the magnetic signal to be measured may stem from a modulation of a magnetic field carried by a magnet associated with the sensor system itself. The modulation of the magnetic field comes about, for example, by varying the distance between the Hall sensor and the magnet. In this case, the magnetic signal is used just as a means to measure a non-magnetic quantity. Accordingly, a Hall magnetic sensor may be applied as a key component in a magnetometer or in a contact-less sensor for linear position, angular position, velocity, rotation speed, electrical current, electrical power, etc [3–6, 106]. We present below a few representative examples of applications of Hall magnetic sensors. More information on the most important applications can be found in the literature of the vendors of Hall magnetic sensors. See, for example, [107–109].

5.9.1 Measuring magnetic fields

For testing and characterizing magnetic fields produced by electro- and permanent magnets, in the range of about 10^{-4} T to a few tens of tesla, Hall sensors are typically used. For example, they are used to test the electromagnets applied in particle accelerators, nuclear magnetic resonance (NMR) imaging systems, NMR spectroscopy systems, and permanent magnets used in magnetic position sensors, electro-motors, and so on. Miniaturized Hall devices (§5.5) are particularly suitable for high spatial resolution magnetic measurement. Applications of micro-Hall devices include precision magnetic field mapping [110], characterization of ferromagnetic nano-particles [111], detection of magnetic micro-beads [112], detection of NMR and ESR (electron spin resonance) [113].

According to the usual terminology in the magnetic field measurement community, instruments suitable for measuring a magnetic field B are called *magnetic field transducers* and *teslameters* (or *gaussmeters*). A Hall magnetic field transducer is an instrument that provides at its output a convenient-level voltage, proportional to the measured magnetic field. In addition, a teslameter usually contains a display, which shows the numerical value of the measured magnetic field. Such an instrument may measure 1, 2 or 3 components of a magnetic field; then we say that this is a 1-, 2- or 3-axis transducer or teslameter.

The market for teslameters and transducers is small and very diversified. This forces manufacturers to adopt a design concept with a minimum of

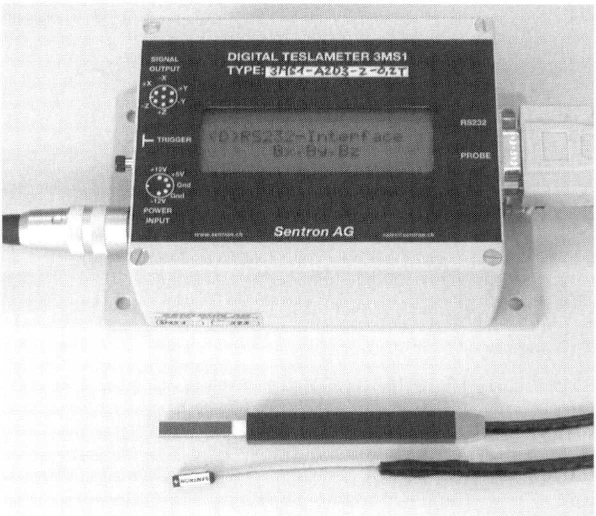

Figure 5.53. A three axis digital teslameter with two different Hall probes. The probes contain the single-chip three-axis Hall magnetic sensor described in §5.4.4. A probe measures simultaneously all three components of a magnetic field at exactly the same spot. The instrument amplifies the signal and converts it into digital form. It provides the possibility of automatic data acquisition by a host computer via the RS232 serial port (courtesy of Sentron AG [108]).

application-specific parts. Typically, the only specific part of a teslameter is a discrete Hall magnetic sensor packaged in an appropriate holder. When used for magnetometry, this part is called a *Hall probe*. The rest of the system, i.e. biasing circuits, amplifier, means for reducing offset and temperature cross-sensitivity, analogue-to-digital converters and display, and others, is made of commercially available general-purpose electronic parts, which are assembled in a separate box. The Hall probe is connected to the electronic box by a long flexible cable. Figure 5.53 shows an example of a teslameter.

As far as I know, the best reported absolute accuracies in the measurement of high magnetic fields achieved with Hall probes is ± 100 ppm (parts per million) [114]. One of the serious problems in the field of high magnetic field measurements is the planar Hall effect (§3.3.9). By properly combining at least two Hall elements in a Hall probe, the planar Hall effect can be strongly reduced [115].

5.9.2 Mechanical displacement transducers

Hall magnetic sensors have by far the highest economical impact in the field of mechanical displacement transducers. They are replacing little by little more traditional mechanical transducers, such as contact- and read-switches and

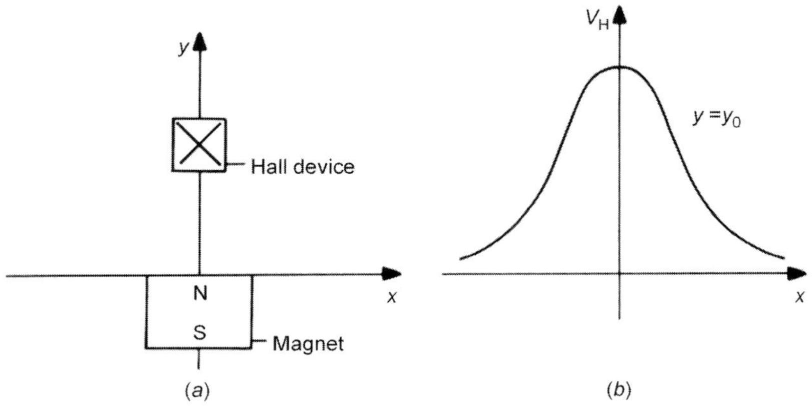

Figure 5.54. A Hall device used in a simple displacement transducer. (a) The basic configuration. (b) The Hall voltage as a function of the relative displacement magnet–Hall device along the x axis (adapted from [116]).

resistive potentiometers. The main reason why magnetic sensors are increasingly used in this area of application is the fact that they are compact, reliable, wear-free, insensitive to pollution and dust, and inexpensive.

The principle of a simple displacement transducer involving a magnetic sensor is illustrated in figure 5.54. The transducer consists of a combination of a permanent magnet and a magnetic sensor, such as a Hall device. Usually the magnet is fastened to the object whose displacement is to be sensed, and the Hall device is stationary. A contemporary permanent magnet of a few millimetres in size gives at its pole surface a magnetic field of about 100 mT. The magnetic field of the magnet seen by the Hall sensor depends on the distance between the magnet and the Hall device: $B_\perp = f(x, y, z)$. If this function is known and the displacement is limited, by measuring the magnetic induction one may find the displacement x, y or z. A problem with this arrangement is the high non-linearity of the function $B_\perp = f(x, y, z)$. Therefore, this arrangement is usually used only in simple proximity on–off sensors, called *Hall switches*.

Better performance of a displacement transducer are obtained if the magnet (or a magnetic circuit) is made to produce a simple convenient function of the displacement. Figure 5.55 shows two such examples based on a single axis Hall sensor. In the *linear displacement* (translation) sensor, the magnetic field 'seen' by the Hall sensor varies approximately linearly with the horizontal displacement of the magnetets. In the *angular position sensor*, the perpendicular component of the magnetic field varies as cos(angle) as the magnetic circuit rotates around the Hall sensor.

Inherent to such systems is the problem that the position information may vary over time. The instability is due to ageing effects and to variation

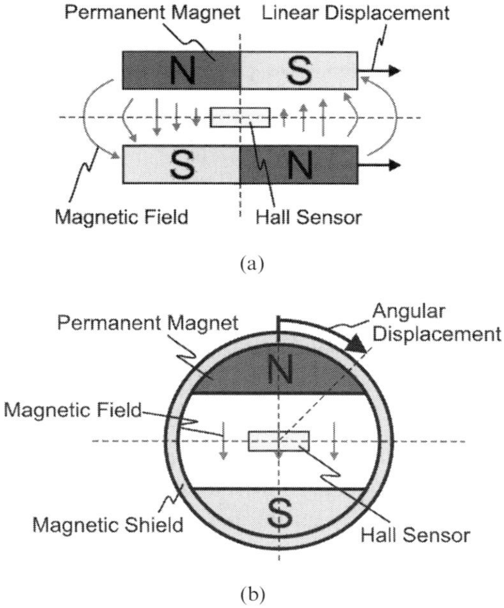

Figure 5.55. Conventional Hall sensor applied for linear position sensing (a) and angular position sensing (b). The Hall sensors are sensitive to a magnetic field parallel to their vertical axis.

in temperature, which causes a drift of magnetization of the magnet and a drift in the sensitivity of the Hall sensor. Therefore a position measurement system must be separately temperature compensated and frequently recalibrated to give precise position information.

These drawbacks are eliminated by using the field direction instead of the field amplitude as a meter for position. Although the measured field strength of a permanent magnet changes, the directions of the flux lines in the volume around the magnet hardly do. Therefore, a much more robust detection principle is based on measuring the field direction instead of the field amplitude. This becomes possible by measuring two orthogonal components of the magnetic field instead of just one, using an integrated two-axis Hall sensor, such as that described in §5.7.3.

Angular position measurement

The two-axis IMC sensor, shown in figures 5.48 and 5.49, is positioned under a magnet mounted on the shaft end of a rotating axis (figure 5.56). The magnet is magnetized diametrically so that, by rotating the shaft, the field through the sensor also rotates. The generated output voltages of the

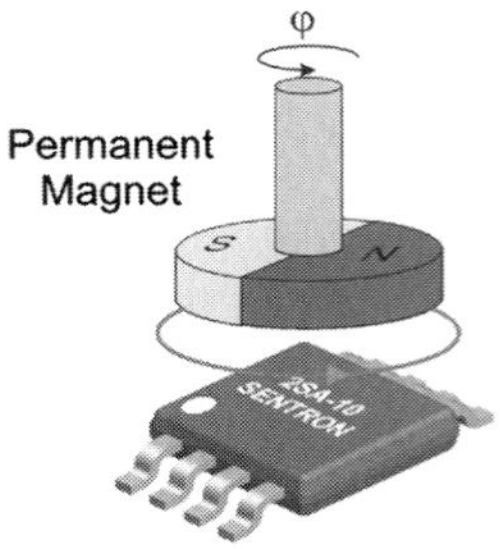

Figure 5.56. A small rotating magnet and the two-axis Hall sensor, shown in figures 5.48 and 5.49, form a very accurate angular position sensor [108].

sensor V_x and V_y represent the sine and cosine of the magnetic field direction in the sensor plane. The two output voltages can be transformed into the angle information via the arctangent function.

Even though both signal amplitudes V_x and V_y depend upon temperature and vertical distance between sensor and magnet, the angle information, which is derived from the ratio of V_x and V_y does not. Measurement results show an absolute accuracy of better than 0.5° with a resolution of better than 0.1° over a full 360° of rotation (figure 5.57).

Linear position sensor

A combination of a two-axis Hall magnetic sensor and a magnet can be used to realize a small, low-cost and robust displacement sensor [117]. Here the two-axis sensor is positioned underneath a small permanent magnet of cylindrical shape. The magnetization axis of the magnet is the rotation-symmetry axis of the cylinder, perpendicular to the sensor plane (figure 5.58).

When the magnet moves parallel to the plane, the magnetic field component parallel to the sensor chip surface rotates. The output signal V_x behaves like $X^* B_0/r^2$ and the output signal V_y like $Y^* B_0/r^2$. Here X^* and Y^* are the coordinates of the magnet with respect to the sensor, B_0 is the parallel component of the magnetic field at the position (0, 0), and r is the projection of the distance between the magnet and the sensor. If the magnet now moves parallel to the x axis, the coordinate Y is constant. The field values measured by the sensor are shown in figure 5.59. At $X^* = 0$, V_x changes sign and V_y has its maximum. If we now build the ratio V_x/V_y, this ratio is a very linear measure of the position X^* of the magnet. No further mathematical transformation has to be applied. Measurement results demonstrate an absolute error of less than 0.5% for ±4 mm travel using a 6 mm diameter permanent magnet.

By using a second two-axis sensor, the principle can be extended to a very linear two-dimensional position sensor.

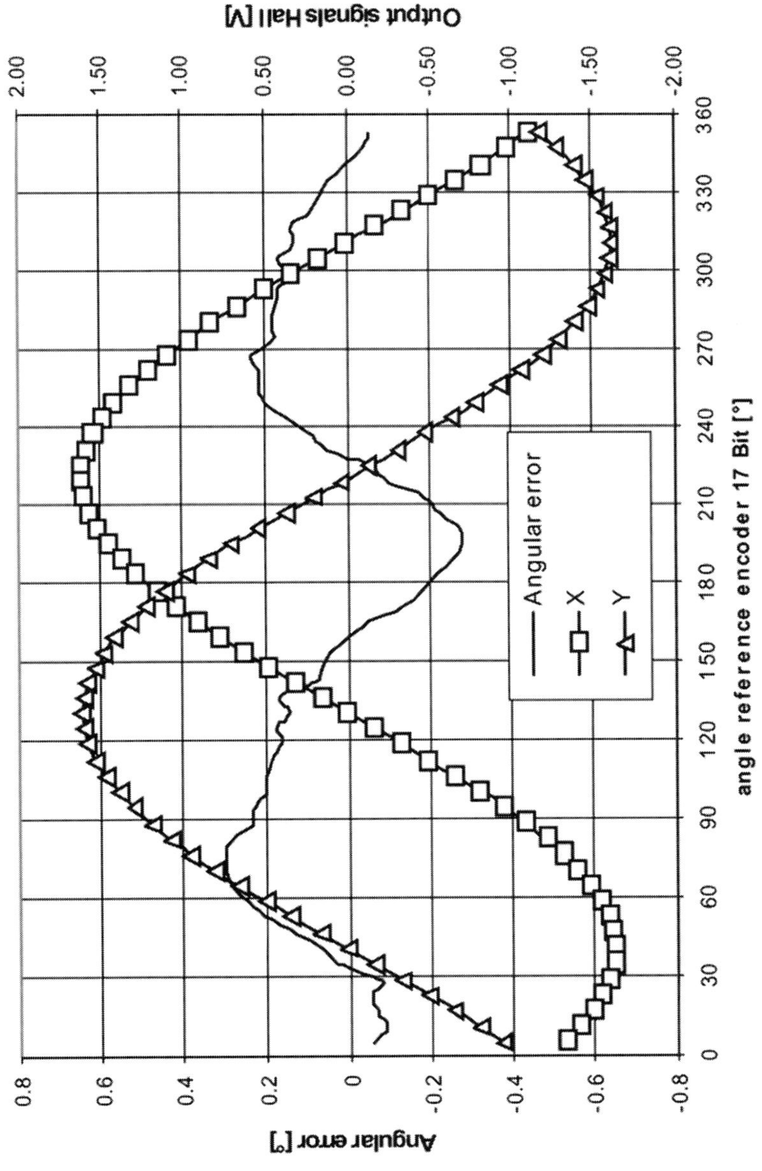

Figure 5.57. The angle error of the angle sensor shown in figure 5.56 is less than 0.3° over the full 360° rotation.

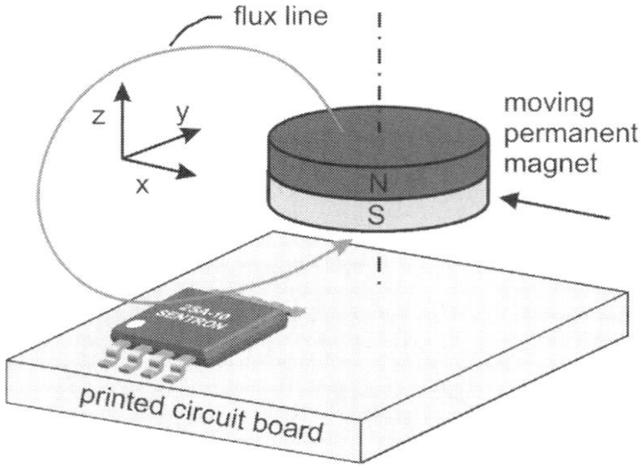

Figure 5.58. A small magnet and the two-axis Hall sensor (figures 5.48 and 5.49) form a very accurate linear position sensor.

5.9.3 Measuring electric current, power and energy

A magnetic field and magnetic induction are inevitably associated with an electric current. For example, the magnetic field about a long straight current-carrying cylindrical wire in air is given by

$$B = \frac{\mu_0 I}{2\pi R}. \tag{5.79}$$

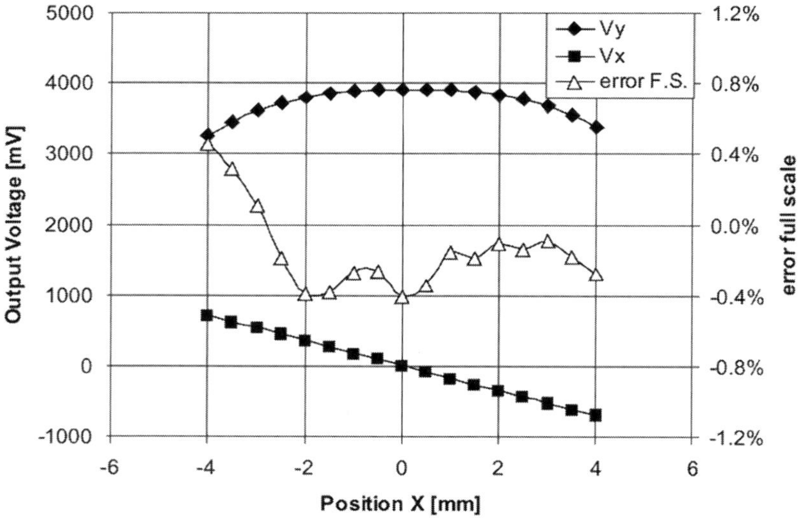

Figure 5.59. By using a 6 mm diameter magnet in the sensor shown in figure 5.58, linear position is measured with better than 0.5% accuracy over a range from −4 mm to +4 mm.

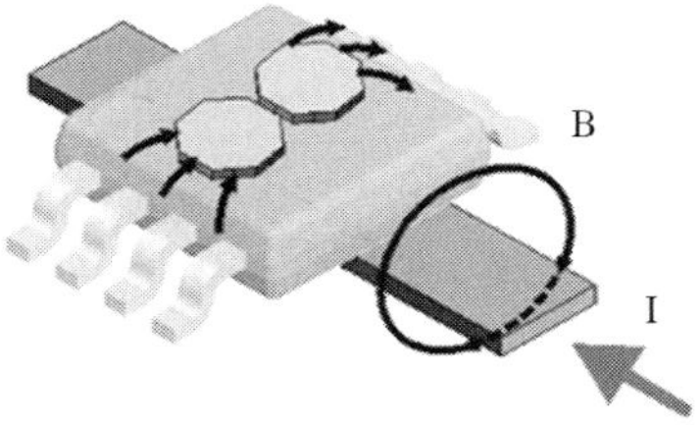

Figure 5.60. A flat stripe-like current conductor, e.g. a current track on the PCB (printed circuit board) underneath the chip, generates a magnetic field, which is collected by the integrated magnetic flux concentrator of the sensor from figure 5.44. Such an implementation is particularly adapted for currents up to 20 A [108].

where μ_0 is the permeability of air and R is the distance from the wire axis. For $I = 1\,\text{A}$ and $R = 2\,\text{mm}$, this gives $B = 10^{-4}\,\text{T}$. Thus the measuring of the magnetic induction is equivalent to the measuring of the current itself. Hence the measuring of the current in a conductor can be done without interrupting the conductor and without any galvanic contact between the current conductor and the sensor.

Current sensors

The single axis Hall magnetic sensor with integrated magnetic concentrator (IMC) shown in figure 5.44 (§5.7.3) is very convenient for the implementation of this current measurement principle. Thanks to its sensitivity-axis direction parallel to the chip surface and its high sensitivity, it is enough to place the sensor above the current-carrying line on a printed circuit board, as shown in figure 5.60.

If the applied Hall device is not sensitive enough and/or in order to get a higher measurement precision, a *magnetic circuit* around the current-carrying wire is needed. Figure 5.61 schematically shows the usual arrangement for such a measurement: a current conductor, enclosed by a magnetic yoke, and a Hall-plate magnetic sensor placed in the narrow air gap of the yoke. The yoke is made of a soft ferromagnetic material with a very high permeability.

The Ampere law here attains the form

$$i = \oint_{\mathrm{C}} H\,\mathrm{d}l = H_0 l_0 + H_y l_y. \tag{5.80}$$

where i denotes the current in the conductor, $\boldsymbol{H}$ the magnetic field, and the integration is done over the circular contour C; H_0 and l_0 are the magnetic fields in the air gap and the gap width, respectively, and H_y and l_y are the magnetic field and the contour length in the yoke, respectively. By substituting here $H = B/\mu_0\mu_r$, μ_0 being the permeability of free space and μ_r the

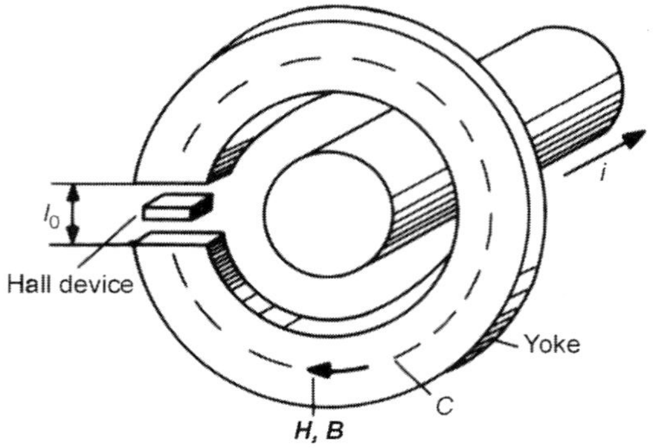

Figure 5.61. The usual basic arrangement for the electrical measurements using a Hall plate. The Hall plate is fitted in the air gap of the magnetic yoke, which encloses a current-carrying conductor. Then $V_H \sim B \sim i$.

relative permeability of each of the materials (yoke and air, respectively), we have

$$B \simeq \frac{\mu_0}{l_0} i. \tag{5.81}$$

Here we assumed that $l_0 \gg l_y/\mu_{ry}$. For example, the current of 1 A produces in the air gap of 1 mm the magnetic field of $B \simeq 1.256$ mT. Substituting equation (5.81) into (5.2) yields

$$V_H = S_I \frac{\mu_0}{l_0} Ii. \tag{5.82}$$

Thus if the bias current of the Hall device I is held constant, the Hall voltage is proportional to the measured current i.

The above two examples of current sensors are called *open-loop sensors*. Higher precision and wider band-width can be achieved with the concept of *closed-* (magnetic feed-back) *-loop current sensor* shown in figure 5.62 [118]. The current to be measured, I_1, passes through the input windings n_1. The Hall magnetic sensor senses the produced magnetic field. Its output voltage is amplified and transformed into the feedback current I_2 which is forced through the feedback windings n_2. The feedback magnetic field compensates the magnetic field produced by the input current. In the steady state, the Hall magnetic sensor sees almost no magnetic field in the air gap. Hence, the output voltage V_{out} is proportional to the input current I_1:

$$V_{out} = R \frac{n_1}{n_2} I_1. \tag{5.83}$$

Note that the output voltage depends neither on the sensitivity of the Hall sensor nor on the non-linearity of the magnetic core. Current sensors based on this principle have been commercially available for many years [119].

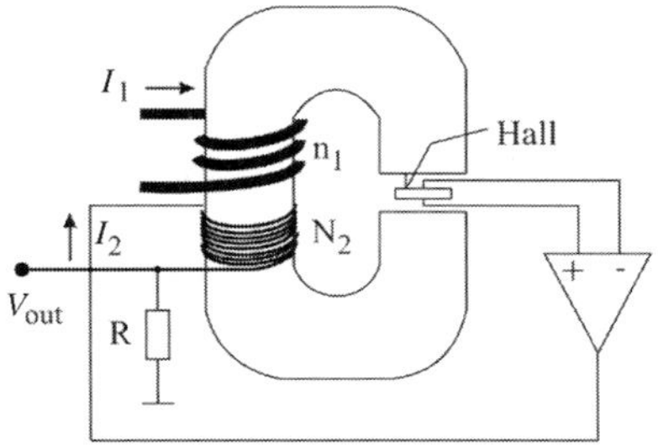

Figure 5.62. A Hall current sensor based on the magnetic feedback principle [119]. The sensor converts an instantaneous value of the measured current I_1 into the proportional value of the output voltage V_{out}.

Electric power sensor

A Hall device is a natural four-quadrant multiplier. An elegant method for measuring electric power is based on this property: if the bias current of the Hall device in figure 5.61 is made proportional to the line voltage v, that is $I \sim v$, then according to (5.82) $V_H \sim v \cdot i$. Suppose the line voltage is given by $v = V_m \cos \omega t$ and the current in the thick conductor by $i = I_m \cos(\omega t + f)$. The Hall voltage is then given by

$$V_H = C_m V_m I_m \cos \varphi + C_m V_m I_m \cos(2\omega t + \varphi) \tag{5.84}$$

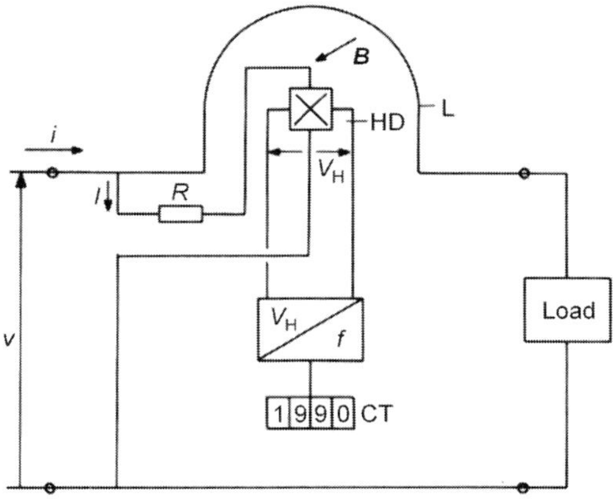

Figure 5.63. Schematic diagram of a single-phase electricity meter (watt-hour meter) based on a Hall device. L denotes a current-carrying conductor, which produces a magnetic induction ***B*** around it. A Hall device HD yields the output voltage V_H which is proportional to the product $i \cdot B$, and thus also $V_H \sim iv$. A voltage-frequency converter (V_H/f) converts the Hall voltage V_H into a frequency f; and a counter CT counts the energy pulses and displays the energy being consumed.

where C_m is the meter transduction constant. The first term on the right-hand side of this equation is a d.c. voltage proportional to active power, and the second term is an a.c. voltage, which can be filtered out.

Figure 5.63 shows a simple single-phase electricity meter (watt–hour meter) based on this principle. The bias current of the Hall device is directly derived from the line voltage using a high-value resistor R. Thus $I = v/(R + R_{in})$, R_{in} being the input resistance of the Hall device.

Substituting this into (5.82), we obtain the watt-meter transduction constant in the form

$$C_m \simeq \frac{S_I}{R} \frac{\mu_0}{l_0}. \qquad (5.85)$$

where we have assumed $R \gg R_{in}$.

The block V_H/f in figure 5.63 denotes a voltage-frequency converter, and CT a counter. In fact, the counter integrates the output signal of the Hall device (5.84). It also works as a low-pass filter. Thus the state of the counter indicates the electric energy used up by the load.

References

[1] Boll R and Overshott K J (eds) 1989 *Magnetic Sensors* vol. 5 of the series *Sensors* ed. W Gopel, J Hesse and J N Zemel (Weinheim: VCH)

[2] Pearson G L 1948 A magnetic field strength meter employing the Hall effect in germanium *Rev. Sci. Instrum.* **19** 263–5.

[3] Kuhrt F and Lippmann H I 1968 *Hallgeneratoren* (Berlin: Springer)

[4] Weiss H 1969 *Physik und Anwendung galvanomagnetischer Hauelemente* (Braunschweig: Vieweg); 1969 *Structure and Application of Galvanomagnetic Devices* (Oxford: Pergamon)

[5] Wieder H H 1971 *Hall Generators and Magnetoresistors* (London: Pion)

[6] Homeriki O K *Poluprovodnhikovye Preobrazovateli Magnitnogo Polja* (Moscow: Energoatomizdat)

[7] Bosch G 1968 A Hall device in an integrated circuit *Solid-State Electron.* **11** 712–4.

[8] Maupin J T and Geske M L 1980 The Hall effect in silicon circuits. In: C L Chien and C R Westgate (eds) *The Hall Effect and its Applications* (New York: Plenum) pp 421–45.

[9] Randhawa G S 1981 Monolithic integrated Hall devices in silicon circuits *Microelectron. J.* **12** 24–9

[10] Popovic R S, Schott Ch, Shibasaki I, Biard J R and Foster R B 2001 Hall-effect magnetic sensors Chapter 5. In: Ripka P (ed.) *Magnetic Sensors and Magnetometers* (Boston: Artech House)

[11] Baltes H P and Popovic R S 1986 Integrated semiconductor magnetic field sensors *Proc. IEEE* **74** 1107–32

[12] Popovic R S 1989 Hall-effect devices *Sens. Actuators* **17** 39–53

[13] Popovic R S and Heidenreich W 1989 Galvanomagnetic semiconductor sensors. In: *Sensors* vol. 5 ed. R Boll and K I Oveshott *Magnetic Sensors* (Weinheim: VCH) pp 43–96

[14] Middelhoek S and Noorlag D J W 1981 Silicon micro-transducers *J. Phys. E: Sci. Instrum.* **14** 1343–52

[15] Popovic R S and Hälg B 1988 Nonlinearity in Hall devices and its compensation *Solid-State Electron.* **31** 681–8

[16] Hara T T, Mihara M, Toyoda N and Zama M 1982 Highly linear GaAs Hall devices fabricated by ion implantation *IEEE Trans. Electron Devices* **ED-29** 78–82

[17] Bjorklund G 1978 Improved design of Hall plates for integrated circuitry *IEEE Trans. Electron Devices* **ED-25** 541–4

[18] Kanda Y and Migitaka M 1976 Effect of mechanical stress on the offset voltage of Hall devices in Si IC *Phys. Status Solidi* (a) **35** K115–8

[19] Kanda Y and Migitaka M 1976 Design consideration for Hall devices in Si IC *Phys. Status Solidi* (a) **38** K41–4

[20] Popovic R S 1984 A CMOS Hall effect integrated circuit *Proc. 12th Yugoslav Conf. on Microelectronics, MIEL '84, Nis* vol. 1 (EZS Ljubljana) pp 299–307

[21] Kleinpenning T G M 1983 Design of an ac micro-gauss sensor *Sens. Actuators* **4** 3–9

[22] Popovic R S 1987 Nonlinearity in integrated Hall devices and its compensation *4th Int. Conf. on Solid-State Sensors and Actuators (Transducers '87), Tokyo, Dig. Tech. Pap.* pp 539–42

[23] Mittleman D M, Cunningham J, Nuss M C and Geva M 1997 Noncontact semiconductor wafer characterization with the terahertz Hall effect *Appl. Phys. Lett.* **71** (1) 16–8.

[24] Sze S M (ed) 1983 *VLSI Technology* (New York: McGraw-Hill)

[25] Folberth O G, Madelung O and Weiss H 1954 Die elektrischen Eigenschaften von Indiumarsenid II *Z. Naturforsch.* **9a** 954

[26] Gallagher R C and Corak W S 1966 A metal-oxide-semiconductor (MOS) Hall element *Solid-State Electron.* **9** 571–80

[27] Middelhoek S and Hoogerwerf A C 1985 Smart sensors: when and where *Sens. Actuators* **8** 39–48

[28] Ichiro Shibasaki, Section 5.2 in reference 10

[29] Ghandhi S K 1994 *VLSI Fabrication Principles* (New York: Wiley)

[30] Veendrick H 2000 *Deep-Submicron CMOS ICs* (London: Kluwer)

[31] Kanda Y, Migitaka M, Yamamoto H, Morozumi H, Okabe G and Okazaki S 1982 Silicon Hall-effect power ICs for brushless motors *IEEE Trans. Electron Devices* **ED-29** 151–4

[32] Improved Hall devices find new uses, orthogonal coupling yields sensitive products with reduced voltage offsets and low drift *Electron. Wkly* April 29 pp 59–61

[33] Thanailakis A and Cohen E 1969 Epitaxial gallium arsenide as Hall element *Solid-State Electron.* **12** 997–1000

[34] Pettenpaul E, Flossmann W, Heidenreich W, Huber J, von Borcke U and Wiedlich H 1982 Implanted GaAs Hall device family for analog and digital applications *Siemens Forsch. Entwickl. Her.* **11** 22–7

[35] Liu W 1999 *Fundamentals of III–V Devices* (New York: Wiley)

[36] Mosser V, Kobbi F, Conteras S, Mercy J M, Callen O, Robert J L, Aboulhouda S, Chevrier J and Adam D 1997 Low-cost 2DEG magnetic sensor with metrological performances for magnetic field and current sensing *Proc. TRANSDUCERS 97, International conference on Solid-State Sensors and Actuators*, Chicago, 16–19 June 1997, 401–4

[37] Hadab Y, Mosser V, Kobbi F and Pond R 2000 Reliability and stability of GaAs-based pseudomorphic quantum wells for high-precision power metering *Microelectronics Reliability* **40** 1443–7

[38] Popovic R S 2000 Not-plate-like Hall magnetic sensors and their applications *Sensors and Actuators* **85** 9–17

[39] Popovic R S 1984 The vertical Hall-effect device *IEEE Electron Device Lett.* **EDL-5** 357–8

[40] Huiser A M and Baltes H P 1984 Numerical modeling of vertical Hall-effect devices *IEEE Electron Device Lett.* **EDL-5** 482–4

[41] For details on the commercially available non-plate-like Hall magnetic sensors, see: www.sentron.ch

[42] Falk U 1990 A symmetrical vertical Hall-effect device *Transducers '89* vol. 2 ed. S Middelhoek (Lausanne: Elsevier Sequoia)

[43] Burger F, Besse P A and Popovic R S 1998 New fully integrated 3-D silicon Hall sensor for precise angular position measurements *Sensors and Actuators* **A67** pp 72–6

[44] Maenaka K, Ohgusu T, Ishida M and Nakamura T 1987 Novel vertical Hall cells in standard bipolar technology *Electron. Lett.* **23** 1104–5

[45] Reference 10, figure 5.25

[46] Schurig E, Schott C, Besse P A and Popovic R S 2002 CMOS integrated vertical Hall sensor with low offset *Proc. 16th Eurosensors Conference,* Prague, Czech Republic, pp 868–71, 15–18 September 2002

[47] Blanchard H, Chiesi L, Racz R and Popovic R S 1996 Cylindrical Hall device *Proc. IEDM* San Francisco pp 541–4

[48] Schott C and Popovic R S Integrated 3D Hall magnetic field sensor *Transducers '99*, Sendai, Japan, 7–10 June 1999 Vol 1 pp 168–71

[49] Sze S M 2002 *Semiconductor Devices* (New York: Wiley) §6.3.2

[50] Thiaville A, Belliard L, Majer D, Zeldov E and Miltat J 1997 Measurement of the stray field emanating from magnetic force microscope tips by Hall effect microsensors *J. Appl. Phys.* **82** 3182–91

[51] Brennan K F and Brown A S 2002 *Theory of Modern Electronic Semiconductor Devices* (New York: Wiley) pp 105–9

[52] Boero G, Demierre M, Besse PA and Popovic R S 2003 Micro-Hall devices: performance, technologies and applications *Sensors and Actuators* **A106** 314–20

[53] Janossy B, Haddab Y, Villiot J M and Popovic R S 1988 Hot carrier Hall devices in CMOS technology *Sensors and Actuators* **A71** 172–8

[54] Kanayama T, Hiroshima H and Komuro M 1988 Miniature Hall sensor fabricated with maskless ion implantation *J. Vac. Sci. Technol.* **B6** 1010–3

[55] Sugiyama Y and Kataoka S 1985 S/N studies of micro-Hall sensors made of single crystal InSb and GaAs *Sensors and Actuators* **8** 29–38

[56] Sugiyama Y, Sano K, Sugaya T and Nakagawa T 1995 in Micro-Hall devices with high resolution made of d-doped InAlAs/InGaAs pseudomorphic heterostructures, in: *Proceedings of Transducers '95 & Eurosensors IX*, Stockholm, Sweden, 25–29 June 1995, pp 225–8

[57] Schweinbock T, Weiss D, Lipinski M and Eberl K 2000 Scanning Hall probe microscopy with shear force distance control *J. Appl. Phys.* **87** 6496–8

[58] Sandhu A, Masuda H, Oral A and Bending S J 2001 Room temperature magnetic imaging of magnetic storage media and garnet epilayers in the presence of external magnetic fields using sub-micron GaAs SHPM *J. Cryst. Growth* **227–8** 899–905

[59] Johnson M, Bennet B R, Yang M J, Miller M M and Shanabrook B V 1997 Hybrid Hall effect device *Appl. Phys. Lett.* **71** 974–6

[60] Sandhu A, Masuda H, Kurosawa K, Oral A and Bending S J 2001 Bismuth nano-Hall probes fabricated by focused ion beam milling for direct magnetic imaging by room temperature scanning Hall probe microscopy *Electron. Lett.* **37** 1335–6

[61] Geim A K, Dubonos S V, Lok J G S, Grigorieva I V, Maan J C, Hansen L T and Lindelof P E 1997 Ballistic Hall micromagnetometry *Appl. Phys. Lett.* **71** 2379–81

[62] Randjelovic Z 2000 *Low-Power High-Sensitivity Integrated Hall Magnetic Sensor Microsystems* (Konstanz: Hartung-Gorre)

[63] Wilson B and Jones B E 1982 Feed-forward temperature compensation for Hall effect devices *J. Phys. E: Sci. Instrum.* **15** 364–6

[64] Ref 4, pp 99–102, and references therein

[65] Thomson W 1856 On the effects of magnetisation on the electric conductivity of metals *Philos. Trans. R. Soc.* **A146** 736–51

[66] Wieder H H 1971 Hall *Generators and Magnetoresistors* (London: Pion) pp 30–2

[67] Taranow S G *et al* 1973 Method for the compensation of the nonequipotential voltage in the Hall voltage and means for its realization German Patent Appl. 2333080

[68] Maupin J T and Geske M L 1980 The Hall effect in silicon circuits. In: C L Chien and C R Westgate (eds) *The Hall Effect and its Applications* (New York: Plenum) pp 421–45

[69] Daniil P and Cohen E 1982 Low field Hall effect magnetometry *J. Appl. Phys.* **53** 8257–9

[70] Matsui K, Tanaka S and Kobayashi T 1981 GaAs Hall generator application to a current and watt meter *Proc. 1st Sensor Symp.* ed. S Kataoka (Tokyo: Institute of Electrical Engineers of Japan) pp 37–40

[71] Enz Ch and Temes G 1996 Circuit techniques for reducing the effects of op-amp imperfections: auto-zeroing, correlated double sampling, and copper stabilization *IEEE J. Solid-State Circuits* **29** 601–10

[72] Bilotti A, Monreal G and Vig R 1997 Monolithic magnetic Hall sensor using dynamic quadrature offset cancellation *IEEE J. Solid-State Circuits* **32** 829–36

[73] Bakker A, Thiele K and Huijsing J 2000 A CMOS nested chopper instrumentation amplifier with 100 nV offset *Proceedings of International Solid-State Circuit Conference ISSCC 2000*, Paper TA 9.4

[74] Trontelj J 1999 Optimization of integrated magnetic sensor by mixed signal processing *Proc. of 16th IEEE Instrumentation and Measurement Technology Conf.* (IMTC99), pp 299–302

[75] Popovic R S, Randjelovic Z and Manic D 2001 Integrated Hall-effect magnetic sensors *Sensors and Actuators* **A91** 46–50

[76] Demierre M, Pesenti S, Frunchi J, Besse P A and Popovic R S 2002 Reference magnetic actuator for self-calibration of a very small Hall sensor array *Sensors and Actuators* **A97-98** 39–46

[77] Demierre M, Pesenti S and Popovic R S 2002 Self calibration of a CMOS twin Hall microsystem using an integrated coil *Eurosensors XVI, September 15–18, 2002*, Prague, Czech Republic 573–4

[78] Popovic R S, Drljaca P M, Schott C and Racz R 2001 Integrated Hall sensor/flux concentrator microsystems, Invited Lecture *37th International Conference on Microelectronics, Devices And Materials, MIDEM 01*, Bohinj, Slovenia, October 2001

[79] See, for example: Boll R (ed) 1979 *Soft Magnetic Materials* (London: Heyden), §4.2.2
[80] O'Handley R C 2000 *Modern Magnetic Materials* (New York: Wiley) §2.3
[81] Barkhausen H 1919 Zwei mit Hilfe der neuen Verstärker entdeckte Erscheinungen *Physik. Zeitschrift* **XX** 401–3
[82] Drljaca P M, Kejik P, Vincent F and Popovic R S 2003 Low noise CMOS micro-fluxgate magnetometer *Transducers '03* Boston (MA), USA 8–12 June
[83] Ross L M *et al* 1955 Applications of InSb *J. Electronics* **2** 223
[84] Hieronymus H *et al* 1957 Über die Messung kleinster magnetischer felder mit Hallgeneratoren *Siemens Z* **31** 404–9
[85] Epstein M *et al* 1961 Magnetic field pickup for low-frequency radio-interference measuring sets *IRE Transactions on Electron Devices* pp 70–7
[86] Nazarov P A *et al* 1981 On selection of magnetic circuit for Hall effect devices *Electric Technology USSR* **1** 69–77
[87] Marchant A H 1984 Hall-effect devices for relay and other applications *Electrotechnology* pp 122–5
[88] Shibasaki I 1989 High sensitivity Hall Element by vacuum deposition *Technical Digest of the 8th Sensor Symposium*, Japan, pp 221–4
[89] InSb and InAs Hall Elements from Asahi Chemical *Compound Semiconductor* September/October 1996 pp 38–40
[90] See Fig 6 in ref [12]
[91] Popovic R S *et al* US patents 5,942,895 and 6,184,679
[92] Drljaca P M *et al* 2002 Design of planar magnetic concentrators for high sensitivity Hall devices *Sensors and Actuators* **97** 10–4
[93] Popovic R S and Schott C 2002 Hall ASICs with integrated magnetic concentrators *Proceedings Sensors Expo and Conference* Boston (Ma) USA, 23–26 September 2002
[94] Swiss Patent Application 2000 1645/00
[95] Popovic R S *et al* 2001 A new CMOS Hall angular position sensor *Technisches Messen* **68** 286–91
[96] Vorthmann E A 1980 Packaging Hall effect devices. In: C L Chien and C R Westgate (eds) *The Hall Effect and its Applications* (New York: Plenum)
[97] Heidenreich W 1989 Aufuautechniken fur Halbleiter-Magnetfeldsensoren *Technisches Messen* **56** pp 436–43
[98] Schneider G 1988 Non-metal hermetic encapsulation of a hybrid circuit *Microelectron. Reliab.* **28** 75–92
[99] Manic D 2000 *Drift in Silicon Integrated Sensors and Circuits Due to Thermomechanical Stress* (Konstanz: Hartung-Gorre)
[100] Schleiser K M, Keneman S A and Mooney R T 1982 Piezoresistivity effects in plastic-encapsulated integrated circuits *RCA Rev.* **43** 590–607
[101] Steiner R, Maier C, Mayer M, Bellekom S and Baltes H 1999 Influence of mechanical stress on the offset voltage of Hall devices operated with spinning current method *J. Micromechanical Systems* **8** 466–72
[102] Manic D, Petr J and Popovic R S 2000 Temperature cross-sensitivity of Hall plates in submicron CMOS technology *Sensors and Actuators* **A85** 244–8
[103] Manic D, Petr J and Popovic R S 2000 Short- and long-term stability problems in Hall plates in plastic *Proceedings of the IEEE International Reliability Physics Symposium (IRPS 2000)* April 2000, San Jose, USA, pp 225–30

[104] Hälg B 1988 Piezo-Hall coefficients of n-type silicon *J. Appl. Phys.* **64** 276–82

[105] Hälg B and Popovic R S 1990 How to liberate integrated sensors from encapsulation stress *Transducers '89* vol. 2 ed. S Middelhoek (Lausanne: Elsevier Sequoia)

[106] Ramsden E 2001 *Hall Effect Sensors* (Cleveland, Ohio: Advannstar)

[107] Melexis—Applications and Databook, Section 3—Applications

[108] Sentron AG: www.sentron.ch (Application notes and Technical papers)

[109] GMW: www.gmw.com

[110] Minden H T and Leonard M F 1979 A micron-size Hall probe for precision magnetic mapping *J. Appl. Phys.* **50** 2945–7

[111] Li Y, Xiong P, von Molnar S, Wirth S, Ohno Y and Ohno H 2002 Hall magnetometry on a single iron nanoparticle *Appl. Phys. Lett.* **80** 1–3

[112] Besse P A, Boero G, Demierre M, Pott V and Popovic R S 2002 Detection of a single magnetic microbead using a miniaturized silicon Hall sensor *Appl. Phys. Lett.* **80** 4199–201

[113] Boero G, Besse P A and Popovic R S 2001 Hall detection of magnetic resonance *Appl. Phys. Lett.* **79** 1498–500

[114] Schott C, Popovic R S, Alberti S and Tran M Q 1999 High accuracy magnetic field measurements with a Hall probe *Rev. Sci. Instrum.* **70** 2703–7

[115] Reference [107], Application Note 003

[116] Schott C, Popovic R S and Racz R 2003 Magnetic position measurement: from amplitude to direction, Paper 1.1 *Proceedings Sensor 2003 11th International Conference* 13–15 May 2003, Nuremberg, Germany, pp 181–6

[117] Schott C, Racz R, Betschart F and Popovic R S 2002 New two-axis magnetic position sensor, Paper 54.6 *Proc. 1st Annual IEEE Sensors Conference*, Orlando, Florida, USA, 12–14 June 2002

[118] Popovic R S and Flanagan J A 1997 Sensor microsystems *Microelectron. Reliab.* **37** 1401–9

[119] LEM current sensors. Product catalogue, LEM SA, Geneva, Switzerland

Chapter 6

Hall devices as means for characterizing semiconductors

The Hall voltage measured at the output of a Hall device depends on the properties of the applied material, device geometry, biasing and magnetic field (see §4.2.3). If the latter three quantities are known, measurement of the Hall voltage enables us to determine the pertinent transport properties of carriers in the material the device is made of. In this case, a Hall device is applied as a means for characterizing material. This was the very first and, for many decades, the only application of Hall devices. Currently, the standard methods for characterizing semiconductor materials inevitably include measurements on Hall devices. The subject has been treated in detail in two monographs [1, 2] and also in a chapter of a more recent book [3]. In the first part of this chapter (§§6.1 and 6.2), we shall review these classical methods for characterizing semiconductor materials using the Hall effect. In the last part (§6.3), we shall present a new method for characterizing semiconductor device packages using Hall devices.

6.1 Extraction of material characteristics

We shall first dwell on the questions: what should we measure on a Hall device, which material parameters can be extracted from the measurement results, and how should we do it? For simplicity, we assume throughout this section the Hall voltage mode of operation of the test Hall devices. The issue of the measurement methods will be discussed in §6.2.

6.1.1 Homogeneous plates and layers

Consider a Hall plate made of a material with uniform properties. For example, such a plate may be cut from a single-crystal ingot, or can be fabricated using the homogeneously doped epitaxial layer. Suppose that such a uniform Hall plate is biased by a current I and exposed to a perpendicular magnetic induction $B_\perp = B$. Then the Hall voltage measured at the output terminals of the device carries information on some material properties. According to (4.90)

$$\frac{V_H}{G_H I B} = \frac{R_H}{t} = R_{HS} \tag{6.1}$$

where G_H denotes the geometrical correction factor (§4.2.6), R_H the Hall coefficient (§3.4.6), and t is the thickness of the plate. The ratio $R_H/t = R_{HS}$ is called the *sheet Hall coefficient*.

The Hall coefficient of strongly extrinsic material is given by (3.308):

$$R_H = \text{sign}[e]\frac{r_H}{qn} \tag{6.2}$$

where sign[e] denotes the sign of majority carriers, r_H is the Hall factor (§3.4.1), q is the magnitude of the electron charge, and n is the concentration of majority carriers.

By inspecting the above two equations, we can see that by measuring the Hall voltage of a device of an unknown material, we can immediately determine the following. (i) The *sign of majority carriers*, for given directions of current and magnetic induction, the polarities of the Hall voltages in a p-type and n-type sample are opposite. This is illustrated in figure 3.1 of §3.1.1. At the boundary where positive charge accumulates, more positive potential occurs. (ii) The *carrier concentration*: from (6.1) and (6.2)

$$nt = \frac{r_H}{q|R_{HS}|} \qquad n = \frac{r_H G_H}{t}\left|\frac{IB}{qV_H}\right|. \tag{6.3}$$

In the exhaustion temperature range, the carrier concentration equals the doping density (see (2.44) and (2.50)). Recall also that the nt product equals the number of carriers per unit plate area, with dimensions of m^{-2}.

The *Hall factor* r_H is usually assumed to equal unity, but this is true only at very high inductions (see (3.312)). The exact value of the Hall factor can be determined by measuring the Hall voltage at a low magnetic induction (B_0) and a very high induction (B_∞). With the aid of (6.1) and (6.2), we have

$$r_H = \frac{V_{H0}}{V_{H\infty}}\frac{B_\infty}{B_0} \tag{6.4}$$

where V_{H0} and $V_{H\infty}$ are the Hall voltages at B_0 and B_∞, respectively. Here we have assumed that both the current and the geometrical correction factor remain constant during the measurements. The latter implies the application of a very long (or point-contact) Hall device (see figure 4.17).

In a mixed conductor, the Hall coefficient R_H is given by (3.206). If one of the conditions $bx = \mu_n n/\mu_p p \gg 1$ or $bx \ll 1$ is fulfilled, the above relations can still be applied with a good accuracy.

An elegant method for measuring the *band gap* of a semiconductor is based on the determination of the temperature dependence of the Hall coefficient. A Hall device sample should be made of p-type material (we assume $\mu_n > \mu_p$, μ_n and μ_p being the mobility of electrons and holes, respectively). By measuring the Hall voltage at various temperatures, one can determine the temperature T_0 at which the Hall coefficient drops to zero (see figure 5.5 and equation (3.207)). At this temperature:

$$\frac{r_{Hp}}{r_{Hn}}\left(\frac{\mu_p}{\mu_n}\right)^2\frac{p(T_0)}{n(T_0)} = 1. \tag{6.5}$$

If $\mu_n \gg \mu_p$, then $p(T_0) \gg n(T_0)$ also. Then to a first approximation, $p(T_0) \simeq N_A$ (2.51) and $n(T_0) \simeq (n_i(T_0))^2/N_A$ (2.52). Substituting these relations into (6.8), we find the intrinsic carrier concentration at the temperature T_0:

$$(n_i(T_0))^2 \simeq N_A^2\left(\frac{\mu_p}{\mu_n}\right)^2\frac{r_{Hp}}{r_{Hn}}. \tag{6.6}$$

From (2.31) or (2.32), we can now readily find the band gap E_g.

The *carrier mobility* can be determined by making use of the sheet resistance, the ratio of Hall voltage to supply voltage, the magnetoresistance, and the current deflection effect.

The *sheet resistance* of a homogeneous conductive layer is defined as [4]

$$R_S = \rho/t \tag{6.7}$$

with ρ being the material resistivity (see (2.91) and (2.95)) and t the thickness of the layer. Physically, the sheet resistance equals the resistance of a square of the layer, irrespective of the dimensions of the square. The sheet resistance of a strongly extrinsic homogeneous layer is given by

$$R_S = 1/\mu q n t \tag{6.8}$$

where μ denotes the conductivity mobility of majority carriers. From (6.3) and (6.8), we find the *Hall mobility* as

$$\mu_H = r_H\mu = R_{HS}/R_S \tag{6.9}$$

where we have also made use of (3.283).

The Hall voltage of a Hall device, made of extrinsic material, biased by a voltage V, and exposed to a weak perpendicular magnetic induction B, is given by (4.92). Accordingly, the Hall mobility of the majority carriers is given by

$$\mu_H = \frac{l}{w}\frac{1}{G_H}\frac{V_H}{V}\frac{1}{B}. \tag{6.10}$$

Here l/w denotes the length-to-width ratio of the equivalent rectangular device (see (4.78) and (4.79)).

Note that the ratio V_H/V is related to the *Hall angle*: according to (4.93), $V_H/V \sim \mu_H B$, and from (3.152), $\Theta_H = \arctan(\mu_H B)$. Therefore, the measurement of the V_H/V ratio yields information on the Hall angle. The Hall angle can also be measured with the aid of the current deflection effect. The application of the current deflection effect will be explained later on, in relation to the corresponding measurement methods.

The Hall mobility can also be determined by measuring the magneto-resistance. Usually, two specimens of the same material are used: one very short, such as a Corbino disc, and the other very long (or point contact). The Hall mobility is then given by (4.73)

$$\mu_H = \left[\left(\frac{\Delta R_B}{R_B(0)} - \frac{\Delta R_b}{R_b(0)} \right) \Big/ \left(1 + \frac{\Delta R_b}{R_b(0)} \right) \right]^{1/2} \frac{1}{B} \tag{6.11}$$

where $\Delta R_B / R_B(0)$ denotes the Corbino magnetoresistance (4.128) and $\Delta R_b / R_b(0)$ denotes the physical magnetoresistance. Alternatively, the Hall mobility can also be found from

$$\mu_H = \left[\left(\frac{R_B(B)}{R_B(0)} \Big/ \frac{R_b(B)}{R_b(0)} \right) - 1 \right]^{1/2} \frac{1}{B} \tag{6.12}$$

where R_B denotes the resistance of a Corbino disc and R_b the resistance of a point-contact device.

The *conductivity mobility* of carriers can be now determined by

$$\mu = \mu_H / r_H \tag{6.13}$$

with r_H being given by (6.4).

In summary, with the aid of a Hall device fabricated from an unknown electrically conductive material, we can experimentally determine the type of the majority carriers in the material, their concentration and mobility, and the Hall factor. By repeating measurements at various magnetic fields, the magnetic field dependences of these quantities can also be found. Thus all the galvanomagnetic transport coefficients (see §3.4) of carriers in a material can be measured. If the material under investigation is a semiconductor and properly doped, its band-gap energy can also be found.

6.1.2 Inhomogenous layers

Consider the Hall effect in a one-dimensionally inhomogeneous semi-conductor layer, such as an ion-implanted layer. We choose the coordinate system so that the plane $y = 0$ leans on the top surface of the layer, and the inhomogeneity appears only along the y axis (see figure 6.1). We assume strongly extrinsic material with one kind of carrier. The carrier concentration and transport coefficients depend on the depth in the layer: $n(y)$, $\mu(y)$, $\mu_H(y)$, etc. The external electric field is parallel to the plane $y = 0$, and the magnetic induction $\boldsymbol{B}$ is parallel to the y axis. Both $\boldsymbol{E}$ and $\boldsymbol{B}$ are weak.

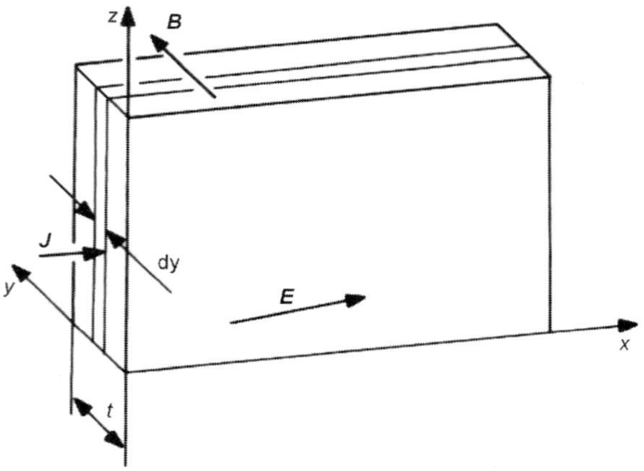

Figure 6.1. An inhomogeneous semiconductor layer. The layer properties, such as concentration and mobility of carriers, are functions of the depth (y) in the layer. t is the layer thickness, and dy denotes the thickness of a lamina parallel to the top surface of the layer ($y = 0$).

Imagine that the layer is subdivided into laminae of infinitesimal thickness dy. The laminae are 'connected' in parallel, and each has the same electric field $\boldsymbol{E}$ and a different current density $\boldsymbol{J}(y)$. We may introduce the average current density over the layer by

$$(\boldsymbol{J}) = \frac{1}{t}\int_0^t \boldsymbol{J}(y)\,dy \tag{6.14}$$

where t is the thickness of the layer. Obviously, if $\boldsymbol{J} \neq f(y)$ as in a homogeneous layer, $(\boldsymbol{J}) = \boldsymbol{J}$. Let us now derive equations relating the vectors $\boldsymbol{E}$, $(\boldsymbol{J})$ and $\boldsymbol{B}$, in analogy with the basic galvanomagnetic equations (3.149) and (3.175). (Note the similarity of the present problem to that of the Hall effect in a mixed conductor, treated in §3.3.7.)

The relation between the vectors $\boldsymbol{E}$, $\boldsymbol{J}$ and $\boldsymbol{B}$ in each of the laminae is given by (3.149):

$$\boldsymbol{J} = \sigma_B \boldsymbol{E} + \sigma_B \mu_H [\boldsymbol{E} \times \boldsymbol{B}]. \tag{6.15}$$

By substituting this equation into (6.14), we obtain

$$(\boldsymbol{J}) = (\sigma_B)\boldsymbol{E} + (\sigma_B)(\mu_H)[\boldsymbol{E} \times \boldsymbol{B}] \tag{6.16}$$

where we have introduced the notation

$$(\sigma_B) = \frac{1}{t}\int_0^t \sigma_B(y)\,dy \tag{6.17}$$

$$(\mu_H) = \int_0^t \sigma_B(y)\mu_H(y)\,dy \Big/ \int_0^t \sigma_B(y)\,dy. \tag{6.18}$$

The quantity (σ_B) is the *average value of the conductivity* σ_B over the layer depth, and (μ_H) is the *average value of the Hall mobility* over the layer conductivity. Since we assume $B \approx 0$ and extrinsic conditions, σ_B is given by (3.146). Then

$$(\sigma_B) \simeq \frac{1}{t}\int_0^t qn\mu \, dy \tag{6.19}$$

$$(\mu_H) \simeq \int_0^t n\mu\mu_H \, dy \bigg/ \int_0^t n\mu \, dy. \tag{6.20}$$

In analogy with equations (3.155) and (3.156), the solution of (6.16) with respect to $\boldsymbol{E}$ is given by

$$\boldsymbol{E} = \frac{1}{(\sigma_B)[1 + (\mu_H)^2 B^2]}(\boldsymbol{J}) - \frac{(\mu_H)}{(\sigma_B)[1 + (\mu_H)^2 B^2]}[(\boldsymbol{J}) \times \boldsymbol{B}]. \tag{6.21}$$

In analogy with (3.175) this can be rewritten as

$$\boldsymbol{E} = (\rho_b)(\boldsymbol{J}) - (R_H)[(\boldsymbol{J}) \times \boldsymbol{B}] \tag{6.22}$$

where (ρ_b) and (R_H) are the averages over the layer of the resistivity $\rho_b(y)$ and the Hall coefficient $R_H(y)$, respectively. At a low magnetic induction

$$(\rho_b) \simeq \frac{1}{\sigma_B} = t \bigg/ \int_0^t qn\mu \, dy \tag{6.23}$$

$$(R_H) \simeq (\mu_H)/(\sigma_B) = t\int_0^t n\mu\mu_H \, dy \bigg/ q\left(\int_0^t n\mu \, dy\right)^2. \tag{6.24}$$

By comparing equations (6.16) with (3.149), and (6.22) with (3.175), we notice that the basic equations describing the galvanomagnetic effects in a homogeneous plate and an inhomogeneous layer are fully analogous. The analogous quantities are: the electric fields $\boldsymbol{E}$; the current density $\boldsymbol{J}$ and the average current density $(\boldsymbol{J})$; the Hall mobility μ_H and the average Hall mobility (μ_H); the conductivity σ_B and the average conductivity (σ_B); etc. On the other hand, all the other basic equations and boundary conditions (§§3.3.1 and 3.3.4) hold for both homogeneous and inhomogeneous Hall plates. Therefore, the electrical characteristics of an inhomogeneous Hall plate can be expressed by the formulae pertinent to a homogeneous Hall plate of equal geometry, the only difference being that the 'normal' transport coefficients must be replaced by the corresponding average values. Consequently, by evaluating an inhomogeneous layer using the methods described in §6.1.1, we can obtain as results the corresponding average values of the transport coefficients.

Here are two examples.

By measuring the Hall voltage on a current-driven Hall plate, we can determine the *sheet Hall coefficient*. By replacing R_H by (R_H) in (6.1), we obtain

$$\frac{1}{G_{\mathrm{H}} I B} V_{\mathrm{H}} = \frac{(R_{\mathrm{H}})}{t} = (R_{\mathrm{HS}}) \tag{6.25}$$

where (R_{HS}) denotes the average sheet Hall coefficient. According to (6.24), the average sheet Hall coefficient is now given by

$$(R_{\mathrm{HS}}) = \int_0^t n\mu\mu_{\mathrm{H}}\, \mathrm{d}y \Big/ q\left(\int_0^t n\mu\, \mathrm{d}y\right)^2. \tag{6.26}$$

Using the value of the sheet Hall coefficient, we may find the mobility-weighted average of the area density of carriers (N_{S}). If we define (N_{S}) by

$$(N_{\mathrm{S}}) = (r_{\mathrm{H}})\left(\int_0^t n\mu\, \mathrm{d}y\right)^2 \Big/ \int_0^t n\mu|\mu_{\mathrm{H}}|\, \mathrm{d}y \tag{6.27}$$

then from (6.26) and (6.27) it follows that

$$(N_{\mathrm{S}}) = (r_{\mathrm{H}})/q(R_{\mathrm{HS}}). \tag{6.28}$$

This is analogous to the first equation (6.3). If we assume that the mobilities vary little over the depth of the layer, (6.27) reduces to

$$(N_{\mathrm{S}}) \simeq \int_0^t n\, \mathrm{d}y. \tag{6.29}$$

This is identical with N_{S} according to (5.4). If the layer under investigation is an *ion-implanted layer*, (N_{S}) gives approximately the area concentration of activated dopants. If all dopant atoms are activated (ionized), then (N_{S}) approximately equals the *implanted dose*.

According to (6.7) and (6.23), the sheet resistance of an extrinsic inhomogeneous layer is given by

$$R_{\mathrm{S}} = 1 \Big/ \int_0^t qn\mu\, \mathrm{d}y. \tag{6.30}$$

Let us now find the ratio $R_{\mathrm{HS}}/R_{\mathrm{S}}$. Using (6.26) and (6.30), and in view of (6.20), it is

$$(\mu_{\mathrm{H}}) = \frac{(R_{\mathrm{HS}})}{R_{\mathrm{S}}}. \tag{6.31}$$

This equation is analogous to (6.9). Thus we can determine the average value of the Hall mobility of carriers in an inhomogeneous layer by measuring the sheet Hall coefficient and the sheet resistance of the layer.

In summary, simple measurements on a Hall device with a one-dimensionally inhomogeneous active layer can deliver both the quantity (N_{S}), which approximately equals the area carrier concentration in the layer (number of charge carriers per unit area), and the average values of the transport coefficients of the carriers in the layer.

6.1.3 Measurement of profiles

The distribution function of a physical quantity over the depth of an inhomogeneous layer is called the profile of this quantity. In semiconductor technology, impurity profiles in doped semiconductor layers are of great importance. A summary of various techniques for measuring profiles in semiconductors can be found in [5]. We shall consider here the measurement of profiles with the aid of the Hall effect [2].

Suppose we measure the sheet resistance and the sheet Hall coefficient on an inhomogeneous conductive layer, such as that shown in figure 6.1. The measurement results are R_{S1} and R_{HS1} respectively. Then we remove from the surface a lamina of thickness $\Delta y1$ and remeasure the remaining portion of the specimen. Let us denote these new measurement results by R_{S2} and R_{HS2}, respectively. With the aid of these four measured values, we can determine the average concentration of carriers and their mobility in the removed lamina, as we show below.

According to (6.26) and (6.30), we have:

$$(R_{HS1}) = \int_0^t n\mu\mu_H \, dy \Big/ q\left(\int_0^t n\mu \, dy\right)^2$$
$$R_{S1} = 1 \Big/ \int_0^t qn\mu \, dy \tag{6.32}$$

$$(R_{HS2}) = \int_{\Delta y1}^t n\mu\mu_H \, dy \Big/ q\left(\int_{\Delta y1}^t n\mu \, dy\right)^2$$
$$R_{S2} = 1 \Big/ \int_{\Delta y1}^t qn\mu \, dy. \tag{6.33}$$

Then we find

$$A = \frac{R_{HS1}}{R_{S1}^2} - \frac{R_{HS2}}{R_{S2}^2} = qn\mu\mu_H \Delta y1 \tag{6.34}$$

$$B = \frac{1}{R_{S1}} - \frac{1}{R_{S2}} = q\mu n \Delta y1 \tag{6.35}$$

and finally

$$\mu_H = \frac{A}{B} \tag{6.36}$$

$$\frac{r_H}{qn\Delta y1} = \frac{A}{B^2} \qquad n = \frac{r_H}{q\Delta y1}\frac{B^2}{A}. \tag{6.37}$$

These are the required average values of the Hall mobility and the carrier concentration over the depth (0, $\Delta y1$). By repeating the procedure, namely by removing another lamina of thickness $\Delta y2$ and remeasuring R_{HS} and

R_S, we can determine the average values of μ_H and n over the depth $(\Delta y1, \Delta y2)$. Proceeding sequentially in this manner, we can construct the profiles $\mu_H(y)$ and $n(y)$ throughout the analysed layer.

Several techniques have been developed for removing thin laminae of material: chemical etching, anodic oxidation followed by oxide stripping, and ion milling. The whole process of profiling can be automated [6, 7].

Instead of actually removing a lamina, in a semiconductor sample a lamina can be made inactive by depleting the charge carriers out of it. To this end, the test device must incorporate a Schottky contact, a p–n junction, or an MOS structure. Recall that the depletion-layer width of a p–n junction depends on the applied voltage (see (2.116) and (2.117)). The width of the depleted region W can be measured by measuring the small-signal capacitance of the structure. For example, in the case of a Schottky contact or a p–n junction,

$$W = \varepsilon_s \varepsilon_0 / C \tag{6.38}$$

where C is the depletion-layer capacitance per unit area, and the other notation is the same as in (2.116).

6.2 Measurement methods for characterizing material

As we have seen in previous sections, in order to determine the type, concentration and mobility of charge carriers in a material, we need to measure some of the following quantities pertinent to this material: resistivity, Hall coefficient, physical magnetoresistance, Corbino magnetoresistance, and the Hall angle. We shall now discuss the methods for measuring these quantities.

To avoid errors due to parasitic effects, and ease the extraction of relevant material characteristics from the row measurement results, the measurements are normally done on a well-defined specimen of the material to be characterized. The specimen has to be fitted by appropriate electrical contacts. For metals, small-band-gap semiconductors and highly doped semiconductors, simple pressure contacts are usually adequate. A contact metal/large-band-gap low-doped semiconductor usually exhibits a Schottky barrier [8]. To make a good ohmic contact on such a semiconductor, the semiconductor surface must be heavily doped. A doping depth of some 10 nm usually suffices [9].

6.2.1 Measurements on a long slab

The simplest form of a sample suitable for the galvanomagnetic measurements is shown in figure 6.2: it is a long slab fitted with five small-area contacts. A current I is forced along the slab through the contacts A and C, the longitudinal voltage drop is measured between the contacts E and

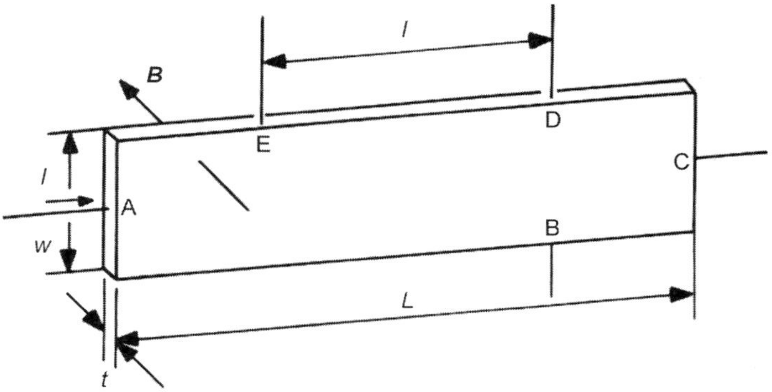

Figure 6.2. Specimen for characterizing material in the form of a long strip, fitted with five small-area contacts. To simplify the extraction of material characteristics from the measured results, the contacts should be small, the sample long ($L \gg w$, $L \gg l$), and the contacts good.

D, and the Hall voltage is measured between the contacts B and D. In such an arrangement, the voltage drop between the contacts E and D does not depend on the contact resistance, and the geometrical correction factor (§4.2.6) approximately equals unity. The resistance of the slab between the contacts E and D is given by

$$R_{\mathrm{ED}} = \rho_{\mathrm{b}} \frac{l}{wt} \tag{6.39}$$

where ρ_{b} is the resistivity (§3.4.5), and l, w and t are the dimensions defined in figure 6.2. The voltage drop between the contacts E and D is given by

$$V_{\mathrm{ED}} = R_{\mathrm{ED}} I. \tag{6.40}$$

Hence, by measuring I and V_{ED}, the resistivity ρ_{b} can be readily determined. The sheet resistance, equation (6.7) or (6.30), is given by

$$R_{\mathrm{S}} = \frac{\rho_{\mathrm{b}}}{t} = \frac{w}{l} \frac{V_{\mathrm{ED}}}{I}. \tag{6.41}$$

The voltage between the terminals B and D is given by the sum of the Hall voltage V_{H}, the offset voltage V_{off} and the noise voltage V_{N} (see (4.86)). To achieve a high signal-to-noise ratio (see §4.3.6), the sample has to be symmetrical, the contacts good, and both bias current and magnetic induction high enough. The offset voltage can be compensated by a potentiometer (§5.6.3). Alternatively, the offset voltage can be eliminated by measuring the output voltage V_{DB} at two opposite directions of a magnetic induction B. Since $V_{\mathrm{H}}(B) = -V_{\mathrm{H}}(-B)$, and $V_{\mathrm{off}}(B) = V_{\mathrm{off}}(-B)$, we obtain

$$V_{\mathrm{DB}}(B) - V_{\mathrm{DB}}(-B) = 2V_{\mathrm{H}}(B). \tag{6.42}$$

Note, however, that this method of offset elimination works well enough only if the offset voltage is small and independent of the direction of the magnetic induction. The latter condition might not be fulfilled if the magnetoconcentration effect is present in the sample (see §3.5.2).

According to (6.1) and (6.25), the *sheet Hall coefficient* is readily obtained from

$$R_{HS} = \frac{R_H}{t} = \frac{V_H}{IB} \tag{6.43}$$

since in a long sample $G_H = 1$.

Instead of a simple parallelepiped, the sample can also be made in the form of a bridge, as shown in figure 4.15(b). This form is especially useful for large-band-gap and high-resistivity semiconductors, since it allows the application of larger-area contacts. Nevertheless, a bridge exhibits a geometrical correction factor close to unity, and the above simple equations hold.

An early refined circuit for precise measurements of the ratios V_{ED}/I and V_{DB}/I is described in [10]. Nowadays, much better circuit components are available, and the processing of the voltage and current signals generally presents no difficulty.

6.2.2 Van der Pauw method

Van der Pauw showed in 1958 that the resistivity and the Hall coefficient can be measured on a flat sample of arbitrary shape, if the following conditions are fulfilled [11, 12]:

(i) the contacts are positioned at the circumference of the sample;
(ii) the contacts are sufficiently small;
(iii) the sample is homogeneous and uniform in thickness;
(iv) the area of the sample is singly connected, that is the sample does not contain isolated holes.

Such a sample is shown in figure 6.3, where A, B, C and D denote small contacts.

If we force a current I between two non-neighbouring contacts, say A and C, then in the presence of a magnetic induction the Hall voltage occurs at the other two non-neighbouring contacts, B and D. Recall that we have already considered the Hall voltage mode of operation of a flat plate of arbitrary shape (see figure 4.26). The Hall voltage generated in such a structure is given by (4.89). Obviously, it does not depend on the plate shape. Therefore, the Hall coefficient of any flat plate fitted with point contacts can be measured in exactly the same manner as that of a long slab. In particular, equations (6.42) and (6.43) also hold.

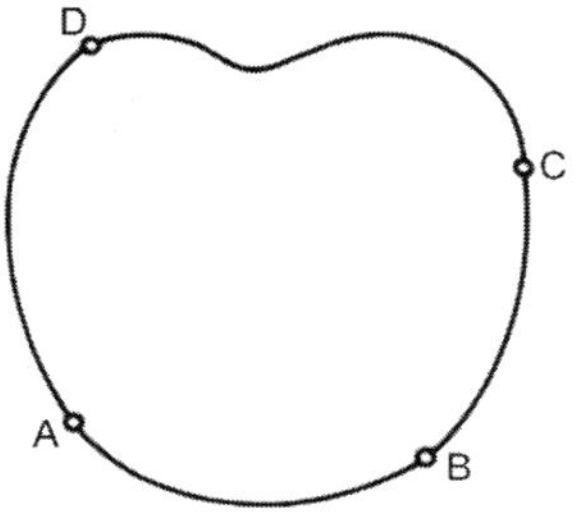

Figure 6.3. A van der Pauw sample for characterizing material. This is a thin plate of arbitrary shape, fitted with four small contacts arbitrarily placed at its periphery.

To determine the sheet resistance of the sample, according to van der Pauw, two measurements have to be made (see figure 6.4). First, a current I is forced through the sample between two neighbouring contacts, say A and B in figure 6.4(a), and the voltage is measured between the other two contacts, D and C. Then the measurement is repeated for other pairs of contacts (b), and the ratios

$$R_{AB,CD} = \frac{V_D - V_C}{I_{AB}} = \frac{V_{DC}}{I_{AB}} \qquad R_{BC,DA} = \frac{V_{DA}}{I_{BC}} \tag{6.44}$$

are calculated. We shall call these ratios the trans-resistances. Van der Pauw showed that the following relation holds:

$$\exp\left(-\pi\frac{R_{AB,CD}}{R_S}\right) + \exp\left(-\pi\frac{R_{BC,DA}}{R_S}\right) = 1 \tag{6.45}$$

where R_S denotes the *sheet resistance*. This equation can be put in the following form:

$$R_S = \frac{\pi}{\ln 2}\frac{R_{AB,CD} + R_{BC,DA}}{2} f\left(\frac{R_{AB,CD}}{R_{BC,DA}}\right) \tag{6.46}$$

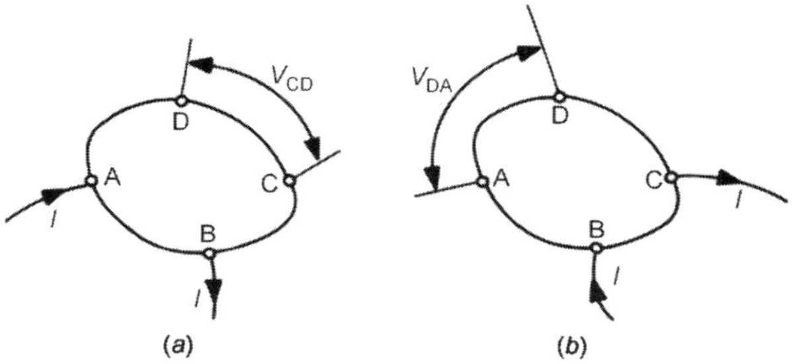

Figure 6.4. The two measurements required to determine the sheet resistance of a layer according to the van der Pauw method. I denotes the current forced through the sample, and V is the voltage which occurs between the other two contacts.

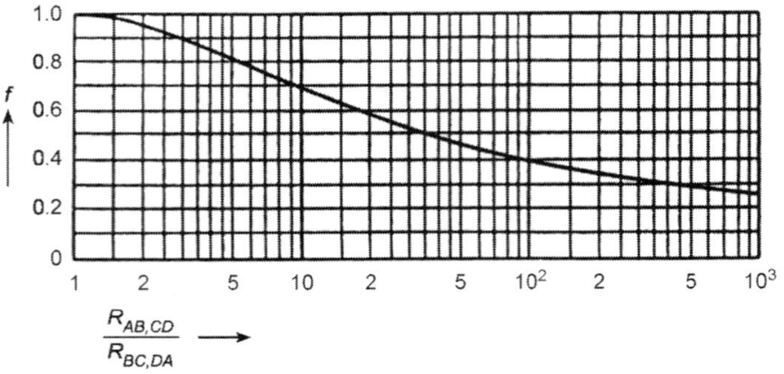

Figure 6.5. The function $f(R_{\mathrm{AB,CD}}/R_{\mathrm{BC,DA}})$, figuring in equation (6.46) (reprinted from [11]).

where f denotes a function of the trans-resistance ratio. This function is plotted in figure 6.5. Thus if the two trans-resistances are known, with the aid of equation (6.46) the sheet resistance of the sample can easily be determined.

In practice, structures which are invariant for a rotation of 90° (figure 4.15(c)–(h)) are preferably used. In such a structure, offset is small, and the Hall voltage can be accurately extracted from the output voltage (see (6.42)). Moreover, the two transverse resistances are then approximately equal, $f \approx 1$, and (6.46) reduces to

$$R_{\mathrm{S}} \simeq \frac{\pi}{\ln 2} R_{\mathrm{T}} \simeq 4.53 R_{\mathrm{T}} \qquad R_{\mathrm{T}} = \frac{R_{\mathrm{AB,CD}} + R_{\mathrm{BC,DA}}}{2}. \tag{6.47}$$

Here R_{T} denotes the average value of the two trans-resistances.

If the contacts are of finite size and not positioned at the boundary of the sample, equations (6.43)–(6.47) are not exact. Table 6.1 illustrates the errors introduced in the experimentally determined values of resistivity and Hall coefficient due to the shown non-idealities of a contact in a circular sample.

For a rotation symmetrical structure with finite-sized contacts, an expression analogous to (6.47) is valid [13]:

$$R_{\mathrm{S}} = C(\lambda) R_{\mathrm{T}} \tag{6.48}$$

Here λ denotes the ratio of the sum of the lengths of the contacts to the length of the plate boundary (see (4.57)) and $\mathrm{C}(\lambda)$ is a geometrical coefficient. In view of (6.47), at $\lambda \to 0$ (point contacts), $C(\lambda) \to \pi/\ln 2$. The function $C(\lambda)$ has been determined for several symmetrical shapes. For a circular structure (see figure 4.15(f)) the function $C(\lambda)$ is plotted in figure 6.6 [13]. In this case, $\lambda = \theta/\pi$. At small values of λ, the geometrical coefficient $C(\lambda)$ may be approximated by

$$C(\lambda) \simeq 4.53236 + 1.00837\lambda^2. \tag{6.49}$$

This formula is accurate to within 1% if $\lambda < 0.49$ [13].

Table 6.1. The relative errors $\Delta\rho/\rho = \Delta R_S/R_S$ and $\Delta R_H/R_H$ in the values of the resistivity ρ and the Hall coefficient R_H caused by the final size or the displacement (l) of a contact in a circular sample. (Adapted from [12].)

		$\Delta\rho/\rho$	$\Delta R_H/R_H$
(a)	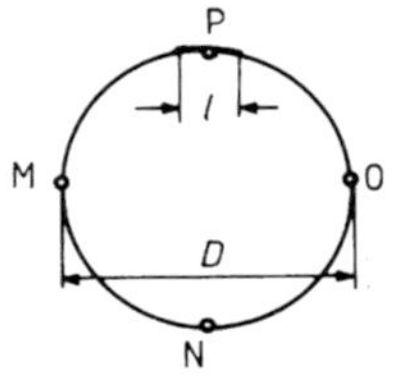	$\approx \dfrac{-l^2}{16D^2 \ln 2}$	$\approx \dfrac{-2l}{\pi^2 D}$
(b)	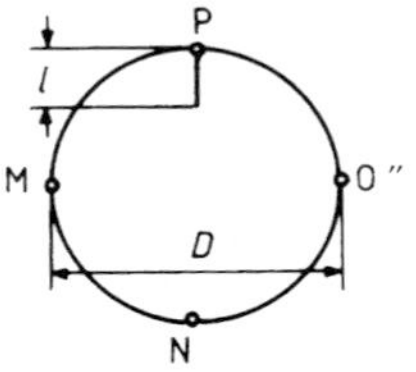	$\approx \dfrac{-l^2}{4D^2 \ln 2}$	$\approx \dfrac{-4l}{\pi^2 D}$
(c)	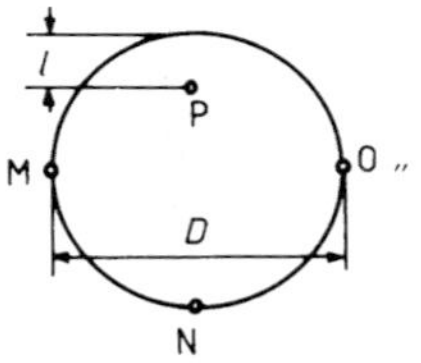	$\approx \dfrac{-l^2}{2D^2 \ln 2}$	$\approx \dfrac{-2l}{\pi D}$

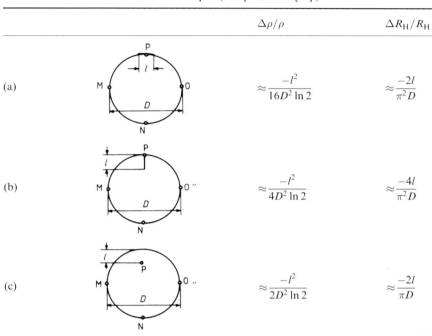

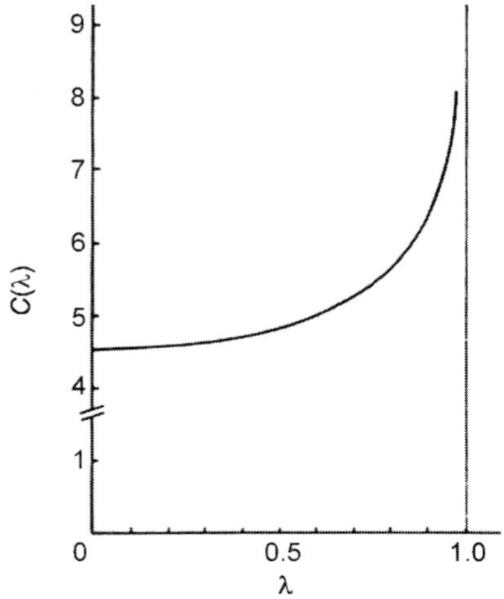

Figure 6.6. The coefficient $C(\lambda)$ figuring in equation (6.48), for a circular sample with four equal finite contacts (adapted from [13]).

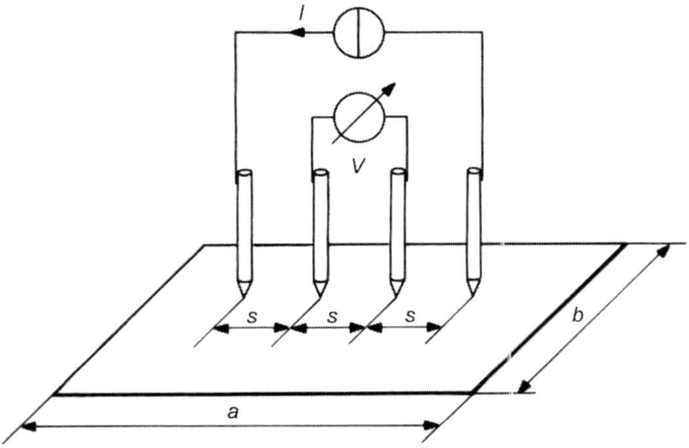

Figure 6.7. Four-point probe method for measuring the resistivity of a thin conductive layer.

Van der Pauw noted that the influence of the finite contacts can be virtually eliminated by using a clover-leaf-shaped sample, shown in figure 4.15(c). Clover-leaf-shaped samples have since been referred to as van der Pauw samples.

6.2.3 Four-point probe methods

In a four-point probe method, the four contacts needed for the resistivity and Hall coefficient measurements are provided by pressing four needle-like probes on the layer to be characterized. Therefore, the preparation of the test sample can be greatly simplified.

Resistivity

A four-point probe method for measuring resistivity was proposed by Valdes in 1954 [14]. His method has been used ever since as the standard method for resistivity measurement in the semiconductor industry. It is illustrated in figure 6.7. The four equally spaced collinear probes are pressed on to the layer to be measured. A current I is forced through the outer probes, and the voltage drop V across the inner probes is measured. The sheet resistance of the layer is then given by

$$R_{\mathrm{S}} = C\frac{V}{I} \qquad (6.50)$$

where C is a numerical coefficient. The coefficient C depends on the ratio of the probe distance s to the specimen dimensions. For a very thin and infinite layer, $C = \pi/\ln 2 \approx 4.5324$. For a rectangular specimen with probes pressed

Table 6.2. Correction factor C (see equation (6.50)) for the measurement of sheet resistance with a four-point probe on a rectangular sample. a and b are the sample dimensions, and s is the distance between the probes (see figure 6.7). It is assumed that $t/s \rightarrow 0$, t being the thickness of the sample (after [16]).

b/s	$a/b = 1$	$a/b = 2$	$a/b = 3$	$a/b \geq 4$
2.0		1.9454	1.9475	1.9475
2.5		2.3532	2.3541	2.3541
3.0	2.4575	2.7000	2.7005	2.7005
4.0	3.1137	3.2246	3.2248	3.2248
5.0	3.5098	3.5749	3.5750	3.5750
7.5	4.0095	4.0361	4.0362	4.0362
10.0	4.2209	4.2357	4.2357	4.2357
15.0	4.3882	4.3947	4.3947	4.3947
20.0	4.4516	4.4553	4.4553	4.4553
40.0	4.5210	4.5129	4.5129	4.5129
∞	4.5324	4.5324	4.5324	4.5324

in the middle as shown in figure 6.7, the values of the coefficient C are listed in table 6.2 [16].

Obviously, the four point contacts needed in the van der Pauw method can be provided by pressing four probes at the periphery of a sample. For example, the probes can be pressed at the periphery of a circular sample [15].

Hall measurements

A method has been devised for measuring resistivity and the Hall coefficient with the aid of a four-point probe pressed directly on to a non-structured wafer [17]. The arrangement is illustrated in figure 6.8. The four contacts needed for the measurements are provided by pressing the four collinear probes on the layer to be measured, in a similar manner to the conventional four-point probe resistivity measurement (see figure 6.7). The measurement procedure is identical to that in the van der Pauw method.

To understand how this measurement method works, imagine that we want to measure resistivity and the Hall coefficient using the van der Pauw method. As we noted before, it is convenient for the test sample to be rotation symmetrical. But what is really important is not geometrical, but electrical, symmetry. Therefore, with good reason, we may choose the geometry of the test sample as shown in figure 6.9(a). This is a plate with a partially straight boundary, on which the four point contacts A–D are placed. According to examples 1 and 2 of §4.2.3, such a plate is equivalent to the circular symmetrical Hall plate shown in figure 6.9(b), if the following conditions are fulfilled (see figure 6.9(a)):

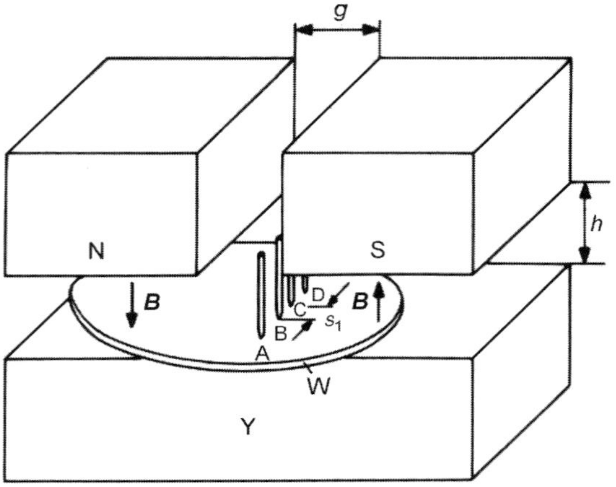

Figure 6.8. The four-point probe method for measuring resistivity and the Hall coefficient of a non-structured wafer. The wafer to be measured (W) is exposed to the magnetic induction ***B*** between the poles of a magnet (N, S) and a soft ferromagnetic yoke (Y). The contacts are provided by pressing the four probes (A, B, C, D) on to the wafer. To simplify the pertinent mathematical relations, the air gaps h and g should be small relative to the probe spacing: $s_1 \gg h$ and $s_1 \gg g$.

(i) The contacts are positioned at the straight boundary so that

$$\begin{aligned} s_1 &= 2x_{12} = 2U \tan \pi/8 \simeq 0.83U \\ s_2 &= x_{45} - x_{12} = U(\cot \pi/8 - \tan \pi/8) \simeq 2U \end{aligned} \tag{6.51}$$

where U denotes a length unit.

(ii) The distance d between any contact and the closest other boundary is $d > 10U$.

The first condition follows from (4.34) if we put $\theta \to 0$ (see figure 4.10); then $x_1 = x_2 = x_{12} = \tan \pi/8$ and $x_4 = x_5 = x_{45} = \cot \pi/8$. The second condition is required to keep the influence of the other boundaries on the electrical symmetry small (see (4.43)).

If the above conditions are fulfilled, the sample shown in figure 6.9(a) is suitable for the application of the van der Pauw measurement method. Figure 6.10 illustrates the measurements of the Hall coefficient (a) and resistivity (b). The two measurements (a) at B and $-B$ yield the Hall voltage (6.42), and the sheet Hall coefficient is then given by (6.43). The two measurements (b) yield the two trans-resistances (6.44), and the sheet resistance is then given by (6.47).

Let us now consider a sandwich of two identical straight boundary samples. Let the contacts be arranged as explained above, and each pair of corresponding contacts shorted (see figure 6.11(a)). Since the bias current is now divided into two slices, the van der Pauw measurements yield only

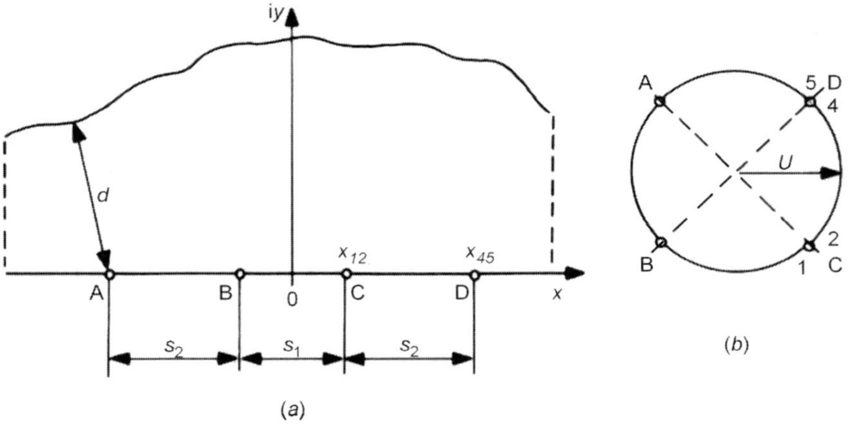

Figure 6.9. (a) A *van der Pauw sample* with the four point contacts (A, B, C, D) placed at the *straight boundary*. (b) A circular sample with four point contacts, invariant for a rotation of $\pi/2$. If the conditions (6.51) are fulfilled, these two samples are galvanomagnetically equivalent.

half of the voltages measured before. Therefore, instead of (6.43) and (6.47), for one slice we have

$$R_{HS} = 2V_H/IB \tag{6.52}$$

$$R_S \simeq \frac{\pi}{\ln 2}\frac{2V_{CD}}{I_{AB}}. \tag{6.53}$$

Imagine now that we start rotating the top slice around the straight boundary, together with its magnetic induction vector (see figure 6.11(b)). During

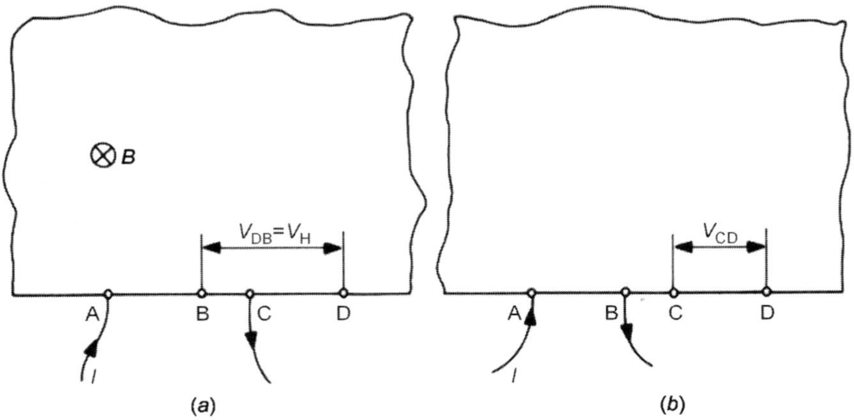

Figure 6.10. Schematic representation of the van der Pauw measurements on the straight boundary sample from figure 6.9(a): (a) measuring the Hall coefficient; (b) measuring the sheet resistance.

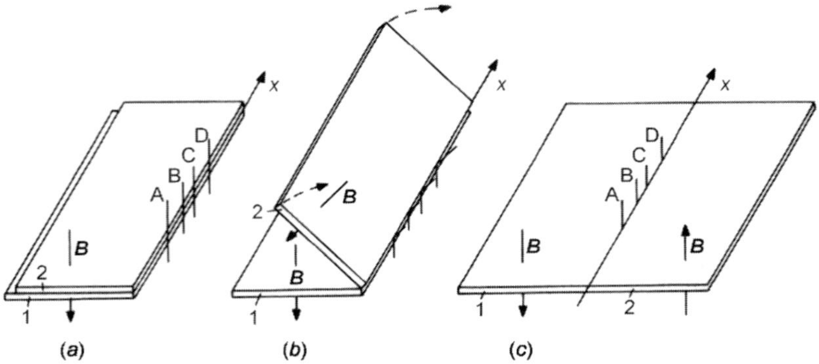

Figure 6.11. Imaginary transformation of a sample with contacts at a straight boundary (a) into a sample with contacts in the middle of the sample (c). (a) Two identical straight-boundary samples, 1 and 2, connected in parallel by the contacts A, B, C, D. (b) The sample 2 rotates around the x axis. The corresponding magnetic induction $\boldsymbol{B}$ also rotates. (c) The two samples 1 and 2 are now in the same plane. The x axis is the line where the magnetic induction $\boldsymbol{B}$ suddenly changes direction.

the rotation the current and voltage distributions in the samples do not alter. After completing a rotation of 180°, the two slices lie in the same plane (figure 6.11(c)). Since the electrical potentials at the common boundary have remained identical, we may 'remove' the boundary by merging the slices along the boundary. Thus we obtain one single continuous plate. The former boundary of the electrically conductive regions of the two samples is now merely the line where the magnetic field changes its sign.

Obviously, an isolating boundary of a plate and the line in a plate where a magnetic induction suddenly reverses its direction are fully equivalent. It is convenient to introduce the notion of a galvanomagnetic discontinuity line, which includes both these cases. Then we can describe the arrangements shown in figures 6.11(a) and (c) by stating that the four contacts are placed on a straight galvanomagnetic discontinuity line.

Since the arrangement of figure 6.11(c) is equivalent to that of figure 6.11(a), it can also be used for measuring resistivity and the Hall coefficient. In particular, equations (6.52) and (6.53) also hold. Figure 6.8 shows schematically an implementation of the arrangement from figure 6.11(c). Therefore, this configuration enables measurements of resistivity and the Hall coefficient to be made on an unstructured wafer.

6.2.4 Measurements on a large-contact sample

Measurements on a small-contact sample and on a large-contact sample can provide equivalent information for characterizing a material. However, if the

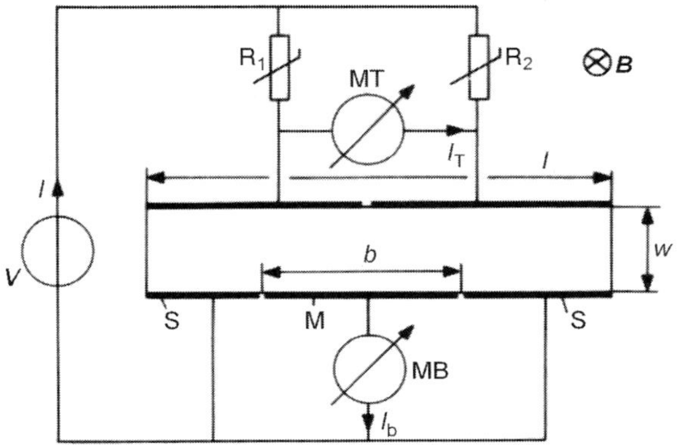

Figure 6.12. Arrangement for measuring resistivity and Hall mobility using a large-contact sample. V denotes the biasing voltage source, R_1 and R_2 are adjustable resistors, and MT and MB are low-resistance ammeters (after [18]).

material under investigation is of a very high resistivity, measurements on a large-contact device are usually much easier to make [18, 19].

For a reliable measurement, the bias current in the sample must be high enough. In order to increase the bias current in a high-resistivity sample, one may increase the supply voltage. But the maximum supply voltage is usually limited. For example, a severe limitation of the applicable voltage comes about because of breakdown phenomena in the sample. Alternatively, one may increase the current in the material under test by choosing a short- and large-contact sample. A large-contact sample is preferable for operation in the Hall current mode (see §4.4). In addition, the Hall current mode of operation enables us to avoid the interference of the surface leakage currents with the measurements. In a high-resistivity material, parasitic surface currents may be much larger than the bulk current.

Let us consider the galvanomagnetic measurements on a large-contact sample operated in the *Hall current mode* [18, 19].

A suitable configuration for measuring sample resistivity and Hall current is shown in figure 6.12 [18]. The sample has the form of a parallelepiped of dimensions l, w and t, with $l \gg w$ (note that we keep the same notation as in figure 4.33). The dimensions of the contacts and the isolation slits between them are chosen so that the device approximates well an infinitely-large-contact device. The resistances of the resistors R_1 and R_2 are much smaller than the sample resistance; and the resistances of the galvanometers are much smaller than R_1 or R_2. Thus both upper and lower sample boundaries are at constant potentials. The lower side contacts S shield the lower middle contact M from the surface leakage currents.

In a measurement in the absence of a magnetic field, the resistors R_1 and R_2 are first adjusted so that the transverse current I_T vanishes. The measured resistance is given by

$$R(0) = \frac{V}{I_b} = \rho \frac{w}{bt} \tag{6.54}$$

where I_b denotes the bulk current over the area bt and ρ is the sample resistivity. After exposing the device to a magnetic field, the galvanometer MT shows a transverse current

$$I_T(B) = I_H/2. \tag{6.55}$$

The Hall current I_H is now given by

$$I_H = \mu_H \frac{w}{b} I_b B \tag{6.56}$$

which is obtained from (4.124) by replacing there $2I_0/l$ by I_b/b (compare figures 4.33(d) and 6.12). In addition, by again reading the values of V and I_b, the resistance $R(B)$ can be found. The quantities $R(0)$ and $R(B)$ determine the Corbino magnetoresistance (see §3.1.3).

In deriving the above expressions, we assumed an extrinsic material. If this is not the case, the mobility μ_H has to be replaced by the average mobility $\bar{\mu}_H$ (3.177).

6.2.5 The Hall mobility of minority carriers

Buehler and Pensak proposed a method for directly measuring the Hall mobility of minority carriers [20]. The measurement set-up for measuring the Hall mobility of holes in an n-type semiconductor is shown in figure 6.13. In this

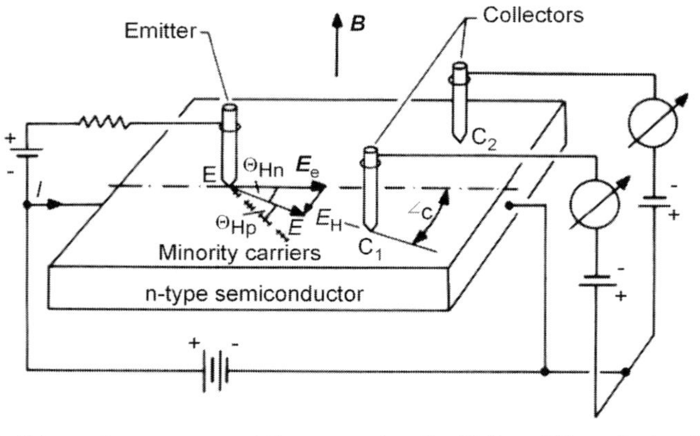

Figure 6.13. Schematic arrangement for measuring the Hall mobility of minority carriers. The minority carriers (holes), injected by the emitter E, drift along the specimen under the influence of the resultant electric field $\boldsymbol{E}$ and the magnetic induction $\boldsymbol{B}$. The collector current depends on the density of minority carriers passing under the collector probes C_1 or C_2 (after [20]).

method, a long slab of n-type semiconductor material is used simultaneously as a Hall plate and as the base region of a point-contact transistor. As in a conventional Hall device, a bias current I is forced through the slab. The needles E, C_1 and C_2 can be used either as point probes to measure the electric potential at the plate surface, or as the emitter (E) and collectors (C_1 and C_2) of a point-contact transistor.

Let us first consider briefly the operation of the structure as a Hall plate. The external electric field $\boldsymbol{E}_e$ in the sample is collinear with the sample's longitudinal axis. As a result of the Hall effect, the total electric field tilts for the Hall angle with respect to the external field (see §3.1.1). This Hall angle is predominantly determined by majority carriers. According to (3.152),

$$\tan \Theta_{\mathrm{Hn}} = \mu_{\mathrm{Hn}} B. \tag{6.57}$$

The Hall angle of majority carriers can easily be found by measuring the longitudinal and transverse components of the electric field with the aid of the three probes. According to figure 3.2, the Hall angle is given by

$$|\tan \Theta_{\mathrm{Hn}}| = |\boldsymbol{E}_{\mathrm{H}}|/|\boldsymbol{E}_{\mathrm{e}}| \tag{6.58}$$

where $\boldsymbol{E}_{\mathrm{H}}$ is the Hall electric field, which is the transverse component of the total field.

Assume now that we apply the contact E as an emitter, which injects minority carriers into the slab. If the external electric field is high, we may neglect the diffusion terms in the current density (3.324). Then the injected carriers make a well-defined minority carrier beam. The carriers in the beam see no boundaries, and thus cannot take part in creating the Hall electric field. The transport of minority carriers in the beam is thus subject to the current deflection effect (see §3.1.2). Owing to the current deflection effect, the beam is tilted with respect to the total electric field in the sample. This tilting angle equals the Hall angle of minority carriers:

$$\Theta_{\mathrm{Hp}} = \arctan(\mu_{\mathrm{Hp}} B). \tag{6.59}$$

The total deflection angle of the minority carrier beam with respect to the sample longitudinal axis is then

$$\phi_{\mathrm{p}} = -\Theta_{\mathrm{Hn}} + \Theta_{\mathrm{Hp}}. \tag{6.60}$$

Recall that, by convention, $\mu_{\mathrm{Hn}} < 0$ and $\mu_{\mathrm{Hp}} > 0$. Hence the negative sign in (6.60): the two angles add up.

The total deflection angle can be determined in the following way. When reverse biased with respect to the base, the contacts C_1 and C_2 operate as collectors. A collector current depends on the position of the collectors with respect to the minority carrier beam. If the magnetic field varies, so do the deflection angle ϕ_{p} and the collector current. If we determine the magnetic induction B_{m} at which a collector current attains a maximum value, then

$$\phi_{\mathrm{p}}(B_{\mathrm{m}}) = \phi_{\mathrm{c}} \tag{6.61}$$

where ϕ_c denotes the angle enclosed between the sample axis and a collector C_1 (see figure 6.13).

From equations (6.58)–(6.61), we may now determine the Hall angle and the Hall mobility of minority carriers. For small Hall angles the total deflection angle of the minority carrier beam is

$$\phi_p \simeq -\mu_{Hn} B_m + \mu_{Hp} B_m. \tag{6.62}$$

Then the Hall mobility of minority carriers is given by

$$\mu_{Hp} = \frac{1}{B_m}\left(\phi_c - \frac{|\boldsymbol{E}_H|}{|\boldsymbol{E}_e|}\right) \tag{6.63}$$

where we have made use of (6.58).

6.3 Measuring packaging stress

In plastic encapsulated microcircuits, mechanical stress is induced during the encapsulation process. The largest packaging stresses are due to the mismatch of the thermal expansion coefficients between the die and the moulding material. Packaging stresses can degrade silicon IC performance. The piezo-resistance effect can influence silicon resistors [21] and MOSFET characteristics [22]. Piezo-junction effect [23] can influence band-gap references and silicon temperature sensors. Both the piezo-resistance effect and the piezo-Hall effect [24] affect Hall sensors (see §5.8.1). If the stresses are not stable in time, they may lead to instability of the above mentioned electron devices.

In order to understand the problem, it is necessary to measure the stress in plastic packages during the encapsulation process. Different test chips based on piezo-resistor arrays were proposed for the stress mapping on a silicon die [25]. Silicon resistors are strongly temperature dependent. Thus, temperature fluctuations can cause serious errors in the experimental values of in plane normal stresses. Only the difference of in plane normal stresses and in plane shear stress can be measured in a temperature-compensated manner.

In this section, we present a method for measuring the in plane normal stresses in a silicon die based on the *piezo-Hall effect* [26]. In terms of relative change of appropriate physical quantities (magnetic sensitivity for the piezo-Hall method and resistance for the piezo-resistance method), the piezo-Hall effect is as strong as the piezo-resistance effect (see §3.5.5). However, the current related magnetic sensitivity of a Hall device is roughly 10 times less dependent on temperature ($\alpha_H = 500\,\text{ppm}/^\circ\text{C}$) than the resistance of the piezoresistors in the similar n-type silicon layers ($\alpha_R = 6700\,\text{ppm}/^\circ\text{C}$). Consequently, 10 times more accurate in plane normal stress sensors can be realized using Hall plates [27].

6.3.1 Measurement method

The relative change of the Hall coefficient R_H in the (100) silicon plane with $\boldsymbol{B} \parallel \langle 100 \rangle$, $\boldsymbol{J} \parallel \langle 010 \rangle$ or $\langle 010 \rangle$, under mechanical stress is given by [24]

$$\frac{\Delta R_H}{R_H} = P_{12}(\sigma_{XX} + \sigma_{YY}) \tag{6.64}$$

where $\boldsymbol{B}$ is the magnetic induction, $\boldsymbol{J}$ is the current density, P_{12} is the piezo-Hall coefficient ($P_{12} = 40 \times 10^{-11}\ \mathrm{Pa}^{-1}$) [24]. When stress $\sigma \perp B \parallel \langle 100 \rangle$, σ_{XX} is in plane normal stress component in the x direction ($\langle 011 \rangle$, long edge of the chip, see figure 6.14) and σ_{YY} is the stress component in the y direction. The stress σ_{ZZ} in the direction normal to the chip surface and the shear stress σ_{XY} are negligible compared with the stresses σ_{XX} and σ_{YY}. This assumption is based on the results summarized in [25].

In order to measure the *encapsulation stress*, one must first measure the current-related magnetic sensitivity of a Hall plate incorporated in a non-encapsulated test chip. Once the test chip is encapsulated in the IC plastic package, the magnetic sensitivity should be measured again. The current-related magnetic sensitivity of a Hall plate is given by $S_I = G_H |R_H| / t$ (see §5.1.1). After temperature cycles, the same measurements can be repeated under the same conditions. The changes of the sensitivity ΔS_I can be determined with a high accuracy. A change of the sensitivity is related to a change of the mechanical stress. The change of the stress is given by

$$\Delta(\sigma_{XX} + \sigma_{YY}) = \frac{1}{P_{12}} \frac{\Delta S_I}{S_I}. \tag{6.65}$$

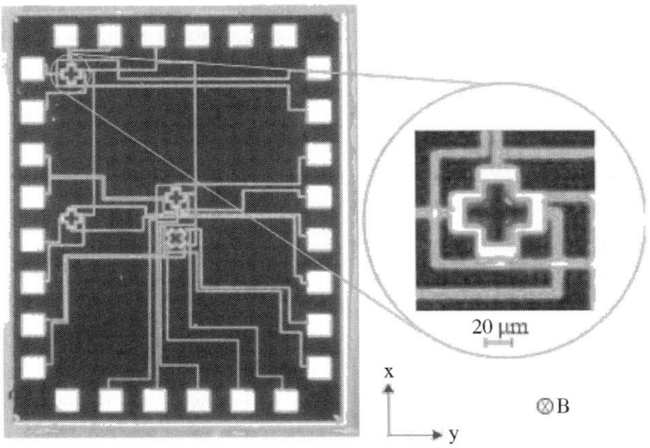

Figure 6.14. CMOS Hall stability test chip. The size of the chip is 1.6 × 2 mm. Four Hall elements (cross-shapes) exist on the chip: two at the chip centre (one in $\langle 110 \rangle$ and another rotated for 45°, i.e in the $\langle 010 \rangle$ direction), one in a corner and one at the chip side. On the right-hand side, a zoom-in on a Hall element is shown. The coordinate directions are as follows: $y \parallel \langle 110 \rangle$ and $x \perp \langle 110 \rangle$) (reprinted from [26]).

To verify the method, stress drift measurements based on the piezo-Hall effect were performed on the Hall plates made in a conventional n-well CMOS IC process. The chip is shown in figure 6.14. This chip was packaged in a conventional plastic TSSOP (thin shrink small outline package). All measurements were carried out at the magnetic field of 1 T controlled by an NMR teslameter. The connection-commutation technique was applied for offset reduction. An air-stream system was utilized for temperature cycling tests. The Hall plate sensitivity was always measured at 20 °C ($\Delta T \leq \pm 0.1$ °C). The overall accuracy of the current-related magnetic sensitivity was better than 200 ppm. The corresponding stress drift accuracy was ±0.25 MPa. Temperature cycling (−55 °C, 125 °C) with 15 min at the extreme temperatures was performed. Sensitivity and time-drift were measured after temperature cycling and the stress drift was calculated using equation (6.65).

6.3.2 Examples of the measurement results

The variations in the die stress after a temperature cycle are presented in figure 6.15. The stress variations are always related to the initial stress state (the stress state before any temperature cycling).

A shift of −14 MPa after the first temperature cycle is observed. This stress behaviour is a result of the viscoelastic properties of the package materials. There was no essential change after the first half-cycle (15 min at −55 °C). The most important change was at the high temperature because of the much faster relaxation of the material. The stress decreased with time (measurement b1–b5 in figure 6.15). After the second temperature cycle, the stress variation was −19 MPa. Since the normal in plane stresses in the IC plastic packages are compressive, i.e. with negative sign, the magnitude of the stress is increasing with the number of temperature cycles.

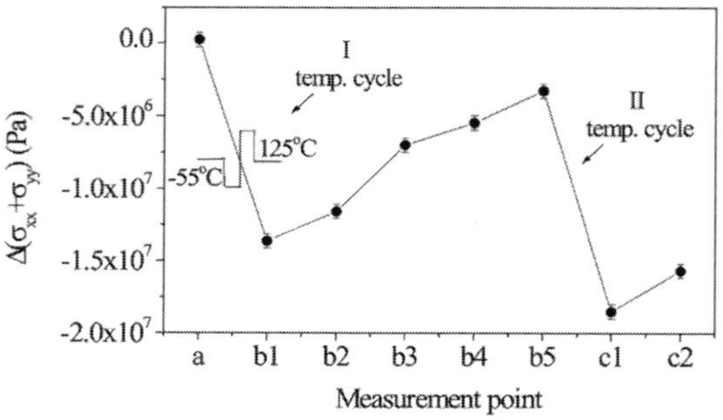

Figure 6.15. The change in in-plane normal stresses (measured always at 20 °C) after 15 min at −55 °C (a), after 15 min at 125 °C (b1–b5) and after the second temperature cycle (c1, c2). Note that the horizontal axis is not on a time scale. The measurement points from b1 to b5 are presented on a time scale in figure 6.16 (reprinted from [26]).

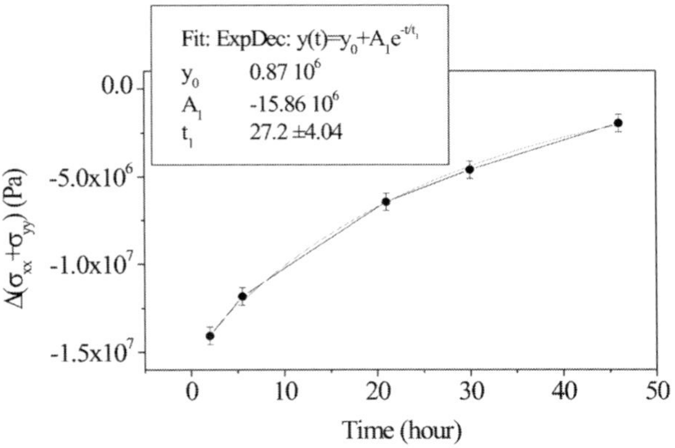

Figure 6.16. Time-drift of the sum of the in plane normal packaging stresses (measured at 20 °C) after one temperature cycle (from the measurement points b1–b5 in figure 6.15) (reprinted from [26]).

Time-drift of the stress after one temperature cycle is presented in figure 6.16. The relaxation process can be well represented with an exponential decay function (the equation is shown in figure 6.16). The time constant of the relaxation is 27 hours.

In figure 6.17 a lower stress shift can be observed for the Hall plate in the chip corner than for that at the chip centre. Note that the stress is lower in the corners but stress gradients are higher and offset problems may arise. Since the magnetic sensitivity of a Hall device is affected by the absolute stress values (not by stress gradients), design for stability can include the placement of the Hall device in a less stressed area, e.g. closer to a corner.

Another measurement shows that the Hall cell along the ⟨010⟩ direction is equally affected by the piezo-Hall effect as the cell in the ⟨011⟩ crystallographic direction. This is in accordance with theory [24]. In the (100)

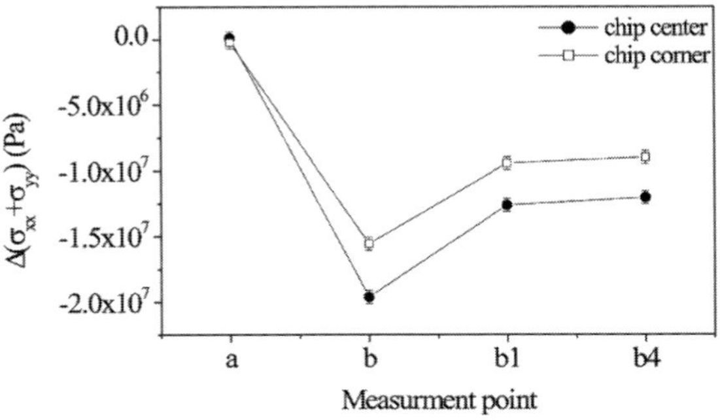

Figure 6.17. The change of the in plane normal stresses (measured at 20 °C) after 15 min at −55 °C (a), after 15 min at 125 °C (b1–b4) at two positions on the chip (chip centre and chip corner). Note that the horizontal axis is not on a time scale (reprinted from [26]).

crystallographic plane all current flow directions are equivalent from the piezo-Hall effect point of view.

References

[1] Kuchis E V 1974 *Metody Issledovanija Effekta Holla* (Moscow: Sovetskoe Radio)

[2] Wieder H H 1979 *Laboratory Notes on Electrical and Galvanomagnetic Measurements* (Amsterdam: Elsevier)

[3] Schroder D K 1998 *Semiconductor Material and Device Characterization* (New York: Wiley), sections 2.5, 8.3 and 8.4

[4] Ghandi S K 1983 *VLSI Fabrication Principles* (New York: Wiley) sections 4.10.2 and 5.7.1

[5] Sze S M (ed.) 1983 *VLSI Technology* (New York: McGraw-Hill) pp 186–93

[6] Galloni R and Sardo A 1983 Fully automatic apparatus for the determination of doping profiles in Si by electrical measurements and anodic stripping *Rev. Sci. Instrum.* **54** 369–73

[7] Blight S R, Nicholls R E, Sangha S P S, Kirby P B, Teale L, Hiscock S P and Stewart C P 1988 Automated Hall profiling system for the characterisation of semiconductors at room and liquid nitrogen temperatures *J. Phys. E: Sci. Instrum.* **21** 470–9

[8] Rhoderick E H and Williams R H 1988 *Metal-Semiconductor Contacts* (Oxford: Clarendon)

[9] Popovic R S 1978 Metal-N-type semiconductor ohmic contact with a shallow N+ surface layer *Solid-State Electron.* **21** 1133–8

[10] Dauphinee T M and Mooser E 1955 Apparatus for measuring resistivity and Hall coefficient of semiconductors *Rev. Sci. Instrum.* **26** 660–4

[11] van der Pauw L J 1958 A method of measuring specific resistivity and Hall effect of discs of arbitrary shape *Philips Res. Rep.* **13** 1–9

[12] van der Pauw L J 1958/9 Messung des spezifischen Widerstandes und Hall-Koeffizienten an Scheibchen beliebiger Form *Philips Tech. Rundsch.* **20** 230–4

[13] Versnel W 1978 Analysis of symmetrical van der Pauw structures with finite contacts *Solid-State Electron.* **21** 1261–8

[14] Valdes L B 1954 Resistivity measurements on germanium for transistors *Proc. IRE* 42420–7

[15] Lange J 1964 Method for Hall mobility and resistivity measurements on thin layers *J. Appl. Phys.* **35** 2659–64

[16] Smits F M 1958 Measurement of sheet resistivities with four-point probe *Bell Syst. Tech. J.* 37711–18

[17] Popovic R 1990 Vorrichtung und Verfahren zum Messen characterisierender Eigenschaften eines Halbleiermaterials *Swiss Patent Appl.* 00317/90–3

[18] Dobrovol'skii V N and Krolovets A N 1983 Hall current and its use in investigations of semiconductors (review) *Sov. Phys.-Semicond.* **17** 1–7

[19] Dobrovol'skii V N and Krolovets A N 1979 Determination of the electrical resistivity and Hall coefficient of samples dual to van der Pauw samples *Sov. Phys.-Semicond.* **13** 227–8

[20] Buehler M G and Pensak L 1964 Minority carrier Hall mobility *Solid-State Electron.* **74** 31–8

[21] Kanda Y 1982 A graphical representation of the piezoresistance coefficients in silicon *IEEE Trans. on EDs* **29** 64–70

[22] Hamada A *et al* 1991 A new aspect of mechanical stress effect in scaled MOS devices *IEEE Trans. on EDs* **38**(4) 895–900

[23] Kanda Y 1967 Effects of stress on germanium and silicon p–n junctions *Jpn. J. Appl. Phys.* **6** 475–86

[24] Halg B 1988 Piezo-Hall coefficients of n-type silicon *J. Appl. Phys.* **64** 276–82

[25] Zou Y *et al* 1998 Three-dimensional die surface stress measurements in delaminated and non-delaminated plastic packages *Proc. 48th Elec. Components and Technology Conf.* 1223–34

[26] Manic D, Petr J and Popovic RS 2001 Die stress drift measurement in IC plastic packages using the piezo-Hall effect *Microelectronics Reliability* **41** 767–71

[27] Manic D 2000 *Drift in Silicon Integrated Sensors and Circuits due to Thermo-Mechanical Stresses* (Konstantz: Hartung-Gorre)

Chapter 7

Magnetic-sensitive field-effect and bipolar devices

This chapter is devoted to the magnetic-sensitive versions of otherwise non-magnetic semiconductor devices, such as diodes, MOS transistors, bipolar transistors and thyristors. These devices are known as magnetodiodes, MOS Hall devices, magnetotransistors and carrier-domain magnetometers. Like conventional Hall devices, these devices may be applied as magnetic sensors, or as tools for exploring carrier transport phenomena.

In principle, any semiconductor device is sensitive to a magnetic field. In the presence of a magnetic field, the Hall effect takes place in any substance carrying current, and thus also in the active region of a semiconductor device. Via the Hall effect, the magnetic field modulates the current and voltage distributions in the device. This modulation shows up in the device external characteristics. However, the magnetic sensitivity of 'normal' devices is usually negligibly small. To become useful for practical applications, the magnetic sensitivity must be boosted by choosing an appropriate device structure and operating conditions.

In conventional Hall devices, the Hall effect is accessible in its 'pure' form. In the devices discussed in this chapter, the Hall effect coexists with other physical phenomena pertinent to 'non-magnetic' semiconductor devices, such as the field effect, injection of carriers, and conductivity modulation. The Hall effect may combine with these phenomena in various ways and thus yield results, which appear as completely new effects (see, for example, the magnetoconcentration effect, §3.5.2). This makes the analysis and optimization of devices, such as magnetotransistors, much more difficult than that of a Hall plate. But this very fact has also motivated a lot of research in the field. In the hope of devising a magnetic sensor, which might outperform a Hall plate, many interesting Hall effect devices related to diodes, transistors and thyristors have been proposed.

The aim of this chapter is to provide a simple and clear picture of the physical effects basic to the operation of non-conventional Hall effect devices. Accordingly, we shall discuss only some selected examples of the devices proposed up to now. Exhaustive reviews of the work in the field and performances achieved can be found in the review-type papers [1–3].

7.1 MOS Hall effect devices

One of the conditions for the high magnetic sensitivity of a Hall plate is a low area carrier density in the plate (see (5.6)). A very low area carrier density can be obtained if the channel of a MOSFET is exploited as the active region of a Hall device [4]. (MOSFET stands for metal-oxide semiconductor field-effect transistor: see e.g. [5].) The idea of implementing a Hall device in the form of an MOS device looks especially attractive for sensor applications: such a device is readily integrable with MOS bias and signal-processing circuits. Unfortunately, an MOS Hall effect device also has a few serious drawbacks: the carrier mobility in the channel amounts to only half its value in the bulk [5]; the $1/f$ noise of an MOS device is usually several orders of magnitude higher than that of a comparable bulk device [6]; and the surface device might be insufficiently stable. We shall review here a few example of magnetic-sensitive MOSFETs, which are sometimes called *MagFETs*.

7.1.1 MOS Hall plates

A rectangular Hall plate realized as a modified MOS transistor is shown in figure 7.1. The channel (Ch) of the MOS transistor is used as an extremely

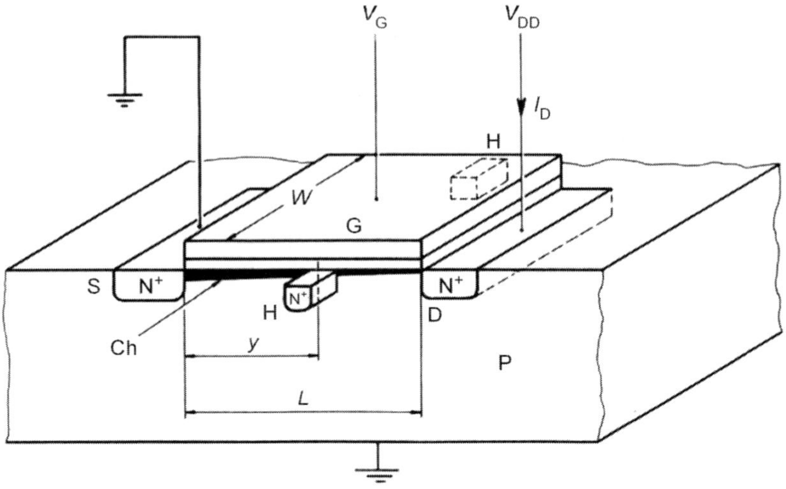

Figure 7.1. MOS Hall plate: the channel (Ch) of a MOSFET is exploited as an extremely thin conductive plate. Two Hall voltage probes H are added to the usual MOSFET structure. (Adapted from [1]. © 1986 IEEE.)

thin Hall plate, the source (S) and drain (D) as the biasing contacts, and the two additional heavily doped regions (H) are provided for sensing the Hall voltage (compare with figure 5.7). The two sense regions are fabricated simultaneously with the source and drain regions. The channel length is L, its width is W, and the sense probes are positioned at a distance y from the source.

For future reference, it is useful to recall the main properties of a MOSFET [5]. We assume an enhancement-type n-channel device, with earthed source and substrate. We shall denote the drain-to-source voltage by V_D, the gate-to-source voltage by V_G, and the threshold voltage by V_T.

At very low drain voltages, when

$$V_D \ll V_G - V_T \tag{7.1}$$

the area density of carriers in the channel is approximately constant over the channel. This charge density is given by

$$Q_{ch} \simeq C_{ox}(V_G - V_T) \tag{7.2}$$

where C_{ox} denotes the gate oxide capacitance per unit area. Then the drain current is given by

$$I \simeq \frac{W}{L} \mu_{ch} C_{ox} (V_G - V_T) V_D \tag{7.3}$$

where μ_{ch} denotes the drift mobility of carriers in the channel. This is the linear region of operation of a MOSFET.

At higher drain voltages, the carrier charge density in the channel continuously decreases with increasing distance from the source. The drain current is generally given by

$$I \simeq \frac{W}{L} \mu_{ch} C_{ox} [(V_G - V_T) V_D - \tfrac{1}{2} V_D^2] \quad \text{for } V_D \leq V_G - V_T. \tag{7.4}$$

When the drain voltage reaches the value

$$V_{Dsat} \simeq V_G - V_T \tag{7.5}$$

the carrier density at the drain end of the channel reduces virtually to zero. This is the pinch-off point. Beyond the pinch-off point, the drain current stays essentially constant. The saturated drain current is given by I_{Dsat} (V_{Dsat}) according to (7.4) and (7.5).

Let us now consider the operation of an MOS Hall device in the Hall voltage mode. An MOS Hall plate working in the linear region is equivalent to a conventional Hall plate. The only difference is that, in a conventional Hall plate, the charge carriers in the plate are provided by the material itself; and in an MOS Hall plate, the charge carriers are due to the surface field effect. Substituting in (4.91) the qnt product by Q_{ch} and I by I_D, we obtain the Hall voltage of the MOS Hall plate in the form

$$V_H = G_H \frac{r_H}{Q_{ch}} I_D B_\perp \tag{7.6}$$

with Q_{ch} being given by (7.2). Recall that G_H denotes the geometrical correction factor and r_H the Hall factor.

Let us estimate the numerical value of the Hall voltage (7.6). The oxide capacitance per unit area is given by $C_{ox} = \varepsilon_r \varepsilon_0 / t_{ox}$ with ε_r being the relative permittivity of the gate oxide, ε_0 the permittivity of free space, and t_{ox} is the oxide thickness. For SiO_2, $\varepsilon_r \simeq 4$. Thus for $t_{ox} \simeq 50\,\text{nm}$, $C_{ox} \simeq 7 \times 10^{-4}\,\text{F m}^{-2}$. For $V_T = 1\,\text{V}$ and $V_G = 5\,\text{V}$, the charge density in the channel (7.2) is $Q_{ch} \simeq 2.8 \times 10^{-3}\,\text{cm}^{-2}$. This corresponds to an area carrier density of $N_s \simeq 1.75 \times 10^{12}\,\text{cm}^{-2}$. Assuming $G_H r_H \simeq 1$, equation (7.6) yields $V_H \simeq 280\,\text{mV mA}^{-1}\,\text{T}^{-1}$.

By substituting (7.2) and (7.3) into (7.6), we obtain

$$V_H \simeq \mu_{ch} \frac{W}{L} G_H V_D B_\perp. \tag{7.7}$$

This equation is analogous to equation (4.93) for a conventional Hall plate.

At a higher drain voltage, the channel charge in the MOS Hall plate decreases with an increase in the distance from the source (y). The lower the channel charge, the higher the Hall electric field. Therefore, if we could neglect the short-circuiting effect by the supply electrodes, the Hall voltage would steadily increase by moving the sense contacts towards the drain (see figure 7.2) [7]. In addition, the channel sheet resistance also increases with an increase in distance from the source. Therefore, the short-circuiting effect by the supply electrodes is more intensive in the vicinity of the source than in the vicinity of the drain. The consequence is that the Hall voltage varies with the position of the sense contacts along the channel length, as shown in figure 7.2: the curve is not symmetrical with respect to the mid-channel as in the conventional Hall plate (see figure 4.7); the maximum value of the Hall voltage occurs closer to the drain.

Beyond the pinch-off point, the short-circuiting effect due to the drain electrode almost vanishes altogether. This is because the carriers in the drain depletion region move by the saturated velocity. Therefore, the differential resistance between a point at the drain end of the channel and the drain is virtually infinitely high. Now, the maximum Hall voltage can be obtained if the sense electrodes are positioned right at the end of the channel, where $y \simeq L$. However, if $y = L$, the offset voltage also tends to be high. A practical optimum position of the sense contacts has been found to be $0.7 \leq y/L \leq 0.8$ [8, 9].

I have demonstrated that the short-circuiting effect in an MOS Hall plate can be completely eliminated [10]. To this end, the source and drain regions of a short Hall plate were replaced by two distributed current

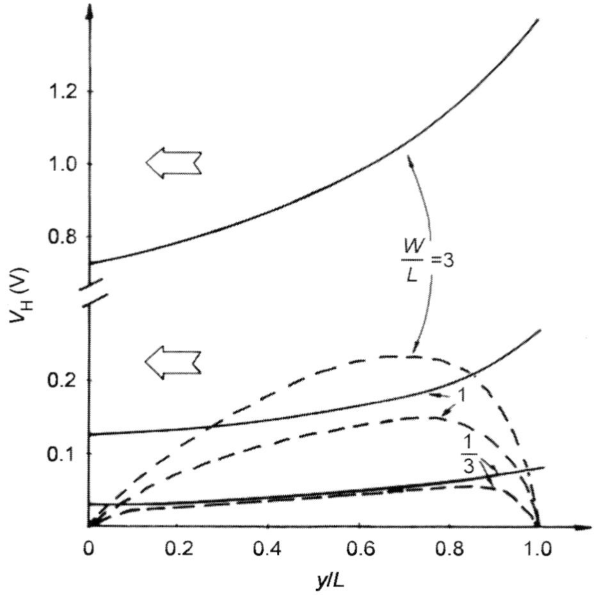

Figure 7.2. Distributions of the Hall voltage along the channel length of an MOS Hall plate (see figure 7.1). Numerical simulation. The parameters are: $W = 60\,\mu m$, $\mu_{Hn} = 0.06\,m^2\,V^{-1}\,s^{-1}$, $V_T = 0.55\,V$, $V_D = V_G = 10\,V$, $B = 1\,T$. The labels show the width-to-length ratio (W/L) of the channel. Broken lines, conventional MOS Hall device, such as the one shown in figure 7.1; full lines, the short-circuiting effect by the source and drain regions has been eliminated with the aid of the method illustrated in figure 7.3. (Adapted from [7].)

sources (see figure 7.3). Owing to the high differential resistance of the current sources, the Hall plate behaves as if it were infinitely long. The actual short length of the channel allows the realization of a high longitudinal electric field, and thus also a very high Hall field (see equations (3.339)–(3.341)).

7.1.2 Split-drain MOS devices

A split-drain MOS device is a MOSFET with two or three adjacent drain regions replacing the conventional single-drain region. A split-drain MOSFET with two drains D1 and D2 is shown in figure 7.4. The two drains share the total channel current. If a split-drain device is exposed to a perpendicular magnetic field, the current lines in the device skew, as illustrated in figure 7.5 [11]. The current deflection causes an imbalance in the two drain currents. The relative sensitivity of a split-drain device to a

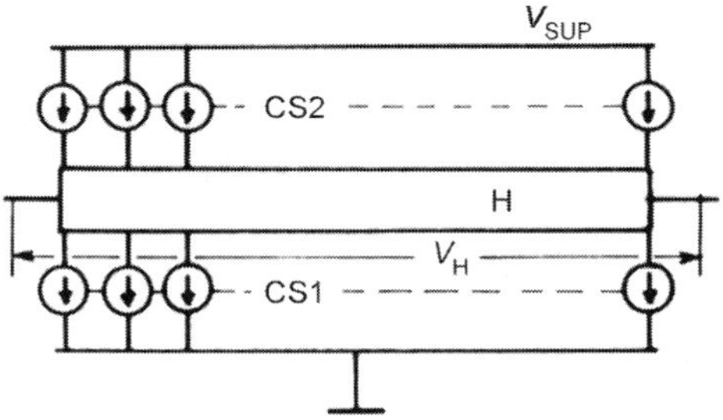

Figure 7.3. Schematic representation of a short Hall device (H), supplied by two arrays of the current sources (CS1 and CS2). The current sources inject the bias current directly into the active region of the Hall device: there are no current contacts. Since there are no current contacts, the short-circuiting effect does not occur. Such a device exhibits a geometrical correction factor $G_H \simeq 1$, irrespective of its aspect ratio. (Adapted from [10].)

magnetic field is defined in analogy to (5.7) as

$$S_I = \left| \frac{1}{I_D} \frac{\Delta I_D}{B_\perp} \right| \tag{7.8}$$

with the unit 'per tesla' ($AA^{-1}T^{-1} = T^{-1}$). Here I_D denotes the total drain current, $I_D = I_{D1} + I_{D2}$, and ΔI_D the change in each drain current due to a magnetic induction $\boldsymbol{B}$:

$$\Delta I_D = I_{D1}(B) - I_{D1}(0) = I_{D2}(B) - I_{D2}(0). \tag{7.9}$$

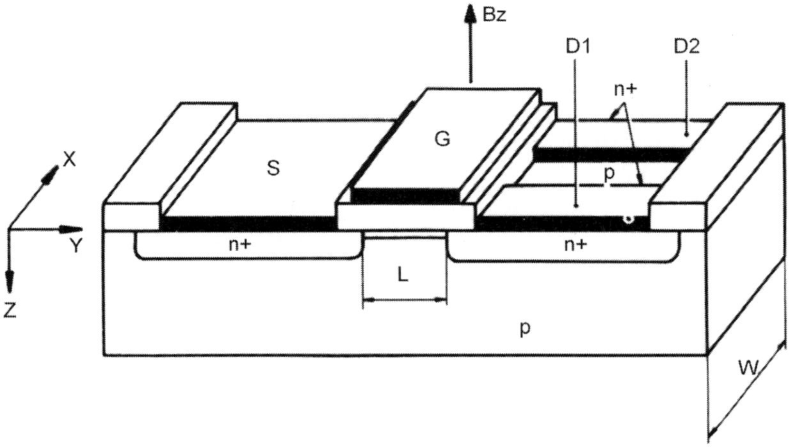

Figure 7.4. Split-drain magnetic-sensitive MOSFET (MAGFET). Magnetic induction perpendicular to the channel plane produces a current redistribution between the two drains Dl, D2.

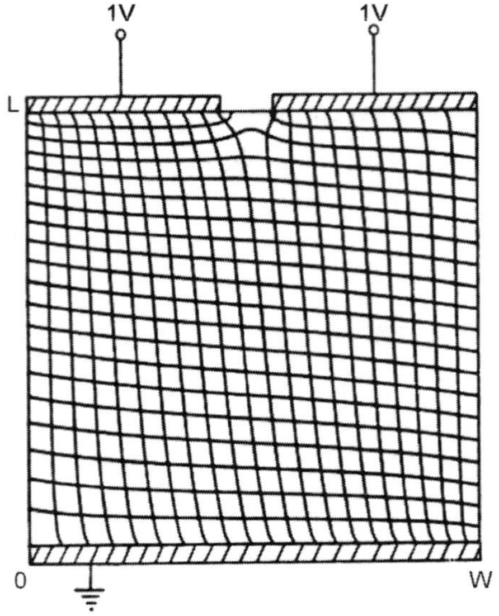

Figure 7.5. Result of a numerical modelling of a split-drain MAGFET operation. The earthed bottom contact denotes the source, and the top contacts are the two drains. The lines connecting the contacts are the current lines. As a result of the magnetic field, the current lines are deflected towards the left drain. Thus the current of the left drain increases at the expense of the right drain.

Let us find the relative sensitivity (7.8) of a rectangular split-drain MOSFET, such as that shown in figure 7.4. We shall assume that the device operates in the linear range and that the drains are kept at the same drain voltage V_D. Then, this split-drain device is a Hall device which operates in the Hall current mode, ΔI_D being related to the Hall current. To determine the Hall current, we shall make use of the properties of dual Hall devices, discussed in §4.4.2

Figure 7.6 shows schematically the split-drain device (a) and the corresponding dual Hall device operating in the Hall voltage mode (b) (compare with figure 4.34). Since the split-drain device has only three contacts, so also has its dual device. (We have met a Hall device with only one sense contact before: see figure 4.15(i).) Let us assume that the slit between the drains is very small; then the sense contact of the device (b) is nearly a point contact. The output signal in the device (b) equals half of the Hall voltage. Analogously, in the device (a), $\Delta I_D = I_H/2$. Then with the aid of equation (4.125),

$$\Delta I_D = \frac{I_H}{2} = \frac{1}{2}\mu_{ch}\frac{w}{l}G_H I_D B_\perp \tag{7.10}$$

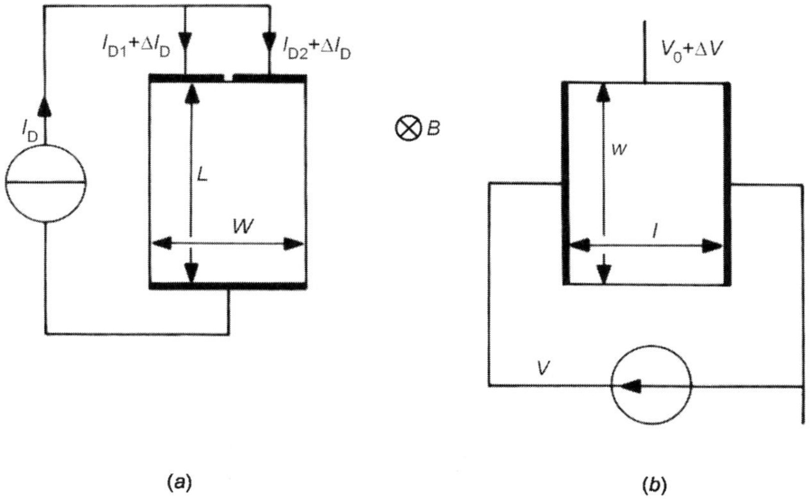

Figure 7.6 (a) Split-drain device operated in the Hall current mode. Its dual device is the one-sense-contact Hall device (b). The device (b) operates in the Hall voltage mode.

with $w/l = L/W$ (see figure 7.6). Recall the notation: μ_{Hch} denotes the Hall mobility of the carriers in the channel, and G_{H} is the geometrical correction factor of the device (b). By substituting (7.10) into (7.8), we find the relative sensitivity of the split-drain device, namely

$$S_I = \frac{1}{2}\mu_{\mathrm{ch}}\frac{L}{W}G_{\mathrm{H}}. \tag{7.11}$$

According to figure 4.14, $(w/l)G_{\mathrm{H}} \to 0.74$ when $w/l \to \infty$. Therefore, the maximum relative sensitivity of a split-drain device is given by

$$S_{I_{\max}} \simeq 0.37\mu_{\mathrm{ch}} \quad \text{for } L/W \to \infty. \tag{7.12}$$

When a split-drain device operates outside the linear range, its magnetic sensitivity cannot be expressed in such a simple form. Though there are experimental indications that the magnetic sensitivity does not change much when the operating regime of the device goes from linear to saturation.

A simple circuit for converting the drain current variations into a differential voltage is shown in figure 7.7. If the split-drain MOSFET works in saturation, then the drain output resistances R_{D} are very high. Suppose $R_{\mathrm{D}} \gg R$, R being the load resistance. The differential output voltage is then given by

$$\Delta V_{\mathrm{D}} = 2\Delta I_{\mathrm{D}}R = \mu_{\mathrm{ch}}\frac{L}{W}G_{\mathrm{H}}RI_{\mathrm{D}}B_{\perp} \tag{7.13}$$

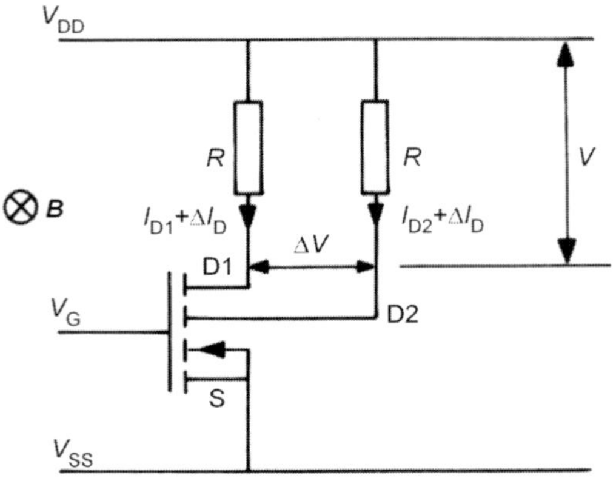

Figure 7.7. A circuit with a split-drain magnetic-sensitive MOSFET. The imbalance in the drain currents caused by a magnetic field brings about the differential voltage ΔV between the two drains.

where we have made use of (7.10). The product $RI_D = V_R$ denotes the voltage drop over the load resistors. It is interesting to note that by substituting this into (7.13) we obtain

$$\Delta V_D = \mu_{ch} \frac{w}{l} G_H V_R B_\perp \tag{7.14}$$

where we have again put $L/W = w/l$. This equation has the same form as equation (7.7), pertinent to a conventional MOS Hall device. If V_D in (7.7) and V_R in (7.13) denote the available voltage in a circuit, we may conclude that a conventional MOS Hall plate and a split-drain MOSFET are equivalent with respect to magnetic sensitivity.

However, *split-drain MOS Hall* devices also lend themselves to application in a more efficient current–voltage conversion *circuit*. Figure 7.8 shows an example. The circuit consists of a pair of complementary split-drain MOSFETS in a CMOS differential amplifier-like configuration [12]. To understand the operation of the circuit, first compare figure 7.7 and figure 7.8(a). Obviously, the load resistors (figure 7.7) now play the role of the p-channel split-drain device (figure 7.8(a)). Note further that a magnetic field leads to changes in the drain currents of both split-drain devices (I_n', I_n'' and I_p', I_p'' in figure 7.8(a)). The drain with increasing current of one transistor is connected to the drain with decreasing current of the complementary transistor, and vice versa. As a result of this cross-coupling and the dynamic load technique, small imbalances in the drain currents lead to large variations in the output voltage (see figure 7.9).

The salient feature of this circuit is that it performs the following four operations simultaneously: working point control, sensing of the magnetic

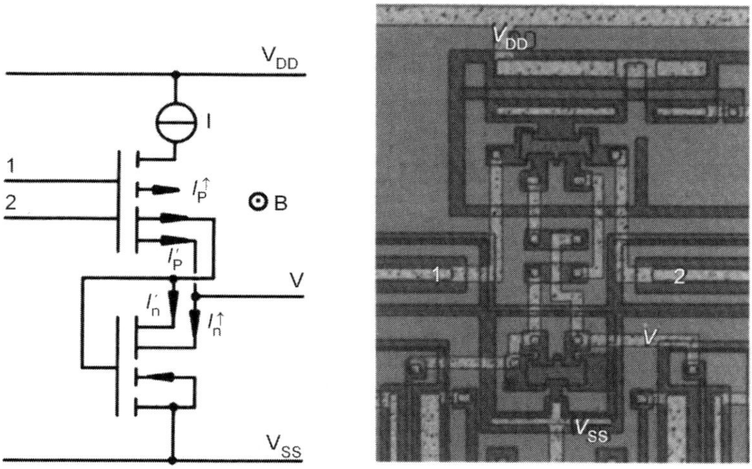

(a) (b)

Figure 7.8. CMOS magnetic-sensitive circuit: (a) schematic, (b) photograph of a detail. The essential part of the circuit consists of two complementary split-drain MAGFET devices. They replace the usual two pairs of MOS transistors in a CMOS differential amplifier. (Adapted from [12]. © 1983 IEEE.)

field, provision of gain, and differential to single-ended voltage conversion. The circuit has been implemented in a commercially available CMOS technology. A magnetic sensitivity of $1.2\,\mathrm{V\,T^{-1}}$ at 10 V supply and 100 μA current consumption has been achieved.

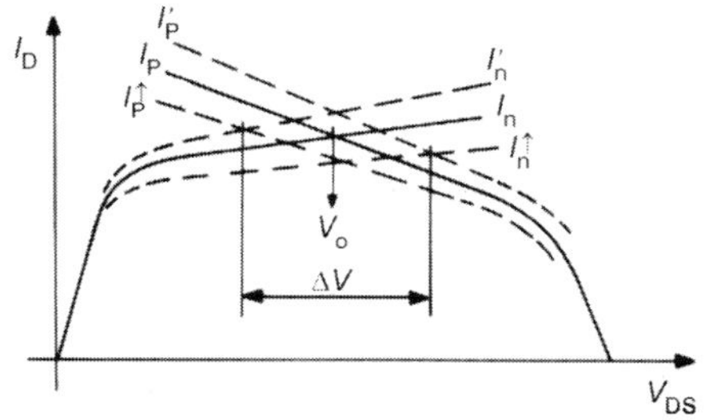

Figure 7.9. Drain current (I_D) against drain-source voltage (V_{DS}) characteristics of the split-drain complementary MOS transistors. I_n and I_p correspond to the n-channel and p-channel devices, respectively (see figure 7.8(a)). Full lines: $\boldsymbol{B} = 0$. Then $I'_n = I''_n = I_n$ and $I'_p = I''_p = I_p$. V_0 is the corresponding output voltage. Broken lines: $\boldsymbol{B} \neq 0$. The magnetic field produces an imbalance in the drain currents, and thus $I'_n \neq I''_n$ and $I'_p \neq I''_p$. ΔV is the output voltage shift caused by this imbalance in currents. (Adapted from [12]. © 1983 IEEE.)

7.2 Magnetodiodes

A magnetodiode is a two-terminal Hall effect device similar to a conventional bipolar diode. The voltage–current characteristic of a magnetodiode is sensitive to magnetic field. The physical effect responsible for the magnetic sensitivity of magnetodiodes is called the magnetodiode effect. The magnetodiode effect was discoverd by Stafeev in 1958 [13, 14]. A monograph covering much of the older work on magnetodiodes was published in 1975 [15]. More recent works are reviewed in [16, 1].

7.2.1 The magnetodiode effect

The magnetodiode effect comes about as a result of the cooperation of the Hall effect and a few other effects pertinent to a p–n junction diode. The basic effects are: the conductivity modulation due to a high injection level (see (2.124)); the current deflection (§3.1.2); and the magnetoconcentration effect (§3.5.2). Details of the magnetodiode effect depend on the diode structure and operating conditions. We shall consider two typical examples: magnetodiodes with volume and surface carrier recombination.

The *magnetodiode with volume recombination* is a relatively long-base pin diode with a well passivated surface (see figure 7.10). The length of the quasi-intrinsic (i) region L_i is chosen so as to be slightly smaller than the diffusion length of carriers (§2.5.2): $L_i \leq L_n$ or $L_i \leq L_p$. When the diode is forward biased, carriers are injected into the i-region and recombine there. Figure 7.10 shows schematically the current density lines in the diode. In the absence of a magnetic field (a), the current lines are straight; in the presence of a magnetic field (b), both holes and electrons are deflected towards the same insulating boundary (see also figure 3.1) and the current lines become curved. Consequently the current lines become longer: the longer the current lines, the larger the effective base length L.

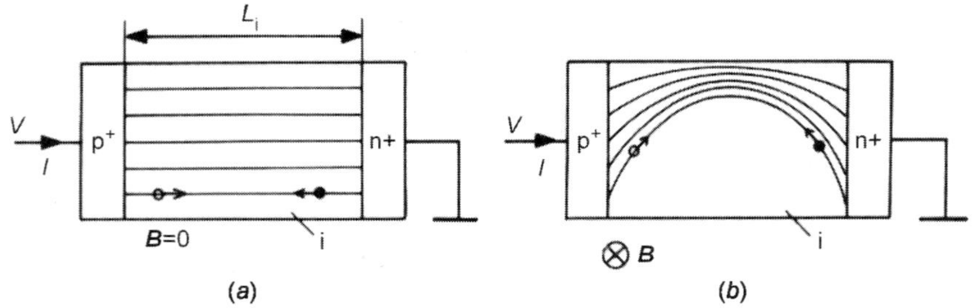

Figure 7.10. Current density lines in a magnetodiode with volume recombination. (a) No magnetic field; (b) a magnetic field perpendicular to the drawing plane is present. Then both electrons (•) and holes (∘) are deflected towards the same boundary of the slab, and the current lines get longer.

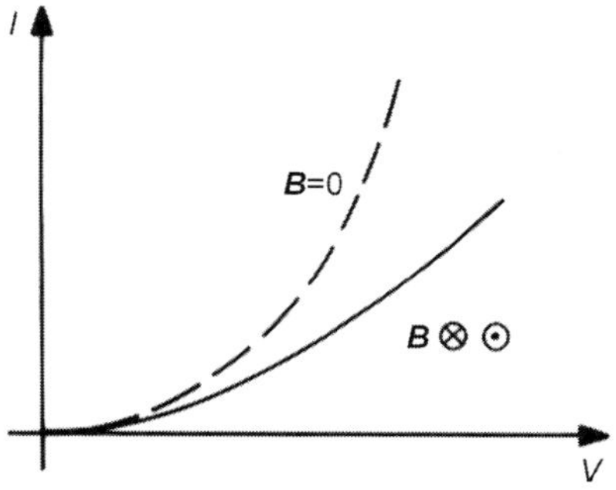

Figure 7.11. Current–voltage characteristics of a magnetodiode with volume recombination. At a constant bias voltage, a magnetic field causes a decrease in diode current, irrespective of the sign of the magnetic field.

Thus a magnetic field effects a change from a short diode ($\boldsymbol{B} = 0$) to a longer diode ($\boldsymbol{B} \neq 0$). According to equations (2.118) and (2.119), this change causes a decrease in the diode current. The corresponding change in the voltage–current characteristic of the diode is depicted in figure 7.11: the voltage–current characteristic is magnetic-sensitive. The same things happen for the opposite direction of the magnetic field. Hence, such a magnetodiode is not sensitive to the sign of a magnetic field.

The magnetic sensitivity of a magnetodiode is higher at a *high injection level*. There are two reasons for this. First, at a high injection level, the base region (i) becomes really quasi-intrinsic, irrespective of some doping (see equation (2.122)). In an intrinsic material, the Hall field is small and the current deflection more intensive, irrespective of the device geometry (see §3.3.7). Second, at a high injection level, the *conductivity modulation* takes place (see equations (2.123) and (2.124)). The conductivity modulation effect supports the magnetic sensitivity of a magnetodiode in the following way.

The current in the magnetodiode decreases in the presence of a magnetic field. The lower the current, the lower the injection level and the higher the resistance of the i-region. Then the larger portion of the diode voltage drops across the i-region, the voltage across the injecting junctions decreases, and the diode current decreases. Thus at a high injection level, a magnetic field triggers a cumulative process of current reduction, which greatly boosts the magnetic sensitivity of the magnetodiode.

The structure of a *magnetodiode with surface recombination* is shown in figure 7.12(a). This is again a pin diode as above. However, the length of the i-region is now much smaller than the carrier diffusion length. Thus the volume recombination is of no importance. In addition, the two opposite surfaces of the i-region, S_1 and S_2, are prepared so as to get two very different surface recombination velocities, s_1 and s_2, respectively. Suppose $s_1 = 0$ (no

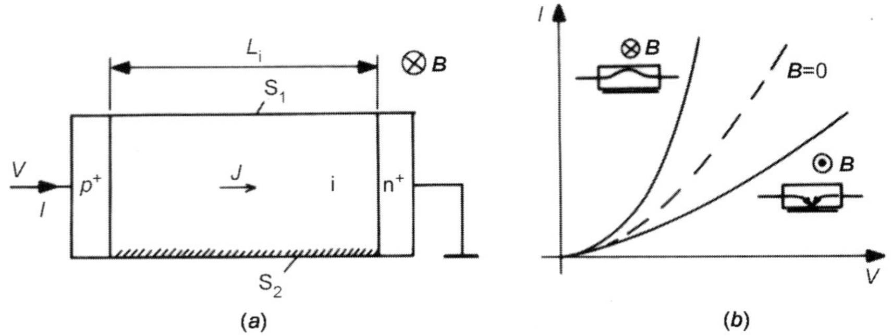

Figure 7.12. Magnetodiode with surface recombination. (a) Device structure. This is a long pin diode. One insulating boundary (S_1) has a very low surface recombination velocity, and the other (S_2) a very high surface recombination velocity. The device is sensitive to a magnetic field perpendicular to the current density (***J***) and parallel to the surfaces S_1 and S_2. (b) Current–voltage characteristics. The insets illustrate the carrier trajectories in the presence of a magnetic field.

recombination) and $s_2 \rightarrow \infty$. In §3.5.2, we described the magnetoconcentration effect in a similar structure. The operation of the magnetodiode from figure 7.12(a) relies on the cooperation of that magnetoconcentration effect (see figure 3.33) and the conductivity modulation, which was explained above. Here a high injection level is necessary, since the magnetoconcentration effect is intensive only at quasi-intrinsic conditions.

The voltage–current characteristics of the surface recombination diode are displayed in figure 7.12(b). If the carriers are deflected towards the high-recombination surface S_2, their concentration in the diode decreases (figure 3.33(b)), and the current also decreases. If the carriers are deflected towards the low-recombination surface S_1, their concentration in the diode increases (figure 3.33(c)), and the diode current also increases. Therefore, this magnetodiode is sensitive to the sign of the magnetic field.

7.2.2 Structures and technology

Magnetodiodes have been made of various semiconductors, including germanium, silicon and GaAs. Early devices were discrete. The magnetodiodes with surface recombination seem to have attracted more development effort. The difference in the surface recombination velocity has been achieved by polishing and etching one surface and grinding the other.

Figure 7.13 shows a magnetodiode realized in SOS (silicon-on-sapphire) integrated circuit technology [17, 18]. Owing to the crystal defects at the silicon–sapphire interface, this interface exhibits a high recombination velocity. The top surface is thermally oxidized and has a very small surface recombination velocity.

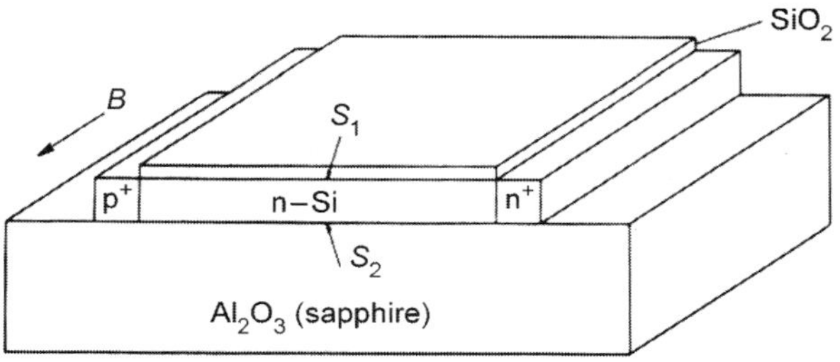

Figure 7.13. Magnetodiode with surface recombination made using SOS technology (SOS: silicon-on-sapphire). The notation is the same as in figure 7.12(a).

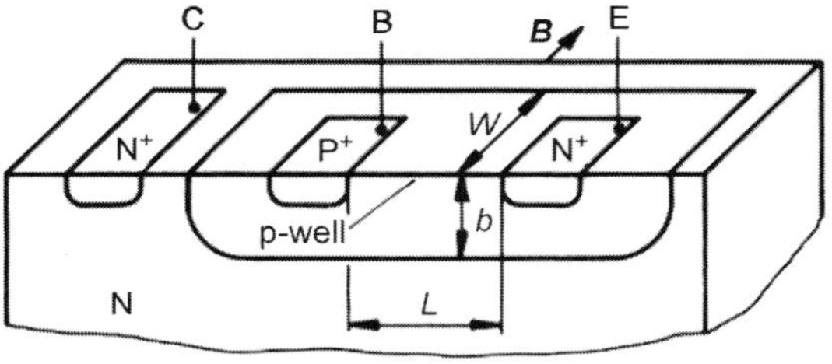

Figure 7.14. Magnetodiode with a reverse-biased p–n junction (p-well/N-substrate) replacing the high-recombination surface. The regions B and E correspond to the regions p^+ and n^+ in the basic magnetodiode structure (figure 7.12(a)). (Adapted from [20]. © 1984 IEEE.)

Figure 7.14 shows a magnetodiode where a reverse-biased p–n junction plays a role similar to that of the high-recombination surface: here the recombination of carriers is replaced by collection [19, 20]. Magnetodiodes of this type have been manufactured using a standard CMOS integrated circuit process [20]. The p-well is used as the quasi-intrinsic region, and the junction p-well/substrate (N) is the collector junction. The device has the structure of a bipolar transistor, the contacts E, B, C corresponding to emitter, base, collector. However, the collector is kept at a constant bias voltage, and the useful signal is the modulation of the emitter-base forward voltage–current characteristics.

Typical magnetic sensitivities of magnetodiodes are about $5\,\mathrm{V\,mA^{-1}\,T^{-1}}$ at bias currents of 1 mA to 10 mA. The noise characteristics are similar to those of Hall plates [18].

7.3 Magnetotransistors

A magnetotransistor is a bipolar junction transistor (BJT) whose structure and operating conditions are optimized with respect to the magnetic sensitivity of its collector current. The very first indication of what we would now call a magnetotransistor mechanism was a report by Suhl and Shockley in 1949 [21]. In 1950, Brown modulated the current gain of a transistor by a magnetic field [22]. Melngailis and Rediker were the first to propose using a transistor as a magnetic sensor [23]. Since then, many works have been published on various versions of magnetic-sensitive bipolar junction transistors (*MagBJTs*) [1–3].

The magnetic sensitivity of a magnetotransistor is usually defined in analogy to (5.7) and (7.8) as

$$S_I = \left| \frac{1}{I_c} \frac{\Delta I_c}{B} \right|. \tag{7.15}$$

Here, I_c denotes the collector current, and ΔI_c is the change in the collector current due to a magnetic induction $\boldsymbol{B}$:

$$\Delta I_c = I_c(B) - I_c(0). \tag{7.16}$$

The sensitivities of magnetotransistors reported hitherto cover a surprisingly wide range: from 10^{-2} to over $10\,\mathrm{T}^{-1}$. This large spread of sensitivities indicates that the operation of various magnetotransistors is based on different effects. Indeed, the Hall effect may interfere with the action of a bipolar transistor in many ways and give rise to different end effects. The following three major end effects may be distinguished:

(i) *The current deflection effect*. This is essentially the same effect that we studied in large-contact Hall plates.
(ii) *The injection modulation*. The Hall voltage generated in the base region of a magnetotransistor modulates the emitter-base voltage, and thus also the carrier injection.
(iii) *The magnetodiode effect*. The emitter-base diode of a transistor may function as a magnetodiode. This leads to a magnetic sensitivity of collector current.

In principle, these three effects coexist and cooperate in any magnetotransistor. This makes the operation of some magnetotransistors rather involved. However, there are also magnetotransistor structures in which, under appropriate operating conditions, one of these three effects strongly prevails. Let us have a look at a few such examples.

7.3.1 The current deflection mechanism

Recall that the current deflection effect is one way for the Hall effect to appear. It occurs whenever the generation of the Hall field is prevented: for example, by short-circuiting, as in a short sample; or by compensating the space charge, as in mixed conductors.

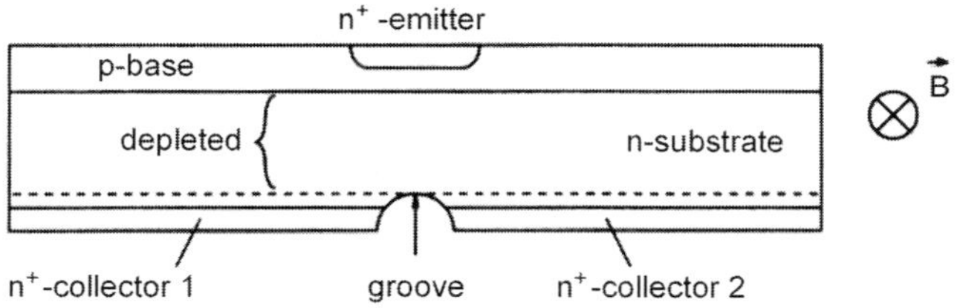

Figure 7.15. Cross section of split-collector magnetotransistor. A magnetic induction perpendicular to the drawing plane causes the current injected from the emitter to deflect. This gives rise to an imbalance in the collector currents. (Adapted from [24].)

One of the earliest magnetotransistors, devised by Flynn [24], operated on the current deflection principle. This is a *split-collector n–p–n transistor* (see figure 7.15). The larger part of the collector region is low doped, and depletes upon reverse biasing of the collector-base junction. The electrons injected from the emitter traverse the base, and then drift downwards, towards the collectors. If a magnetic field is present, the electrons will be deflected. This gives rise to an imbalance in the two collector currents.

This magnetotransistor is reminiscent of the split-drain MOSFET Hall effect device (see §7.1.2): the depletion region between the base and the two collectors of the magnetotransistor in figure 7.15 is analogous to the channel region of the split-drain device in figure 7.4. The role of the emitter and base regions of the magnetotransistor is only to inject carriers into the depletion layer. These two regions do not essentially participate in the magnetic sensitivity of the transistor. They are too strongly doped and too short for this.

A version of the split-collector magnetotransistor, which is compatible with bipolar integrated circuit technology, is shown in figure 7.16 [25]. Here a part of the low-doped epitaxial layer (n^-) serves as the collector region, and the split-collector 'contacts' are realized by splitting the buried layer. The operation is similar to that of Flynn's device described above.

The magnetic sensitivity of this magnetotransistor can be estimated by making use of its similarity to the split-drain MOS Hall device (see figure 7.4). In analogy with (7.11), we may write.

$$S_I \simeq \frac{1}{2}\mu_{\mathrm{Hn}}\frac{L}{W_{\mathrm{E}}} \tag{7.17}$$

with S_I being defined by (7.15). Here we put $G_{\mathrm{H}} \simeq 1$, μ_{Hn} denotes the average Hall mobility of electrons on their way between the emitter and the collectors, L is the emitter–collector distance, and W_{E} is the emitter width (see figure 7.16). Measured sensitivities range from 0.03 to $0.05\,\mathrm{T}^{-1}$. Equation (7.17) yields a roughly equal value.

Using magnetotransistor structures similar to that shown in figure 7.16, two-axial [25, 26] and three-axial [27] magnetic field sensors have also been realized.

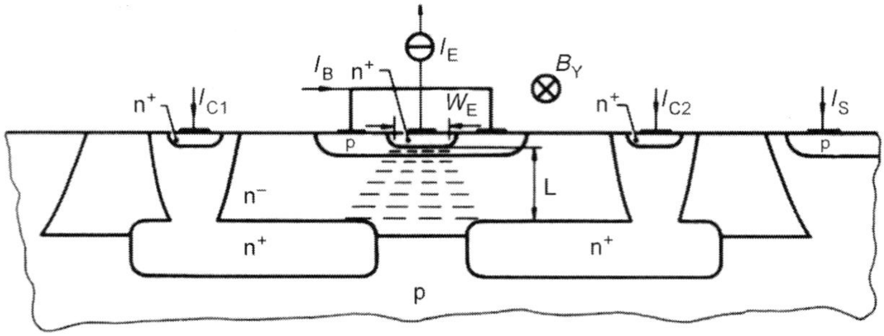

Figure 7.16. Cross section of a split-collector magnetotransistor made in bipolar integrated circuit technology. Magnetic induction $\boldsymbol{B}_y$ causes a deflection of the current 'beam' and thus an asymmetry in collector currents I_{C1}, I_{C2}. (Reprinted from [25]. © 1982 IEEE).

Another example of a magnetotransistor operating on the current deflection principle is illustrated in figure 7.17 [28]. This is a drift-aided lateral double-collector p–n–p transistor. The emitter and both collectors are embedded into a plate-like n-type base region. The base region may be part of the n-type epitaxial layer, so that the device is compatible with bipolar integrated circuits. The two base contacts B_1 and B_2 are used to apply the bias voltage needed to establish the lateral acceleration electric field in the base region. Obviously, the base region makes a Hall plate with no sense contacts.

The structure of this magnetotransistor is similar to the arrangement for measuring the Hall mobility of minority carriers shown in figure 6.13. Moreover, the principles of operation of the two devices are identical. Briefly, the emitter injects holes into the base region. Under the influence of the accelerating field in the base, these holes form a minority carrier

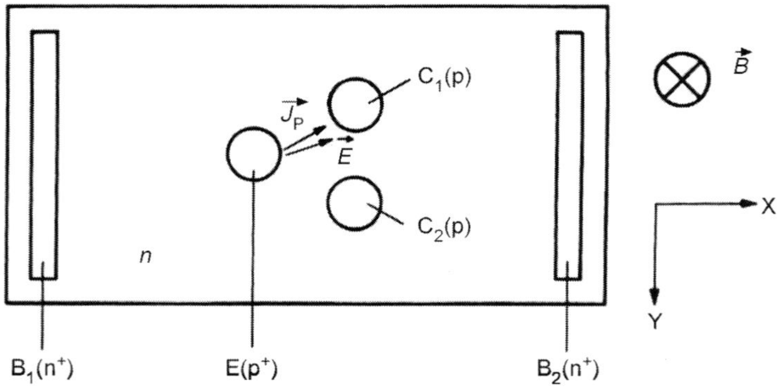

Figure 7.17. Top view of a drift-aided double-collector magnetotransistor [28]. Magnetic induction $\boldsymbol{B}$ causes the rotation of the electric field $\boldsymbol{E}$ with respect to the x axis and the deflection of the current density $\boldsymbol{J}$ with respect to $\boldsymbol{E}$ (compare with figure 6.13).

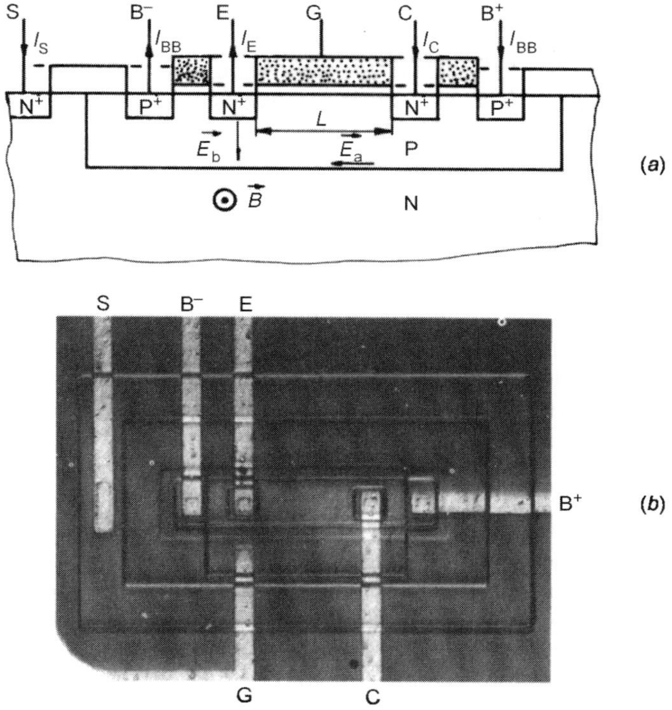

Figure 7.18. Lateral magnetotransistor in CMOS technology. (a) Cross section. A magnetic induction $\boldsymbol{B}$ perpendicular to the drawing plane modulates the distribution of the emitter current I_E among I_C and I_S. The modulation in the collector current I_C is used as the sensor signal. (b) Photograph (top view). (Adapted from [29]. © 1986 IEEE.)

beam. In the presence of a normal magnetic field, the electric field in the base region rotates for the Hall angle of majority carriers Θ_{Hn}, and the hole beam tilts with respect to the electric field for the Hall angle Θ_{Hp} (see equation (6.62)). Therefore, the total angle of deflection of the minority carrier beam is given by

$$\Theta_{tot} \simeq -\mu_{Hn} B + \mu_{Hp} B \tag{7.18}$$

where we have assumed a weak magnetic induction. Since $\mu_{Hn} < 0$ and $\mu_{Hp} > 0$, the two angles add up. The deflection of the beam leads to an imbalance in the two collector currents.

The next example, shown in figure 7.18, is a *magnetotransistor compatible with CMOS* integrated circuits [29]. This device has the structure of a long-channel MOS transistor, but operates as a lateral bipolar transistor with a drift-aided field in the base region. The device is situated in a p-well, serving as the base region of the transistor. Two base contacts, B^+

and B^-, allow the application of the accelerating voltage for the minority carriers injected into the base region. The two n^+ regions, separated by the lateral base-length distance L, serve as the emitter E and the primary collector C. The substrate S works as the secondary collector. As will be shown later, the action of this secondary collector, which is a mere parasitic one in a 'non-magnetic' lateral transistor, plays a crucial role in the magnetic sensitivity of this device.

Consider qualitatively how the device operates. Let us assume that it is biased adequately for the forward active operation: the emitter-base junction forward biased, both collector-base junctions reverse biased, and the base contact B^+ more positive than B^-. Owing to the accelerating field $\boldsymbol{E}_a$ in the base region, the electrons are injected into the base region mostly by the emitter right-hand side-wall, drift mainly along the base length, and are collected by collector C, producing collector current I_C. Some of them, however, diffuse downwards, are collected by the secondary collector S, and thus produce the substrate current I_S. The accelerating field helps a favourable ratio to be obtained between the useful current I_C and the parasitic current I_S.

If we apply a magnetic field $\boldsymbol{B}$ perpendicularly to the plane of figure 7.18(a), it will cause a change in the ratio I_C/I_S. The physical mechanism giving rise to this effect is the deflections of the electron paths. When the field is directed into the plane of the figure, the electrons are deflected towards the substrate junction and therefore fewer electrons contribute to the collector current (see figure 7.19).

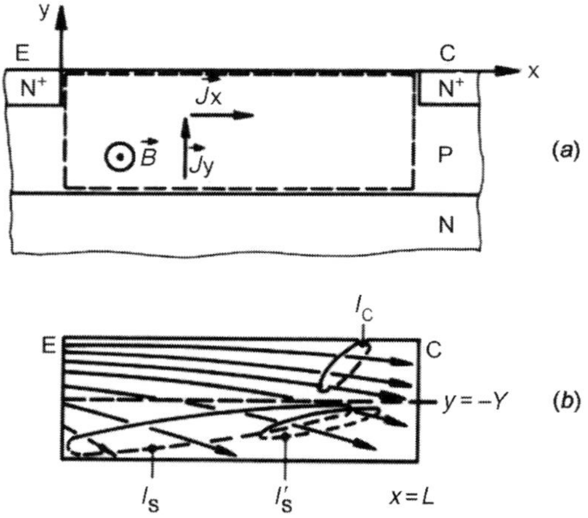

Figure 7.19. The base region of the magnetotransistor from figure 7.18. (a) Cross section; (b) current density lines. The lines that stay in the region $y > -Y$ make up the collector current I_C; all other lines make up the substrate current I_S. Depending on magnetic induction $\boldsymbol{B}$, the current I'_S may add to I_C or I_S. (Adapted from [29]. © 1986 IEEE.)

When the field direction is reversed, the electron flux lines are deflected in the opposite direction, towards the device surface; consequently, the collector current increases. But this will happen only if there is no appreciable recombination at the silicon–oxide interface. The MOS structure on top of the base region serves to ensure this condition. Application of a negative voltage to the gate enriches the base-region surface and pushes the minority carriers away from the interface. This effect is very beneficial to MOS transistor operation in the bipolar mode [6].

A theoretical analysis [29] yields the following results for the magnetic sensitivity (7.15) of this CMOS magnetotransistor. If the acceleration field $\boldsymbol{E}_a$ in the base region is very small, so that the electrons move essentially by diffusion, the sensitivity is given by

$$S_I \simeq |\mu_{Hn}| \frac{L}{Y}. \tag{7.19}$$

In the opposite case, namely for a strong acceleration field,

$$S_I \simeq (\mu_{Hp} - \mu_{Hn}) \frac{L}{Y}. \tag{7.20}$$

Here μ_{Hp} and μ_{Hn} are the Hall mobilities of electrons and holes in the p-well, L is the distance between the emitter and the collector, and Y is a geometrical parameter given approximately by

$$y_{jn} < Y < y_{jp}. \tag{7.21}$$

Here y_{jn} and y_{jp} denote the junction depths of the collector region and the p-well, respectively.

Note that with a weak accelerating electric field in the base, this magnetotransistor operates in a similar way as the devices shown in figures 7.15 and 7.16. The expressions for sensitivity are also similar: compare equations (7.17) and (7.19). With a high accelerating field, this magnetotransistor functions similarly as the device shown in figure 7.17. The higher sensitivity (7.20) is obviously due to double-deflection phenomena (see equation (7.18)).

The experimental results corroborate the theoretical predictions (7.19) and (7.20): the magnetic sensitivity is proportional to the lateral base length L, and a change between two sensitivity values occurs when the base bias current increases.

A few other versions of CMOS lateral magnetotransistors have been proposed and tested [30, 31]. For one of these devices a very high magnetic sensitivity of more than $10\,\mathrm{T}^{-1}$ has been reported [31].

7.3.2 Injection modulation

The injection modulation operating mechanism of magnetotransistors shows up in its clearest form in the structure presented in figure 7.20 [32]. This is a

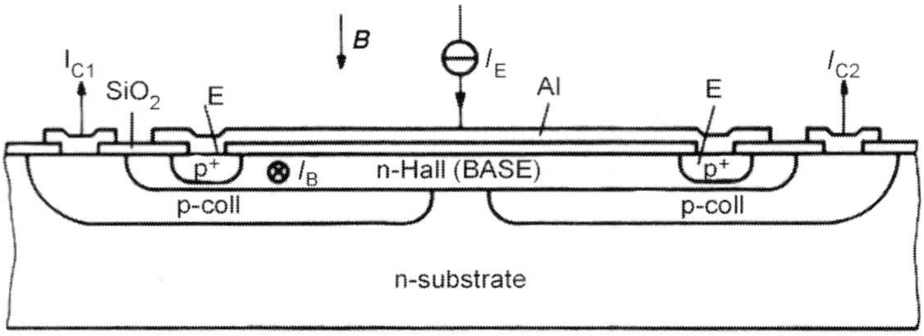

Figure 7.20. Merged combination of a Hall plate and two bipolar transistors: cross section perpendicular to the current density in the Hall plate (n-Hall). The Hall voltage generated in the Hall plate modulates the base-emitter bias voltages of the two transistors. The modulation of the junction bias voltages causes a modulation of the current injection, and thus also a modulation of the collector currents I_{C1} and I_{C2}. (Reprinted from [32]. © 1972 IEEE).

merged combination of a horizontal Hall plate and a pair of vertical bipolar transistors. Two p^+ emitters are diffused into the Hall plate instead of the usual sense contacts. Two separate n-collectors embrace the Hall plate, isolating it almost completely from the n-substrate. The Hall plate serves as the common base of the two vertical p–n–p transistors. In operation, the Hall plate biasing current I_B flows along the base region: its direction is perpendicular to the drawing plane of figure 7.20. With $\boldsymbol{B}$ perpendicular to the chip surface, a Hall voltage appears across the base region. If the two emitters are kept at the same potential, the Hall voltage generated in the base region acts as the differential emitter-base voltage of the transistor pair. If the two transistors are properly biased for forward active operation, the differential emitter-base voltage will be directly converted into a difference in the two collector currents.

The injection modulation by magnetic field may come about in a transistor if the following two conditions are fulfilled. (i) The base region is properly structured and biased so as to operate efficiently as a Hall device in the Hall voltage mode. (We mention only the base region, since the emitter region is normally heavily doped, and the Hall voltage generated in it is negligible.) (ii) The transistor is adequately structured and biased so as to enable the generated Hall voltage to modulate the emitter-base bias voltage, and thus the collector current. Obviously, these two conditions are fulfilled in the structure presented above (figure 7.20), and not in the magnetotransistor shown in figure 7.16.

We shall now give an example which illustrates how a change in the operating conditions of a transistor may affect the fulfilment of the conditions for the appearance of the injection modulation magnetotransistor effect.

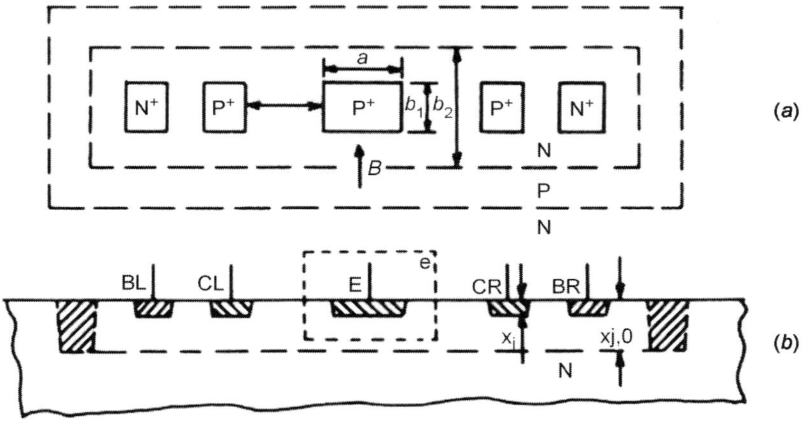

Figure 7.21. Double-collector lateral p–n–p magnetotransistor: (a) top view; (b) cross section. (Adapted from [33].)

The structure of the device we are going to discuss is shown in figure 7.21 [33]. This is a dual-collector lateral p–n–p magnetotransistor. The five heavily doped regions (the emitter E, the two collectors CL and CR, and the two base contacts BL and BR) are symmetrically arranged at the surface of an n-type silicon substrate. To confine the current into a plane perpendicular to the magnetic field and thus increase the magnetic sensitivity, the active device region is surrounded by a deep p-type ring.

The important feature of this magnetotransistor is its short-circuited emitter (see figure 7.22). In case (a), the p^+n emitter-base junctions are connected in parallel with Schottky diodes, and in case (b) by an ohmic contact. The efficiency of such a short-circuited emitter depends strongly on the emitter-base biasing voltage (or the total emitter current).

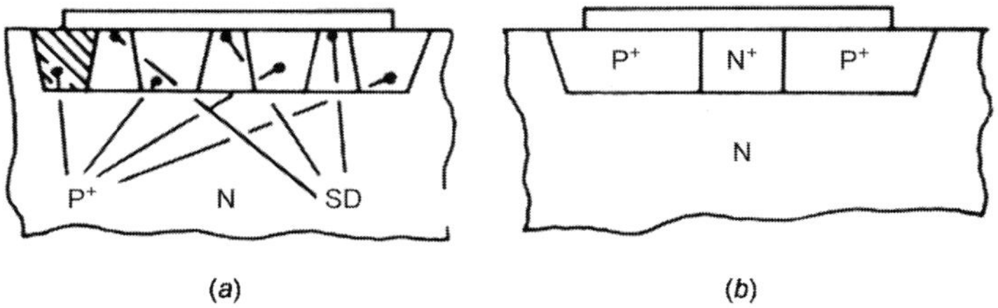

Figure 7.22. Low-injection-efficiency emitters (section e from figure 7.21). (a) An array of small-area p^+n junctions is connected in parallel with intermediary Schottky diodes (SD). (b) A split p^+n junction short-circuited by an ohmic contact (metal–n^+n structure). (Adapted from [33].)

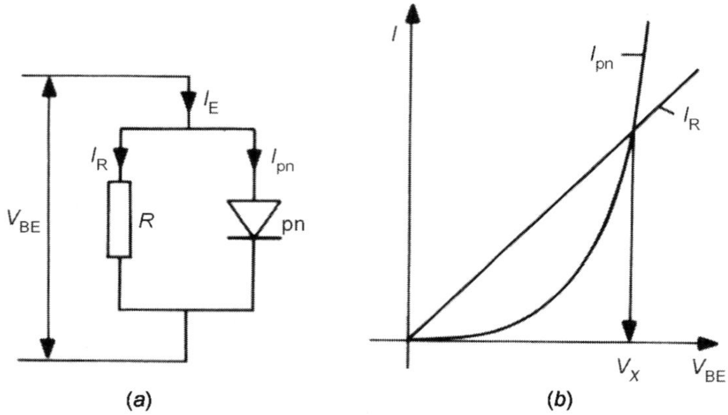

Figure 7.23. (a) Equivalent circuit of the short-circuited emitter from figure 7.22(b). pn denotes the p^+n junction, and R is the spreading resistance of the base region in the vicinity of the ohmic contact. Both branches (pn and R) are biased by the same base-emitter voltage V_{BE}. The emitter current I_E is divided into the injection current I_{pn} and the idle current I_R. (b) The current–voltage characteristics of the circuit (a). At $V_{BE} = V_x$, $I_{pn} = I_R$.

The emitter efficiency of a p^+n emitter-base junction is defined as [34]

$$\gamma = I_p/I_E \tag{7.22}$$

where I_p denotes the hole current injected from the emitter into the base and I_E is the total emitter current. Usually, $\gamma \geq 0.99$ over a broad range of the emitter current.

Consider the emitter efficiency of the short-circuited emitter shown in figure 7.22(b). With reference to figure 7.23, we see that, at a small emitter current, the emitter efficiency is very low: $\gamma \to 0$ (since at $V_{EB} \ll V_x$, $I_{pn} \ll I_R \simeq I_E$). At a high emitter current, the emitter efficiency tends to the value pertinent to the junction without short circuit: $\gamma \to \gamma_{pn}$ (since at $V_{EB} \ll V_x$, $I_R \ll I_{pn} \simeq I_E$). At $V_{EB} \simeq V_x$, the emitter efficiency changes very fast with a change in the emitter current.

The magnetotransistor functions as follows. Suppose the transistor is properly biased for forward active operation, and the two base contacts are at the same potential. In the absence of a magnetic field, the two collector currents are equal (we neglect offset). If a magnetic field as shown in figure 7.21(a) is established, an imbalance between the left and right collector current arises. Under certain biasing conditions, this asymmetry is due to the injection modulation effect.

Let us consider the Hall voltage generated in the base region, between the two points A and B situated close to the side-walls (opposite to the collectors) of the emitter (see figure 7.24). Neglecting the short-circuiting effect, the Hall voltage is equal to the line integral of the Hall electric field over some line AB,

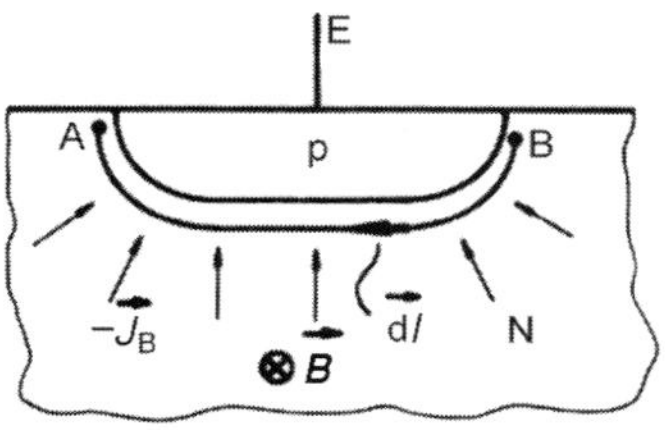

Figure 7.24. Generation of the Hall voltage around an emitter-base junction. The arrows ***J*** denote the majority carrier density, and the line AB is the integration path. (Reprinted from [33].)

namely

$$V_H \simeq \int_A^B \boldsymbol{E}_H \, d\boldsymbol{l}. \tag{7.23}$$

Here d***l*** denotes the line element, and

$$\boldsymbol{E}_H = -R_H[\boldsymbol{J} \times \boldsymbol{B}] \tag{7.24}$$

the local Hall field (see equation (3.165); R_H denotes the Hall coefficient (§3.4.6), ***J*** the total current density, and ***B*** the magnetic induction. Under the assumption that, close to the emitter-base junction, most of the current flows in a direction perpendicular to ***B*** and d***l***, we can write

$$V_H \simeq -R_H B \frac{1}{t} \int_A^B Jt \, dl \tag{7.25}$$

$$|V_H| \simeq \frac{|R_H|}{t} I_E B. \tag{7.26}$$

Here I_E stands for the total emitter current, and t is an effective thickness of the base region. With reference to figure 7.21(a), we assess this thickness to be

$$b_1 < t < b_2. \tag{7.27}$$

We note that equation (7.26) is identical in form to the corresponding equation holding for a point-contact Hall device (4.89).

The collector current of a bipolar transistor is generally given by [34]

$$I_C \simeq I_S \exp\left(\frac{qV_{EB}}{kT}\right) \tag{7.28}$$

where I_S is a transistor parameter, V_{EB} is the emitter-base voltage, and the other notation is as usual. The Hall voltage (7.26) modulated the emitter-base voltage.

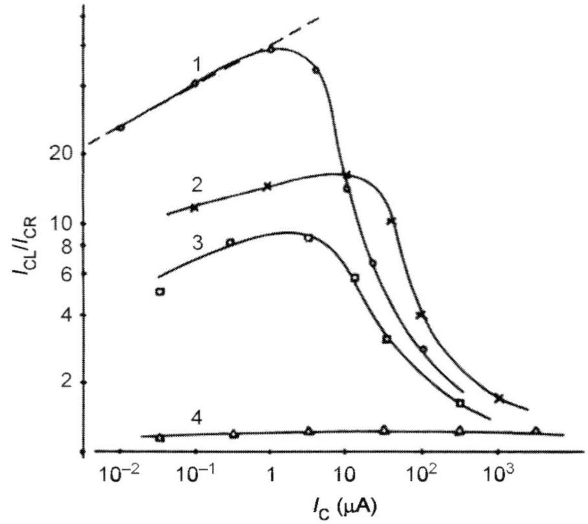

Figure 7.25. Ratio of the collector currents of the magnetotransistors shown in figure 7.21 as a function of average collector current at $\boldsymbol{B} = 1$ T. Curves 1, 2, 3: low-efficiency emitters; curve 4: conventional emitter; curves 1 and 2: with P-ring; curves 3 and 4: without ring; broken line: theory with $t = 37\,\mu$m. (Adapted from [33].)

Thus the currents of the left and the right collectors are given by

$$I_{CL} \simeq I_S \exp\left(\frac{q(V_{EB} + \frac{1}{2}V_H)}{kT}\right)$$
$$I_{CR} \simeq I_S \exp\left(\frac{q(V_{EB} + \frac{1}{2}V_H)}{kT}\right). \tag{7.29}$$

The ratio of these two collector currents is given by

$$\frac{I_{CL}}{I_{CR}} = \exp\left(\frac{qV_H}{kT}\right). \tag{7.30}$$

Figure 7.25 shows experimentally measured ratios of the two collector currents of various test devices, at the magnetic induction $B = 1$ T. Equation (7.29) is plotted as the broken straight line. Obviously, at low collector currents the injection modulation theory agrees very well with experimental results. But what happens at higher collector currents?

At a small collector current, the injection level in the base region is low, and the appropriate formula for the Hall coefficient figuring in (7.25) is given by (3.293):

$$|R_H| = r_H/qn. \tag{7.31}$$

Thus $R_H \neq f(I_E)$ and $V_H \sim I_E$ (see (7.26)).

To reach a high collector current, the emitter-base voltage V_{EB} has to be increased. Then the emitter efficiency also increases. At $V_{EB} \gg V_x$, the injection level in the base region eventually becomes high (see equations (2.122)–(2.124)). Then in the base region the concentrations of electrons and holes become almost equal, $n \simeq p$, and the Hall coefficient reduces to the value given by (3.206):

$$\bar{R}_H \simeq \frac{1-b}{1+b}\frac{1}{qp}. \tag{7.32}$$

Here we put $x = n/p \simeq 1$, $s = r_{Hn}/r_{Hp} \simeq 1$ and $b = \mu_n/\mu_p$. Also, according to (2.123) and (2.124),

$$I_E \sim p. \tag{7.33}$$

By substituting (7.32) and (7.33) into (7.26), we obtain

$$V_H \simeq \text{constant} \times B \quad \text{for } V_{EB} \gg V_x. \tag{7.34}$$

Therefore, the Hall voltage would no longer depend on the emitter current. But in addition, at $V_{EB} \gg V_x$, the resistance of the emitter-base junction starts to decrease. Thus the short-circuiting effect becomes active, and the Hall voltage deteriorates further on.

The result of these two effects is a fast decay in the magnetic sensitivity of the magnetotransistor at higher currents, as displayed in figure 7.25. At higher currents, all curves tend to join the curve 4, which belongs to a magnetotransistor with a normal non-shorted emitter. The small but persistent magnetic sensitivity of this transistor is obviously not due to the injection modulation effect.

In the base region of a 'normal' transistor, close to a normal high-efficiency emitter, the Hall electric field cannot develop. There the majority carrier current is too small, and the Hall field due to the minority carrier current is always well short-circuited. Numerical simulations [35] and experimental results [36] have been reported that indicate the presence of a weak Hall field along a high-efficiency emitter of a magnetotransistor. However, this field is too minute to play any appreciable role in the magnetic sensitivity of such a magnetotransistor.

7.3.3 The magnetodiode effect in magnetotransistors

The magnetodiode effect may become the dominant sensitivity mechanism of a magnetotransistor operating at high currents. As we have seen in §7.2.1, the magnetodiode effect is intensive only under high-injection conditions.

We shall explain the influence of the magnetodiode effect on the magnetic sensitivity of a transistor collector current using figure 7.26 [1]. This figure shows a double-collector double-base-contact transistor, similar

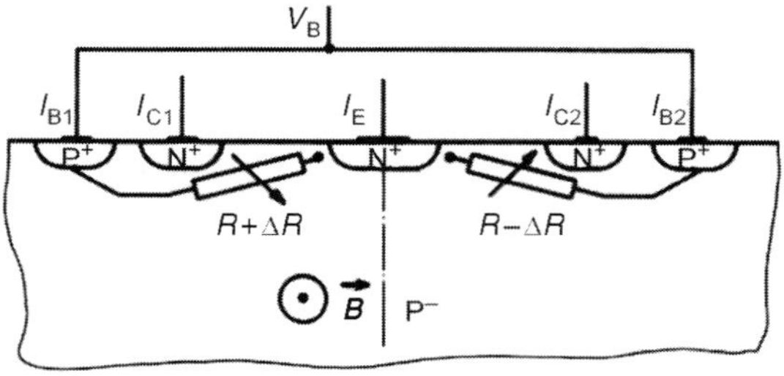

Figure 7.26. Magnetodiode effect as operating principle of a double-collector magnetotransistor with low-doped base region (P^-). Magnetic induction ***B*** causes asymmetrical carrier injection, leading to asymmetrical modulation $\pm\Delta R$ of the base-spreading resistances R, resulting in collector current imbalance. (Reprinted from [1]. © 1986 IEEE.)

to that shown in figure 7.21, but with a normal emitter. The resistors in the base region indicate that the material resistivity in the low-doped base region may be modulated as a result of the magnetodiode effect.

Suppose the device operates at a high injection level in the base region. Then a magnetic field perpendicular to the plane of the drawing gives rise to an increase in carrier concentration at the right-hand side and a decrease at the left-hand side (see also figure 7.10). An increase in carrier concentration causes a decrease in the resistivity of the corresponding base region, and vice versa. The lower the base resistance, the higher the base current at this side. Accordingly, an imbalance in the two base currents I_{B1}, I_{B2} arises. These result in an asymmetrical bias of the emitter-base junction, asymmetrical minority carrier injection, and finally a further increase of the imbalance ΔR, and so on. The asymmetrical minority carrier injection, however, leads to a corresponding imbalance in the collector currents. In this way, the magnetodiode effect participates in the magnetic sensitivity of a magnetotransistor.

7.3.4 Magnetotransistor figures of merit

Magnetotransistors are usually intended to be used for sensing magnetic fields. To enable a comparison with other magnetic sensors, it is useful to characterize magnetotransistors in a similar way as conventional Hall devices (see §5.1). We shall now do so briefly.

Sensitivity

A magnetotransistor may be regarded as a modulating transducer, converting a magnetic field signal into a current signal. This current signal is the change in the collector current I_C caused by the induction $\boldsymbol{B}$:

$$\Delta I_C = I_C(B) - I_C(0). \tag{7.35}$$

The absolute sensitivity of a magnetotransistor is defined, in analogy with (5.1), as

$$S_A = \left| \frac{\Delta I_C}{B} \right|. \tag{7.36}$$

The relative sensitivity of a magnetotransistor is usually defined as

$$S_I = \frac{S_A}{I_C} = \left| \frac{1}{I_C} \frac{\Delta I_C}{B} \right| \tag{7.37}$$

with the unit 'per tesla' ($AA^{-1}\,T^{-1} = T^{-1}$). For a double-collector transistor, the definitions (7.36) and (7.37) also apply if we put

$$I_C = I_{C1} + I_{C2} \tag{7.38}$$

$$\Delta I_C = (I_{C1}(B) - I_{C2}(B)) - (I_{C1}(0) - I_{C2}(0)) \tag{7.39}$$

where I_{C1} and I_{C1} denote the two collector currents.

We have used this definition throughout this section (see (7.15)). As we have seen, various magnetotransistors exhibit very different sensitivity values, ranging from $0.03\,T^{-1}$ to more than $10\,T^{-1}$. However, sensors with a high S_I value require additional bias current, which is much larger than the collector current. This additional current also participates in producing parasitic effects, such as offset and noise. Therefore, it seems reasonable to introduce the relative sensitivity defined as

$$S_{It} = \left| \frac{1}{I_{tot}} \frac{\Delta I_C}{B} \right| \tag{7.40}$$

where I_{tot} denotes the total current supplied to the magnetotransistor. For example, for the magnetotransistor shown in figure 7.18, $I_{tot} = I_{BB} + I_E$. In this magnetotransistor, typically $I_{tot} \geq 10 I_C$. When judged according to (7.40), the relative sensitivities of the magnetotransistors reported hitherto look less impressive. Their S_{It} values tend to approach the relative sensitivities of the Hall plates made of the same materials. For silicon magnetotransistors, a typical value is $S_{It} \simeq 0.1\,T^{-1}$.

Offset-equivalent magnetic field

For a double-collector magnetotransistor, such as that shown in figure 7.21, the offset collector current is

$$\Delta I_{Coff} = I_{CL}(0) + I_{CR}(0). \tag{7.41}$$

Here $I_{CL}(0)$ and $I_{CR}(0)$ denote the currents of the left and right collector, respectively, at zero magnetic field. The causes of offset are essentially the same as those responsible for offset in Hall plates, that is material non-uniformity, misalignment and piezo-effects.

The offset-equivalent magnetic field of a double-collector magnetotransistor may be defined, in analogy with (5.10), as

$$B_{\text{off}} = \frac{\Delta I_{\text{Coff}}}{S_A} = \frac{\Delta I_{\text{Coff}}}{S_I I_C}. \tag{7.42}$$

The offset-equivalent magnetic field values for magnetotransistors and Hall devices fabricated in a similar way are comparable. For a silicon magnetotransistor, $B_{\text{off}} \simeq 10\,\text{mT}$ is a good figure. It has been demonstrated that the offset of a magnetotransistor can also be electrically controlled [37].

Noise-equivalent magnetic field

The noise affecting the collector current of a magnetotransistor is shot noise and $1/f$ noise (see §2.5.3).

The mean square noise-equivalent magnetic field (NEMF) of a magnetotransistor can be defined in analogy with (5.14) as

$$\langle B_N^2 \rangle = \frac{\int_{f1}^{f2} S_{NI}(f)\,df}{(S_I I_C)^2}. \tag{7.43}$$

Here S_{NI} is the noise current spectral density in the collector current, (f_1, f_2) is the frequency range, and S_I is given by (7.37). For example, for the lateral magnetotransistor shown in figure 7.18, it has been found [38] that at higher frequencies (say, $f_1 > 100\,\text{Hz}$)

$$\langle B_N^2 \rangle \lesssim 2q \left(\frac{Y}{L}\right)^2 \frac{1}{\mu_{Hn} I_C} \Delta f. \tag{7.44}$$

This follows from (7.43), after substituting there equation (2.129) for the shot noise, and expression (7.19) for the sensitivity. The parameters Y and L are defined by (7.21) and in figure 7.18.

Figure 7.27 shows experimentally measured collector current *noise spectral density* in this magnetotransistor. The measured values in the white-spectrum range correspond to the theoretical values for the shot noise. Figure 7.28 shows the NEMF spectral density, defined by (7.43), corresponding to the two curves from figure 7.27. For comparison, values for two different Hall devices are also shown. Obviously, this magnetotransistor is superior to Hall devices with respect to noise at very low frequencies. In particular, it features a NEMF of less than $0.4 \times 10^{-6}\,\text{T}\,\text{Hz}^{-1/2}$ at 10 Hz and total current supply of 0.6 mA.

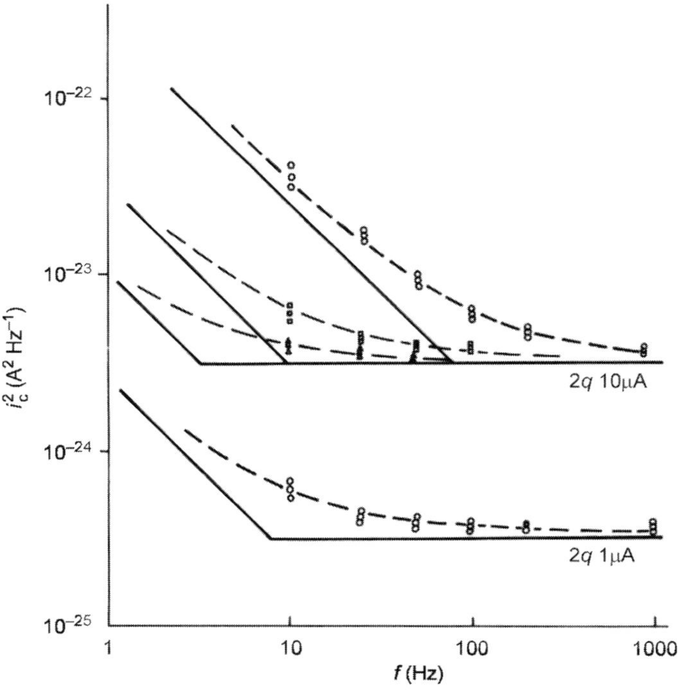

Figure 7.27. Collector current noise spectral density against frequency of magnetotransistors shown in figure 7.18. Experimental. The horizontal lines show the theoretical values of the shot noise at the collector currents 10 and 1 μA. In the $1/f$ noise region, a large spread of results for different samples has been observed [38].

In a dual-collector magnetotransistor with large current gain, a strong positive correlation between noise in the two collector currents has been found [39, 40]. Consequently, in spite of the low magnetic sensitivity of the magnetotransistor, very large signal-to-noise ratios are obtained, resulting in a high magnetic field resolution. A NEMF of about 0.19×10^{-6} T $Hz^{-1/2}$ at 10 Hz and a collector current $I_C < 1$ mA has been obtained.

Cross-sensitivity

The definition (5.17) also holds for the cross-sensitivity of a magnetotransistor. To the author's knowledge, only the temperature dependence of magnetotransistor sensitivity has been studied. By way of example, figure 7.29 shows the temperature dependence of sensitivity for the magnetotransistor shown in figure 7.18. Since $S_1 \sim \mu_{Hn}$ (7.19), and $\mu_{Hn} \sim \mu_n \sim T^{-2.4}$ (figure 2.9), a dependence of $S_I \sim T^{-2.4}$ is expected. The measurement corroborated this conclusion.

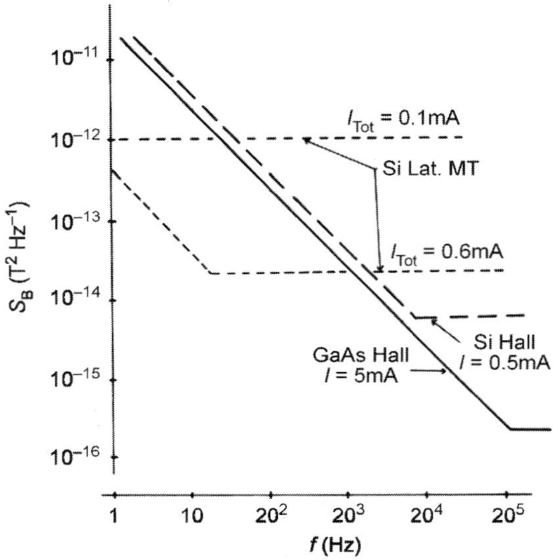

Figure 7.28. Noise-equivalent magnetic field (NEMF) spectral density against frequency for a magnetotransistor (figure 7.18) and two Hall devices [38].

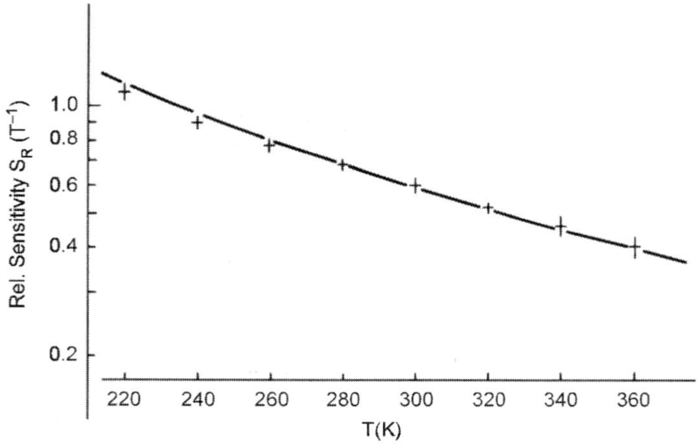

Figure 7.29. Temperature dependence of the relative magnetic sensitivity of a magnetotransistor with $L = 30\,\mu m$. The structure of the magnetotransistor is shown in figure 7.18. Points represent experiment, $I_{BB} = 0$, and I_C is between 10 nA and 10 μA. The line represents theory. (Reprinted from [29]. © 1986 IEEE.)

Non-linearity

The non-linearity of a magnetotransistor may be defined in analogy with (5.21) as

$$NL = \frac{\Delta I_C(B) - \Delta I_{C0}(B)}{\Delta I_{C0}(B)}. \tag{7.45}$$

Here $\Delta I_C(B)$ denotes the actual output signal (7.35) at an induction $\boldsymbol{B}$, and ΔI_{C0} is the best linear fit to the measured values.

Non-linearity strongly depends on the underlying operating principle. There is experimental evidence that magnetotransistors based on current deflection might be as linear as Hall plates [25,31]. Magnetotransistors based on the other two operating principles are inherently non-linear.

Stability

To the author's knowledge, the stability of magnetotransistors has not been studied. It may be speculated that a magnetotransistor cannot be more stable than either a Hall plate or a differential transistor pair fabricated in the same technology.

7.3.5 Magnetotransistor circuits

Most magnetotransistors lend themselves very well as circuit elements. Usually, a conventional transistor in a circuit can be directly substituted by a magnetotransistor. Then such a circuit becomes magnetic-sensitive.

Figure 7.30 shows the simplest magnetotransistor circuits. For the single transistor (a), the useful signal is the variation of the collector voltage V_C due to a magnetic field:

$$\Delta V_C = V_C(B) - V_C(0). \tag{7.46}$$

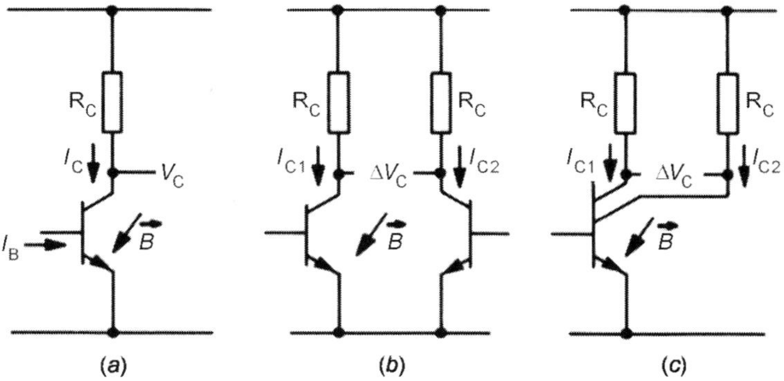

Figure 7.30. Simple magnetotransistor circuits. Magnetic induction causes variation of collector currents and, hence, collector voltages. (Reprinted from [1]. © 1986 IEEE.)

The relative sensitivity of the magnetotransistor circuit is given by

$$S_C = \left| \frac{1}{V_C} \frac{\Delta V_C}{B} \right| \tag{7.47}$$

with the unit 'per tesla' (T^{-1}).

At a sufficiently large collector output resistance r_C, such that $r_C \gg R_C$, it holds that $\Delta V_C = R_C \Delta I_C$. Then (7.47) reduces to

$$S_C = \left| \frac{1}{I_C} \frac{\Delta I_C}{B} \right| \tag{7.48}$$

which is identical to the relative sensitivity of the magnetotransistor itself (see equation (7.37)). Equation (7.48) also applies for the magnetotransistor pair circuit (figure 7.30(b)) and the circuit with the double-collector magnetotransistor (figure 7.30(c)), if we take into account the definitions (7.38) and (7.39).

Several attempts have been made to build sophisticated integrated circuits that incorporate magnetotransistors: see, for example, [41–44]. The aim of these efforts has been the development of a smart magnetic sensor that should perform better than a conventional circuit based on a Hall device. As yet, none of them has. Nevertheless, here is an example.

Figure 7.31 [42] shows a CMOS circuit which is reminiscent of an operational amplifier with n-channel MOS transistors operated in the bipolar mode [6]. The two unconventional six-terminal devices in the middle of the circuit are the magnetotransistors. These are the lateral double-base contact n–p–n magnetotransistors, such as that shown in figure 7.18. The six terminals are equally designated in figures 7.18 and 7.31. Assume that the input terminals IN^+ and IN^- are at the common potential, and there is no magnetic field. Then the two symmetrical branches of the amplifier have equal currents. In particular, $I_{EL} \simeq I_{ER}$ and $I_{CL} \simeq I_{CR}$. The symmetry may be disturbed either by a potential difference at the inputs, or by a magnetic field. A potential difference at the inputs produces an imbalance in the emitter currents I_{EL} and I_{ER}. Since $I_{CL,R} = \alpha I_{EL,R}$, α being the common-base current gain of the magnetotransistors, the imbalance in the emitter currents I_{EL} and I_{ER} also appears in the collector currents I_{CL} and I_{CR}. On the other hand, a magnetic field directly affects the collector currents I_{CL} and I_{CR}. As a result of the imbalance in the collector currents, a voltage change arises at the output of the circuit OUT1.

The circuit in figure 7.31 has been implemented using a commercially available p-well CMOS process. With the voltage supply $V_{DS} = 10\,\mathrm{V}$ and current consumption of about 1 mA, the circuit shows a magnetic sensitivity $S_A \simeq 100\,\mathrm{V\,T^{-1}}$, an offset-equivalent induction $B_{off} < 0.1\,\mathrm{T}$ and a temperature coefficient of sensitivity $TC_s \simeq -0.006\,\mathrm{K^{-1}}$. The electrical input proved useful for correcting offset and applying a feedback signal.

The application of an electrical feedback signal to a magnetic-sensitive device has been exploited in circuits called *magnetooperational amplifiers* [41–43].

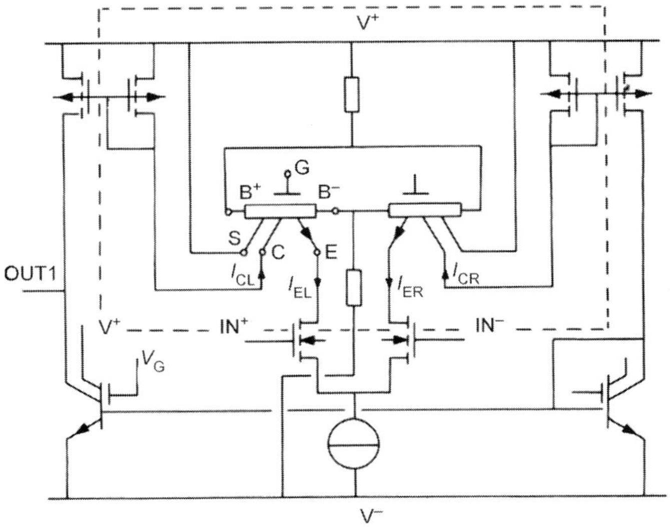

(*a*)

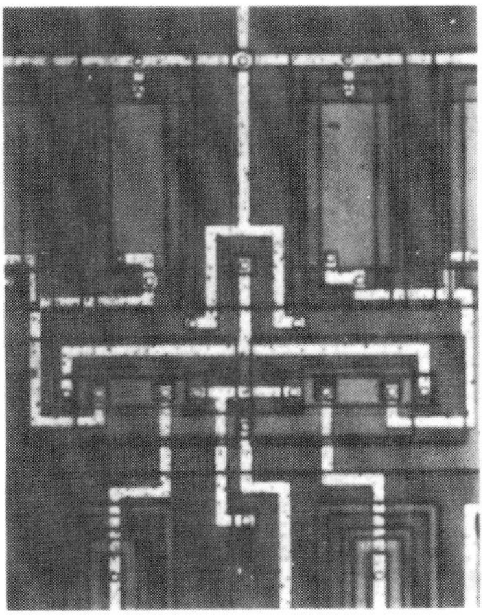

(*b*)

Figure 7.31. Schematic of a magnetic-sensitive circuit in CMOS technology. The magnetic-sensitive elements are the two magnetotransistors shown in figure 7.18 (a) schematic; (b) photograph of the part enclosed in the box (a) [42].

A circuit block diagram of a magnetooperational amplifier is shown in figure 7.32. Here S denotes the sensor element for electric and magnetic signals, A is a conventional amplifier and (R_1, R_2) is a voltage divider. In a magneto-operational amplifier, an electrical feedback signal balances the disturbance

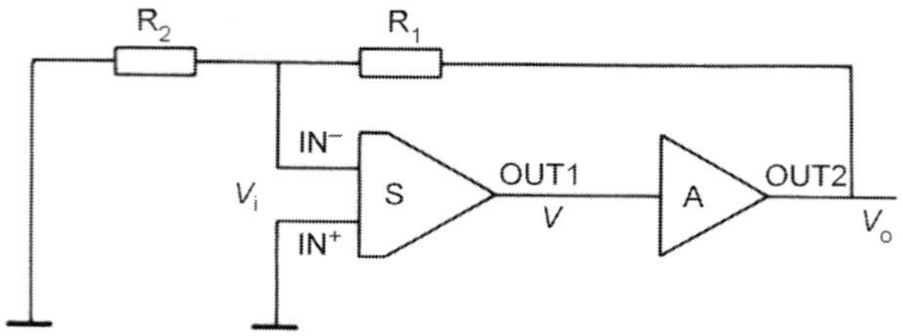

Figure 7.32. Block diagram of the magnetooperational amplifier. S denotes an element sensitive to both magnetic and electric signals (such as the circuit shown in figure 7.31), and A is a conventional voltage amplifier. The electrical feedback signal V_i balances the disturbance in S produced by a magnetic signal. (After [41].)

in the input sensor made by a magnetic signal. In this way, the magnetic sensitivity of the input sensor can be stabilized.

Let us consider how the magnetooperational amplifier from figure 7.32 operates [41]. Suppose the sensor element delivers at its output OUT1 a voltage V, given by

$$V = S_B B - S_E V_i \tag{7.49}$$

where S_B denotes the absolute sensitivity for the magnetic signal, $\boldsymbol{B}$ is the magnetic field (magnetic signal), S_E is the absolute sensitivity for the electric signal, and V_i is the input voltage at the sensor element (electric signal). The output voltage of the whole circuit is given by

$$V_0 = AV \tag{7.50}$$

where A denotes the amplifier gain. The feedback voltage is given by

$$V_i = KV_0 \qquad K = \frac{R_2}{R_1 + R_2}. \tag{7.51}$$

From equations (7.49)–(7.51), we find

$$B = \frac{1 + S_E KA}{S_B A} V_0. \tag{7.52}$$

If $S_E KA \gg 1$, this reduces to

$$V_0 \simeq \frac{S_B}{S_E} \frac{1}{K} B. \tag{7.53}$$

If the magnetic and electric sensitivities of the element S depend on the same material parameter (such as carrier mobility), the ratio S_B/S_E should be much more stable than S_B alone.

7.4 Carrier-domain magnetic-sensitive devices

A carrier domain in a semiconductor is a region of high non-equilibrium carrier density. Owing to the charge neutrality condition (2.36), the concentrations of electrons and holes in a carrier domain are approximately equal. Therefore, a carrier domain is an electron–hole–plasma domain. Another term in use for the carrier domain is current filament. The formation of a carrier domain has been demonstrated in pin structures [45, 23], p–n–p or n–p–n structures [23, 46–47], and in p–n–p–n structures [48]. The important feature of a carrier domain is that, at least in one dimension, it has no structural boundaries. Therefore, the domain can easily migrate in this dimension.

Carrier-domain magnetic-sensitive devices exploit a peculiarity of the Hall effect in the carrier domain. A carrier domain has no insulating boundaries. Therefore, in the presence of a magnetic field, the current deflection takes place in the domain. Since both electrons and holes are deflected in the same direction, the entire carrier domain eventually travels through the semiconductor. By detecting the migration of the domain, one may obtain information on the magnetic field. A carrier-domain device used as a magnetic sensor is also-called a carrier-domain magnetometer [1].

7.4.1 Three-layer carrier-domain devices

Melngailis and Rediker were the first to demonstrate the formation of a carrier domain in a semiconductor [45]. The experimental device was an n^+pp^+ InSb diode with the p-base about 1 mm thick (see figure 7.33). The diode operated with a forward bias, at a temperature below 100 K.

For low injection levels, as a consequence of trapping, the electron lifetime in the base region is very short, of the order of 10^{-9} s. However, if a large number of electrons are injected into the base, the traps eventually become saturated, the electron lifetime increases and reaches a value of 10^{-7} to 10^{-6} s. At a current density higher than about 0.2 A cm^{-2}, a carrier domain forms. The traps become saturated first near the n^+p junction, and

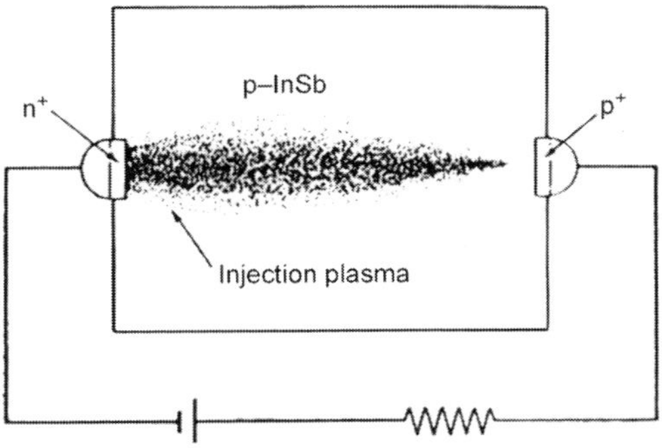

Figure 7.33. Carrier domain formed by injection plasma in a n^+pp^+ InSb diode. (Adapted from [23].)

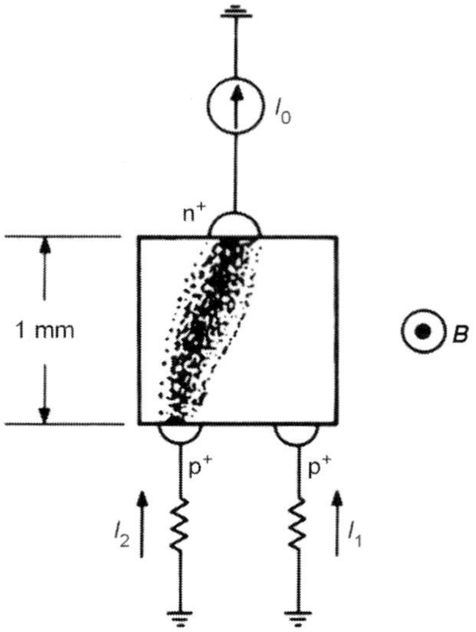

Figure 7.34. Dual-base-contact diode with a carrier domain between the n^+p junction and one of the p^+ contacts. In the shown configuration, $I_0 \simeq I_2 \gg I_1$, and the current depends little on the magnetic induction $\boldsymbol{B}$. Upon removing the magnetic field, the domain remains stable in the same position. A magnetic induction of opposite polarity and strength of about 1 mT is required to switch the domain to the other contact. The device is bistable and can operate as a magnetically triggerable flip-flop. (After [23].)

a region of high conductivity is formed. With an increase in diode current, the high-conductivity plasma domain grows in length, and eventually a high-current density filament forms (see figure 7.33). Inside the filament the electron diffusion length is of the order of a millimetre, while outside it the electron diffusion length has its original value of about 0.1 mm.

This great difference in electron diffusion lengths provides the spatial positive feedback mechanism, which keeps the carrier domain localized. An electron injected into a domain highly populated with fellow electrons survives for a long time, engages (but does not marry) a hole, travels a long distance and thus helps the newly injected electrons survive. An electron injected outside the populated area dies (recombines) very soon. Everybody would.

A magnetic field perpendicular to the current filament causes its deflection. This deflection can be exploited in various ways, two of them being illustrated in figures 7.34 and 7.35 [23].

Another three-layer carrier-domain Hall effect device is shown in figure 7.36 [49]. The device has the structure of a circular lateral bipolar n–p–n transistor surrounded by four voltage-probing contacts S_1–S_4. The

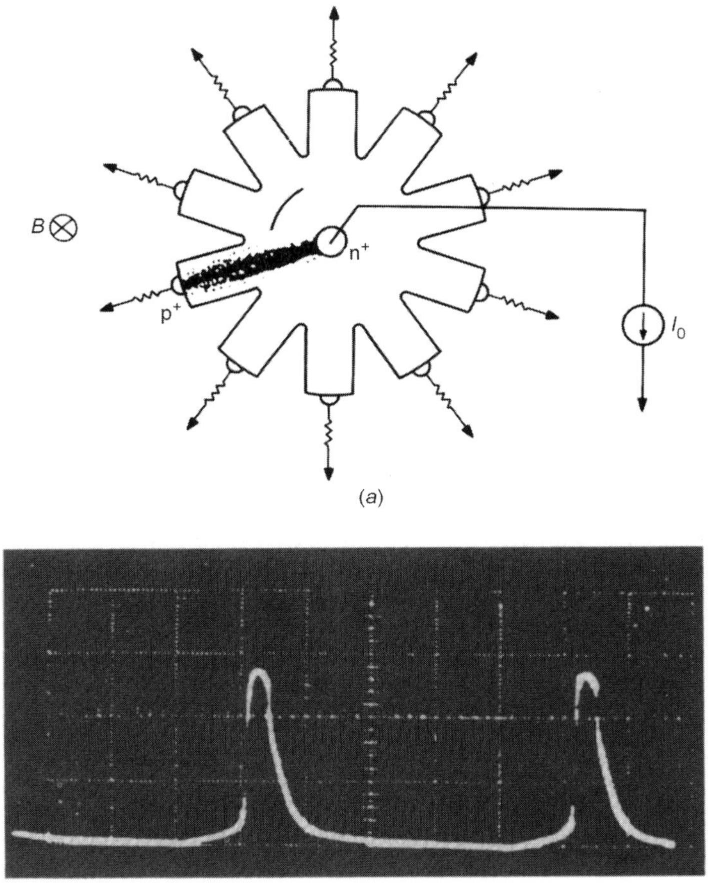

(b)

Figure 7.35. Carrier-domain rotational stepping switch. (a) Device structure. n^+p junction is centrally located and surrounded by 10 branches with p^+ contacts. (b) Current as a function of time in one branch of the switch exposed to a constant magnetic field. (Reprinted from [23].)

transistor operates in the collector–emitter breakdown regime with short-circuited emitter and base contacts (E and B). The internal feedback involved in the transistor breakdown mechanism confines the current domain to a narrow sector of the base region.

The experimental device had been made using a standard silicon planar process. The diameter of the collector region was 500 μm and the width of the base region 8 μm. The device can operate at room temperature (but not for very long). The

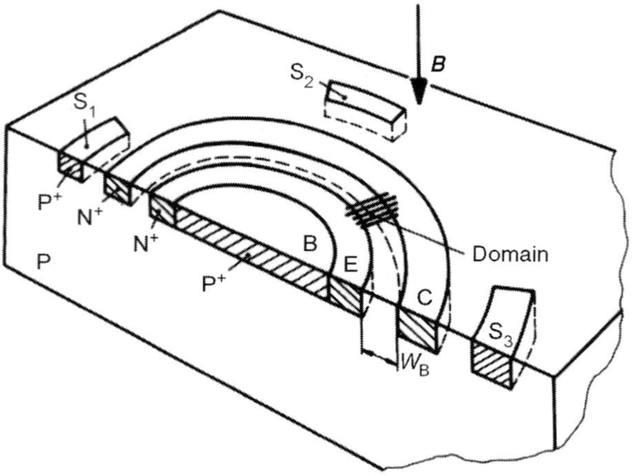

Figure 7.36. Circular three-layer carrier-domain magnetic-sensitive device. The carrier domain occurs in the base region of a circular, lateral n–p–n transistor operating in the breakdown regime. E is the emitter, B is the base contact and C is the collector. The frequency of domain rotation is modulated by magnetic induction $\boldsymbol{B}$ and detected at contacts S_1 to S_4. (Reprinted from [49].)

carrier domain appears for sufficiently large supply current (about 30 mA), which is supplied by the external current source. Once formed, the domain rotates spontaneously with a frequency $f \simeq 280$ kHz at $B = 0$, the direction of the rotation being incidental. The domain rotation is detected via voltage pulses appearing at the contacts S_1–S_4 whenever the domain passes by.

The frequency of the carrier-domain rotation is modulated by a magnetic field perpendicular to the device planar surface. The rotation frequency is enhanced or lowered, depending on the relative directions of B and the domain rotation. The sensitivity is about 250 kHz T^{-1} around zero magnetic induction.

7.4.2 Four-layer carrier-domain devices

A four-layer p–n–p–n device, also known as a Shockley diode, has two stable states at forward bias (see figure 7.37): the 'off' state with very high impedance, and the 'on' state with a low impedance [50]. To keep the device in the 'on' state, a certain minimal current must flow through the device. Thanks to this condition, a carrier domain can be created in such a device.

Let us consider what happens in a large-area or large-circumference p–n–p–n device working in the 'on' state at various currents. Figure 7.38(a) shows a four-layer device biased by a large current. The vertical lines indicate an approximately uniform distribution of the current over the device area. If we now start

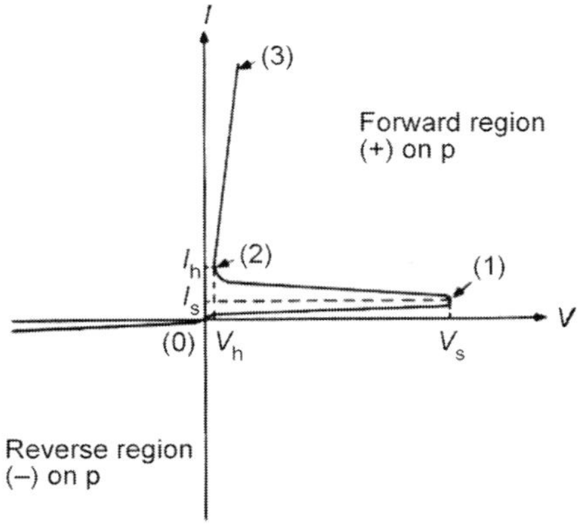

Figure 7.37. Current–voltage characteristics of a p–n–p–n device. In the forward region (external p-type layer positive biased), 0–1 is the forward blocking region; V_s and I_s are the switching voltage and switching current, respectively; 1–2 is the negative resistance region; V_h and I_h are the holding voltage and holding current, respectively; 2–3 is the forward conduction region. (Adapted from [50].)

decreasing the bias current I, at the beginning only the current density decreases. Note now that a curve similar to that shown in figure 7.37 also holds for current density. Eventually the average current density in our device drops below the holding current density (point (2) in figure 7.37). Because of the negative resistance in the portion (1)–(2) of the characteristics, the current density can

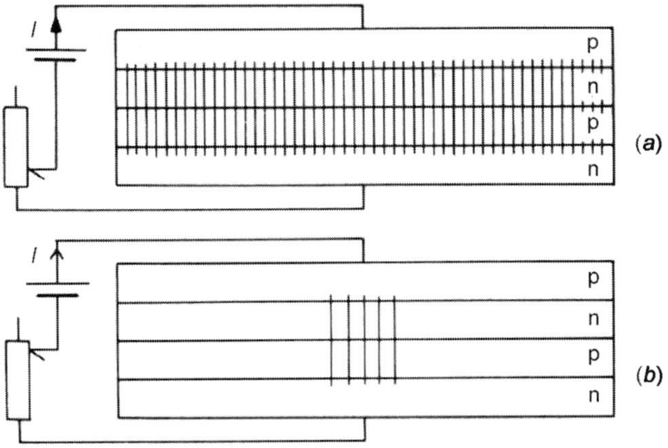

Figure 7.38. Large-area p–n–p–n device working at two different forward current levels. The vertical lines represent current density. (a) High current. The whole device is in the 'on' state. (b) Low current. Most of the available current concentrates in the small region of the structure, keeping it in the 'on' state. There a carrier domain forms. The rest of the structure has turned into the 'off' state.

> no longer remain constant over the device area. Instead, the current tends to concentrate in one area, whereas the other area starts losing current. The lower the current density, the higher the voltage. If the current supply is further reduced, the voltage at the current-depleted part of the structure will eventually reach the switching voltage V_s. Then this part of the device comes into the 'off' state. The rest of the structure takes over the whole current, and so the carrier domain forms (see figure 7.38(b)). If a magnetic field perpendicular to the drawing plane of figure 7.38(b) is established, the carrier domain will move left or right.

Based on this principle, a few carrier-domain magnetic sensors have been proposed. Here are two examples.

Figure 7.39 shows a circular four-layer carrier-domain device [51, 52]. A carrier domain forms with the current flow in the radial direction. A magnetic field perpendicular to the device surface makes the domain travel around the circumference of the structure. The frequency of this rotation is proportional to the magnetic field. The actual domain position is detected by monitoring currents at the segmented outer collectors SC.

Experiments [52–54] show a linear dependence of the domain rotation frequency on the magnetic induction, but only at inductions of $B \geq 0.2\,\mathrm{T}$. The domain rotation stops if B drops below a threshold value, which is

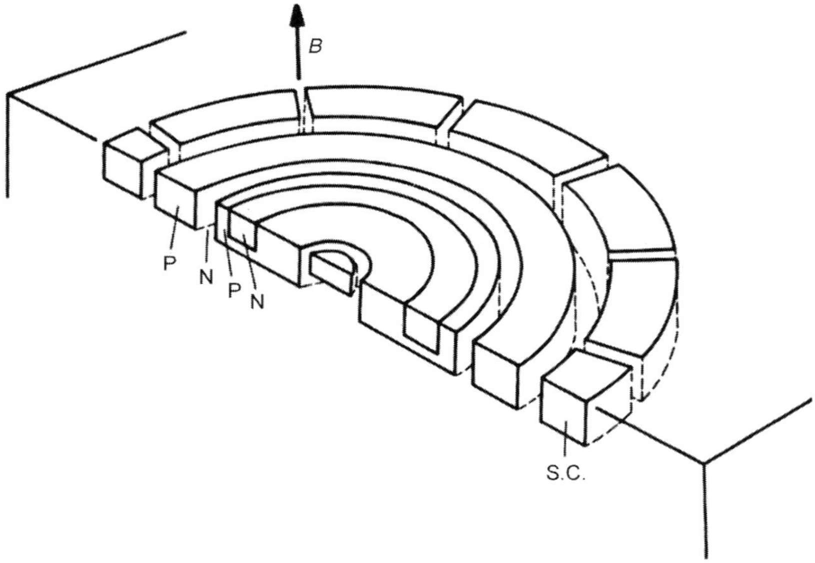

Figure 7.39. Circular four-layer carrier-domain magnetometer. The four layers are labelled N, P, N, P. The carrier domain rotates under the influence of a perpendicular magnetic field $\boldsymbol{B}$. Then current pulses appear at segmented collectors SC with frequency proportional to B [51, 52].

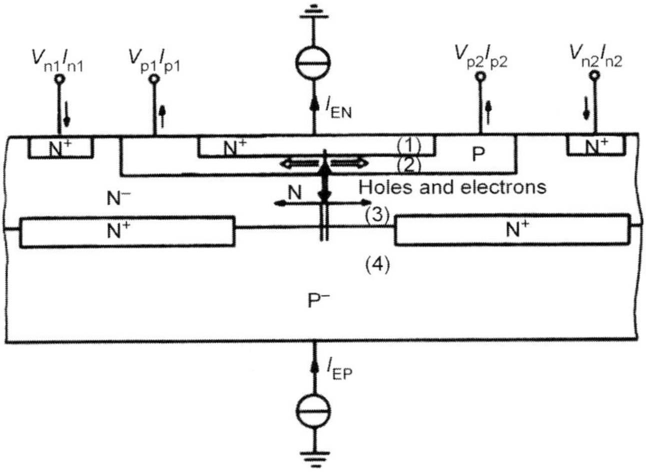

Figure 7.40. Cross section of vertical four-layer carrier-domain magnetometer [55]. A magnetic field perpendicular to the drawing plane causes a displacement of the domain and thus an imbalance in the currents I_{p1}, I_{p2} and I_{n1}, I_{n2}.

between 0.1 and 0.4 T. It is believed that the domain may stick to a preferred location as a consequence of geometrical imperfections due to alignment tolerances. Sensitivities are between 10 and 100 kHz T^{-1} at about 10 mA supply current.

The four-layer carrier-domain device shown in figure 7.40 [55] has a structure similar to that of the split-collector magnetotransistor shown in figure 7.16. The basic difference between the two devices is in their modes of operation. In the magnetotransistor, the junction (3)–(4) is always reverse biased, and serves only as insulation. In the carrier-domain device, this junction must also be forward biased in order to support the carrier domain.

When the device is biased as shown in figure 7.40, a carrier domain forms at an optimal place near the symmetry line. The optimum, or equilibrium, point exists because of the lateral currents and corresponding voltage drops in the two base regions (2) and (3).

If the domain moves out of the place of equilibrium, a restoring force brings the domain back to equilibrium. A magnetic field perpendicular to the plane of figure 7.40 produces a displacement of the domain. If, for example, the domain moves to the right, both right-hand base currents I_{p2} and I_{n2} increase, while both left-hand base currents I_{p1} and I_{n1} decrease. This current modulation indicates the domain displacement and hence the presence of the magnetic field. On the other hand, the asymmetry produced by the domain displacement eventually brings about the restoring force that prevents further displacement of the domain.

At the total current supply of 10 mA, and small magnetic fields, the current difference $I_{n1} - I_{n2}$ amounts to $3\,\mathrm{mA\,T^{-1}}$. Regarding this device as a magnetotransistor, we may calculate the relative sensitivity according to equation (7.40). Thus we obtain the value $S_{It} \simeq 0.3\,\mathrm{T^{-1}}$, which is among the highest sensitivity figures reported so far for silicon magnetic sensors.

References

[1] Baltes H P and Popovic R S 1986 Integrated semiconductor magnetic field sensors *Proc. IEEE* **74** 1107–32

[2] Kordic S 1986 Integrated silicon magnetic field sensors *Sens. Actuators* **10** 347–78

[3] Baltes H P and Nathan A 1989 Integrated magnetic sensors. In: *Sensor Handbook 1* ed. T Grandke and W H Ko (Weinheim: VCH) ch 7 pp 195–215

[4] Gallagher R C and Corak W S 1966 A metal-oxide-semiconductor (MOS) Hall element *Solid-State Electron.* **9** 571–80

[5] Sze S M 1981 *Physics of Semiconductor Devices* (New York: Wiley) ch 8

[6] Vittoz E A 1983 MOS transistors operated in lateral bipolar mode and their application in CMOS technology *IEEE Solid-State Circuits* **SC-18** 273–9

[7] Popovic R S 1985 Numerical analysis of MOS magnetic field sensors *Solid-State Electron.* **28** 711–16

[8] Mohan Rao G R and Carr W N 1971 Magnetic sensitivity of a MAGFET of uniform channel current density *Solid-State Electron.* **14** 995–1001

[9] Yagi A and Sato S 1976 Magnetic and electrical properties of n-channel MOS Hall-effect device *Jpn. J. Appl. Phys.* **15** 655–61

[10] Popovic R S 1984 A MOS Hall device free from short-circuit effect *Sens. Actuators* **5** 253–62

[11] Nathan A, Huiser A M J and Baltes H P 1985 Two-dimensional numerical modeling of magnetic field sensors in CMOS technology *IEEE Trans. Electron. Devices* **ED-31** 1212–19

[12] Popovic R S and Baltes H P 1983 A CMOS magnetic field sensor *IEEE Solid-State Circuits* **SC-18** 426–8

[13] Stafeev V I 1959 Modulation of diffusion length as a new principle of operation of semiconductor devices *Sov. Phys.-Solid State* **1** 763

[14] Karakushan E I and Stafeev V 11961 Magnetodiodes *Sov. Phys.-Solid State* **3** 493

[15] Stafeev V I and Karakushan E 11975 *Magnetodiody* (Moscow: Nauka)

[16] Cristoloveanu S 1979 L'effet magnétodiode et son application aux capteurs magnétiques de haute sensibilité *L'Onde Electrique* **59** 68–74

[17] Lutes O S, Nussbaum P S and Aadland O S 1980 Sensitivity limits in SOS magnetodiodes *IEEE Trans. Electron. Devices* **ED-27** 2156–7

[18] Chovet A, Cristoloveanu S, Mohaghegh A and Dandache A 1983 Noise limitations of magnetodiodes *Sens. Actuators* **4** 147–53

[19] Fujikawa K and Takamiya S 1975 Magnetic-to-electric conversion semi-conductor device *US Patent* 3911468

[20] Popovic R S, Baltes H P and Rudolf F 1984 An integrated silicon magnetic field sensor using the magnetodiode principle *IEEE Trans. Electron. Devices* **ED-31** 286–91

[21] Suhl H and Shockley W 1949 Concentrating holes and electrons by magnetic fields *Phys. Rev.* **75** 1617–18

[22] Bradner Brown C 1950 High-frequency operation of transistors *Electronics* **23** 81–3

[23] Melngailis I and Rediker R H 1962 The madistor-a magnetically controlled semiconductor plasma device *Proc. IRE* **50** 2428–35

[24] Flynn J B 1970 Silicon depletion layer magnetometer *J. Appl. Phys.* **41** 2750–1

[25] Zieren V and Duyndam B P M 1982 Magnetic-field-sensitive multicollector n–p–n transistors *IEEE Trans. Electron. Devices* **ED-19** 83–90; Zieren V 1983 Integrated silicon multicollector magnetotransistors *PhD Dissertation* Delft University of Technology

[26] Zieren V and Middlehoek S 1982 Magnetic-field vector sensor based on a two-collector transistor structure *Sens. Actuators* **2** 251–61

[27] Kordic S 1986 Integrated 3-D magnetic sensor based on an n–p–n transistor *IEEE Electron. Device Lett.* **EDL-7** 196–8

[28] Davies L W and Wells M S 1970 Magneto-transistor incorporated in a bipolar IC *Proc. ICMCST, Sydney* pp 34–5

[29] Popovic R S and Widmer R 1986 Magnetotransistor in CMOS technology *IEEE Trans. Electron. Devices* **ED-33** 1334–40

[30] Popovic R S and Baltes H P 1983 A bipolar magnetotransistor in CMOS technology *Proc. 11th Yugoslav Conf. on Microelectronics, MIEL, Zagreb* (Ljubljana: EZS) pp 299–306

[31] Ristic Lj, Smy T and Baltes H P 1989 A lateral magnetotransistor structure with a linear response to the magnetic field *IEEE Trans. Electron. Devices* **36** 1076–86

[32] Takamiya S and Fujikawa K 1972 Differential amplification magnetic sensors *IEEE Trans. Electron. Devices* **ED-19** 1085–90

[33] Popovic R S and Baltes H P 1983 Dual-collector magnetotransistor optimized with respect to injection modulation *Sens. Actuators* **4** 155–63

[34] Ref. 5, ch 3

[35] Nathan A, Maenaka K, Allegretto W, Baltes H P and Nakamura T 1989 The Hall effect in silicon magnetotransistors *IEEE Trans. Electron. Devices* **ED-36** 108–17

[36] Nathan A and Baltes H P 1990 Rotation of base region Hall field in magnetotransistors *Sens. Actuators* **A22** 758–61

[37] Ristic Lj, Smy T and Baltes H P 1988 A magnetotransistor structure with offset elimination *Sens. Mater.* **2** 83–92

[38] Popovic R S and Widmer R 1984 Sensitivity and noise of a lateral bipolar magnetotransistor in CMOS technology *IEDM Tech. Dig.* pp 568–71

[39] Nathan A, Baltes H P, Briglio D R and Doan M 1989 Noise correlation in dual-collector magnetotransistors *IEEE Trans. Electron. Devices* **ED-36** 1073–5

[40] Nathan A and Baltes H P 1989 How to achieve nanotesla resolution with integrated silicon magnetotransistors *IEDM, Washington, DC, 3–6 December 1989, Tech. Dig.* pp 511–14

[41] Popovic R 1987 Electrical circuit which is linearly responsive to changes in magnetic field intensity *US Patent* 4683429

[42] Popovic R S 1984 The magnetotransistor-operation and CMOS circuit application *IEEE Switzerland Fall Meet. on Silicon Sensors, Bern*, 16 October

[43] Maenaka K, Okada H and Nakamura T 1989 Universal magneto-operational amplifier (MOP) *Transducers '89, Abstracts, Montreux*, pp 236–7

[44] Krause A 1986 An integrated magnetic field sensor based on magnetotransistor *Proc. ESSCIRC '86, Delft, The Netherlands, 16–18 September 1986* pp 193–5

[45] Melngailis I and Rediker R H 1962 Negative resistance InSb diodes with large magnetic-field effects *J. Appl. Phys.* **33** 1892–3

[46] Gilbert B 1971 New planar distributed devices based on a domain principle *IEEE ISSCC Tech. Dig.* p 166

[47] Popovic R S and Baltes H P 1983 New oscillation effect in a semiconductor device *Europhysics Conf Abstracts* vol. 7b (European Physical Society) p 223

[48] Persky G and Bartelink D J 1974 Controlled current filaments in PNIPN structures with application to magnetic field detection *Bell Syst. Tech. J.* **53** 467–502

[49] Popovic R S and Baltes H P 1983 A new carrier-domain magnetometer *Sens. Actuators* **4** 229–36

[50] Ref. 5, ch 4

[51] Gilbert B 1976 Novel magnetic field sensors using carrier domain rotation: proposed device design *Electron. Leu.* **12** 608–10

[52] Manley M H and Bloodworth G G 1976 Novel magnetic field sensor using carrier domain rotation: operation and practical performance *Electron. Leu.* **12** 610–11

[53] Manley M H and Bloodworth G G 1978 The carrier-domain magnetometer: a novel silicon magnetic field sensor *Solid-State Electron. Devices* **2** 176–84

[54] Manley M H and Bloodworth G G 1983 The design and operation of a second-generation carrier-domain magnetometer device *Radio Electron. Eng.* **53** 125–32

[55] Golcolea J I, Muller R S and Smith J E 1984 Highly sensitive silicon carrier domain magnetometer *Sens. Actuators* **5** 147–67

Chapter 8

Comparing Hall sensors with other galvanomagnetic sensors

After having seen in this book various semiconductor devices based on the Hall effect (in a broader sense, we include here also the magnetoresistance effect), we may ask ourselves: what is the relative importance of each of these devices? How do these devices compare with other galvanomagnetic sensors, such as thin-film ferromagnetic magnetoresistors (which we did not treat in this book)? And how might this picture look in the years to come [1]? In this concluding little chapter of the book, I would like to share with you, the curious reader, my personal opinion on these issues.

We shall discuss here only magnetic field sensors, which are similar to Hall sensors: that means solid-state magnetic sensors, operating around room temperature, low-cost, and suitable for many industrial and laboratory applications. Only some galvanomagnetic sensors fit these criteria. In our discussion we shall leave aside many other magnetic sensors, such as search coil, flux-gate, SQUID, magnetooptical, and NMR magnetic sensors. In order to make the comparison of various galvanomagnetic devices that meet the above criteria, I chose as the major additional criterion *the impact* of these sensors both in scientific research and in industry. Thereby I will try to adopt a marketing point of view. Accordingly, it is not relevant what we, researchers in the field of magnetic sensors, think about the beauty of an effect or a device concept. What determines the impact of a magnetic sensor is what our peers in other research fields and in industry think about the potential of this device to solve their magnetic sensing problems.

8.1 Hall elements versus other Hall effect devices

In this book, we treated in chapters 4 and 7 various semiconductor devices based on the Hall effect. However, in the practice-oriented chapters 5 and 6, we discussed only the applications of Hall elements working in the Hall voltage mode of operation. This choice corresponds to the probability to find such cases in the real world. Let me explain why it is so.

8.1.1 Hall elements: Hall voltage mode versus Hall current mode and magnetoresistance mode of operation

In virtually all industrial applications of magnetic sensors, such as position sensing and current sensing, the most important requirements are a low threshold magnetic field for d.c. and low-frequency magnetic fields and the stability of the magnetic sensitivity over a broad temperature range. We can express these requirements in terms of the basic characteristics of Hall magnetic sensors: offset-equivalent magnetic field (§5.1.2), $1/f$ noise-equivalent magnetic field (§5.1.3), signal-to-noise ratio (4.3.6), and temperature cross-sensitivity (§5.1.4).

For an a.c. magnetic signal above the $1/f$-noise region, the signal-to-noise ratios of equivalent Hall plates working in the Hall voltage mode (§4.3), in the Hall current mode (§4.4), and in the magnetoresistance mode (§4.5) are similar [2]. However, for a d.c. and low-frequency magnetic signal, this is not necessarily so: by applying the connection-commutation technique to a Hall element working in the Hall voltage mode, one can strongly reduce offset and $1/f$ noise. An equivalent technique applicable in the Hall current mode and in the magnetoresistance mode does not exist (yet).

We have also seen that the current-related magnetic sensitivity of an extrinsic Hall plate is not very temperature-dependent (5.1.4). On the other hand, both the current-deflection effect (§4.4.2) and the magnetoresistance effect (4.5.1) depend directly on the mobility of charge carriers, which is strongly temperature-dependent (§2.4.1).

These two facts explain the predominance of the use of the Hall elements in the Hall voltage mode of operation over the other two modes.

Are there exceptions of this general rule? Yes, I think there are. Here are two examples.

If a magnetic switch (5.9.2) is made using a high-mobility galvanomagnetic plate, it might be preferable to operate the plate in the magnetoresistance mode: the quadratic magnetoresistor response to a magnetic field may help to achieve a higher resolution at high magnetic fields.

The second example is the immerging field of very small magnetic field sensors (§5.5). Recall that a Hall device operating in the conventional Hall voltage mode requires four terminals; the one operating in the current-deflection mode, like a split-drain MagFET (§7.1.2), requires three terminals; and a magnetoresistor requires only two terminals. The smaller the number of terminals, the easier the ultimate miniaturization of the device, and the less parasitic effects in its high-frequency operation (§4.6). Moreover, to the first approximation, in the Hall current mode of operation, there is no variation of the terminal potentials, and so no influence of the parasitic capacitances.

Does this mean that, due to the general trend in microelectronics towards miniaturization, in the future the Hall voltage mode may lose its dominant position? I don't think so. Hall plates working in the Hall voltage mode have been for many years absolutely dominant in most applications. This fact places them very high on a learning curve, a place difficult to achieve with a competing magnetic sensor concept, even if it is in principle better.

8.1.2 Hall Elements versus magnetodiodes, MagFETs, MagBJTs, . . .

A reader who had a look at the first edition of this book [3], might notice that chapter 7 on magnetic-sensitive diodes, field-effect transistors (FETs), bipolar junction transistors (BJTs) and thyristors is modified for this edition only by a slight reduction of its content. The reason for not formally updating chapter 7 is not the lack of new publications on these devices. This is simply because no new publication on the subject changed my impression of little potential for an impact [1].

Why have these, otherwise intellectually challenging, devices stayed without any impact? And this in spite of the fact that some of them, like MagFETs and MagBJTs, are so perfectly compatible with the mainstream microelectronics technology? I think that the following two facts are the main reason for this.

If you analyse any known 'MagDevice', you find in the best case that either its performance is equivalent to that of a Hall plate (see, for example, (7.14) in §7.1.2) or that it behaves as a combination of a Hall plate and a transistor (see, for example, §7.3.2). In view of the above-mentioned learning-curve effect, this is simply not enough to challenge the dominant position of the Hall plates.

The other reason is the existence of the connection-commutation technique for the reduction of offset and $1/f$ noise of Hall plates, and the non-existence of an equivalent technique applicable on MagDiodes, MagFETs, MagBJTs and other 'MagDevices'.

There is, however, a slight chance that someone finds a narrow niche application where a MagDevice is a better choice than a conventional Hall device. For example, if the current consumption in a system is limited, it might be optimal that one and the same device, such as a magnetotransistor, performs two functions, i.e. magnetic sensing and signal amplification. This might bring a better signal-to-noise ratio than that of a combination Hall plate/transistor. But I have not seen such niche applications yet.

So I kept chapter 7 in this edition of the book not because the MagDevices are now an alternative to the Hall elements described in chapters 4 and 5. I kept it only for reference purposes: to help young researchers avoid reinvent a wheel, and to stimulate them to address the real problems and look for the right applications.

8.2 Hall sensors versus ferromagnetic magnetoresistors

The only other solid-state magnetic field sensors with a practical importance comparable with that of Hall devices are the magnetic sensors based on the magnetoresistance effect in thin ferromagnetic films. We shall now briefly review the main properties of commercially available Hall magnetic sensors and of thin-film ferromagnetic magnetoresistors. Then we shall compare their characteristics and discuss their respective positions in the landscape of industrial applications.

8.2.1 Performance of integrated Hall magnetic sensors

By way of example, we shall now briefly look at the characteristics of a few leading commercially-available Hall ASIC single-axis magnetic field sensors. The first two (Hall-1 and Hall-2) are conventional integrated linear Hall sensors, similar to that described in §5.6.5. They are made in BiCMOS and CMOS technology, respectively. The third one (IMC/Hall) is the CMOS integrated Hall ASIC combined with integrated magnetic concentrators (IMC), described in §7.7.3, figure 5.44. The characteristics of these Hall sensors are summarized in table 8.1.

We see that the integration of magnetic concentrators (IMC) brings about a higher magnetic sensitivity (b) and lower equivalent offset and noise ((f)–(i)); this results in a better magnetic field resolution ((j)–(l)). The magnetic gain associated with IMC makes possible also the realization of a larger bandwidth (c). Recall also that conventional Hall magnetic sensors respond to a magnetic field normal to the chip surface, whereas an IMC/Hall sensor responds to a magnetic field parallel with the chip surface.

8.2.2 Performance of ferromagnetic magnetoresistors

As we mentioned in the introduction of this book (§1.1.2), the magnetoresistance effect in ferromagnetic materials was discovered almost 150 years ago, long before the Hall effect. Today this effect is called anisotropic magnetoresistance effect, and is used commercially in the form of magnetic field sensors called *anisotropic magnetoresistors* (AMRs). An AMR is basically a planar thin-film resistor, made of a ferromagnetic film of about 100 nm in thickness. It responds to a magnetic field parallel to the chip surface (see figure 1.3). The typical magnetic sensitivity is 10 mV/mT for a bias voltage of 1 V. This corresponds to a voltage-related sensitivity (5.7)

Table 8.1. Characteristics of some integrated Hall single axis magnetic sensors [4].

Technology → Sensor →	BiCMOS Hall-1	CMOS Hall-2	IMC/CMOS IMC/Hall	Unit
Characteristics (*1) ↓				
(a) Full scale field range (FS)	±40	±14 (*2)	7 (*2)	mT
(b) Sensitivity (typical)	50	140 (*2)	300 (*2)	mV/mT
(c) Bandwidth (BW)	30	0.13	120	kHz
(d) Linearity error within FS	0.1	1	0.5	%FS
(e) Hysteresis error (±10 mT)	0	0	20	μT
(f) Equivalent offset at 300 K	3	0.06 (*3)	0.45 (*3)	%FS
(g) Offset TemCoef. error	16	3	0.66 (*4)	μT/K
(h) $1/f$ noise density at 1 Hz	0.3 (*5)	(*5), (*6)	0.3 (*5)	$\mu T/\sqrt{Hz}$
(i) White noise density (*7)	0.2 (*5)	1.3 (*5)	0.03 (*5)	$\mu T/\sqrt{Hz}$
Resolution:				
(j) For a quasi-d.c. field (*8)	167	67	31	μT
(k) For low frequency fields (*9)	7	37	4.5	μT
(l) For high frequency fields (*10)	0.2	1.3	0.03	μT

Notes (*):
(1) At the supply voltage of 5 V.
(2) Programmable. The figure shows the optimal value for high resolution.
(3) Quiescent output voltage is programmable. The figure shows the rest offset.
(4) With the best programmed parameters.
(5) Our own measurement.
(6) Covered by the white noise.
(7) Equivalent noise field at a sufficiently high frequency, beyond the $1/f$ region.
(8) After zeroing the initial offset, within $\Delta T = 10$ K: (j) = (e) + (g)ΔT + (k).
(9) (k) = Peak to peak equivalent noise field: 7 × (rms noise field, $0.1 < f < 10$ Hz).
(10) For BW $\ll$ 1 Hz, we take (l) $\simeq$ (i)$\sqrt{1\,\text{Hz}}$. Otherwise, (l) is proportional to $\sqrt{\text{BW}}$.

of about 10 V/V T, which is about 100 times better than that of the best silicon Hall plate.

Why is the magnetoresistive effect in a thin ferromagnetic film so much stronger than the Hall effect in semiconductors?

The explanation of this enigma is that in MRs, the high sensitivity is not only the merit of the appropriate galvanomagnetic effect. Instead, the origin of the high sensitivity is the result of a synergy of two distinct effects [5]: the one is, in our terminology, a strong magnetic flux concentration effect, which we explained (in another context) in §5.7.1; and the other is a mediocre inherent magnetoresistive effect, which can be described by equation (3.244) of §3.3.10.

Table 8.2. Characteristics of some AMR and GMR single axis magnetic sensors [8].

Characteristics (*1)	AMR-1	AMR-2	GMR-1	Unit
(a) Full scale field range (FS)	0.6 (*2)	0.6 (*2)	7 (*3)	mT
(b) Sensitivity (typical)	50	64	27.5	mV/mT
(c) Bandwidth (BW)	5	1	1	MHz
(d) Linearity error within FS	1.4	6	2	%FS
(e) Hysteresis error within FS	0.3	0.5	3	%FS
(f) Equivalent offset at 300 K	30	19	10	%FS
(g) Offset TemCoef. error (*4)	0.1	0.47	0.22	μT/K
(h) $1/f$ noise density at 1 Hz	1	n.a	16 (*5)	$nT/\sqrt{Hz}$
(i) White noise density (*6)	0.1	0.1	0.2	$nT/\sqrt{Hz}$
Resolution:				
(j) For a quasi-d.c. field (*7)	3	30	210	μT
(k) For low frequency fields (*8)	8.5	n.a	140	nT
(l) For high frequency fields (*9)	0.1	0.1	0.2	nT

Notes (*) :
(1) At the supply voltage of 5 V.
(2) If exposed to a field higher than 2 mT, the sensor has to be reset.
(3) Maximum rating: 30 mT.
(4) Without Set/Reset offset reduction.
(5) Our own measurement.
(6) Equivalent noise field at a sufficiently high frequency, beyond the $1/f$ region.
(7) After zeroing the initial offset, without Set/Reset offset reduction, within $\Delta T = 10$ K: (j) = 0.01 (e) FS + (g) ΔT + (k).
(8) (k) = Peak to peak equivalent noise field: 7 × (rms noise field, $0.1 < f < 10$ Hz).
(9) For BW ≪ 1 Hz, we take (l) ≃ (i)$\sqrt{1\text{ Hz}}$. Otherwise, (l) is proportional to $\sqrt{\text{BW}}$.

Another much used MR has the structure of a sandwich of thin ferromagnetic and non-ferromagnetic films. Its operation principle is based on the relatively recently discovered *giant magnetoresistance effect* (GMR). Paradoxally, a typical GMR sensor is less magnetic-sensitive than an AMR, but has much higher saturation field. There is a sizeable literature on AMRs and GMRs. See, for example, monographs [5]–[7] and references herein.

In table 8.2 we list the main characteristics of some commercially available general-purpose AMRs and GMR magnetic sensors.

The full-scale measurement ranges of AMR and GMR sensors are limited by a magnetic saturation of the corresponding ferromagnetic films. Related with their structure is also the hysteresis phenomenon. A very unpleasant feature of AMR is the possible reversal of the internal magnetization of the magnetoresistor film (flipping), which causes a reversal of the sign of the output signal. A magnetic field shock (disturbing field) may demagnetize the film and so deteriorate the sensitivity of the sensor. These problems can be efficiently eliminated by magnetizing back and forth the film using an associated electromagnet. In the notes to table 8.2, we refer to this function as Set/Reset offset reduction. For GMR, a strong magnetic shock is a more serious problem: if a GMR is exposed to a strong-enough magnetic field (destroying field), it will be irreparably damaged.

8.2.3 Integrated Hall sensors versus AMRs and GMRs

Let us now compare the characteristics and application areas of commercially available integrated Hall, AMR and GMR magnetic sensors. We shall consider only general-purpose sensors described in §§8.2.1 and 8.2.2, leaving aside sensors for very specific applications such as reading heads for magnetic discs. In the first part of the following discussion I will briefly describe the state of the art in the field of magnetic sensors in the year 2002, that is, just before the advent of the IMC/Hall technology (§5.7.3). Later we shall see how this technology starts to change the landscape of Hall, AMR and GMR sensors.

Hall plates have a relatively small magnetic sensitivity. This has two important consequences: first, detecting small magnetic fields is difficult; and second, a Hall sensor system may require a big electronic gain, which limits the achievable bandwidth of the sensor system. On the other hand, Hall sensors show no saturation effects at high magnetic fields; and, due to their compatibility with microelectronics technology, they can be integrated with interface electronics, which result in a low-cost smart magnetic sensor microsystem. Therefore, integrated Hall sensors are preferably used at higher magnetic fields. It was generally believed that Hall sensors are applicable only at magnetic fields well above 1 mT, and the measured magnetic field had to be perpendicular to the sensor chip.

AMR magnetic sensors have high resolution and high bandwidth, but they saturate at a rather small magnetic field (less than 1 mT) and they may require a complex resetting procedure. GMR magnetic sensors also have high resolution and high bandwidth, can operate also in the millitesla range, but have a high hysteresis and can be destroyed by a not-so-high magnetic field. Both AMR and GMR sensors respond to a magnetic field parallel with the MR layer.

Generally speaking, ferromagnetic MR sensors were considered as highly sensitive and good for small magnetic fields, whereas Hall sensors were considered as less sensitive and good only for high magnetic fields.

Obviously, according to the state of the art in 2002, there was a complex gap between the characteristics and application areas of Hall magnetic sensors and ferromagnetic magnetoresistors. The major gap existed between the applicable magnetic field ranges: the gap was at about 1 mT. The other related gaps appeared in the achievable bandwidth of the sensor system and in the system costs. Still another difference existed in the relative orientation of the measured magnetic field vector with respect to the sensor chip: for MRs, parallel; for Halls, perpendicular.

Over the last few years we developed the concept of Hall magnetic field sensors based on an integrated combination of Hall elements and magnetic flux concentrators (IMC Hall sensors, §5.7.3). The first IMC Hall ASICs are now commercially available [4, 9]. The IMC functions as a passive magnetic amplifier, boosting the performance of the Hall sensor. So the resolution and the bandwidth of an IMC Hall sensor approach those of ferromagnetic MR sensors. Moreover, seen from outside, an IMC Hall sensor responds to a magnetic field parallel with the chip surface, much as an MR sensor does.

To facilitate the comparison of the above sensors, we present now the key performance data of all sensors listed in tables 8.1 and 8.2 in one diagram (figure 8.1 [4]).

By inspecting the diagram, we conclude that AMRs and GMRs are still much better than modern Hall ASIC sensors only in the field of high-frequency

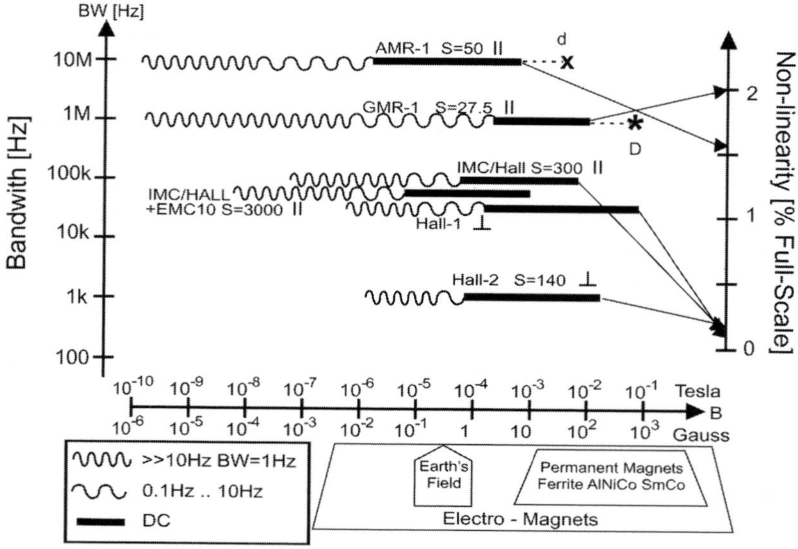

Figure 8.1. Key performance data of some AMR, GMR and integrated Hall magnetic field sensors. MR performance are without Set/Reset offset reduction. Notation (see the inset at bottom left): The vertical position of the a.c. and d.c. lines indicates the bandwidth of the sensor; The left-side ends of the a.c. and d.c. lines denote the detectivity limits (resolution) for d.c. field, low frequency field and high frequency field, respectively. The high field limit of the full lines denotes the full scale range. d, disturbing field; D, destroying field; S, sensitivity at the supply voltage of 5 V; ||, sensitive to a field parallel with the chip surface; ⊥, sensitive to a field perpendicular to the chip surface; +EMC10, with an additional, external, magnetic flux concentrator, providing a magnetic gain of 10 [4].

a.c. magnetic measurements: MRs are preferred for both higher resolution and larger bandwidth. But at lower frequencies, the difference becomes smaller. For d.c. fields, an AMR or GMR are good only if some kind of chopper stabilization (switching, set/reset) is applied (not shown in figure 8.1); otherwise, the resolution of the IMC Hall ASIC is much better than that of GMR and approaches the resolution of a non-switched AMR. Therefore, the IMC Hall sensor ASICs are about to bridge the gap between AMR, GMR and traditional silicon integrated Hall magnetic sensors.

In conclusion, the stereotype opinion 'ferromagnetic MR sensors are much better for small magnetic fields than Hall magnetic sensors' starts to lose its ground. Today (2003), for small quasi-d.c. magnetic fields, a magnetic sensor based on the combination Hall ASIC/Integrated Magnetic Concentrator (IMC) is superior than a GMR and it is approaching the performance of AMR magnetic sensors.

8.2.4 Looking into a crystal ball

Let me now try to predict what will happen in the field of magnetic field sensors over the next dozen years. I will base my predictions only on the facts that I am aware of now. I will not consider the consequences of future genial inventions and discoveries.

The system costs of a magnetic sensor system based on an IMC Hall ASIC tends to be lower than that of a similar sensor system based on an AMR or GMR. From this and from the above results we may infer that, in the near future, the application field of integrated Hall magnetic sensors shall broaden at the expense of AMRs and GMRs.

Shall this trend continue in the future? This depends on the development of the performance/price ratio pertinent to each family of magnetic field sensors. But paradoxally, this shall not depend so much on the cleverness and effort of us, working on research and development of magnetic sensors. Much more, the race will depend on the trends in the mainstream technologies, on the peripheries of which float the technologies of magnetic sensors.

If we except the applications for reading heads for computer magnetic disc memories, we may estimate the total world market for Hall, AMR and GMR sensors to be less than $1 billion (2003). On the other hand, Hall, AMR and GMR magnetic sensors are low-cost high-technology products. Making something high-tech at low cost is possible only if it is compatible with a mainstream high technology. And indeed, Hall ASICs are compatible with the microelectronics technology, and AMRs and GMRs are compatible with computer magnetic disc technology.

It seems very probable that the microelectronics technology will continue to be dominated by silicon CMOS technology at least until the year 2015. That means that the basics for the current Hall ASIC technology will be readily available over this period. If the microelectronics industry drifts more towards high-mobility semiconductors, then it will be even better for Hall applications. In any case, I think that the Hall ASIC technology will continue to be present on the market place. People shall learn to cope even better with the problems of offset, thermal drift and limited bandwidth. Accordingly, the basic performance of integrated Hall magnetic sensors shall continue to be incrementally improved. In addition, more electronic functions (such as analogue to digital conversion, look-up tables, even complete micro-controllers), shall be integrated on the chip. This will further boost the performance/price ratio of integrated Hall magnetic sensors.

It is less clear what will happen on the magnetic memory market. If 'magnetoelectronics' [10], also-called 'spintronics' [11], produce results worth large scale commercialization, then AMR, GMR and related magnetic sensors shall get a very strong wind in their sails. Among other things, low-cost 'smart' versions of these sensors, integrated with interface electronics, will appear on the market. If magnetoelectronics does not succeed commercially, but the magnetic disc industry keeps its big importance, then also AMRs and GMRs shall keep their big niche markets: high detectivity at high frequency and/or very small size (e.g. spin-valves [10]). But if the semiconductor memories start to replace magnetic discs, then AMRs and GMRs may face hard times.

References

[1] Here I will briefly revisit and update the relevant part of our paper: Popovic R S, Flanagan J A and Besse P-A 1996 The future of magnetic sensors *Sensors and Actuators* **A56** 39–55
[2] Boero G, Demierre M, Besse P-A and Popovic R S 2003 Micro-Hall devices: performance, technologies and applications *Sensors and Actuators* **A63**
[3] Popovic R S 1991 *Hall Effect Devices* (Boston: IOP Publishing)
[4] Popovic R S and Schott C 2002 Hall ASICs with integrated magnetic concentrators, *Proceedings Sensors Expo & Conference*, Boston, MA, USA, 23–26 September 2002
[5] O'Handley R C 2000 *Modern Magnetic Materials* (New York: Wiley)
[6] Ripka P (ed.) 2001 *Magnetic Sensors and Magnetometers* (Boston: Artech House) ch 4
[7] Tumanski S 2001 *Thin Film Magnetoresitive Sensors* (Bristol: IOP Publishing)
[8] Popovic R S, Drljaca P M and Schott C 2002 Bridging the gap between AMR, GMR and Hall magnetic sensors *Proc. 23rd Int. Conference on Microelectronics* (MIEL 2002), Vol. 1, Nis, Yugoslavia, 12–15 May, 2002, pp 55–58
[9] www.sentron.ch; www.gmw.com
[10] de Boeck J and Borghs G 1999 Magnetoelectronics *Physics World* April, pp 27–32
[11] Grundler D 2002 Spintronics *Physics World* April, pp 39–43

Appendix A

International system of units

Quantity	Unit	Symbol	Dimensions
Length	metre	m	
Mass	kilogram	kg	
Time	second	s	
Temperature	kelvin	K	
Current	ampere	A	
Frequency	hertz	Hz	s^{-1}
Force	newton	N	$kg\ m\ s^{-2}$
Pressure	pascal	Pa	$N\ m^{-2}$
Energy	joule	J	N m
Power	watt	W	$J\ s^{-1}$
Electric charge	coulomb	C	A s
Potential	volt	V	$J\ C^{-1}$
Conductance	siemens	S	$A\ V^{-1}$
Resistance	ohm	Ω	$V\ A^{-1}$
Capacitance	farad	F	$C\ V^{-1}$
Magnetic flux	weber	Wb	V s
Magnetic induction	tesla	T	$Wb\ m^{-2}$
Inductance	henry	H	$Wb\ A^{-1}$

Appendix B

Physical constants

Quantity	Symbol	Value
Angstrom unit	Å	1 Å = 10^{-4} μm = 10^{-8} cm
Avogadro constant	N_{AVO}	6.02204×10^{23} mol^{-1}
Bohr radius	a_B	0.52917 Å
Boltzmann constant	k	1.38066×10^{-23} J K^{-1} (R/N_{AVO})
Elementary charge	q	1.60218×10^{-19} C
Electron rest mass	m_0	0.91095×10^{-30} kg
Electron volt	eV	1 eV = 1.60218×10^{-19} J
Gas constant	R	1.98719 cal mol^{-1} K^{-1}
Permeability in vacuum	μ_0	1.25663×10^{-6} H m^{-1} $(4\pi \times 10^{-7})$
Permittivity in vacuum	ε_0	8.85418×10^{-12} F m^{-1} $(1/\mu_0 c^2)$
Planck constant	h	6.62617×10^{-34} J s
Reduced Planck constant	$\hbar$	1.05458×10^{-34} J s $(h/2\pi)$
Proton rest mass	M_p	1.67264×10^{-27} kg
Speed of light on vacuum	c	2.99792×10^{8} m s^{-1}
Standard atmosphere		1.01325×10^{5} N m^{-2}
Thermal voltage at 300 K	kT/q	0.0259 V
Wavelength of 1 eV quantum	λ	1.23977 μm

Appendix C

Properties of some semiconductors

Semiconductor		Band gap (eV)		Mobility at 300 K[1] ($cm^2\ V^{-1}\ s^{-1}$)		Relative effective mass m^*/m		
		300 K	0 K	Electrons	Holes	Electrons	Holes	$\varepsilon_s/\varepsilon_0$
Element	C	5.47	5.48	1 800	1 200	0.2	0.25	5.7
	Ge	0.66	0.74	3 900	1 900	0.64*l*	0.04*li*	16.0
						0.082*t*	0.28*he*	
	Si	1.12	1.17	1 500	450	0.98*l*	0.16*li*	11.9
						0.19*t*	0.49*he*	
	Sn		0.082	1 400	1 200			
IV–IV	α-SiC	2.996	3.03	400	50	0.60	1.00	10.0
III–V	AlSb	1.58	1.68	200	420	0.12	0.98	14.4
	GaN	3.36	3.50	380		0.19	0.60	12.2
	GaSb	0.72	0.81	5 000	850	0.042	0.40	15.7
	GaAs	1.42	1.52	8 500	400	0.067	0.082	13.1
	GaP	2.26	2.34	110	75	0.82	0.60	11.1
	InSb	0.17	0.23	80 000	1 250	0.0145	0.40	17.7
	InAs	0.36	0.42	33 000	460	0.023	0.40	14.6
	InP	1.35	1.42	4 600	150	0.077	0.64	12.4
II–IV	CdS	2.42	2.56	340	50	0.21	0.80	5.4
	CdSe	1.70	1.85	800		0.13	0.45	10.0
	CdTe	1.56		1 015	100			10.2
	ZnO	3.35	3.42	200	180	0.27		9.0
	ZnS	3.68	3.84	165	5	0.40		5.2
IV–VI	PbS	0.41	0.286	600	700	0.25	0.25	17.0
	PbTe	0.31	0.19	60 000	4 000	0.17	0.20	30.0

(1) Drift mobilities obtained in the purest and most perfect materials. *m*, electron rest mass in vacuum; *l*, longitudinal; *li*, light hole; *t*, transverse; *he*, heavy hole. (After Sze S M 1981 *Physics of Semiconductor Devices* (New York: Wiley) p 849.)

Appendix D

Magnetic quantities, units, relations and values

Quantity	Symbol	SI unit	CGS unit	Conversion of units
Magnetic field intensity, or Magnetizing field	H	ampere per metre (A/m)	oersted (Oe)	$1\ \mathrm{A/m} = 1.2566 \times 10^{-2}$ Oe $1\ \mathrm{Oe} = 79.578$ A/m
Magnetic induction, or Magnetic flux density, or Magnetic field	B	tesla (T) $\mathrm{T} = \mathrm{Wb/m^2} = \mathrm{Vs/m^2}$	gauss (G)	$1\ \mathrm{T} = 10^4$ G $1\ \mathrm{G} = 10^{-4}$ T
Magnetic flux	Φ	weber (Wb) Wb = Vs	maxwell (Mx)	$1\ \mathrm{Wb} = 10^8$ Mx $1\ \mathrm{Mx} = 10^{-8}$ Wb

Relations between *H*, *B* and Φ

$$\boldsymbol{B} = \mu \boldsymbol{H};\ \text{in vacuum, } \mu = \mu_0 = 4\pi \times 10^{-7}\ \mathrm{T/(A/m)};$$

Therefore, in vacuum, $H = 80\ \mathrm{A/m} \approx 1$ Oe corresponds to $B = 10^{-4}\ \mathrm{T} = 1$ G.

$$\Phi = \int_A \boldsymbol{B}\, \mathrm{d}\boldsymbol{A};\ \text{A: surface.}$$

How strong a magnetic field can be

Human brain	10^{-14}–10^{-12} T
Earth's magnetic field	6×10^{-5} T at the Earth's poles
Current-carrying wire	10^{-3} T for $I = 100$ A, at the distance 2 cm from the wire axis
Small bar magnet	0.01–0.1 T at the magnet surface
Electromagnet with an iron core	≤ 2 T
Superconducting magnet	5–20 T
Solenoid with a short current pulse	≤ 100 T

Appendix E

Mathematical relations

I. Vector indentities

$$[\boldsymbol{a} \times [\boldsymbol{b} \times \boldsymbol{c}]] = \boldsymbol{b}(\boldsymbol{a} \cdot \boldsymbol{c}) - \boldsymbol{c}(\boldsymbol{a} \cdot \boldsymbol{b}) \tag{E.1}$$

$$[[\boldsymbol{a}\times]\boldsymbol{b} \times \boldsymbol{c}] = \boldsymbol{b}(\boldsymbol{a} \cdot \boldsymbol{c}) - \boldsymbol{a}(\boldsymbol{c} \cdot \boldsymbol{b}) \tag{E.2}$$

$$[\boldsymbol{a} \times \boldsymbol{b}] \cdot [\boldsymbol{c} \times \boldsymbol{d}] = (\boldsymbol{a} \cdot \boldsymbol{c})(\boldsymbol{b} \cdot \boldsymbol{d}) - (\boldsymbol{a} \cdot \boldsymbol{d})(\boldsymbol{b} \cdot \boldsymbol{c}). \tag{E.3}$$

II. Vector equations

II.1. Let us solve the equation

$$\boldsymbol{x} = \boldsymbol{a} + [\boldsymbol{x} \times \boldsymbol{b}] \tag{E.4}$$

The scalar and the vector product of this equation by $\boldsymbol{b}$ are:

$$\boldsymbol{x} \cdot \boldsymbol{b} = \boldsymbol{a} \cdot \boldsymbol{b} \tag{E.5}$$

$$[\boldsymbol{x} \times \boldsymbol{b}] = [\boldsymbol{a} \times \boldsymbol{b}] + [[\boldsymbol{x} \times \boldsymbol{b}] \times \boldsymbol{b}]. \tag{E.6}$$

With the aid of the identity (E.2), we can transform (E.6) into

$$[\boldsymbol{x} \times \boldsymbol{b}] = [\boldsymbol{a} \times \boldsymbol{b}] + \boldsymbol{b}[\boldsymbol{x} \cdot \boldsymbol{b}] - \boldsymbol{x}(\boldsymbol{b} \cdot \boldsymbol{b}). \tag{E.7}$$

The solution of (E.4) is now given by

$$\boldsymbol{x} = \frac{\boldsymbol{a} + [\boldsymbol{a} \times \boldsymbol{b}] + \boldsymbol{b}(\boldsymbol{a} \cdot \boldsymbol{b})}{1 + (\boldsymbol{b})^2}. \tag{E.8}$$

II2. Let us find the inverse function ($\boldsymbol{y} = \varphi(\boldsymbol{x})$) of the function

$$\boldsymbol{x} = f(\boldsymbol{y}) = a\boldsymbol{y} + b[\boldsymbol{x} \times \boldsymbol{x}] + c(\boldsymbol{y} \cdot \boldsymbol{z})\boldsymbol{z}. \tag{E.9}$$

To this end, we shall first transform (E.9) as follows:

$$\boldsymbol{y} = \frac{1}{a}\boldsymbol{x} - \frac{b}{a}[\boldsymbol{y} \times \boldsymbol{z}] - \frac{c}{a}(\boldsymbol{y} \cdot \boldsymbol{z})\boldsymbol{z}. \tag{E.10}$$

Now the scalar and the vector products of (E.10) by z give respectively

$$y \cdot z = \frac{a}{a\left(1 + \frac{c}{a} z^2\right)} x \cdot z \tag{E.11}$$

$$[y \times z] = \frac{1}{a}[x \times z] - \frac{b}{a}[z(y \cdot z) - y \cdot z^2] \tag{E.12}$$

where we applied the identity (E.2).

From (E.10)–(E.12) we obtain the required inverse function $y = \varphi(x)$:

$$y = \frac{1}{1 + \frac{b^2}{a^2} z^2} \left\{ \frac{1}{a} x - \frac{b}{a^2}[x \times z] + \frac{\frac{b^2}{a^2} - \frac{c}{a}}{a\left(1 + \frac{c}{a} z^2\right)} (x \cdot z) z \right\}. \tag{E.13}$$

We can write this equation in the compact form similar to that of (E.6):

$$y = \alpha x - \beta[x \times z] + \gamma(x \cdot z) z. \tag{E.14}$$

Index